# Nonlinear Fracture Mechanics for Engineers

# Nonlinear Fracture Mechanics for Engineers

**Ashok Saxena**

CRC Press
Boca Raton   Boston   London   New York   Washington, D.C.

TA
409
S29
1998

**Library of Congress Cataloging-in-Publication Data**

Saxena, A. (Ashok)
    Nonlinear fracture mechanics for engineers / by Ashok Saxena.
        p.   cm.
    Includes bibliographical references and index.
    ISBN 0-8493-9496-1 (alk. paper)
    1. Fracture mechanics. 2. Nonlinear mechanics. I. Title.
TA409.S29 1997
620.1'126—dc21                                                                                                 97-37838
                                                                                                                                                            CIP

    This book contains information obtained from authentic and highly regarded sources. Reprinted material is quoted with permission, and sources are indicated. A wide variety of references are listed. Reasonable efforts have been made to publish reliable data and information, but the author and the publisher cannot assume responsibility for the validity of all materials or for the consequences of their use.

    Neither this book nor any part may be reproduced or transmitted in any form or by any means, electronic or mechanical, including photocopying, microfilming, and recording, or by any information storage or retrieval system, without prior permission in writing from the publisher.

    The consent of CRC Press LLC does not extend to copying for general distribution, for promotion, for creating new works, or for resale. Specific permission must be obtained in writing from CRC Press LLC for such copying.

    Direct all inquiries to CRC Press LLC, 2000 Corporate Blvd., N.W., Boca Raton, Florida 33431.

    **Trademark Notice:** Product or corporate names may be trademarks or registered trademarks, and are only used for identification and explanation, without intent to infringe.

© 1998 by CRC Press LLC

No claim to original U.S. Government works
International Standard Book Number 0-8493-9496-1
Library of Congress Card Number 97-37838
Printed in the United States of America  1 2 3 4 5 6 7 8 9 0
Printed on acid-free paper

# AUTHOR VITAE

Dr. Ashok Saxena is currently Professor and Chair of the School of Materials Science and Engineering at the Georgia Institute of Technology in Atlanta, Georgia. Dr. Saxena was born on August 28, 1948 in Jhansi, India. In 1970, he received his B.Tech. degree in Mechanical Engineering from the Indian Institute of Technology in Kanpur. Subsequently, he received his M.S. and Ph.D degrees in Materials Science and Metallurgical Engineering from the University of Cincinnati in 1972 and 1974, respectively. Following that Dr. Saxena joined the Research and Development Center of National Steel Co. in Weirton, W. Va. as a Research Metallurgist. In 1976, he moved to join the Research and Development Center of Westinghouse Electric Co. in Pittsburgh, Pa. where he held the title of Fellow Scientist in the Materials Science Division until leaving in 1985. From 1980 to 1981, he served as a Reader in the School of Materials Science and Technology at the Banaras Hindu University in Varanasi, India. Dr. Saxena moved to Georgia Tech. in the June of 1985 as Professor of Materials Engineering and was appointed the Chair of the School in 1993. From 1991 to 1993, he served as the Director of the Composites Education and Research Center at Georgia Tech.

Dr. Saxena's early research interests were in the area of the effect of microstructure on the fracture toughness and fatigue crack growth behavior of structural alloys. He made several notable contributions in the area of near-threshold fatigue crack growth testing and was one of the prime contributors to the development of the ASTM standard E-647 for fatigue crack growth testing. In 1978, he began research in the area of time-dependent fracture mechanics. He lead the ASTM effort to develop a standard test method for conducting creep crack growth testing and data analysis, E-1457, issued in 1992. In 1987, he developed a software package, PCPIPE, for integrity analysis of high temperature steam pipes which is used in several organizations world-wide to establish safe inspection intervals in reheat piping in fossil power plants. Dr. Saxena has authored and co-authored over a hundred research papers, numerous other technical reports, and has co-edited four ASTM special technical publications in the area of fracture mechanics. In addition, he has co-authored a text book entitled "The Science and Design of Engineering Materials" and another book entitled "Nickel-Titanium Instruments Applications in Endodontics".

Dr. Saxena is a Fellow of ASTM, a Fellow of ASM International, and a member of TMS, ASEE, and Sigma Xi and Alpha Sigma Mu honor societies. He has also received several research awards. In 1992, he received the George Irwin Medal from ASTM for his contributions to the area of creep fracture. In 1993, he received Georgia Tech's Outstanding Research Author Award and 1994 he received the ASTM Award of Merit and the honorary title of Fellow.

He routinely consults in the area of fracture mechanics for several large and small corporations. This book is based on ten years of experience in teaching linear and nonlinear fracture mechanics at the graduate level to classes comprised of students from several engineering disciplines.

# PREFACE

In the past fifteen years, several excellent books on fundamentals of Fracture Mechanics have been written focusing on the relatively mature subject of linear elastic fracture mechanics (LEFM). The treatment of nonlinear fracture mechanics in these books lacks depth because the subject is considerably more complex and must be rightfully treated in a separate book. More recently, some books providing an advanced treatment of fracture mechanics have also emerged. These books provide an excellent treatment of elastic-plastic fracture mechanics, and to a somewhat lesser degree, a treatment of time-dependent fracture mechanics, in a mathematically rigorous frame work. Thus, these books are very useful for those who approach fracture mechanics from a fundamental solid mechanics viewpoint.

The field of fracture mechanics is highly multidisciplinary and is studied and used by engineers from several disciplines. For example, engineers in basic materials industry such as steel, aluminum, plastics, and composites use fracture mechanics based test methods to evaluate and characterize materials. Engineers in aerospace, automotive, marine equipment, power generation, petro-chemical, and construction industries use fracture mechanics based methods for design and life prediction, for remaining life prediction, for establishment of safe inspection intervals, for developing inspection criteria, and last but not least, for understanding field failures of structural components. Thus, it is also an engineering tool used by design and maintenance engineers in a variety of industries. Keeping this in mind, I feel that a book focusing on the engineering aspects of nonlinear fracture mechanics is needed to complement the currently available books. In writing this book, I have attempted to fulfill this perceived need.

When I first arrived in Georgia Tech. in 1985 after spending several years in industry, I immediately embarked on developing a graduate course in nonlinear fracture mechanics to supplement an existing course which was exclusively focused on LEFM. Initially, my objective for this advanced course was to teach nonlinear fracture mechanics to those Materials Science and Engineering graduate students who would be conducting research in this field. Soon, the number of students in the course grew and it began attracting students from other disciplines such as Mechanical, Engineering Mechanics, Aerospace, and Civil Engineering. Catering to such a diverse student group required some changes to be made in my approach to teaching the subject. For example, the course had to include more coverage of the component life prediction related issues, in addition to materials testing and evaluation. Over the years, I have come to realize that these changes have helped tremendously in keeping the interest of all students alive in this subject. Exposing materials science students to life prediction methods based on fracture mechanics has been very useful.

This book has evolved from the extensive class notes and example problems that have been formulated for this course over the past ten years. Several example problems are included to reinforce the concepts being discussed. Homework problems are provided for each chapter to further improve the understanding of the students. Several universities have LEFM courses available as part of their undergraduate and graduate curricula. However, relatively few have a course entitled nonlinear fracture mechanics or advanced fracture mechanics. It is my hope that this book will help to spur similar courses in other universities. It should also be a useful book for teaching advanced fracture mechanics concepts to practicing engineers in continuing education courses and also as a valuable reference source.

A first course in fracture mechanics is a prerequisite for understanding the topics covered in this book. The emphasis in this book is on the concepts of nonlinear fracture mechanics and the mathematics has been simplified as much as possible to emphasize physical concepts. The practical applications of the field in materials testing and evaluation and in life prediction models for structural components are the primary focus of the book.

I should also mention that the field is dynamic and, in several areas, it is still evolving. Therefore, in some places the description may appear speculative. I have tried to point that out by labeling it as such, as much as possible. I have also tried to emphasize the simplifying assumptions in the analytical framework used in deriving the various fracture mechanics parameters. These simplifying assumptions in the underlying theories dictate the restrictions that must be applied during the use of nonlinear fracture mechanics in practical applications. These limitations are emphasized through out the book.

The field of fracture mechanics has been extremely fortunate to have attracted some of the most creative engineering minds. Achieving a high degree of rigor in their approaches to even very complex problems has always been a priority with these people. This was the primary reason why such rapid progress occurred in the field and also for its simultaneous wide spread acceptance as a practical engineering tool. In forty years, the field has progressed from its very elementary roots to essentially a well developed and widely accepted approach for solving complex fracture problems. I have had the good fortune of coming in personal contact with several of the leaders in this field who have provided the critical break-throughs to solve a yet new class of fracture problems. I am indebted to these colleagues because I feel that they have directly and indirectly contributed to the book; directly by helping me in preparation of the manuscript by providing me feedback and suggestions and indirectly by shaping my own personal views on this complex subject over the past twenty years.

At the risk of forgetting to include some names, I would like to acknowledge the following: all my colleagues from my days in Westinghouse Electric Co., in particular John D. Landes, Hugo A. Ernst, Peter K. Liaw, W.G. Clark, Jr., Norman E. Dowling, W.A. Logsdon, E.T. Wessel, S.J. Hudak, Jr., Don E. McCabe, T.T. Shih, W.R. Brose, J.A. Begley, and G.A. Clarke for their friendship and several very stimulating discussions.

I also wish to acknowledge my friends associated with the Mechanical Properties Research Laboratory (MPRL) at Georgia Tech: S.D. Antolovich, Richard C. Brown, David L. McDowell, and W.S. Johnson, as well as all my past and current students, who provided me with strong support in completing the book. I am indebted to Lisa Novak and B. Carter Hamilton, Richard H. Norris, Lewis Zion, and Christopher L. Muhlstein for their assistance in preparing the manuscript.

My deepest gratitude goes to my wife, Madhu, our children, Rahul and Anjali, who gave me unlimited freedom during weekends and evenings so I could complete this time-intensive project, and to my mother and late father for all the encouragement and support they have provided me throughout my life.

Ashok Saxena
July, 1997

# TABLE OF CONTENTS

## Chapter 1  Overview — 1

1.1  Introduction — 1
1.2  Classification of Fracture Mechanics Regimes — 5
1.3  History of Developments in Fracture Mechanics — 8
1.4  References — 13

## Chapter 2  Review of Solid Mechanics — 17

2.1  Stress — 17
2.2  Strain — 20
2.3  Elasticity — 22
2.4  Plasticity — 25
2.5  Consideration of Creep — 30
2.6  Component Analysis in the Plastic Regime — 35
2.7  Fully Plastic/Limit Loads — 41
2.8  Summary — 42
2.9  References — 43
2.10  Exercise Problems — 43

## Chapter 3  Review of Linear Elastic Fracture Mechanics — 45

3.1  Basic Concepts — 45
3.2  Crack Tip Plasticity — 54
3.3  Compliance Relationships — 58
3.4  Fracture Toughness and Predicting Fracture in Components — 60
3.5  Subcritical Crack Growth — 64
3.6  Limitations of LEFM — 70
3.7  Summary — 75
3.8  References — 76
3.9  Exercise Problems — 78

## Chapter 4  Analysis of Cracks Under Elastic-Plastic Conditions — 81

4.1  Introduction — 81
4.2  Rice's J-Integral — 81
4.3  J-Integral, Crack Tip Stress Fields, and Crack Tip Opening Diplacement — 88
4.4  J-Integral as a Fracture Parameter and its Limitations — 92
4.5  Summary — 102
4.6  References — 103
4.7  Exercise Problems — 104

## Chapter 5  Methods of Estimating J-Integral — 107

| | | |
|---|---|---|
| 5.1 | Analytical Solutions | 107 |
| 5.2 | Determiniation of J in Test Specimens | 111 |
| 5.3 | J for Growing Cracks | 122 |
| 5.4 | Numerically Obtained J-Solutions | 125 |
| 5.5 | Summary | 142 |
| 5.6 | References | 142 |
| 5.7 | Exercise Problems | 143 |

## Chapter 6  Crack Growth Resistance Curves — 147

| | | |
|---|---|---|
| 6.1 | Fracture Parameters under Elastic-Plastic Loading | 147 |
| 6.2 | Experimental Methods for Determining Stable Crack Growth and Fracture | 149 |
| 6.3 | Special Considerations for Weldments | 162 |
| 6.4 | Summary | 170 |
| 6.5 | References | 170 |
| 6.6 | Exercise Problems | 172 |

## Chapter 7  Instability, Dynamic Fracture, and Crack Arrest — 175

| | | |
|---|---|---|
| 7.1 | Fracture Instability | 175 |
| 7.2 | Fracture under Dynamic Conditions | 187 |
| 7.3 | Crack Arrest | 201 |
| 7.4 | Test Methods for Dynamic Fracture and Crack Arrest | 206 |
| 7.5 | Summary | 213 |
| 7.6 | References | 215 |
| 7.7 | Exercise Problems | 218 |

## Chapter 8  Constraint Effects and Microscopic Aspects of Fracture — 221

| | | |
|---|---|---|
| 8.1 | Higher Order Terms of Asymptotic Series | 221 |
| 8.2 | Cleavage Fracture | 234 |
| 8.3 | Ductile Fracture | 246 |
| 8.4 | Ductile-Brittle Transition | 257 |
| 8.5 | Summary | 260 |
| 8.6 | References | 261 |
| 8.7 | Exercise Problems | 263 |

## Chapter 9  Fatigue Crack Growth under Large-Scale Plasticity — 267

| | | |
|---|---|---|
| 9.1 | Crack Tip Cyclic Plasticity, Damage, and Crack Closure | 267 |
| 9.2 | $\Delta J$-Integral | 281 |
| 9.3 | Test Methods for Characterizing FCGR under Large Plasticity Conditions | 290 |
| 9.4 | Behavior of Small Cracks | 292 |
| 9.5 | Summary | 302 |
| 9.6 | References | 303 |
| 9.7 | Exercise Problems | 306 |

## Chapter 10  Analysis of Cracks in Creeping Materials — 309

| | | |
|---|---|---|
| 10.1 | Stress Analysis of Cracks under Steady-State Creep | 310 |
| 10.2 | Analysis of Cracks under Small-Scale and Transition Creep | 319 |
| 10.3 | Consideration of Primary Creep | 335 |
| 10.4 | Effects of Crack Growth on the Crack Tip Stress Fields | 345 |
| 10.5 | Crack Growth in Creep-Brittle Materials | 348 |
| 10.6 | Summary | 358 |
| 10.7 | References | 359 |
| 10.8 | Exercise Problems | 361 |

## Chapter 11  Creep Crack Growth — 363

| | | |
|---|---|---|
| 11.1 | Test Methods for Characterizing Creep Crack Growth | 364 |
| 11.2 | Microscopic Aspects of Creep Crack Growth | 377 |
| 11.3 | Greep Crack Growth in Weldments | 384 |
| 11.4 | Summary | 389 |
| 11.5 | References | 390 |
| 11.6 | Exercise Problems | 392 |

## Chapter 12  Creep-Fatigue Crack Growth — 395

| | | |
|---|---|---|
| 12.1 | Early Approaches for Characterizing Creep-Fatigue Crack Growth Behavior | 396 |
| 12.2 | Time-Dependent Fracture Mechanics Parameters for Creep-Fatigue Crack Growth | 399 |
| 12.3 | Crack Tip Parameters during Creep-Fatigue | 408 |
| 12.4 | Methods of Determining $(C_t)_{avg}$ | 409 |
| 12.5 | Experimental Methods for Characterizing Creep Crack Growth | 413 |
| 12.6 | Creep-Fatigue Crack Growth Correlations | 414 |
| 12.7 | Summary | 421 |
| 12.8 | References | 422 |
| 12.9 | Exercise Problems | 425 |

## Chapter 13  Case Studies — 427

| | | |
|---|---|---|
| 13.1 | Applications of Fracture Mechanics | 427 |
| 13.2 | Fracture Mechanics Analysis Methodology | 428 |
| 13.3 | Case Studies | 429 |
| 13.4 | Summary | 461 |
| 13.5 | References | 461 |

**Appendices**

**Index**

CHAPTER 1

# OVERVIEW

**1.1 Introduction**

Fracture of load-bearing components is always an important consideration for engineers who design, build, operate, and maintain bridges, highways, automobiles, trains, airplanes, powerplants, chemical process equipment, and numerous other large pieces of machinery. Everyone understands the catastrophic consequences of structural failure and that sometimes it is unavoidable because the factors involved in predicting it are very complex. Over the past forty years, the developments in fracture mechanics have contributed immensely to our understanding of fracture and also to our understanding of how to deal with cracks or crack-like defects which escape detection.

A large fraction of failures in structural components occur due to pre-existing defects or defects that initiate rapidly from clusters of non-metallic inclusions or from other imperfections such as casting, forging, and welding defects. On the other hand, several defects also lie dormant in the components and pose no threat of fracture. In fracture mechanics, we are interested in both, the defects that can ultimately cause fracture and those that are benign. Some examples of the use of fracture mechanics are considered next to illustrate the above points.

*1.1.1 Weld Fractures in Liberty Ships*

All welded construction of ships was first introduced during the early days of World War II as a means for expediting ship construction. Of the 2700 Liberty ships built during this era, approximately 400 sustained fractures and several of them severe enough to completely incapacitate the vessel. The most severe cases were ones in which the hull fractured into two halves. Subsequent analysis showed that fractures emanated from defective welds in high stress areas, a situation which was further exasperated by the use of low fracture toughness steel. Once the problems were identified, corrections were made which prevented several potential fractures.

*1.1.2 Failures in Reheat Steam Pipes*

In June 1985, a 60 cm diameter steam pipe carrying steam at a pressure of 4 MPa at a temperature of 538°C suddenly ruptured and sprayed supersaturated steam on people gathered in a nearby lunch room. Several people were killed and many more sustained serious injuries. The rupture occurred along the pipe's longitudinal seam which was welded and propagated through an axial length of 6m along the pipe. The opening of the rupture in the central region was approximately 2m. A very similar failure occurred in another power station six months later. In this case, the weld seam was pointed upwards, toward the roof. The gush of steam blew a large hole in the ceiling and roof but, unlike the previous failure, it did not result in fatalities or serious injuries. In both cases, the cause of fracture was identified to be early crack initiation in the weld region and subsequent propagation by creep damage [1.1]. An example of creep cracks developing in seam welds of a steam pipe is shown in Figure 1.1. These mechanisms are explored quite extensively in the following chapters dealing with time-dependent fracture mechanics.

*1.1.3 Failure of a Steam Turbine Rotor*

On June 19, 1974, a high temperature rotor of a steam turbine located at the Gallation Power Plant in Tennessee burst suddenly during a routine start-up operation. A schematic of the fracture and a picture of the actual fractured pieces of the rotor are shown in Figures 1.2 and 1.3, respectively [1.2]. Fractographic analysis revealed two elliptical flaws in the bore region of the rotor from which the fracture initiated. These flaws initiated from clusters of MnS inclusions by creep-fatigue and subsequently grew

**2** *Nonlinear Fracture Mechanics for Engineers*

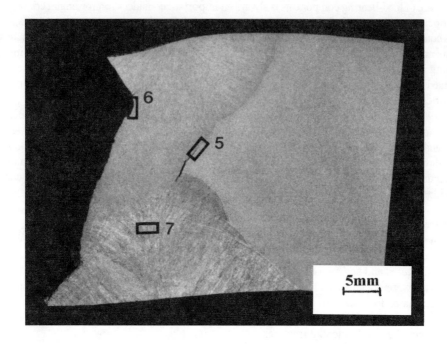

**Figure 1.1** *Photograph of service-generated fracture and cracks in a steam pipe that burst in a power plant in a weld region.*

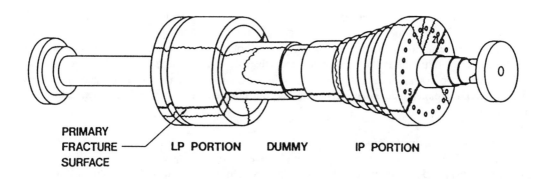

**Figure 1.2** *Schematic diagram showing the fracture in the high temperature Gallatin steam turbine rotor (Ref. 1.2).*

***Figure 1.3*** *Pieces of the fractured Gallatin turbine rotor assembled for studying the failure (Ref. 1.2).*

to the critical size at which time fracture occurred [1.3]. At the time of the sudden fracture, the rotor had experienced 106,000 hours of service. This failure triggered boresonic inspec- tions of turbine rotors at power plants all over the world. It also started a flurry of research in the area of creep-fatigue crack initiation and crack growth. Several concepts described in this book are the result of research conducted on this topic in the aftermath of this failure.

### 1.1.3 Cracks in a Superheater Outlet Steam Header

Figure 1.4 shows a picture of cracks in a superheater outlet steam header uncovered during inspections conducted at several power plants in the 1980s. Inspections were being conducted to assess the condition of headers as part of component life extension programs. The header depicted had been in service at the plant for 25 years. The radius of the holes in the picture are approximately 25mm, thus, it is estimated that some of the cracks in the picture are approaching a size of 10mm. This header carried steam at a pressure of approximately 24MPa and a temperature of 538°C. The outside diameter of the header was approximately 0.4m and the wall thickness was 0.165m. The discovery of these cracks led to a shutdown and replacement of the header. The header was subsequently sectioned and analyzed. One of the conclusions of the analysis was that the deepest cracks had been present for several years. Thus, a substantial portion of the life of the header was spent in crack propagation. The cracks had initiated by thermal-fatigue but creep was involved in their growth. The large size of the cracks found in headers removed from service points to the high crack tolerance of the component. In situations such as these, it becomes essential to establish an inspection criterion which is realistic or a large number of components will be retired prematurely at significant cost to the consumer. Thus, the role of fracture mechanics in this instance is to establish safe inspection criteria and intervals and to aid run/repair/retire decisions.

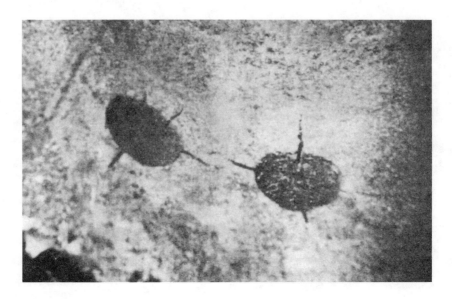

***Figure 1.4*** *Cracks found in a superheater outlet steam header after approximately 25 years of service at a service temperature of 538 °C. Courtesy of Jani and Saxena, copyright TMS, Warrendale, PA.*

### 1.1.4 Cracks in SSTG Turbine Casings

In the 1980s, the U.S. Navy began an extensive inspection program to assess the condition of the ship's steam turbine-generator (SSTG) sets. As part of the program, several steam turbine casings of the type shown in Fig. 1.5 were inspected. Most of the SSTGs inspected had been in operation for 25 years or more and were approaching the end of their original design life. Figure 1.6 shows a schematic and a picture of cracks that were found in the steam passageways in one of the casing inspected [1.4, 1.5]. These cracks were mostly shallow, up to 4mm deep, but some were up to 10mm deep. An analysis of the oxide scale on the cracks removed from a retired casing conclusively showed that these cracks had been present for a high fraction of the service duration. These cracks had initiated and grown by thermal-mechanical fatigue. Similar to the case of headers, it can be concluded from this evidence that the casings have a high tolerance for cracks and that crack propagation occurred over much of the service life of the casings. There was also evidence that the crack growth rate decelerated as the crack grew deeper and some cracks even stopped growing.

An extensive study was conducted to quantify the remaining life, inspection criteria, and interval [1.4, 1.5] for turbine casings. Figure 1.7 shows a plot of the predicted remaining life as a function of crack depth at inspection for cracks found in regions of the casing exposed to 538°C and 427°C. The remaining fatigue life of cracks that are approximately 6mm deep was still in excess of 800 cycles. Since the SSTGs, on average, experience only 50 start-stop cycles per year, this translates to a remaining life of 16 years. Based on these calculations, inspection intervals of 7 to 8 years were recommended.

### 1.1.5 Role of Fracture Mechanics in the Design of Supersonic Transport Aircraft

Figure 1.8 shows the predicted temperature distribution in the various sections of a supersonic transport aircraft being designed to travel at a speed of 2.4 Mach [1.6]. At this speed, aircraft are expected to be able to complete two round trips between the U.S. west coast and the Pacific rim countries during a 24-hour period, thus making a significant difference in the economics of supersonic transport. However, these operating temperatures are sufficiently high for creep to become a consideration in several Al alloys and polymer composites being considered as candidate materials.

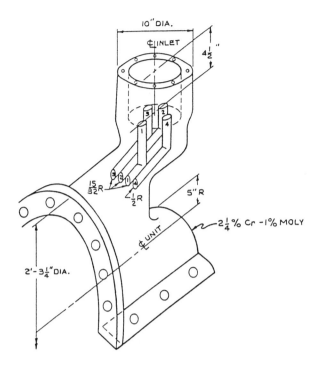

*Figure 1.5* Schematic drawing of a SSTG turbine casing (Ref. 1.5).

This consideration is new to commercial airframe manufacturers. Also, due to the high design and production costs of the aircraft, the long-term durability requirements must be significantly higher than for subsonic aircraft. For example, the requirements being discussed currently specify a certified life of 120,000 hours of flying. Figure 1.9 shows a flow chart of the ingredients of a thorough durability analysis of civil supersonic aircraft structures. As one can see, fracture mechanics analysis and test procedures play a very significant role in such a durability analysis. Therefore, fracture mechanics can be used effectively during design of structural components to avoid service failures and ensure long-term durability.

**1.2 Classification of Fracture Mechanics Regimes**

Fracture mechanics problems have been classified into linear-elastic fracture mechanics (LEFM), elastic-plastic fracture mechanics (EPFM), and time-dependent fracture mechanics (TDFM) regimes. These classifications are based on the dominant operating deformation modes in the cracked bodies as shown in Figure 1.10. When the stress-strain behavior and the load-displacement behavior is linear, LEFM can be used and relevant crack tip parameter is the stress intensity parameter K. In this regime, the plastic zone is small in comparison to the crack size and other pertinent dimensions of the cracked body. When dominantly linear conditions can no longer be ensured due to large-scale plasticity, EPFM is used and the relevant crack tip parameter is the J-integral. Finally, when the stress-strain behavior and the load-displacement behavior is time-dependent due to either dynamic loading or due to time-dependent creep, the concepts of TDFM must be used. When the loading rates are very high, the crack tip stresses are influenced by shear waves and by kinetic energy considerations which are neglected in conventional

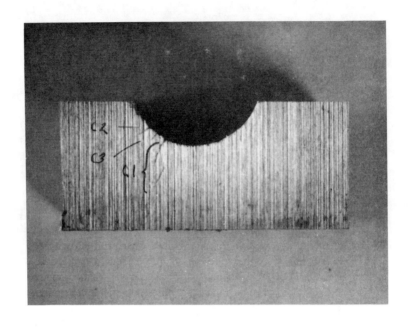

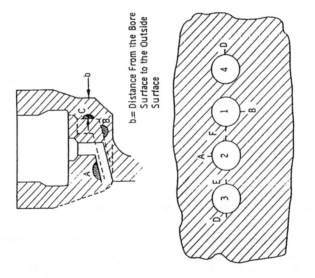

**Figure 1.6** (a) Schematic and (b) actual cracks found in a SSTG turbine casing removed from service after 25 years (Ref. 1.5).

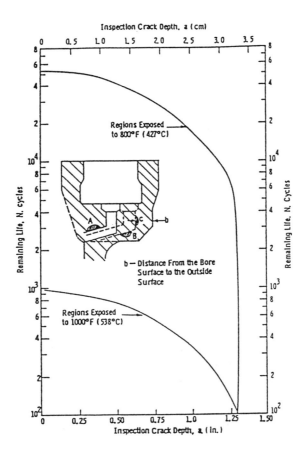

**Figure 1.7** *The predicted relationship between crack size at inspection and remaining life for SSTG turbine casings (Ref. 1.5).*

LEFM. Similarly, when time-dependent creep occurs, the crack tip stress and deformation fields change with time. The crack tip parameters in such cases are the $C_t$ parameter and the $C^*$- integral.

It is important at the outset of our discussion to make a clear distinction between classification of fracture problems on the basis of global deformation mode as presented in the preceding paragraphs and those based on local fracture mechanisms. The terminology for describing the local fracture mechanisms is different. By local, we mean the microscopic fracture mechanisms in a zone very near to the tip of the crack. This region, often referred to as the process zone, is small in comparison to the surrounding zone in which the stresses and strains are characterized by the global crack tip parameters. The terms used to describe the local fracture mechanisms include cleavage fracture, ductile fracture by micro-void coalescence, brittle fracture, intergranular fracture, or transgranular fracture. Cleavage fracture is one that proceeds along well-defined crystallographic planes (Figure 1.11a). Ductile fracture by microvoid coalescence, on the other hand, is characterized by growth of voids by plastic deformation. These voids eventually coalesce and cause fracture (Figure 1.12b). Brittle fracture is said to occur when limited plastic deformation accompanies fracture. Cleavage fractures are usually brittle, but not all brittle fractures are due to cleavage. When fractures propagate through the grains, they are called transgranular and when they propagate along the grain boundaries, they are known as intergranular. Both ductile and brittle fractures

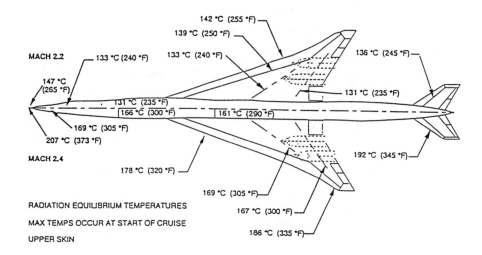

*Figure 1.8* Temperature distributions predicted in the various regions of supersonic transport aircraft traveling at a speed of 2.4 Mach (Ref. 1.6).

can be intergranular or transgranular. Creep fracture occurs due to growth of grain boundary cavities in which case it is intergranular. The appearance of the fracture in that instance is much like the one shown in Figure 1.11b, except cavities form at the grain boundaries.

The selection of the appropriate crack tip parameter is based entirely on the global deformation mechanism(s) which is seemingly independent of the local fracture mechanism(s). This aspect of fracture mechanics has often received severe criticism, somewhat justifiably, from the materials community because the link between the two is not well understood. Current research in micromechanics of fracture processes is likely to make the link between fracture mechanics parameters and local damage mechanisms much clearer.

## 1.3 History of Developments in Fracture Mechanics

In this section, we will trace the significant developments in fracture mechanics and the motivational factors behind them in a somewhat chronological order.

### 1.3.1 Early Theories of Fracture

In 1913, C.E. Inglis [1.7] proposed the concept of stress concentration at geometrical discontinuities, thus providing an explanation for why fractures emanate from cracks, holes, or other defects. Seven years later, A.A. Griffith [1.8] combined Inglis' derivation with his hypothesis about energy exchanges that take place during fracture and derived the concept of critical crack size necessary for brittle fracture. He theoretically demonstrated an inverse relationship between fracture stress and the square root of crack size and was able to experimentally demonstrate this relationship on brittle materials such as glass. However, attempts at applying this theory to metals did not meet with success. In 1939, H.M. Westergaard [1.9] published the results of his analysis demonstrating that the stresses at the tips of cracks in elastic bodies varied as a function of $1/\sqrt{r}$, where r = distance from the crack tip. The significance of this work was not realized until 20 years later. The next significant development in fracture mechanics did not occur until 1948. G.R. Irwin [1.10] and E. Orowan [1.11] independently proposed modifications

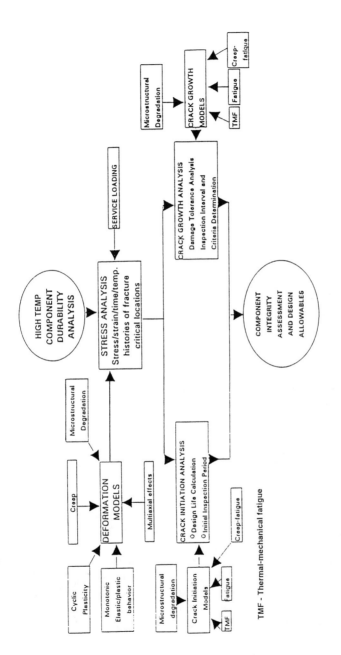

*Figure 1.9 A flow chart showing the various elements of durability analysis for supersonic transport aircraft.*

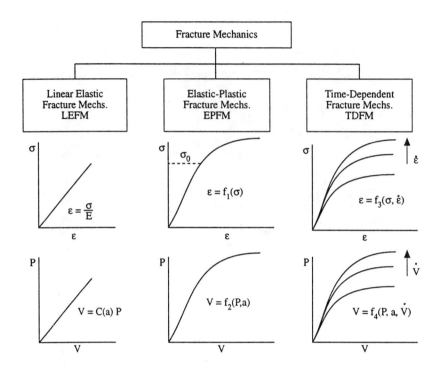

**Figure 1.10** *Classification of fracture mechanics regimes on the basis of dominant deformation modes: σ = stress, ε = strain, ε̇ = strain rate, P = load, V = load-line displacement, and V̇ = load-line displacement rate.*

to Griffith's theory to account for plastic energy that is dissipated during the fracture process. This modification made it possible to apply Griffith's theory to metals. In the same year, N.F. Mott [1.12] published his paper which extended Griffith's analysis by taking kinetic energy into account. He was able to derive expressions for predicting crack speeds. This paper is widely accepted as the first piece of significant research in the field that is now known as dynamic fracture mechanics.

In the subsequent years, the efforts focused on generalizing the modified Griffith's approach to other geometries more suited for engineering applications. In 1956, G.R. Irwin [1.3] proposed the concept of energy release rate or the crack extension force, which successfully met this criterion. Soon after that, Irwin published another landmark piece of research [1.4] in which he was able to use the Westergaard approach to show that the amplitude of stresses and displacements in front of crack tips in elastic solids can be expressed by a single parameter, now widely known as the stress intensity parameter, K. The magnitude of this parameter depends on the remote stress, crack size, and size and geometry of the cracked body. He further demonstrated that K can be uniquely related to the strain energy release rate, thus making the very important connection between the stress and energy based approaches for predicting fracture. Shortly after, M.L. Williams [1.15] used a different technique and proposed a more complete description of the crack tip stress distribution. However, in the region of the crack tip where fracture is believed to occur, William's results essentially agreed with those of Irwin's.

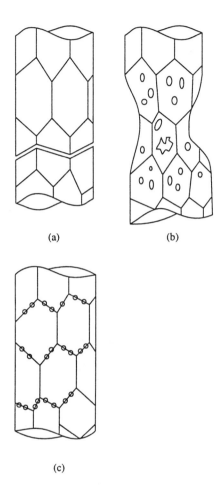

*Figure 1.11* Classification of fracture mechanisms: (a) cleavage fracture, (b) ductile fracture by microvoid growth and coalescence, (c) intergranular creep fracture by growth and coalescence of grain boundary creep cavities, and (d) transgranular creep fracture by growth of voids within the grains by creep deformation.

*1.3.2 Developments in Linear Elastic Fracture Mechanics*

In the mid-1950s, fracture mechanics research received a much needed credibility boost in the aftermath of comet aircraft failures. The aircraft industry, which was using high strength materials such as precipitate hardened Al alloys and high strength steels, realized the merits of the field and joined forces with the U.S. Navy which had already accepted the importance of fracture mechanics research due to Liberty ship failures which occurred much earlier. The period of the early 1960s saw the concept of plane strain fracture toughness, $K_{IC}$, evolve and become widely accepted. The crack tip opening displacement (CTOD) concept originated in the U.K. through the efforts of A.A. Wells at about the same time. The close relationship between CTOD and the K-based approaches has always been recognized. The CTOD approach was used more frequently in Europe while the K-based approach was more popular in the U.S. with both sides recognizing the unique relationship between the two.

In 1960, P.C. Paris and co-workers [1.16] first proposed the relationship between fatigue crack growth rate and the cyclic stress intensity parameter. Although this approach met with initial resistance, it soon became universally accepted as the approach for predicting fatigue crack growth behavior in structures. The nineteen sixties and seventies were decades where linear elastic fracture mechanics (LEFM) flourished and was firmly established as an approach to tackle fracture problems in several industries. It was also fully understood that LEFM behavior is limited to conditions when linear-elastic conditions dominate the cracked body.

### 1.3.3 Developments in Elastic-Plastic Fracture Mechanics

The nuclear power industry was gaining prominence during the nineteen sixties and several ambitious projects were initiated to harness nuclear power safely and more economically. The materials used in the power industry were ductile steels in which fracture was invariably accompanied by extensive plastic deformation, at least when attempts were made to measure the $K_{IC}$ of these materials using laboratory-size specimens. It was clear that if fracture mechanics were to be applied successfully in these applications, the analytical theory had to be extended to include fracture under elastic-plastic and fully-plastic conditions.

In 1968, three papers of considerable significance to the development of elastic-plastic fracture mechanics appeared. J.R. Rice [1.17] idealized plastic deformation as a nonlinear elastic phenomenon for mathematical purposes and was able to generalize the energy release rate for such materials. He expressed this in the terms of a path-independent contour integral that he called J. He noted that this integral could also be derived from one of several conservation integrals proposed earlier by Eshelby [1.18] with no reference to crack problems. Hutchinson [1.19] and Rice and Rosengren [1.20] in the same year derived the relationships between J-integral and the crack tip stress, strain, and displacement fields in a manner similar to how crack tip fields are related to K under linear-elastic conditions. Since J was derived for nonlinear elastic materials, it was considered to have severe limitations for characterizing fracture in elastic-plastic materials. On the other hand, Wells [1.21] had already demonstrated considerable success with CTOD as a fracture parameter even under conditions of significant plasticity.

In the late nineteen sixties, E.T. Wessel, who was one of the pioneers in the development of the ASTM Standard E-399 for measurement of $K_{IC}$, was assembling a research group at Westinghouse Electric Corporation's Research and Development Center to tackle the difficult problem of fracture under elastic-plastic conditions. Wessel had given the responsibility of this difficult task to two of his very young engineers by the names of J.A. Begley and J.D. Landes. Begley and Landes proceeded to apply J-integral for characterizing the initiation of ductile fracture, despite criticism from the mechanics' community. At the time they felt that if their attempts to apply J were unsuccessful, they might be able to find another fracture criterion. In 1972, Landes and Begley [1.22-1.23] published the results of their first study on the use of J to predict the initiation of fracture under elastic-plastic conditions. Following these studies, the field of elastic-plastic fracture mechanics progressed rapidly with continued efforts of Landes and Begley but also due to some land-mark work of C.F. Shih [1.24] and by Hutchinson and Paris [1.25]. The latter two pieces of work provided a rigorous theoretical justification for characterizing stable crack growth using the J-integral. In the late nineteen seventies, P.C. Paris and H.A. Ernst and co-workers [1.26] developed the tearing modulus concept to predict instability following ductile crack growth. About the same time, or actually a little earlier, Dowling and Begley [1.27] proposed the use of the cyclic J-integral for characterizing fatigue crack growth under elastic-plastic and fully-plastic conditions. In the subsequent years, much progress occurred in the development of test methods using the J-integral approach and in the development of methods for estimating J-integral, making elastic-plastic fracture mechanics a viable engineering tool for assessing structural integrity.

### 1.3.4 Developments in Time-Dependent Fracture Mechanics

In the early to mid-seventies, efforts had already begun to extend the concepts of fracture mechanics to crack growth under creep conditions. Some of the early pioneers included Siverns and Price [1.28] and L.A. James [1.29]. Their efforts were directed at extending the use of K for creep and creep-fatigue crack growth. In 1976, Landes and Begley [1.30] and Nikbin, Webster, and Turner [1.31] independently proposed the use of a J-like integral ($C^*$) for characterizing creep crack growth. Subsequent experimental work of Taira and co-workers in Japan [1.32] and Saxena [1.33] in the U.S. confirmed the validity of $C^*$ for characterizing creep crack growth. The primary limitation of $C^*$ was that it applied only to extensive secondary creep conditions. Ohji, Ogura, and Kubo [1.34], Riedel and Rice [1.35], and McCLintock and Bassani [1.36] formulated the problem of small-scale creep which was used subsequently by Saxena [1.37] to define the $C_t$ parameter which can be used to characterize creep crack growth under conditions ranging from small-scale to extensive creep. Next, formulations of $C_t$ and $C^*$ were provided which account for cyclic loading [1.38] and also primary creep deformation [1.39, 1.40].

### 1.3.5 Current Research in Fracture Mechanics

Although much progress has occurred in fracture mechanics over the past 40 years, it still remains an active field of research. Some of the current topics of research are fracture in inhomogeneous and anisotropic materials such as composites, effects of constraint on fracture toughness, creep and creep-fatigue crack growth in creep-brittle materials, crack growth and fracture in weldments, and growth of small cracks and fracture under the conditions of mixed-mode loading. Another limitation of the current capabilities of fracture mechanics is its relatively weak connection with damage mechanisms at the crack tip. While use of global parameters such as K, J, $C^*$, $C_t$, etc. are very useful for predicting crack growth and fracture in engineering components, they do little for advancing the fundamental understanding of the material's resistance to crack growth and fracture. With advances in computer technology and also in experimental techniques for observing fracture mechanisms, this area of research is poised for significant gains.

## 1.4 References

1.1 F.L. Becker, S.M. Walker, and R. Viswanathan, "Guideline for Evaluation of Seam-Welded Steam Pipes", EPRI Report CS-4774, Nov. 1986.

1.2 L.D. Kramer and D.D. Randloph, "Analysis of TVA Gallatin No. 2 Rotor Burst Part I-Metallurgical Considerations", 1976 ASME-MPC Symposium on Creep-Fatigue Interaction, MPC-3, 1976, pp. 1-24.

1.3 D.A. Weisz, "Analysis of TVA Gallatin No. 2 Rotor Burst Part II-Mechanical Analysis", 1976-ASME-MPC Symposium of Creep-Fatigue Interaction, MPC-3, 1976, pp. 25-40.

1.4 W.A. Logsdon, P.K. Liaw, A. Saxena, and V. Hulina, "Residual Life Prediction and Retirement for Cause Criteria for SSTG Upper Casings, Part I: Mechanical and Fracture Mechanics Properties Development", Engineering Fracture Mechanics, Vol. 25, 1986, pp. 259-288.

1.5 A. Saxena, P.K. Liaw, W.A. Logsdon, and V. Hulina, "Residual Life Prediction and Retirement for Cause Criteria for SSTG Upper Casings, Part II, Fracture Mechanics Analysis: Engineering Fracture Mechanics, Vol. 25, 1986, pp. 289-303.

1.6 E.A. Starke et al., "Accelerated Aging of Materials and Structures--the Effects of Long-Term Elevated Temperature Exposure", NMAB-479, National Academy Press, Washington, D.C. 1996.

1.7 C.E. Inglis, "Stresses in Plate Due to the Presence of Cracks and Sharp Corners", Transactions of the Institute of Naval Architects, Vol. 55, 1913, pp. 219-241.

1.8 A.A. Griffith, "The Phenomena of Rupture and Flow in Solids", Philosophical Transactions, Series A, Vol. 221, 1920, pp. 163-198.

1.9 H.M. Westergaard, "Bearing Pressures and Cracks", Journal of Applied Mechanics, Vol. 6, 1939, pp. 49-53.

1.10 G.R. Irwin, "Fracture Dynamics", Fracturing of Metals, American Society for Metals, Cleveland, 1948, pp. 147-166.

1.11 E. Orowan, "Fracture and Strength of Solids", Reports on Progress in Physics, Vol. X11, 1948, pp. 185-232.

1.12 N.F. Mott, "Fracture of Metals: Theoretical Considerations", Engineering, Vol. 165, 1948, pp. 16-18.

1.13 G.R. Irwin, "Onset of Fast Crack Propagation in High Strength Steel and Aluminum Alloys", Sagamore Research Conference Proceedings, Vol. 2, 1956, pp. 289-305.

1.14 G.R. Irwin, "Analysis of Stresses and Strains Near the End of a Crack Traversing a Plate", Journal of Applied Mechanics, Vol. 24, 1957, pp. 361-364.

1.15 M.L. Williams, "On the Stress Distribution at Base of a Stationary Crack", Journal of Applied Mechanics, Vol. 24, 1957, pp. 109-114.

1.16 P.C. Paris, M.P. Gomez, and W.P. Anderson, "A Rational Analytic Theory of Fatigue", The Trend in Engineering, Vol. 13, 1961, pp. 9-14.

1.17 J.R. Rice, "A Path Independent Integral and the Approximate Analysis of Strain Concentration by Notches and Cracks", Journal of Applied Mechanics, Vol. 35, 1968, pp. 379-386.

1.18 J.D. Eshelby, "The Continuum Theory of Lattice Defects", Solid State Physics, Vol. 3, 1956, pp. 79-141.

1.19 J.W. Hutchinson, "Singular Behavior at the End of a Tensile Crack Tip in a Hardening Material", Journal of Mechanics and Physics of Solids", Vol. 16, 1968, pp. 13-31.

1.20 J.R. Rice and G.F. Rosengren, "Plane Strain Deformation Near a Crack Tip in a Power-Law Hardening Material", Journal of Mechanics and Physics of Solids, Vol. 16, 1968, pp. 1-12.

1.21 A.A. Wells, "Unstable Crack Propagation in Metals: Cleavage and Fast Fracture", Proceedings of the Crack Propagation Symposium, Vol. 1, Paper 84, Cranfield, U.K., 1961.

1.22 J.A. Begley and J.D. Landes, "The J-Integral as a Fracture Criterion", Fracture Toughness, ASTM STP 514, American Society for Testing and Materials, 1972, pp. 1-23.

1.23 J.D. Landes and J.A. Begley, "The Effects of Specimen Geometry on $J_{Ic}$", Fracture Toughness, ASTM 514, American Society for Testing and Materials, 1972, pp. 24-39.

1.24  C.F. Shih, "Relationship Between Crack Tip Opening Displacement for Stationary and Extending Cracks", Journal of Mechanics and Physics of Solids, Vol. 29, 1981, pp. 305-326.

1.25  J.W. Hutchinson and P.C. Paris, "Stability Analysis of J-Controlled Crack Growth", Elastic-Plastic Fracture, ASTM STP 668, American Society for Testing and Materials, 1979, pp. 37-64.

1.26  P.C. Paris, H. Tada, A. Zahoor, and H. Ernst, "The Theory of Instability of the Tearing Mode of Elastic-Plastic Crack Growth", Elastic-Plastic Fracture, ASTM STP 668, American Society for Testing and Materials, 1979, pp. 5-36.

1.27  N.E. Dowling and J.A. Begley, "Fatigue Crack Growth During Gross Plasticity and the J-Integral", in Mechanics of Crack Growth, ASTM STP 590, American Society for Testing and Materials, 1976, pp. 82-103.

1.28  M.J. Siverns and A.T. Price, "Crack Propagation Under Creep Conditions in Quenched 2.25Cr-1Mo steel", International Journal of Fracture, Vol. 9, 1973, pp. 199-207.

1.29  L.A. James, "The Effect of Frequency Upon the Fatigue Crack Growth of Type 304 Stainless Steel at 1000F", Stress Analysis of Growth of Cracks, ASTM STP 513, American Society for Testing and Materials, 1972, pp. 218-229.

1.30  J.D. Landes and J.A. Begley, "A Fracture Mechanics Approach to Creep Crack Growth", Mechanics of Crack Growth, ASTM STP 590, American Society for Testing and Materials, 1976, pp. 128-148.

1.31  K.M. Nikbin, G.A. Webster, and C.E. Turner, "Relevance of Nonlinear Fracture Mechanics to Creep Cracking", Cracks and Fracture, ASTM STP 601, 1976, pp. 47-62.

1.32  S. Taira, R. Ohtani, and T. Komatsu, "Application of J-Integral to High Temperature Crack Propagation", Transactions of ASME, Journal of Engineering Materials Technology, Vol. 101, 1979, pp. 163-167.

1.33  A. Saxena, "Evaluation of $C^*$ for Characterization of Creep Crack Growth Behavior of 304 Stainless-Steel", in Fracture Mechanics: Twelfth Conference, ASTM STP 700, American Society for Testing and Materials, 1980, pp. 131-151.

1.34  K. Ohji, K. Ogura, and S. Kubo, "Stress-Strain Field and Modified J-Integral in the Vicinity of the Crack Tip Under Transient Creep Conditions", Japan Society of Mechanical Engineers, No. 790-13, 1979, pp. 18-20 (in Japanese).

1.35  H. Riedel and J.R. Rice," Tensile Cracks in Creeping Solids", Fracture Mechanics: Twelfth Conference, ASTM STP 700, American Society for Testing and Materials, 1980, pp. 112-130.

1.36  J.L. Bassani and F.A. McClintock, "Creep Relaxation of Stress Around a Crack Tip", International Journal of Solids and Structures, Vol. 17, 1981, pp. 79-89.

1.37  A. Saxena, "Creep Crack Growth Under Nonsteady-State Conditions" in Fracture Mechanics: Seventeenth Volume, ASTM STP 905, American Society for Testing and Materials, 1986, pp. 185-201.

1.38  A. Saxena and B. Gieseke, "Transients in Elevated Temperature Crack Growth", Proceedings of MECAMAT, International Seminar on High Temperature Fracture Mechanisms and Mechanics III, EGF-6, 1987, pp. 19-36.

1.39  H. Riedel, "Creep Deformation at Crack Tips in Elastic-Viscoplastic Solids", Journal of Mechanics and Physics of Solids, Vol. 29, 1981, pp. 35-49.

1.40  C. Leung, D.L. McDowell, and A. Saxena, "Consideration of Primary Creep at a Stationary Crack Tip: Implications for the $C_t$ Parameter", International Journal of Fracture, Vol. 36, 1988, pp. 275-289.

# CHAPTER 2

# REVIEW OF SOLID MECHANICS

To understand the material covered in the chapters that follow, a good working knowledge of solid mechanics is essential. The brief review provided in this chapter is to introduce the reader to conventions, symbols, and the specific concepts of solid mechanics used in the remainder of this book. It is anticipated that the reader has already taken some courses in mechanics of deformable bodies and is reasonably knowledgeable in the area. For brevity, the important relationships in solid mechanics are merely stated without any attempts at providing the proofs. Those that desire to learn solid mechanics are referred to other excellent text books on the subject [2.1-2.6].

## 2.1 Stress

Stress is defined as force-per-unit area. Imagine a force $\Delta F$ acting at a point, P, which lies on an area $\Delta A$ as shown in Figure 2.1. Reducing the force into components that are normal and tangential to $\Delta A$ defines the normal stress ($\sigma$) and shear stress ($\tau$) as:

$$\lim \Delta A \to 0, \quad \sigma = \frac{\Delta F_n}{\Delta A} \quad and \quad \tau = \frac{\Delta F_t}{\Delta A} \tag{2.1}$$

where $\Delta F_n$ and $\Delta F_t$ are normal and tangential components of force as shown in Figure 2.1.

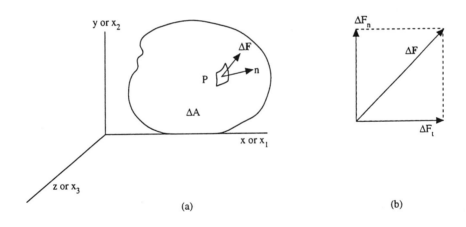

*Figure 2.1* Elemental area $\Delta A$ subjected to (a) total force, $\Delta F$, and (b) normal and tangential components of the force.

In the most general case, each face of a cube may be subjected to an arbitrary force. However, each of these forces may be resolved into components that are parallel to the three coordinate directions. One of these on each face will be normal and the other two will be tangential to yield a total of nine components. If each of these nine components are divided by the area of the face upon which they act, nine components of stress will result as shown in Figure 2.2. The first subscript in describing the stress component refers to the direction normal to the plane and the second subscript to the direction of the stress component. These are shown for the (x, y, z) or the $(x_1, x_2, x_3)$ designations of the coordinate system. $\sigma_{xx}$ (or $\sigma_{11}$) arises from a force acting in the positive x-direction (or $x_1$) on a plane whose normal is also in the positive x direction. When both subscripts are either positive or negative, the stress is considered to be positive or tensile. A positive-negative (or vice-versa) suffix would indicate a negative (or compressive) stress. Also, if the stress state is uniform, a normal stress of the same magnitude must act on the positive x-plane as on the negative x-plane. Since the direction of the stress on the negative x-plane is negative, the stress will be positive (or tensile). This collection of the nine components of stresses is called the stress tensor, designated as $\sigma_{ij}$ and can be expressed as:

$$\sigma_{ij} = \begin{vmatrix} \sigma_{xx} & \sigma_{yx} & \sigma_{zx} \\ \sigma_{xy} & \sigma_{yy} & \sigma_{zy} \\ \sigma_{xz} & \sigma_{yz} & \sigma_{zz} \end{vmatrix} \qquad (2.2)$$

where i, j are iterated over x, y, z, respectively. Two identical subscripts indicate a normal stress while differing subscripts indicate a shear stress. Normal stresses are often designated by a single subscript and shear stress by $\tau$, so:

$$\begin{aligned} \sigma_{xx} &\equiv \sigma_x \\ \sigma_{xy} &\equiv \tau_{xy} \end{aligned} \qquad (2.3)$$

If the element being considered is in equilibrium, it leads to the result that:

$$\sigma_{ij} = \sigma_{ji} \quad \text{or} \quad \sigma_{xy} = \sigma_{yx} \text{ ( etc.)} \qquad (2.4)$$

Thus, only six of the nine stress components are independent. Also, to maintain equilibrium, the following equations must be satisfied:

$$\frac{\partial \sigma_{xx}}{\partial x} + \frac{\partial \sigma_{yx}}{\partial y} + \frac{\partial \sigma_{zx}}{\partial z} = 0 \qquad (2.5a)$$

$$\frac{\partial \sigma_{xy}}{\partial x} + \frac{\partial \sigma_{yy}}{\partial y} + \frac{\partial \sigma_{zy}}{\partial z} = 0 \qquad (2.5b)$$

$$\frac{\partial \sigma_{xy}}{\partial x} + \frac{\partial \sigma_{yz}}{\partial y} + \frac{\partial \sigma_{zz}}{\partial z} = 0 \qquad (2.5c)$$

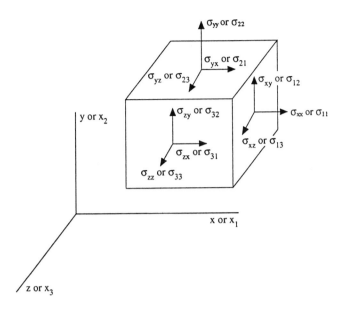

*Figure 2.2* Components of stress acting on an element. Also shown are the positive directions of the various stress components.

In the contracted tensor notations, the above equations can be written as:

$$\frac{\partial \sigma_{ij}}{\partial x_j} = 0 \qquad (2.5d)$$

where i and j vary as 1, 2, 3.

The resultant force-per-unit area is called the traction vector **T**. **T** differs from stress in that it is a vector with a defined direction. The three components of **T** in the x, y, and z directions are $T_1$ (or $T_x$), $T_2$ (or $T_y$), and $T_3$ (or $T_z$) as follows:

$$\mathbf{T} = T_1 \mathbf{i} + T_2 \mathbf{j} + T_3 \mathbf{k} \qquad (2.6)$$

where *i*, *j*, **k** are unit vectors in the x, y, z directions. The various components of the traction vector can be related to the components of stress as:

$$T_1 = \sigma_{11} n_1 + \sigma_{12} n_2 + \sigma_{13} n_3 \tag{2.7a}$$

$$T_2 = \sigma_{21} n_1 + \sigma_{22} n_2 + \sigma_{23} n_3 \tag{2.7b}$$

$$T_3 = \sigma_{31} n_1 + \sigma_{32} n_2 + \sigma_{33} n_3 \tag{2.7c}$$

where $n_1$, $n_2$, and $n_3$ are the three direction cosines associated with the outward normal **n** for the plane of the traction vector. In the short index notation, equation (2.7) can be expressed as:

$$T_i = \sigma_{ij} n_j \tag{2.8}$$

At every point in a body, there exists a set of three orthogonal planes called the principal planes on which the traction vector lies normal to the plane and thus, no shear stresses exist on these planes. The three principal stresses can be determined by solving the following cubic equation:

$$\sigma^3 - I_1 \sigma^2 + I_2 \sigma - I_3 = 0 \tag{2.9}$$

$$I_1 = \sigma_{11} + \sigma_{22} + \sigma_{33}$$

$$I_2 = (\sigma_{11} \sigma_{22} + \sigma_{22} \sigma_{33} + \sigma_{33} \sigma_{11}) - \sigma_{12}^2 - \sigma_{23}^2 - \sigma_{31}^2$$

$$I_3 = \begin{vmatrix} \sigma_{11} & \sigma_{12} & \sigma_{13} \\ \sigma_{21} & \sigma_{22} & \sigma_{23} \\ \sigma_{31} & \sigma_{32} & \sigma_{33} \end{vmatrix}$$

$I_1$, $I_2$, and $I_3$ are also known as stress invariants because their values do not vary with the choice of coordinate system. The state of plane stress is defined when the nonzero components of stress are restricted to a plane; in other words, when $\sigma_z = \sigma_{xz} = \sigma_{yz} = 0$. Thus, the nonzero components of stress are $\sigma_x$, $\sigma_y$, and $\sigma_{xy}$.

## 2.2 Strain

Let two points located in a solid being deformed be displaced relative to their original positions. Strain is defined in terms of such displacements in a manner as to exclude the rigid body translation and rotation. For example, if the distance $l_0$ between two points A and B in a solid refers to an initial underformed condition, and A moves to A' and B to B' after application of the load, a state of strain exists if the distance between A' and B', $l \neq l_0$. The strain, $\varepsilon$, is defined as:

$$\varepsilon = \frac{l - l_0}{l_0} = \frac{\Delta l}{l_0} \tag{2.10}$$

If the change in length is large, we may define strain as follows:

$$\varepsilon = \int_0^\varepsilon d\varepsilon = \int_{l_0}^l \frac{dl}{l} = \ln \frac{l}{l_0} \tag{2.11}$$

If we consider displacements of the four corners of a two-dimensional element ABCD in Figure 2.3 instead of a line as in equation (2.10), the two-dimensional strain tensor can be written as:

$$\varepsilon_{11} = \varepsilon_{xx} = \frac{\partial u}{\partial x} \tag{2.12a}$$

$$\varepsilon_{22} = \varepsilon_{yy} = \frac{\partial v}{\partial y} \tag{2.12b}$$

$$\gamma_{xy} = 2\varepsilon_{xy} = 2\varepsilon_{12} = \frac{\partial u}{\partial y} + \frac{\partial v}{\partial x} \tag{2.12c}$$

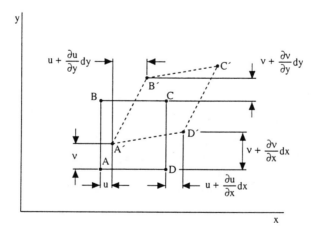

*Figure 2.3* Displacements in a two-dimensional deformed element.

If we consider a 3-dimensional elemental cube, the following additional components of strain can be defined:

$$\varepsilon_{33} = \varepsilon_{zz} = \frac{\partial w}{\partial z} \tag{2.12d}$$

$$\gamma_{xz} = 2\varepsilon_{xz} = 2\varepsilon_{13} = \frac{\partial u}{\partial z} + \frac{\partial w}{\partial x} \tag{2.12e}$$

$$\gamma_{yz} = 2\varepsilon_{yz} = 2\varepsilon_{23} = \frac{\partial v}{\partial z} + \frac{\partial w}{\partial y} \tag{2.12f}$$

where the $u$ and $v$ are displacements as shown in Figure 2.3 and $w$ is the corresponding displacement in the $z$-direction. Equations (2.12) can be collectively written in index notations as:

$$\varepsilon_{ij} = \frac{1}{2}\left( \frac{\partial u_i}{\partial x_j} + \frac{\partial u_j}{\partial x_i} \right) \tag{2.13}$$

From equation (2.13), it also follows that:

$$\varepsilon_{ij} = \varepsilon_{ji} \tag{2.14}$$

Also note the relationship in equations (2.12) between the engineering shear strain and the tensorial shear strain components. The tensorial shear strains are half of the values of the corresponding engineering shear strain components. Thus, all components of the strain tensor may be written as:

$$\varepsilon_{ij} = \begin{vmatrix} \varepsilon_{11} & \varepsilon_{12} & \varepsilon_{13} \\ \varepsilon_{21} & \varepsilon_{22} & \varepsilon_{23} \\ \varepsilon_{31} & \varepsilon_{32} & \varepsilon_{33} \end{vmatrix} \tag{2.15}$$

Similar to principal stresses, principal strains are defined as the normal strains on the planes with zero shear strains and are represented by $\varepsilon_1$, $\varepsilon_2$, $\varepsilon_3$. It can also be rigorously shown that the planes of principal stresses and principal strains are the same.

The state of plane strain is defined as the condition when the strains are limited to a single plane; in other words, when $\varepsilon_{zz} = \gamma_{xz} = \gamma_{yz} = 0$.

## 2.3 Elasticity

The relationships between stress and strain are known as the constitutive equations. For isotropic, homogeneous, and elastic materials, these relationships are defined by the Hooke's Law. For uniaxial tension, the normal strain in the direction of loading is given by:

$$\varepsilon_1 = \frac{\sigma_1}{E} \qquad (2.16a)$$

and the transverse strains (or the other principal strains are given by):

$$\varepsilon_2 = \varepsilon_3 = -\upsilon\varepsilon_1 \qquad (2.16b)$$

where E = elastic modulus (or Young's modulus) and $\upsilon$ = Poisson's ratio. For a 3-dimensional stress state, the relationships between stress and elastic strains are given by:

$$\varepsilon_x = \frac{1}{E}\left[\sigma_x - \upsilon(\sigma_y + \sigma_z)\right] \qquad (2.17a)$$

$$\varepsilon_y = \frac{1}{E}\left[\sigma_y - \upsilon(\sigma_x + \sigma_z)\right] \qquad (2.17b)$$

$$\varepsilon_z = \frac{1}{E}\left[\sigma_z - \upsilon(\sigma_x + \sigma_y)\right] \qquad (2.17c)$$

$$\gamma_{xy} = 2\varepsilon_{12} = \frac{\tau_{xy}}{G} \qquad (2.17d)$$

$$\gamma_{yz} = 2\varepsilon_{23} = \frac{\tau_{yz}}{G} \qquad (2.17e)$$

$$\gamma_{zx} = 2\varepsilon_{31} = \frac{\tau_{zx}}{G} \qquad (2.17)$$

where G is the shear modulus and for an isotropic, homogeneous material, it is given by:

$$G = \frac{E}{2(1 + \upsilon)} \qquad (2.18)$$

For the state of plane strain:

$$\sigma_z = \upsilon(\sigma_x + \sigma_y) \qquad (2.19)$$

### 2.3.1 Elastic Strain Energy

If a bar of length $x$ subjected to a force, $F$, elongates by a distance $dx$, the work done by the force, $F$ is given by $Fdx$. Thus, the work per unit volume is given by:

$$dW = \frac{Fdx}{Ax} = \sigma_1 \, d\varepsilon_1 \tag{2.20}$$

For uniaxial tension, $\sigma_1 = E\varepsilon_1$. Thus:

$$dW = E\varepsilon_1 \, d\varepsilon_1 \tag{2.20a}$$

or

$$W = \frac{E\varepsilon_1^2}{2} = \frac{\sigma_1^2}{2E} \tag{2.21}$$

Since for elastic materials, W is also the stored strain energy-per-unit volume in the solid, it is also known as the strain energy density. For a 3-dimensional state of stress, W is given by:

$$W = \frac{1}{2} \left( \sigma_{xx} \varepsilon_{xx} + \sigma_{yy} \varepsilon_{yy} + \sigma_{zz} \varepsilon_{zz} + \tau_{xy} \gamma_{xy} + \tau_{yz} \gamma_{yz} + \tau_{zx} \gamma_{xz} \right) \tag{2.22a}$$

The above equation in the form of index notation can be written as:

$$W = \frac{1}{2} \sigma_{ij} \varepsilon_{ij} \tag{2.22b}$$

For elastic materials that do not obey the Hooke's law (also known as nonlinear elastic materials), we can write the equivalent of equation (2.20a) as:

$$dW = \sigma_{ij} \, d\varepsilon_{ij}$$

or

$$W = \int_0^{\varepsilon_{ij}} \sigma_{ij} \, d\varepsilon_{ij} \tag{2.23}$$

Thus, if the strain energy density distribution W(x, y, z) is known, we can also write:

$$\sigma_{ij} = \frac{\partial W}{\partial \varepsilon_{ij}} \qquad (2.24)$$

The complementary strain energy density, $W^*$, is defined by:

$$W^* = \sigma_{ij}\, \varepsilon_{ij} - W \qquad (2.25)$$

which leads to the result:

$$\varepsilon_{ij} = \frac{\partial W^*}{\partial \sigma_{ij}} \qquad (2.26)$$

Figure 2.4 shows the strain energy density and complementary strain energy density for linear elastic and nonlinear elastic materials for the uniaxial case. It is also straightforward to see that for linear elasticity $W = W^*$.

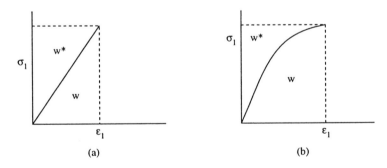

***Figure 2.4*** *Definition of strain energy density, W, and complementary energy density, $W^*$, for a uniaxially loaded body (a) linear-elastic materials and (b) nonlinear-elastic materials.*

## 2.4 Plasticity

### 2.4.1 Uniaxial Stress-Strain Curve

In a uniaxial state of stress, when the applied stress, $\sigma$, exceeds the yield strength of the material, $\sigma_0$, permanent deformation (or plastic deformation) occurs as shown in Figure 2.5. Often, it is not possible to precisely define the critical stress at which plastic deformation commences, therefore, the operational

beyond the yield strength, the material continues to deform plastically until instability is reached. The stress at which instability occurs, $\sigma_u$, is the ultimate tensile strength and the corresponding strain, $\varepsilon_u$, is called the uniform strain. The latter term results from the observation that up to $\varepsilon_u$ the strain is distributed uniformly in the specimen and beyond $\varepsilon = \varepsilon_u$, it is concentrated in the region where a neck develops and eventually fracture occurs. The stress-strain relationship can be described by the so-called Ramberg-Osgood relationship:

$$\varepsilon = \frac{\sigma}{E} + \alpha \varepsilon_0 \left(\frac{\sigma}{\sigma_0}\right)^m \tag{2.27}$$

where $\alpha$ and $m$ are material constants derived from regression of the stress-strain curve and $\varepsilon_0 = \sigma_0 / E$. Both $m$ and $\alpha$ are dimensionless. $m$ is also inverse of the strain hardening exponent which is frequently reported in the literature along with data from tensile tests. Elastic, perfectly-plastic materials are those which have a flat stress-strain curve when $\sigma = \sigma_0$.

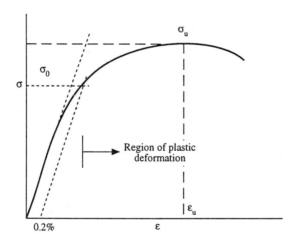

*Figure 2.5* Uniaxial stress-strain curve for a work-hardening material.

### 2.4.2 Von Mises Yield Criterion, Effective Stress, and Effective Strain

Stress states in components are often more complex than in a uniaxial tensile test. Therefore, it is essential to choose a yield criterion for multiaxial stress states. The one used commonly in fracture mechanics is the Von Mises criterion. For a detailed discussion on other yield criteria, the reader should consult other books [2.3-2.5]. The assumptions which accompany the use of the Von Mises criterion are:

- The yield strengths in tension and compression are the same.

- The volume during plastic deformation is conserved, thus the Poisson's ratio is 0.5.
- The mean normal stress (or the hydrostatic stress), $\sigma_m = \frac{1}{3}(\sigma_1 + \sigma_2 + \sigma_3)$ does not participate in the plastic deformation process.

The last assumption implies that yielding must occur when the function $f$ of $(\sigma_1 - \sigma_2)$, $(\sigma_2 - \sigma_3)$ and $(\sigma_3 - \sigma_1)$ is a constant:

$$f\left[(\sigma_1 - \sigma_2), (\sigma_2 - \sigma_3), (\sigma_3 - \sigma_1)\right] = C \tag{2.28}$$

Thus, if a stress state defined by $\sigma_1, \sigma_2, \sigma_3$ causes yielding, an equivalent stress state will also cause yielding, if $\sigma_1' = \sigma_1 - \sigma_m$, $\sigma_2' = \sigma_2 - \sigma_m$, $\sigma_3' = \sigma_3 - \sigma_m$ because the two stress states differ only by the mean normal stress which does not participate in yielding. Mises postulated that yielding occurs when the root mean square of the three maximum values of the shear stresses becomes equal or exceeds a certain constant value:

$$\left[\frac{(\sigma_1 - \sigma_2)^2 + (\sigma_2 - \sigma_3)^2 + (\sigma_3 - \sigma_1)^2}{3}\right]^{1/2} = C_1$$

or

$$(\sigma_1 - \sigma_2)^2 + (\sigma_2 - \sigma_3)^2 + (\sigma_3 - \sigma_1)^2 = C_2 \tag{2.29}$$

Applying the above criterion to the tensile test yields:

$$C_2 = 2\sigma_0^2 \tag{2.30}$$

In the more general form, the criterion can be written as:

$$(\sigma_x - \sigma_y)^2 + (\sigma_y - \sigma_z)^2 + (\sigma_z - \sigma_x)^2 + 6(\tau_{xy}^2 + \tau_{yz}^2 + \tau_{zx}^2) = 2\sigma_0^2 \tag{2.31}$$

Equations (2.29) and (2.30) directly lead to the definition of effective (or equivalent) stress, $\sigma_e$, as being:

$$\sigma_e = \frac{1}{\sqrt{2}}\left[(\sigma_1 - \sigma_2)^2 + (\sigma_2 - \sigma_3)^2 + (\sigma_3 - \sigma_1)^2\right]^{1/2} \tag{2.32}$$

For uniaxial tensile loading, $\sigma_e = \sigma_1$. Effective strain is defined in a manner such that the incremental work per unit volume is:

$$dW_p = \sigma_e d\varepsilon_{pe} = \sigma_1 d\varepsilon_{p1} + \sigma_2 d\varepsilon_{p2} + \sigma_3 d\varepsilon_{p3} \tag{2.33}$$

where the subscript p represents the plastic part of the strain component and strain energy. For the Von Mises criterion, we state without proof that:

$$d\varepsilon_{pe} = \left[\frac{2}{3}(d\varepsilon_{p1}^2 + d\varepsilon_{p2}^2 + d\varepsilon_{p3}^2)\right]^{\frac{1}{2}} \tag{2.34}$$

For the case of proportional increments in strains for which $d\varepsilon_1 : d\varepsilon_2 : d\varepsilon_3$,

$$\varepsilon_{pe} = \left[\frac{2}{3}(\varepsilon_{p1}^2 + \varepsilon_{p2}^2 + \varepsilon_{p3}^2)\right]^{\frac{1}{2}} \tag{2.35}$$

If the strain path is not proportional, the effective strain must be found by integrating equation (2.34) along the paths of strain increments. The total strain can be obtained by adding the equivalent plastic strain to the elastic strain which is given by $\sigma_o / E$.

The above equations imply that the uniaxial stress-strain curve, equation (2.27), is in fact an effective stress-effective strain curve because during plastic deformation in a tensile test, $d\varepsilon_2 = d\varepsilon_3 = -0.5 d\varepsilon_1$, thus we are assured of proportionality in strain increment and from equations (2.32) and (2.35), $\sigma_e = \sigma_1$ and $\varepsilon_e = \varepsilon_1$. In Figure 2.6, the stress-strain behavior of copper for several multiaxial states of stress under proportional loading is compared to the uniaxial stress-strain curve [2.2, 2.7]. These data clearly support the result that the uniaxial stress-strain curve is essentially the effective stress-effective strain curve.

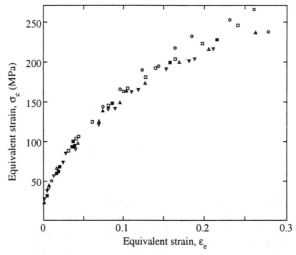

*Figure 2.6* Stress-strain behavior for several multiaxial proportional loading paths, in which $\sigma_2/\sigma_1$ was varied between 0 and 1, compared with the uniaxial stress-strain behavior of copper (Ref. 2.2, 2.7).

## 2.4.3 Flow Rules or Plastic Stress-Strain Equations

The constitutive equations for elastic materials were given in equation (2.17). Similar equations can also be given for plastic behavior and are known as the flow rules:

$$d\varepsilon_1 = \frac{d\varepsilon_e}{d\sigma_e}\left[\sigma_1 - 0.5(\sigma_2 + \sigma_3)\right] \qquad (2.36a)$$

$$d\varepsilon_2 = \frac{d\varepsilon_e}{d\sigma_e}\left[\sigma_2 - 0.5(\sigma_1 + \sigma_3)\right] \qquad (2.36b)$$

$$d\varepsilon_3 = \frac{d\varepsilon_e}{d\sigma_e}\left[\sigma_3 - 0.5(\sigma_3 + \sigma_1)\right] \qquad (2.36c)$$

The above flow rule can be arranged to show that:

$$d\varepsilon_1 : d\varepsilon_2 : d\varepsilon_3 = \left[\sigma_1 - 0.5(\sigma_2 + \sigma_3)\right] : \left[\sigma_2 - 0.5(\sigma_1 + \sigma_3)\right] : \left[\sigma_3 - 0.5(\sigma_1 + \sigma_2)\right] \qquad (2.37)$$

Equation (2.37) shows that if the ratios of principal stresses $\sigma_1/\sigma_2$ and $\sigma_1/\sigma_3$ remain constant, the ratios of the increment in strain are also constant (or vice-versa). For these conditions the increments in strains can be replaced by total strains and equation (2.36) can be written as:

$$\varepsilon_1 = \frac{\varepsilon_e}{\sigma_e}\left[\sigma_1 - 0.5(\sigma_2 + \sigma_3)\right] \qquad (2.38)$$

The equations for $\varepsilon_2$ and $\varepsilon_3$ can be similarly changed. The ratio $\varepsilon_e/\sigma_e$ can be evaluated easily along the stress-strain curve as shown in Figure 2.7, making plasticity analysis much simpler. When the flow rule is expressed in the form of equation (2.38), the resulting theory is known as the deformation theory of plasticity. When the flow rules in the form of equation (2.36) are applied, we call it the incremental theory of plasticity. It is also important to restate that for proportional loading, the incremental and deformation theories yield the same results.

In the subsequent chapters in this book, we will be using the deformation theory of plasticity quite extensively because crack tip plasticity conditions, for the most part, lend themselves nicely to the assumption of proportional loading. However, there is an additional limitation to the use of this theory for metals which must also be clearly understood. For nonlinear-elastic materials, $\sigma_e/\varepsilon_e$ is the same for loading and unloading. Thus, deformation theory of plasticity applies for loading and unloading. Metals, on the other hand, have different loading and unloading stress-strain curves in the plastic regime. Therefore, the value of $\sigma_e/\varepsilon_e$ is not unique if both loading and unloading is involved. However, if we consider only loading, the behavior of metals is the same as nonlinear elastic materials. Thus, we will apply the additional restriction of only monotonic loading, on the use of deformation theory of plasticity. Figure 2.6 showed a plot of equivalent stress vs. equivalent strain for several different ratios of $\sigma_2/\sigma_1$ for copper [2.7]. It appears that the data from all the stress ratios collapse into almost a single trend. These results demonstrate the applicability of the deformation theory of plasticity as described in this section. The same study also showed that using the maximum shear stress criterion, the correlation between the equivalent shear stress and equivalent plastic shear strain is somewhat better than the one based on Von Mises theory of yielding, Figure 2.8. Nevertheless, Von Mises' theory is used extensively in analyzing multiaxial stress states and will also be the primary theory used in the later chapters of this book.

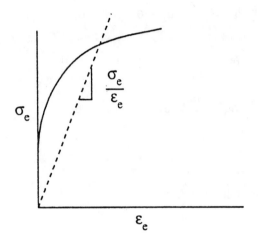

*Figure 2.7* Schematic of the definition of plastic modulus in equation (2.8).

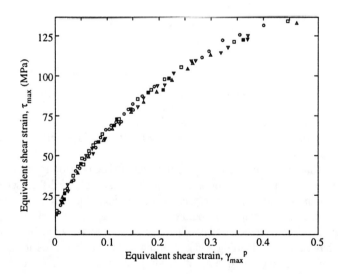

*Figure 2.8* The data in Figure 2.6 for various multiaxial stress states correlated in the form of equivalent plastic shear strain and maximum shear stress (Ref. 2.2, 2.7).

## 2.5 Consideration of Creep

When the temperature of metals increases above approximately $0.35T_m$, where $T_m$ = melting point in degrees Kelvin, time-dependent deformation under stress or creep becomes a consideration. Time-dependent deformation also occurs in polymers at much lower temperatures. In this section, we will cover some basic models for representing creep deformation behavior which are needed to understand the time-dependent fracture mechanics concepts discussed in Chapters 10 to 13. As with the other deformation models, this discussion is also not intended to be a complete treatment of creep. Instead, the emphasis is on providing the reader with some working knowledge of creep and familiarity with the terms. Let us begin with the discussion on uniaxial creep behavior and then consider the effects of multi-axial loading. Figure 2.9 shows a schematic of the typical strain vs. time behavior at constant stress during creep deformation. The deformation behavior can be divided into three regions. In Region I, the strain rate continuously decreases. This region is called the primary creep region in which strain-hardening occurs. Region II is known as the secondary (or steady-state) creep region and is characterized by constant strain rate. Region III, known as the tertiary creep region, is characterized by increasing strain rate and also the region in which extensive damage develops eventually leading to rupture. For engineering purposes, Regions I and II are most relevant.

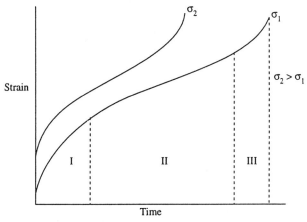

Region I : Primary creep
Region II : Secondary (or steady-state) creep
Region III : Tertiary creep

***Figure 2.9*** *Schematic of the creep deformation curve at constant stress through the various stages.*

### 2.5.1 Models for Representing Creep Deformation

If the constant creep rates in Region II are correlated with stress, the following relationship called the Norton's Law is observed:

$$\dot{\varepsilon} = A\sigma^n \qquad (2.39)$$

Figure 2.10 shows a plot of such data for a 1Cr-1Mo-0.25V steel at several temperatures. The slope of the various lines gives the value of $n$ and the intercept is related to the value of $A$. For materials in which the steady-state creep dominates, the creep behavior is completely specified by equation (2.39). Note the similarity between the plastic term in equation (2.27) and equation (2.39). Here, strain is replaced by strain rate, $m$ by $n$ and the term $\alpha \, \varepsilon_0 / \sigma_0^m$ is replaced by $A$. This analogy is important in analyzing bodies subjected to creep deformation for which a fully-plastic analysis is available. Specifically, the results of the fully-plastic analysis are directly applicable to identical configurations which are dominated by secondary creep.

Primary creep deformation is represented by a strain (or time) hardening law with the following characteristic equation:

$$\dot{\varepsilon}_p = A_1 \, \varepsilon^{-p} \, \sigma^{n_1(1+p)} \tag{2.40}$$

where p, $A_1$, and $n_1$ are regression constants and $\dot{\varepsilon}_p$ is the primary creep rate. At any instant, the primary creep rate may be obtained by subtracting the secondary creep rate from the total creep rate. Equation (2.40) can be integrated and written in the following form:

$$\varepsilon_p = \left[ A_1 \, (1+p) \, t \, \right]^{\frac{1}{1+p}} \sigma^{n_1} \tag{2.41}$$

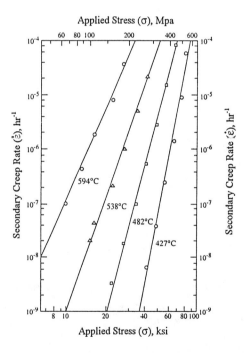

**Figure 2.10** *Steady-state creep rates as a function of temperature for 1Cr-1Mo-0.25V steel at various temperatures.*

If we compare equation (2.41) with equation (2.27), we can draw the following analogies. $n_1$ replaces m and $[A_1 (1 + p)t]^{1/(1+p)}$ replaces the value of $\alpha \varepsilon_0 / \sigma_0^m$. Again, this model offers the advantage of being able to apply existing solutions for fully-plastic bodies to easily obtain solutions to bodies whose behavior is dominated by primary creep. This ability has tremendous value in fracture mechanics. Therefore, the above equation is almost exclusively used to represent the primary creep behavior in this text, even though it is recognized that there are other models which can also be used to represent primary creep behavior in metals. Often, the deformation behavior of components at elevated temperature involves a combination of elastic, primary and secondary creep behaviors. In such instances, the total strain can be written as:

$$\varepsilon = \varepsilon_{el} + \varepsilon_p + \varepsilon_{ss} \qquad (2.42)$$

where $\varepsilon_{el}$ = elastic strain, $\varepsilon_p$ = primary creep strain, and $\varepsilon_{ss}$ is the strain due to steady-state creep. By recognizing that $\varepsilon_{el} = \sigma/E$ and taking the time derivative of equation (2.42), we get:

$$\dot{\varepsilon} = \frac{\dot{\sigma}}{E} + A_1 \varepsilon^{-p} \sigma^{n_1(1+p)} + A\sigma^n \qquad (2.43)$$

or

$$\varepsilon = \frac{\sigma}{E} + \left[ A_1 (1+p)t \right]^{\frac{1}{1+p}} \sigma^{n_1} + A\sigma^n t \qquad (2.44)$$

The above equations are constitutive relationships which include elastic, primary creep and secondary creep behavior.

### 2.5.2 Influence of Temperature on Creep Rates

Since creep is a thermally activated process, it has a characteristic activation energy, $Q_c$ given by:

$$\dot{\varepsilon} = A_0 \exp\left[ -\frac{Q_c}{R}\left( \frac{1}{T} - \frac{1}{T_0} \right) \right] f(\sigma) \qquad (2.45)$$

where R = universal gas constant, T = absolute temperature, $A_0$ = pre-exponent constant at some reference temperature $T_0$, and f ($\sigma$) is a function of stress. For example, f ($\sigma$) = $\sigma^n$ for secondary creep and $\sigma^{n1}$ for primary creep. The value of $Q_c$ can be determined from creep strain rate data at a constant stress at different temperatures. For example, if the natural log of steady-state strain rates at various temperatures are plotted as a function of 1/T, the slope of the curve is related to $Q_c$. An example of this calculation is shown in Figure 2.11 for a 1.25Cr - 0.5Mo steel for data obtained at a stress level of 68.94 MPa. The value of the activation energy is approximately equal to activation energy for self-diffusion. This is based on the observation that the rate of dislocation climb is controlled by vacancy migration which is also the rate-controlling step for self-diffusion. Equation (2.45) may be used to estimate creep rates at various temperatures using the published values of self-diffusion activation energies in place of $Q_c$. A comparison of creep and self-diffusion activation energies for several pure metals was made by Sherby and Miller [2.8] and is reproduced in Figure 2.12. This plot clearly shows the one-to-one relationship between the two. Based on these arguments, the temperature dependence of constants A and

$A_1$ in equations (2.43) and (2.44) can be given by:

$$A = A_0 \exp\left[ -Q_c/R \left( \frac{1}{T} - \frac{1}{T_0} \right) \right] \tag{2.46a}$$

and

$$A_1 = A_{10} \exp\left[ -Q_c/R \left( \frac{1}{T} - \frac{1}{T_0} \right) \right] \tag{2.46b}$$

where $A_0$ and $A_{10}$ are the respective values of A and $A_1$ at the reference temperature, $T_0$. As a first approximation, it may be assumed that the exponents $p$, $n_1$, and $n$ are not dependent on temperature. Although, examining Figure 2.10, it is clear that this is only a rough approximation. It must also be pointed out that the value of $Q_c$ can vary significantly from the activation energy for self-diffusion for alloys because dislocation movement at high temperatures in alloys is not totally dominated by dislocation climb. Also, the value of $Q_c$ is dependent on stress levels because diffusion can be aided by stress thereby making the apparent activation energy less than the activation energy for self-diffusion.

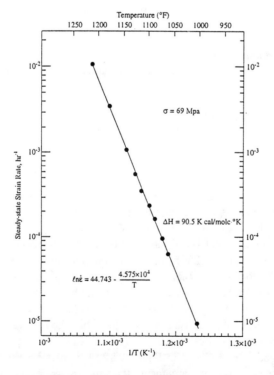

**Figure 2.11** *Steady-state creep rates in 1.25Cr-0.5Mo steel at various temperatures plotted as a function of the inverse of the absolute temperature (1/T).*

### 2.5.3 Creep Under Multiaxial Stress States

The uniaxial creep behavior may be generalized to multiaxial behavior for homogeneous and isotropic bodies by simply considering the effective stress and effective strain rate, $\dot{\varepsilon}_e$, in a manner analogous to the case of plasticity. Thus:

$$\dot{\varepsilon}_e = g(\sigma_e) \qquad (2.47)$$

where the function $g$ is the same for uniaxial and multiaxial cases. Thus, the creep constitutive equations are analogous to the equation for deformation theory of plasticity.

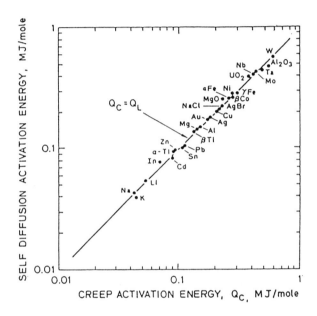

*Figure 2.12* Comparison of the activation energy for creep and self-diffusion for several pure metals, from Sherby and Miller (Ref. 2.8).

### 2.6 Component Analysis in the Plastic Regime

We consider the simple geometry of a rectangular beam subjected to three-point bending as an example of a component. The applied load, P, is increased in steps until large deflections occur and the beam effectively collapses. We are to determine the stress distribution in the beam as a function of the moment, M, in the beam or the applied load, P. Let the beam thickness be B, the height 2H and the length 2L, as shown in Figure 2.13a. We will consider this problem for two types of material behavior (*i*) elastic, perfectly-plastic materials, and (*ii*) Ramberg-Osgood materials.

The normal stress, $\sigma_x$, for linear elastic bending is given by [2.2]:

$$\sigma_x = \frac{My}{I_z} \tag{2.48}$$

where the coordinate system is chosen such that the x-axis coincides with the neutral axis and $I_z$ = bending moment of inertia = $B(2H)^3/12$. We assume in this case that the beam length is sufficiently large that shear stresses are negligible. Substituting for $I_z$ and choosing $y = H$, the maximum tensile stress is given by:

$$\sigma_H = \frac{3M}{2BH^2} \tag{2.49}$$

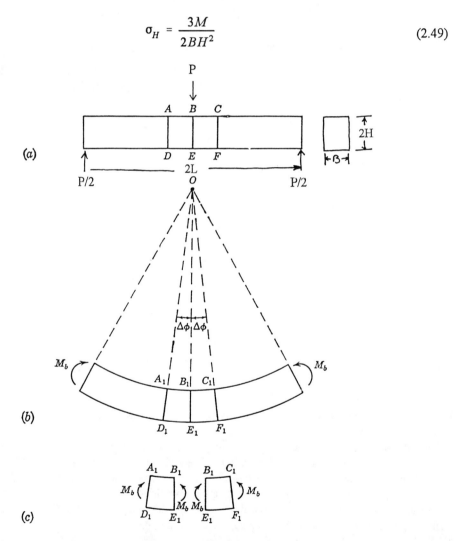

**Figure 2.13** (a) A beam of thickness B, height 2H, and length 2L, subjected to 3-point bending; (b) deformation in the beam as a result of bending showing that planes remain as planes in the deformed state.

Next, we consider that the loads have been increased to high enough values to cause plastic deformation. If we assume that even under plastic deformation, planes remain as planes as illustrated in Figure 2.13b, the strain distribution along the y-axis can be estimated by:

$$\varepsilon_x = \frac{\varepsilon_H}{H} y \qquad (2.50)$$

where $\varepsilon_H$ is the strain at the outer surface of the beam, (y = H). This result is the same as for the elastic case [2.2]. Thus, the stress distribution in the beam is influenced by yielding, but the strain distribution is not. To maintain equilibrium, the force and moment must follow the following equations:

$$PL/2 = M = \int_{-H}^{H} \sigma_x \, By\,dy \qquad (2.51)$$

and

$$\int_{-H}^{H} \sigma_x \, B\,dy = 0 \qquad (2.52)$$

Equation (2.52) results from the consideration that the volumes under tensile and compressive stresses result in equal and opposite forces because there is no external force along the x-axis. In the following analysis, we make a further assumption that the stress-strain behavior of the material is the same in tension as it is in compression. We begin first with the case of elastic, perfectly-plastic material. The yield strength is $\sigma_0$. The stress distribution is schematically shown in Figure 2.14. There is a discontinuity in the stress distribution at the value of $y$ where yielding occurs. Let this value of $y$ be $y_0$. Thus for $|y| \geq y_0$, $\sigma = \sigma_0$ and for $|y| < y_0$ the stress distribution is linear and can be obtained from equation (2.50) as follows:

$$\frac{\sigma_x}{E} = \frac{\varepsilon_H}{H} y \qquad (2.53)$$

Applying equation (2.50) at $y = y_0$ yields:

$$\varepsilon_H = \frac{\varepsilon_0 H}{y_0} \qquad (2.54)$$

thus, equation (2.53) becomes:

$$\sigma_x = \frac{E\varepsilon_0}{y_0} y = \frac{\sigma_0}{y_0} y \qquad (2.55)$$

Since we assume that the stress distribution about the neutral axis is symmetric, equation (2.52) is automatically satisfied. We proceed to substitute equation (2.55) into equation (2.51). Since the stress distribution has a discontinuity, the integration must be performed in two steps as:

$$M = 2B \left[ \int_0^{y_0} \sigma_x y \, dy + \int_{y_0}^{H} \sigma_x y \, dy \right] \quad (2.56)$$

Substituting $\sigma_x = \sigma_0$ in the second term and $\sigma_x$ from equation (2.55) and using equation (2.54) to eliminate $y_0$, we get:

$$M = BH^2 \sigma_0 \left[ 1 - \frac{1}{3} \left( \frac{\sigma_0}{E\varepsilon_H} \right)^2 \right] \quad \text{for} \quad \varepsilon_H > \varepsilon_0 \quad (2.57)$$

A limiting situation occurs when $\varepsilon_H$ becomes very large and

$$M = M_0 = BH^2 \sigma_0 \quad (2.58)$$

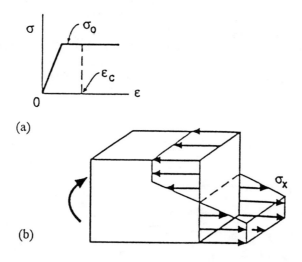

**Figure 2.14** *Schematic of the stress distribution in a beam subjected to bending for the case of elastic, perfectly-plastic material.*

At this point, plastic deformation has spread over the entire cross-section of the beam. Thus, $M_0$ is known as the fully-plastic moment. Initial yielding occurs when $\varepsilon_H = \varepsilon_0$. The moment at that point $M_i$ is given by:

$$M_i = BH^2 \sigma_0 \,(2/3) \qquad (2.59)$$

or

$$M_0/M_i = 1.5$$

A plot of $M/M_0$ and $\varepsilon_H/\varepsilon_0$ is shown in Figure 2.15 along with the accompanying stress distributions at the various values of $M/M_0$. Figure 2.16 schematically shows the development of plasticity in the beam. At $M = M_0$, a plastic hinge develops in the center of the beam as shown in Figure 2.16d.

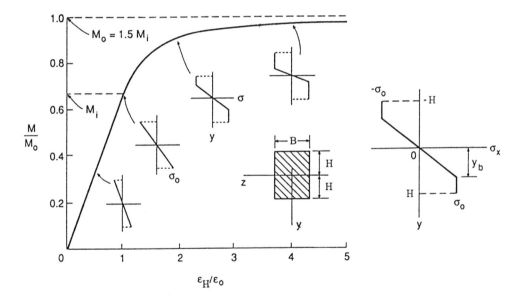

**Figure 2.15** *Plot of $M/M_0$ as a function of the surface strain for an elastic, perfectly-plastic material, adapted from Dowling (Ref. 2.3).*

Next, we will determine the relationship between the moment and the outer fiber strain for a material which plastically deforms by the Ramberg-Osgood relationship, equation (2.27). This relationship does not have a discontinuity, therefore, equation (2.56) becomes:

$$M = 2B \int_0^H \sigma_x y \, dy \qquad (2.60)$$

Substituting for $y$ and $dy$ from equation (2.50), we get:

$$M = 2B\left(\frac{H}{\varepsilon_H}\right)^2 \int_0^{\varepsilon_H} \sigma_x \varepsilon_x d\varepsilon_x \tag{2.61}$$

From equation (2.27), we can substitute for $\varepsilon_x$ and by differentiating equation (2.27), we can substitute for $d\varepsilon_x$. The resulting equation will be:

$$M = 2B\left(\frac{H}{\varepsilon_H}\right)^2 \int_0^{\sigma_H} \sigma_x \left[\frac{\sigma_x}{E} + \alpha\,\varepsilon_0\left(\frac{\sigma_x}{\sigma_0}\right)^m\right]\left[\frac{1}{E} + \frac{\alpha\varepsilon_0}{\sigma_0^m} m\sigma_x^{m-1}\right] d\sigma_x \tag{2.62}$$

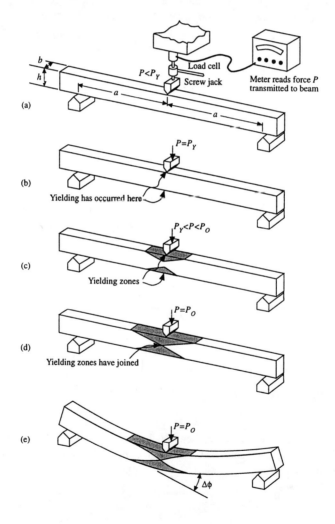

*Figure 2.16* Development of plasticity in a beam loaded in 3-point bending with increasing load levels, adapted from Crandall and Dahl (Ref. 2.2).

The above integral can be easily evaluated to yield [2.3]:

$$M = \frac{2BH^2\sigma_H}{3} \left| \frac{1 + \frac{3m+3}{2m+2}\beta + \frac{3m}{1+2m}\beta^2}{(1+\beta)^2} \right| \quad (2.63)$$

$$\text{where} \quad \beta = \frac{E\alpha\varepsilon_0}{\sigma_0}\left(\frac{\sigma_H}{\sigma_0}\right)^m = \frac{\varepsilon_{Hp}}{\varepsilon_{He}}$$

Where $\varepsilon_{Hp}$ is the plastic part of the outer fiber strain and $\varepsilon_{He}$ is the elastic part of the outer fiber strain. The relationship between M and the outer fiber strain, $\varepsilon_H$, is shown in Figure 2.17. In this case, the moment rises continuously with the outer fiber strain as opposed to reaching a saturation value as in the case of material with no strain hardening. If $\varepsilon_{Hp}$ is small in relation to $\varepsilon_{He}$ ($\beta \to 0$) equation (2.63) reduces to equation (2.49). If, on the other hand, plasticity dominates or $\varepsilon_{Hp} \gg \varepsilon_{He}$, $\beta \to \infty$:

$$M = \frac{2BH^2 m}{(1+2m)}\sigma_H = \frac{2BH^2 m}{(1+2m)}\sigma_0\left(\frac{\varepsilon_H}{\alpha\varepsilon_0}\right)^{\frac{1}{m}}$$

or

$$\varepsilon_H = \alpha\varepsilon_0 \left| \frac{1+2m}{2m}\left(\frac{M}{M_0}\right) \right|^m \quad (2.64)$$

If the beam is operated at elevated temperature in the regime where the strain rate is dominated by secondary creep, the following relationship can be derived:

$$\dot{\varepsilon}_H = A\left[\frac{1+2m}{2m}\left(\frac{M}{BH^2}\right)\right]^n \quad (2.65)$$

The derivation of the above relationship is left as an exercise problem.

If a member is subjected to combined axial load and bending, the neutral axis is shifted from its geometrical centroid. Also, the portions of the cross-section subjected tensile and compressive stresses must yield a net force which equals the applied axial force. Finding a solution to this problem is also left as an exercise.

## 2.7 Fully Plastic/Limit Loads

Equation (2.58) provides an estimate of the bending moment, $M_0$, at which a beam under bending is expected to become fully plastic. This is the highest moment that a beam made from an elastic, perfectly-plastic material can carry. For strain-hardening materials, the beam is capable of sustaining higher moments than $M_0$. For such materials, $M_0$ represents a lower-bound estimate of the moment at

42  *Nonlinear Fracture Mechanics for Engineers*

which the entire beam experiences plastic strains. It is also often referred to as limiting moment. For a center cracked panel of width 2W and crack length, 2a, the limit-load, $P_0$, is given by:

$$P_0 = 2B(W-a)\,\sigma_0 = 2BW(1-a/W)\,\sigma_0 \qquad (2.66)$$

Similarly, the limiting moment or load can be easily estimated for several notched geometries.

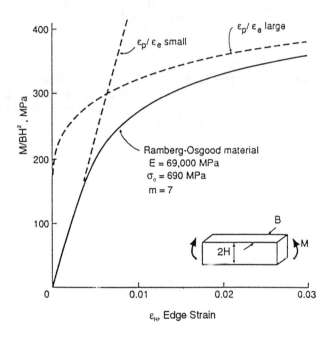

**Figure 2.17** *Plot of $M/M_0$ as a function of outer fiber strain for a material which deforms according to the Ramberg-Osgood relationship, adapted from Dowling (Ref. 2.3).*

## 2.8 Summary

A brief overview of elasticity, plasticity, and creep was presented in this chapter. We began with the definition of the stress and strain tensors and the equations for equilibrium. Both the conventional engineering notations and the index notations were used to represent stress and strain. It is important for the student to become familiar with both notations to derive maximum benefit from reading the rest of the book.

The constitutive equations, or the stress-strain relationships, were considered for elastic, plastic, and creep deformation. For the case of plasticity, only a power-law form for representing work-hardening was considered. This form has the advantage of representing the behavior of most materials accurately and with mathematical simplicity, which is useful in component analysis. Similarly, the creep constitutive laws from primary and secondary creep discussed in the chapter were also of the power-law form. This form allows us to use the analogy to fully-plastic solutions for components to derive the relationships between applied loads and the stress strain and strain-rate distribution in the component under creep conditions. This feature will be used extensively in later chapters.

A brief description and critique of the deformation and incremental theories of plasticity was provided. Both theories yield identical results if constant ratios between the three principal stresses (or proportional loading) can be maintained through the loading histories and no unloading occurs.

An example of a beam in bending was considered to illustrate plasticity analysis in components. For elastic, perfectly-plastic materials, the bending moment reaches a maximum value, $M_0$, at the time the entire cross-section of the beam becomes plastic. This value of the bending moment is known as the fully-plastic or the limiting moment. For work-hardening materials, $M_0$ represents the lower-bound estimate of the moment at which yielding occurs everywhere in the beam.

## 2.9 References

2.1   S.P. Timoshenko and J.M. Gere, "Mechanics of Materials", Third Edition, PWS Publishing Co., Boston, 1990.

2.2   S.H. Crandall and N.C. Dahl, "An Introduction to Mechanics of Solids", McGraw Hill, New York, 1959.

2.3   N.E. Dowling, "Mechanical Behavior of Materials", Prentice Hall, Englewood Cliffs, N.J., 1993.

2.4   W.F. Hosford and R.M. Caddell, "Metal Forming Mechanics and Metallurgy", 2nd edition, Prentice Hall, Englewood Cliffs, N.J., 1993.

2.5   W. Johnson and P.B. Mellor, "Engineering Plasticity", Van Nostrand Reinhold, New York, 1973.

2.6   R. Hill, "Plasticity", Clarendon Press, Oxford, 1950.

2.7   E.A. Davis, "Increase of Stress with Permanent Strain and Stress-Strain Relations in the Plastic State for Copper under Combined Stresses", Transactions of ASME Vol. 65, 1943, pp. A187-A196.

2.8   O.D. Sherby and A.K. Miller, "Combining Phenomenology and Physics in Describing the High Temperature Mechanical Behavior of Crystalline Solids", Transactions of ASME - Journal of Engineering Materials and Technology, Vol. 101, 1979, pp. 387-393.

## 2.10 Exercise Problems

2.1   Consider a stress state $\sigma_x = 70$, $\sigma_y = 35$, $\tau_{xy} = 20$, MPa and $\sigma_z = \tau_{xz} = \tau_{yz} = 0$. Find the principal stresses.

2.2   In Problem 2.1, determine the principal strains and the largest shear strains if $E = 70$ GPa and $\nu = 0.3$.

2.3   Determine the principal stresses for the following stress state:

$$\sigma_{ij} = \begin{vmatrix} 70 & -20 & 30 \\ -20 & 35 & 15 \\ 30 & 15 & 50 \end{vmatrix}$$

**44**   *Nonlinear Fracture Mechanics for Engineers*

2.4   Using Figure 2.3, derive equations (2.12 a to c).

2.5   A circle of 10mm diameter is printed on a thin sheet of metal prior to being shaped into an automotive part. After the shaping operation, the circle turns into an ellipse with its major and minor axes being 15 and 11mm, respectively.

(a) Determine the effective strain in the region of the ellipse and carefully state all of your assumptions.

(b) Assume plane stress prevailed during the deformation process and $\sigma_1 / \sigma_2$ remained constant, find the ratio $\sigma_1 / \sigma_e$.

2.6   For the cases of (a) uniaxial tension and (b) plane strain compression ($\varepsilon_2 = 0$ and $\sigma_3 = 0$), show that the incremental plastic work-per-unit volume as calculated by $dW = \sigma_e \, d\varepsilon_e$ is the same as calculated from $dW = \sigma_1 d\varepsilon_1 + \sigma_2 d\varepsilon_2 + \sigma_3 d\varepsilon_3$.

2.7   Starting from equation (2.62), derive equation (2.63).

2.8   From first principles show that the relationship between the applied moment and strain rate at the outer fiber for a beam in bending for a material controlled by secondary creep is given by equation (2.65). State all your assumptions in the derivation.

2.9   Repeat Problem 2.8 for the case of dominant primary creep and derive the relationship between outer fiber strain rate and applied moment.

2.10   Justify whether equations derived in Problems 2.8 and 2.9 can be combined to derive a relationship between applied bending moment and outer fiber strain rate if the material deformation is controlled by a combination of primary and secondary creep.

2.11   For a standard compact type specimen used in fracture mechanics testing, derive a formula for estimating the limit-load.

2.12   Derive the limit-load equation for a tensile test specimen with a rectangular cross-section ($b \times c$) containing a semicircular crack of radius $a$.

# CHAPTER 3

# REVIEW OF LINEAR ELASTIC FRACTURE MECHANICS

In this chapter, we will review the concepts of linear elastic fracture mechanics (LEFM) which is a widely used analytical tool for predicting fracture under brittle conditions. The limitation to brittle fracture is due to the use of linear elasticity theory as the mechanics framework for LEFM. Thus, for LEFM to be valid, only limited amounts of plasticity can be permitted to accompany the fracture process. This chapter provides only a brief refresher on the subject as a necessary background for the remaining chapters. The main results of the LEFM theory are described without providing proofs which are given in other text books [3.1-3.3] and it is assumed that the reader is already familiar with the topic at this level.

## 3.1 Basic Concepts

There are two primary approaches for predicting brittle fracture under dominantly linear-elastic conditions. The first is based on energy balance which establishes the necessary condition for fracture. The second is based on the amplitude of the crack tip stress intensity reaching a critical level for fracture to occur. The two approaches are described in the following sections. Subsequently, it will be shown that the two approaches are equivalent, thus, avoiding the need to choose between them, and placing the LEFM theory of fracture on firm ground. We will also show that both approaches can admit limited amounts of plasticity which is necessary for them to be applicable to metals and alloys.

### 3.1.1 Energy Balance Approaches to Fracture

This theory is based on the pioneering work of Griffith [3.4] with important modifications by Orowan [3.5] and by Irwin [3.6]. If we consider a crack of area $A$ in a deformable body subjected to an arbitrary load, Figure 3.1a, the energy balance dictates that:

$$\dot{W} = \dot{U} + \dot{K}_E + 2\gamma_s \dot{A} \tag{3.1}$$

where W = work performed by the external loads; U = increase in the elastic energy, $U_e$, of the cracked body plus the energy required to perform the work of plastic deformation, $U_p$, which accompanies fracture; $K_E$ = kinetic energy of the body; $\gamma_s$ = energy per unit surface area required for extending the crack. Note that surface area associated with a crack is twice the crack area. The dots denote the derivatives with time. If we assume that the loads are not time-dependent and the contribution of the kinetic energy is negligible at the onset of fracture, we can write:

$$\frac{\partial}{\partial t} = \frac{\partial A}{\partial t}\frac{\partial}{\partial A} = \dot{A}\frac{\partial}{\partial A} \tag{3.2}$$

Substituting equation (3.2) into (3.1), we get:

$$\frac{\partial W}{\partial A} = \left(\frac{\partial U_e}{\partial A} + \frac{\partial U_p}{\partial A}\right) + 2\gamma_s$$

$$\text{or} \quad \frac{\partial W}{\partial A} - \frac{\partial U_e}{\partial A} = \frac{\partial U_p}{\partial A} + 2\gamma_s \tag{3.3}$$

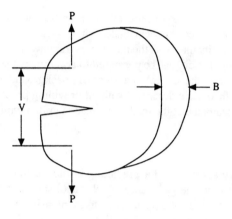

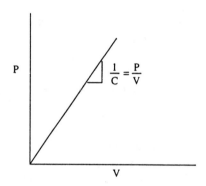

**Figure 3.1** *(a) A cracked body of an arbitrary shape loaded by a point load P resulting in a displacement along the load-line, V. (b) The relationship between load and displacement and the definition of compliance.*

The first term on the right-hand side of equation (3.3) is also an energy dissipative term like $\gamma_s$. Thus, we can represent $\partial U_p / \partial A$ by $2\gamma_p$ and then write $\gamma_s + \gamma_p = \gamma$. We get:

$$\frac{\partial W}{\partial A} - \frac{\partial U_e}{\partial A} = 2\gamma$$

or  $\dfrac{\partial}{\partial A}(W - U_e) = 2\gamma$ \qquad(3.4)

The above equation states that fracture can only occur if the difference between the work done by external forces and the increase in strain energy of the body is sufficient to supply the energy required for fracture. The energy for fracture is the sum of the energy for plastic deformation and the energy needed to form new surfaces. For metals, the surface energy term is negligible in comparison to the plastic energy term. For extremely brittle materials such as glass, the reverse is true. The left-hand side of equation (3.4) is called the Griffith's crack extension force, $\mathcal{G}$, because it can be evaluated independently of the right-hand side. When its value is less than $2\gamma$, it represents the tendency for fracture to occur, and when it equals or exceeds $2\gamma$, fracture is believed to occur.

If we assume that the cracked body in Figure 3.1a remains linear-elastic, the load vs. load-line displacement (V) is linearly related to the applied load, P, as:

$$V = CP \tag{3.5}$$

where C = compliance of the cracked body and is the inverse of stiffness. The compliance is a function of crack size and the geometry of the body but is independent of the applied load or displacement.

If we consider that the crack size increases by an amount equal to $\Delta a$ under the conditions of fixed grips, the incremental work, $\Delta W = 0$ and:

$$\Delta U_e = \frac{d}{B da}\left(\frac{1}{2} PV\right) B \Delta a$$

where B = thickness of the body. It can be shown that:

$$-\frac{\partial U_e}{\partial A} = \mathcal{G} = \frac{1}{B}\frac{\partial U_e}{\partial a} = \frac{P^2}{2B}\frac{dC}{da} \tag{3.6}$$

Note that $C$ is only a function of crack size, therefore:

$$\frac{\partial C}{\partial a} = \frac{dC}{da}$$

For an increase in crack size under the conditions of fixed load, P:

$$\Delta W = PV(B\Delta a)$$

$$\text{and} \quad \Delta U_e = \frac{1}{2} PV(B\Delta a)$$

$$\text{thus,} \quad \mathcal{G} = \frac{1}{B}\frac{d}{da}\left(\frac{1}{2} PV\right) = \frac{P^2}{2B}\frac{dC}{da} \tag{3.7}$$

In both cases of crack growth under fixed grips and fixed load, $\mathcal{G} = (P^2/2B)(dC/da)$ and it is also related to the magnitude of the change in elastic strain energy. Under fixed grip conditions, the energy needed for fracture is balanced by a decrease in strain energy. Under fixed load conditions, the external force must work to supply the energy for fracture as well as an equal amount to raise the strain energy of the body. Due to its relationship with change in strain energy, $\mathcal{G}$ is also known as the strain energy release rate. For an infinite plate subjected to uniform stress and containing a center crack of length 2a, also known as the Griffith problem, the strain energy for plane stress conditions is given by:

$$U_e = U_0 - \frac{\pi a^2 \sigma^2}{E} B \qquad (3.8)$$

where $U_0$ = strain energy of the uncracked body. Since there are two crack tips, the increase in crack area due to increment in crack size, da, at each tip is 2Bda. If we consider fracture to occur under fixed grips, equation (3.4) will yield:

$$\frac{1}{2B} \frac{\partial U_e}{\partial a} = \frac{\pi a \sigma_f^2}{E} = 2\gamma$$

$$\text{or} \quad \sigma_f = \sqrt{\frac{2E\gamma}{\pi a}} \qquad (3.9)$$

where $\sigma_f$ is the value of $\sigma$ at which fracture will occur. Thus, the Griffith's problem can be considered as a special case of the much more general formulation given by Irwin [3.7] in which $\mathcal{G}$ is obtained by equations (3.6) and (3.7).

### 3.1.2 Stress Intensity Parameter Approach

Figure 3.2 shows the three distinct modes in which a cracked body can be loaded. In the crack opening mode, mode I, the crack surfaces separate symmetrically with respect to the $x$ - $y$ and $x$ - $z$ planes. In the sliding mode, mode II, the crack surfaces slide relative to each other symmetrically with respect to $x$ - $y$ and skew-symmetrically with respect to $x$ - $z$. In the tearing mode, mode III, the crack surfaces slide relative to each other skew-symmetrically with respect to both $x$ - $y$ and $x$ - $z$ planes. We will be concerning ourselves mostly with fracture under pure mode I loading because most fracture in metallic components occur under mode I conditions. Further, widely accepted theories for fracture under mode II, III, and mixed-mode conditions are currently not available.

The Westergaard semi-inverse method [3.8] constitutes a simple method for solving certain problems in plane elasticity. The details of this method are described in several books [3.1-33]. Here, we will simply describe the results of implementing the method to solve crack body problems. The credit for recognizing the importance of this approach belongs to Irwin. For mode I cracks, the stresses and displacements in a region near the crack tip are given by (Figure 3.3).

$$\sigma_x = \frac{K_I}{\sqrt{2\pi r}} \cos\frac{\theta}{2} \left(1 - \sin\frac{\theta}{2} \sin\frac{3\theta}{2}\right) \qquad (3.10a)$$

$$\sigma_y = \frac{K_I}{\sqrt{2\pi r}} \cos\frac{\theta}{2} \left(1 + \sin\frac{\theta}{2} \sin\frac{3\theta}{2}\right) \qquad (3.10b)$$

$$\tau_{xy} = \frac{K_I}{\sqrt{2\pi r}} \cos\frac{\theta}{2} \sin\frac{\theta}{2} \cos\frac{3\theta}{2} \qquad (3.10c)$$

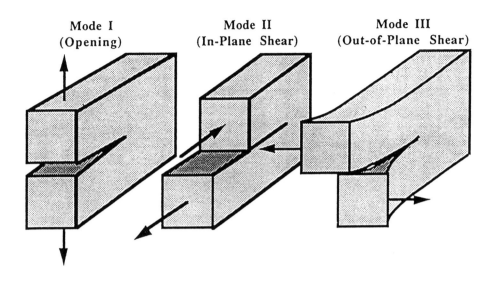

*Figure 3.2* The three modes of loading a cracked body (Ref. 3.2).

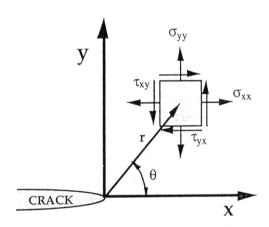

*Figure 3.3* Crack tip coordinate system and the stresses acting on an element ahead of the crack tip (Ref. 3.2).

For plane strain:

$$\sigma_z = \nu(\sigma_x + \sigma_y) \tag{3.10d}$$

but $\sigma_z = 0$ for plane stress; and $\tau_{yz} = \tau_{xz} = 0$ for plane stress and plane strain. The principal stresses are given by:

$$\sigma_1 = \frac{K_I}{\sqrt{2\pi r}} \cos\frac{\theta}{2}\left(1 + \sin\frac{\theta}{2}\right) \tag{3.11a}$$

$$\sigma_2 = \frac{K_I}{\sqrt{2\pi r}} \cos\frac{\theta}{2}\left(1 - \sin\frac{\theta}{2}\right) \tag{3.11b}$$

where $r$ and $\theta$ are as defined in Figure 3.3. The above equations are valid only in the crack tip region in which the stresses are dominated by the singularity. A more complete solution for stress ahead of the crack tip which is not just limited to only the crack tip region was derived by Williams [3.9] and is given by:

$$\sigma_{ij} = \left(\frac{k}{\sqrt{r}}\right) f_{ij}(\theta) + \sum_{m=0}^{\infty} A_m r^{\frac{m}{2}} g_{ij}^{(m)}(\theta) \tag{3.12}$$

where $f_{ij}(\theta)$ and $g_{ij}^{(m)}(\theta)$ are angular functions.

The first term in equation (3.12) is equivalent to the terms in equations (3.10). The higher order terms, as are evident from the nature of equation (3.12), are only important for higher values of $r$. In other words, as $r \to 0$, equations (3.12) and (3.10) yield identical results in that the stresses vary as $1/\sqrt{r}$, regardless of the geometry of the cracked body. The higher order terms, on the other hand, do vary with geometry but near the crack tip make only a second order contribution to the stresses. Further, the first term in Williams' series does not depend on $r$.

Figure 3.4 shows plots of the $\sigma_y$ stress as a function of distance from the crack tip for two cracks, one of which was 0.25mm in size and the other 6.25mm in size, in a 25.0mm wide, single-edge notch panel subjected to uniform stress. The stresses in the panels were adjusted such that the applied K levels in both was 22.2 MPa√m. In this study, Talug and Reifsnider [3.10] determined several coefficients, $A_m$, in equation (3.12) by a numerical technique called boundary collocation. Thus, full crack tip stress field solutions beyond the first singular term were obtained for two cracks. The solid lines in Figures 3.4a and b show the calculation for the full solution for the small crack and the dashed line for the long crack. The Irwin-Westergaard solution, equation 3.10b, is also shown for comparison. The long crack full-field solution and the Irwin-Westergaard solutions are indistinguishable while the full-field solution for small cracks is quite distinct from the other two as the distance from the crack tip increases. When the same results are plotted in Figure 3.4b on the basis of distance normalized by the crack size, the full-field solutions for both small and large cracks begin to diverge at $x/a$ values of 0.15. Thus, for this geometry, the region of dominance of the singular field for which the amplitude is characterized by K, is approximately 15% of the crack size. While the distance from the crack tip as a fraction of crack size for which K dominates may vary somewhat with geometry, the important result here is finding that the region of K-dominance is related to crack size.

The elastic displacement fields near the crack tip are given by:

$$u = \frac{K_I}{2G}\sqrt{\frac{r}{2\pi}}\cos\left(\frac{\theta}{2}\right)\left[\kappa - 1 + 2\sin^2\left(\frac{\theta}{2}\right)\right] \qquad (3.13a)$$

$$v = \frac{K_I}{2G}\sqrt{\frac{r}{2\pi}}\sin\left(\frac{\theta}{2}\right)\left[\kappa - 1 + 2\cos^2\left(\frac{\theta}{2}\right)\right] \qquad (3.13b)$$

where G = shear modulus, $\kappa = 3 - 4\nu$ for plane strain, $\kappa = (3 - \nu)/(1 + \nu)$ for plane stress, and $\nu$ = Poisson's ratio.

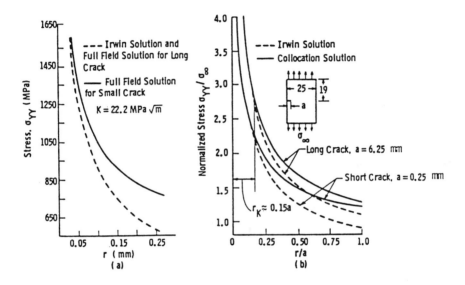

*Figure 3.4* Comparison of full-field and Irwin solutions for long and short cracks at constant K, as a function of (a) distance r from the crack tip and (b) normalized distance, r/a, from the crack tip (Ref. 3.10).

The stress intensity factor, $K_I$, represents the amplitude of the crack tip stress singularity and is dependent on the body geometry, crack size, load level, and loading configuration. The values of $K_I$ (or simply K) for several configurations are cataloged in handbooks and other references to which the readers are referred [3.11 - 3.13]. For some very common geometries and specimens, the expressions for estimating K are provided in Table 3.1. These expressions can be reduced to the following form for geometries subjected to point loads:

$$K = \frac{P}{BW^{1/2}} F(a/W) \qquad (3.14a)$$

and the following form if loadings is expressed in the terms of a characteristic stress:

$$K = \sigma \sqrt{\pi a}\, f(a/W) \qquad (3.14b)$$

where σ represents either a uniform applied stress or, in the case of bending, it can be the outer-fiber stress. W = characteristic length dimension of the cracked body, a = crack size, and B = thickness. F(a/W) or f(a/w) are dimensionless K-calibration functions. By expressing them as a function of a/W, these expressions can be used to determine K for the same crack geometry of any size. For internally pressurized components containing cracks on the inside wall, σ in equation (3.14b) may be replaced by pressure. Also, there may be dimensions other than just W which determine the value of K as seen in Table 3.1b.

*Table 3.1a  K-Calibration Functions for Various Geometries (Refs. 3.11, 3.12).*

| Geometry | F(a/W) |
|---|---|
| Center Crack Tension (CCT) | $\dfrac{1}{2}\sqrt{\pi\left(\dfrac{a}{W}\right)\sec\dfrac{\pi}{2}(a/W)}\left[1-0.25\left(\dfrac{a}{W}\right)^2+.06\left(\dfrac{a}{W}\right)^4\right]$ |
| Compact Type (CT) Specimen, 2H = 1.2W | $\dfrac{(2+aW)}{(1-a/W)^{1.5}}\left(0.886+4.64a/W-13.32(a/W)^2+14.72(a/W)^3-5.6(a/W)^4\right)$ |
| Single Edge Notch Bend (SENB), S=4W | $\dfrac{3\left(\dfrac{S}{W}\right)\left(\dfrac{a}{W}\right)^{\tfrac{1}{2}}\left[1.99-\dfrac{a}{W}(1-a/W)(2.15-3.93(a/W)+2.7(a/W)^2)\right]}{2(1+2a/W)(1-a/W)^{\tfrac{3}{2}}}$ |
| Single Edge Notch Tension (SENT) | $\dfrac{\left(2\tan\dfrac{\pi a}{2W}\right)^{\tfrac{1}{2}}}{\cos\dfrac{\pi a}{2W}}\left[0.752+2.02(a/W)+0.37\left(1-\sin\dfrac{\pi a}{2W}\right)^3\right]$ |
| Double Edge Notch (DEN) | $\dfrac{\dfrac{1}{2}\left(\dfrac{\pi a}{W}\right)^{\tfrac{1}{2}}}{(1-a/W)^{\tfrac{1}{2}}}\left[1.122-0.561(a/W)-0.205(a/W)^2+0.471(a/W)^3-0.190(a/W)^4\right]$ |

*Table 3.1b K Expressions for Typical Component Geometries (Refs. 3.2, 3.13).*

| Geometry | K - expression |
|---|---|
| (cylinder with internal pressure, $R_i$, $R_0$, $a$, $p$) | $K = \dfrac{2pR_0^2}{R_0^2 - R_i^2} \sqrt{\pi a}\, f(a/t, t/R_i) \qquad t = R_0 - R_i$ <br> $f = 1.1 + A\left[4.951(a/t)^2 + 1.092\,9(a/t)^4\right]$ <br> $A = (.125/(t/R_i) - 0.25)^{0.25} \quad$ for $0.1 \leq t/R_i \leq 0.2$ <br> $A = (0.2/(t/R_i) - 1)^{0.25} \quad$ for $0.05 \leq t/R_i \leq 0.1$ |
| (cylinder under axial tension $\sigma$, $R_i$, $R_0$, $a$, $t$) | $K = \sigma \sqrt{\pi a}\, f(a/t, t/R_i) \qquad t = R_0 - R_i$ <br> $= 1.1 + A\left[1.948(a/t)^{1.5} + 0.3342(a/t)^{4.2}\right]$ <br> $= \left[0.125/(t/R_i) - 0.25\right]^{0.25} \quad$ for $0.1 \leq t/R_i \leq 0.2$ <br> $= (0.4/(t/R_i) - 3.0)^{0.25} \quad$ for $0.05 \leq t/R_i \leq 0.$ |
| (surface crack in plate, $2c$, $2t$, $2a$, $2W$, $\sigma$, $\phi$) | $K = \sigma F \dfrac{\sqrt{\pi a}}{Q}$ <br><br> $F = (M_1 + M_2(a/t)^2 + M_3(a/t)^4)\, g\, f_\phi\, f_w$ <br><br> $f_w = \left[\sec\left(\dfrac{\pi c}{2W}\sqrt{\dfrac{a}{t}}\right)\right]^{\frac{1}{2}}$ <br><br> $g = 1 - \dfrac{(a/t)^4 \sqrt{2.6 - 2(a/t)}}{1 + 4(a/t)}\,\lvert\cos\phi\rvert$ <br> $M_2 = .05/(0.11 + (a/c)^{1.5})$ <br> $M_3 = 0.29/(0.23 + (a/c)^{1.5})$ <br><br> for $a/c \leq 1$ $\qquad\qquad$ for $a/c > 1$ <br> $Q = 1 + 1.464(a/c)^{1.65}$ $\quad$ $Q = 1 + 1.464(c/a)^{1.65}$ <br><br> $f_\phi = \left[\left(\dfrac{a}{c}\right)^2 \cos^2\phi + \sin^2\phi\right]^{1/4}$ $\quad$ $f_\phi' = \left[\left(\dfrac{c}{a}\right)^2 \sin^2\phi + \cos^2\phi\right]$ <br><br> $M_1 = 1$ $\qquad\qquad\qquad\qquad$ $M_1 = \sqrt{\dfrac{c}{a}}$ |

### 3.1.3 The Equivalence of $\mathcal{G}$ and $K$

In 1957, Irwin [3.7] showed that:

$$\mathcal{G} = \frac{K^2}{E} \quad \text{for plane stress} \tag{3.15a}$$

and:

$$\mathcal{G} = \frac{K^2}{E}(1-\nu^2) \quad \text{for plane strain} \tag{3.15b}$$

The proof of this relationship is given in other books [3.1-3.3] and is left as an exercise that students should attempt on their own. Relationships similar to equation (3.15) can also be derived for other modes of crack extension such as mode II and III.

For mixed mode conditions, $\mathcal{G}$ for various modes can be algebraically added because energy release rate is a scalar quantity. Combining equations (3.15a) and (3.7), we find the following relationship between K and change in compliance with crack size:

$$\frac{K}{P} BW^{\frac{1}{2}} = F(a/W) = \left(\frac{1}{2} \frac{d(CBE)}{d(a/W)}\right)^{\frac{1}{2}} \tag{3.16}$$

The above equation also provides a method for determining K experimentally by measuring compliance change as a function of crack size for any arbitrary crack/load geometry.

## 3.2 Crack Tip Plasticity

Even the most brittle fractures in metals are accompanied by some plastic deformation. Therefore, any useful theory of fracture must be able to admit some plastic deformation. Recall that for Griffith's theory of fracture to apply to metals, modifications to the original theory were necessary to account for energy dissipated in the form of work of plastic deformation. The crack tip field equation (3.10) predicts that the stresses at the crack tip are infinite. However, in real materials, plastic deformation occurs in the region where the stresses exceed the yield strength of the material. This results in relaxation of stresses within the region of plasticity called the plastic zone. If we assume that the plasticity is contained within a small zone ahead of the crack tip, the stresses and displacements in the crack tip region can be considered as being dominated by the much larger K-controlled zone surrounding it. When plasticity becomes significant, other parameters described in later chapters must be used to characterize crack growth and fracture. To set the limitations for use of LEFM, it is therefore necessary to estimate the plastic zone size. The plastic zone must then be small in comparison to all pertinent length dimensions of the cracked body such as the crack size and the uncracked ligament.

### 3.2.1 Irwin's Plastic Zone Size Calculation

Irwin made a simple estimate of the plastic zone size along $\theta = 0$ for elastic, perfectly-plastic materials. The simplest estimate can be made by substituting $\theta = 0$ in equation (3.10b) or (3.11b) and solving for a distance, $r_y$, at which $\sigma_y = \sigma_0$. This leads to the equation:

$$r_y = \frac{1}{2\pi}\left(\frac{K}{\sigma_0}\right)^2 \tag{3.17}$$

Figure 3.5 shows the distance $r_y$ schematically. This estimate of plastic zone is incorrect because it is based on elastic stress distribution. The elastic-plastic stress distribution is also shown schematically in Figure 3.5. The plastic zone boundary is located at $r_p$ according to this distribution. To satisfy force equilibrium in the y-direction, the areas under the elastic and elastic-plastic stress distributions must be the same. This condition can be met by making $r_p$ such that the following equation is satisfied:

$$\int_0^{r_y} \frac{K}{\sqrt{2\pi x}}\, dx - \sigma_0 r_y = \sigma_0 (r_p - r_y)$$

Solving for $r_p$ gives:

$$r_p = 2 r_y = \frac{1}{\pi}\left(\frac{K}{\sigma_0}\right)^2 \tag{3.18}$$

Equation (3.18) has been derived for nonhardening materials. For hardening materials which follow the Ramberg-Osgood relationship:

$$r_p = \frac{m-1}{m+1}\frac{1}{\pi}\left(\frac{K}{\sigma_0}\right)^2 \tag{3.19}$$

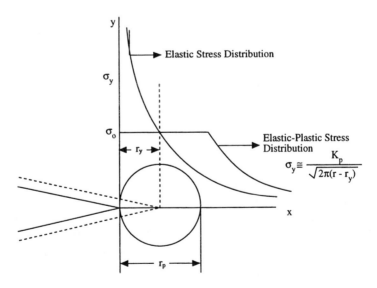

**Figure 3.5** *Elastic and elastic-plastic crack tip stress distribution in front of the crack tip and the plastic zone sized $r_y$ and $r_p$.*

The elastic stress distribution, equation (3.10) can be used to specify stresses beyond the plastic zone. However, these equations must also be modified to account for crack tip plasticity. Irwin suggested that if we consider an effective crack length, $a_{eff}$, such that $a_{eff} = a + r_y$, and move the coordinate origin to the effective crack tip, and adjust the K value to the longer crack size, equation (3.10) will yield the correct magnitude of stresses beyond the plastic zone. This adjustment in crack size can be explained by visualizing that the soft region ahead of the crack tip in the form of the plastic zone is unable to bear the full elastic stress. Note that cracks are unable to transmit any stress at all across their surfaces. Therefore, the plastic region behaves in a manner in between that of an elastic material and a crack. This situation is statically equivalent to a crack which is longer than its physical size by a distance equal to $r_y$. The value of K corrected for the adjusted crack size is known as $K_p$, or the K value corrected for small-scale plasticity. In most cases, the correction is small and can be neglected, but should not be routinely neglected without regard to the material, K-level applied and the size of the uncracked ligament. Thus, for a nonhardening material, the elastic-plastic stress distribution ahead of the crack tip consists of:

$$\sigma_y = \sigma_0 \text{ for } 0 \leq x \leq r_p \text{ and } \sigma_y \approx K_p / \sqrt{2\pi(r-r_y)} \text{ for } x > r_p$$

as shown in Figure 3.5.

Equations (3.18) and (3.19) are for plane stress conditions for which $\sigma_3 = 0$ and yielding occurs when $\sigma_1 = \sigma_0$. If we assume that plane strain conditions exist at the crack tip and $\sigma_3$ is given by equation (3.10d), which for $\theta = 0$ reduces to:

$$\sigma_3 = \sigma_z = 2\nu \left( \frac{K}{\sqrt{2\pi r}} \right) \tag{3.20}$$

Since $\sigma_x$, $\sigma_y$, and $\sigma_z$ are also principal stresses, we can substitute them into Von Mises' effective stress, equation (2.32), and show that for $\nu = 1/3$, $\sigma_1$ must be equal to $3\sigma_0$ for yielding to occur. Substituting $3\sigma_0$ in place of $\sigma_0$ in equation (3.18) gives an estimate of the plane strain plastic zone size which is nine times smaller than the plane stress plastic zone. Irwin [3.14] suggested that plane strain does not occur on the specimen surface and at the crack tip. Consequently, the effective yield strength is only $1.68\,\sigma_0$ instead of $3\sigma_0$ for plane strain. Substituting into equation (3.18) gives:

$$\left. r_p \right|_{pl\,strain} \approx \frac{1}{6\pi} \left( \frac{K}{\sigma_0} \right)^2 \tag{3.21}$$

**Relationship Between K and CTOD**  If we assume the effective crack tip to lie at a distance $r_y$ from the crack tip, we can use the crack tip displacement equation (3.13) to estimate the opening at the actual crack tip. This opening is called the crack tip opening displacement or CTOD and has been used as a fracture criterion. We show that a unique relationship exists between K and CTOD. If we choose $\theta = 180°$ in equation (3.13b), assuming plane stress, and $r = r_y$, we can show that:

$$CTOD = 2v_0 = \frac{4K^2}{\pi E \sigma_0} \tag{3.22}$$

where $v_0$ = the value of $v$ (the y-direction displacement) at the crack tip.

**Shape of the Plastic Zone** The Tresca and Von Mises yield criteria can be substituted into the crack tip field equations to derive the shape of the plastic zone. These are shown in Figure 3.6 a and b according to the two criteria, respectively, for plane stress and plane strain conditions. The extent of the plastic zone along $\theta = 0$ is independent of yield criterion but the shapes are dependent on it. Deriving the equations for the plastic zone shape according to the two yield criteria is left as exercise problems.

The plastic zone size depends strongly on whether plane strain or plane-stress conditions exist. In cracked bodies, plane stress conditions exist in thin sections. In thick sections, plane stress conditions exist on the surface and plane strain conditions exist in the interior. The 3-dimensional plastic zone in a thick section cracked body is schematically shown in Figure 3.7. The state of stress (plane stress vs. plane strain) also depends on the extent of yielding. Large plastic zones cause unconstrained or free yielding. For example, if plastic zone is equal to the plate thickness, unconstrained yielding in the thickness direction will take place causing plane stress conditions to develop.

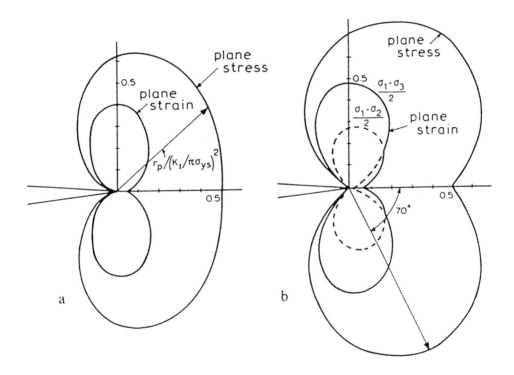

*Figure 3.6 Plastic zone shapes for plane stress and plane strain conditions according to (a) Von Mises criterion and (b) Tresca criterion. Note that the extent of the plastic zone along the x-axis is independent of the yield criterion (Ref. 3.1).*

### 3.3.2 Strip Yield Model

The strip yield model, also known as the Dugdale-Barenblatt model [3.15, 3.16], assumes a long, slender plastic zone in a nonhardening material under plane stress conditions. Thus, by necessity it is only valid for thin sheets. The original analyses were performed for cracks in infinite sheets subjected to uniform stress. In sheets, if we note that slip is concentrated in bands that are ±45° to the crack plane, the slip bands intersect the surface limiting their height. Therefore, the assumption of a slender plastic zone is reasonable. The extent of the plastic zone, $s$, is given by:

$$s = a \sec\left(\frac{\pi}{2}\frac{\sigma}{\sigma_0}\right) - 1 \qquad (3.23)$$

For $\sigma/\sigma_0 \ll 1$, $s$ may be approximated by:

$$s \simeq \frac{\pi}{8}\left(\frac{K}{\sigma_0}\right)^2 \qquad (3.24)$$

Comparing equations (3.24) and (3.18) shows that the results of the two calculations give reasonably comparable results. The derivation of equation (3.24) is left as an exercise problem. For derivation of equation (3.23), the reader is referred to other books [3.2].

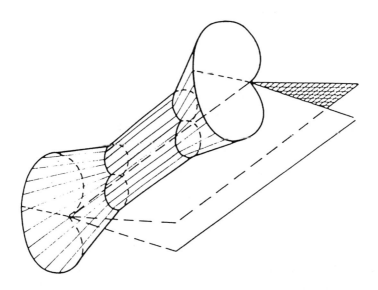

*Figure 3.7* *Three-dimensional plastic zone (Ref. 3.1).*

### 3.3 Compliance Relationships

Elastic compliance measurements play an important role in experimental fracture mechanics. In the days preceding the wide spread use of finite element analysis for determining stress intensity parameters, compliance measurements were used to determine stress intensity parameter, K, through the relationship given in equation (3.16). Now compliance measurements are routinely used for measuring crack size in fracture toughness, fatigue crack growth, stress corrosion cracking, and corrosion-fatigue testing. It is thus worthwhile to review compliance relationships. We will use the compliance relationships for compact type specimen as an example because it is used commonly in fracture toughness and crack growth testing.

We start by recognizing that elastic compliance, C, for any cracked body is given in the functional form by:

$$\frac{V}{P} = C = \frac{1}{BE} f(a/W) \qquad (3.25)$$

In other words, the dimensionless quantity, CBE, is dependent only on a/W provided the planar geometry of the body remains the same. Figure 3.8 shows a plot of CBE as a function of a/W for a compact type specimen and includes experimental and numerical data which agree well with each other. If the point of displacement measurement is other than the load-line as shown in Figure 3.9, the compliance relationships are given by [3.17]:

$$\frac{BEV_x}{P} = \frac{\frac{BEV_0}{P}\left|\frac{x_0}{W} - \frac{x}{W}\right|}{\left[\frac{x_0}{W} + 0.25\right]} \qquad (3.26)$$

where x = distance between the loading line and the point of displacement measurement. x is negative when we move toward the front-face of the specimen from the load-line and positive as we move toward the crack tip. $x_0$ is the location of the point of rotation and is given by [3.17]:

$$\frac{x_0}{W} = 0.09953 + 3.02437(a/W) - 7.95768(a/W)^2 + \\ 13.546(a/W)^3 - 10.6274(a/W)^4 + 3.1133(a/W)^4 \qquad (3.27)$$

Frequently it is necessary to express crack size as a function of compliance. This inverse relationship is expressed by [3.17]:

$$a/W = C_0 + C_1 U_x + C_2 U_x^2 + C_3 U_x^3 + C_4 U_x^4 + C_5 U_x^5 \qquad (3.28)$$

$$\text{where} \quad U_x = \frac{1}{(C_x BE)^{1/2} + 1}$$

where $C_x$ = compliance corresponding the measurement location x. The constants $C_0$, $C_1$, $C_2$ -- are given for various x values from reference [3.17] in Table 3.2. These equations provide ultimate flexibility in using compliance measurements. Similar relationships can be derived for other specimen geometries.

*Table 3.2 Constants in Equation (3.28) for Different Measurement Locations (Ref. 3.17).*

| x/W | $C_0$ | $C_1$ | $C_2$ | $C_3$ | $C_4$ | $C_5$ |
|---|---|---|---|---|---|---|
| 0 | 1.0002 | -4.0632 | 11.242 | -106.04 | 464.33 | -650.68 |
| 0.25 | 1.001 | -4.6695 | 18.46 | -236.82 | 1214.9 | -2143.6 |

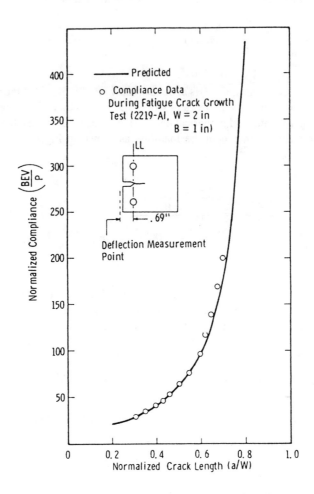

*Figure 3.8* The relationship between nondimensional compliance (CBE) and crack size for compact-type specimens.

### 3.4 Fracture Toughness and Predicting Fracture in Components

Since K uniquely determines (i) the magnitude of stresses in the crack tip region, (ii) the size of the crack tip plastic zone and its shape, and (iii) strain energy available for crack extension, it is attractive as a parameter for characterizing fracture in components. We hypothesize that fracture occurs when K reaches a critical value. In other words, if we measure the value of critical K in a laboratory specimen, we can use it to predict when fracture will occur in a component. The critical value of K at fracture is a material property and is called the fracture toughness. On the other hand, the applied value of K is dependent on the applied load (or stress), the geometry, and crack size. When applied K is equal to or greater than critical K, fracture occurs. The relationship between applied K and critical K is analogous to that between stress and yield strength. Note that critical K and yield strength are material properties, and stress and K are not. In the subsequent discussion, the role of K as a fracture criterion is explored further.

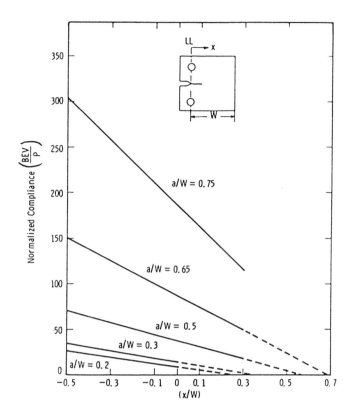

**Figure 3.9** *An arbitrary displacement measurement location on a compact-type specimen for which the compliance relationship can be computed by equation (Ref. 3.26).*

### 3.4.1 Fracture Under Plane Strain Conditions (Thick Sections)

In Section 3.2, we had derived the relationships for estimating crack tip plastic zone sizes for plane stress and plane strain conditions. We had shown that the plastic zone size for plane stress was much larger than that for plane strain. Therefore, if we measure the fracture toughness value under plane strain conditions using a thick specimen and attempt to use the value to predict fracture under plane stress conditions in a thin-sheet or vice-versa, the fracture criterion is not expected to yield satisfactory results because the plastic zone sizes for the two conditions are vastly different at the same applied K levels. Therefore, we need to restate our fracture criterion as follows. Fracture will occur when K reaches a critical value where the critical value has been measured using specimens of the same thickness as those of the component.

From the above discussion, we expect fracture toughness to be dependent on the specimen thickness. The relationship between the critical K and thickness is illustrated in Figure 3.10. The fracture toughness vs. thickness curve has two plateaus, one corresponding to the thickness regime where plane stress conditions occur and one corresponding to thickness where plane strain conditions dominate in the specimen. The two regions are connected by a region in which neither plane stress nor plane strain dominates and therefore the fracture toughness varies with specimen thickness.

The critical value of K corresponding to the lower plateau (or plane strain) is designated as $K_{IC}$ or the plane strain fracture toughness. It represents the lowest fracture toughness value of the material and therefore has great practical significance in applications. It is truly a material constant which can be defined and measured unambiguously. ASTM Standard E-399 [3.18] was the first to standardize the measurement of $K_{IC}$ followed by the British Standard, BS 5447 [3.19]. The method of determining $K_{IC}$ involves selecting a specimen geometry such as a compact specimen or a 3-point bend specimen, developing a sharp precrack ahead of the notch by fatigue loading, and subsequently loading the specimen monotonically until fracture occurs. During the final loading, the specimen load and displacements across the crack surfaces are measured. One of the primary considerations in measuring $K_{IC}$ is meeting several validity requirements. For example, all pertinent dimensions of the specimen, the crack size, the uncracked ligament size, and the thickness must satisfy the following relationship:

$$B, \; W-a, \; a \geq 2.5 \left( \frac{K_Q}{\sigma_0} \right)^2 \tag{3.29}$$

where $K_Q$ is the candidate value of fracture toughness obtained from the test. If the conditions of equation (3.29) are met, $K_Q$ qualifies as a $K_{IC}$ value. If not, the specimen size must be increased and more tests must be performed until the conditions of equation (3.29) are met. The number 2.5 on the right-hand side of equation (3.29) was chosen after examining considerable experimental data. It corresponds to a size which is approximately 50 times the plane strain plastic zone size. Thus, the conditions of linear elasticity and plane strain are completely assured. For details of this test method, the reader is referred to the above referenced standards and other texts [3.2].

Not all applications involve thick plates or large section components. For example, in the aerospace industry, thin plates or sheets are used extensively in the fabrication of shell structures. Therefore, the plane strain fracture toughness, $K_{IC}$, does not represent the correct fracture toughness value for such applications. It is necessary to obtain fracture toughness values of materials to be used in the form of thin plates and sheets.

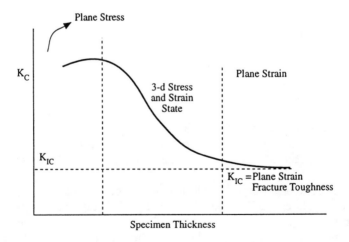

*Figure 3.10* A schematic relationship between fracture toughness and specimen thickness illustrating $K_{IC}$, the plane strain fracture toughness value.

*3.4.2 Fracture in Thin Plates and Sheets*

The considerations involved in characterizing the fracture toughness behavior of thin plates and sheets are somewhat different from the measurement of $K_{IC}$. Due to enhanced plastic deformation in the crack tip region at comparable K levels, instability in the specimen is preceded by stable crack growth. The stable crack growth behavior is characterized by crack growth resistance curves in which the crack extension, $\Delta a$, is correlated with K, as shown schematically in Figure 3.11a. Thus, the entire K-resistance curve represents the fracture toughness behavior of the material as opposed to a single number $K_{IC}$. The instability point in this case is determined by very different considerations.

Figure 3.11b shows the K-resistance curve (or simply the R-curve) on which the applied K vs. crack size curves for various stress levels are also superimposed. For instability to occur, the applied K must exceed the fracture resistance of the material. For the stress level $\sigma_1$ and $\sigma_2$, the crack grows to sizes $a_1$ and $a_2$, respectively, and then the applied K level drops below the resistance curve so no further crack extension can occur. To continue crack extension, a higher load (or stress) level is needed. At stress level $\sigma_3$, the crack extends to $a_3$, beyond which the applied K curve rises with crack extension at a rate above that for the R-curve. The crack grows in a stable fashion until $a_3$ and instability occurs beyond that. The condition for instability can be written as:

$$\frac{dK}{da} \geq \frac{dK_R}{da} \qquad (3.30)$$

and $K > K_R$, where $K_R$ represents the R-curve. For stress level $\sigma_4$, instability occurs immediately. $dK_R/da$ is a material constant for a given specimen thickness, but dK/da depends on geometry, size, and loading configuration.

From the above discussion it is clear that the point of instability is dependent on the geometry, size, and loading configuration for plane stress conditions and is not a material constant. Thus, characterizing fracture toughness involves characterizing the full R-curve behavior. The method for determining R-curves is addressed in ASTM Standard E-561 [3.20]. In this method, a fatigue precracked specimen is loaded under constant displacement rate while the load (P) vs. displacement (V) behavior is recorded as in Figure 3.12. In this case, the load-displacement behavior exhibits considerable nonlinearity. Some nonlinearity occurs due to the presence of plasticity and some due to increase in compliance as the crack length increases. If we choose to calculate crack size by using the instantaneous value of V/P as compliance from relationships of the type given in equation (3.28), the result is expected to yield the plasticity adjusted crack size, $a_{eff}$. Further, if $a_{eff}$ is used to calculate K, the value obtained will be the plasticity adjusted value, $K_p$. Thus, by choosing several points along the load-displacement diagram, the values of $a_{eff}$ and $K_p$ can be calculated. The crack extension can either be measured directly or it can be obtained from unloading compliance. In the latter technique, the specimen is partially unloaded at several points along the load-displacement diagram. The slope of the load-displacement diagram is elastic and it can be related to the instantaneous crack length by substituting the measured compliance in equation (3.28). Thus, we now have the $K_p$ value and the crack extension value at different points along the load-displacement diagram to obtain the K-resistance curve. For the resistance curve to be valid, the extent of plasticity has to be restricted. This is ensured by applying the following restriction:

$$W - a \geq \frac{4}{\pi}\left(\frac{K}{\sigma_0}\right)^2 \qquad (3.31)$$

Thus, all data meeting the above criterion can be used to describe the R-curve.

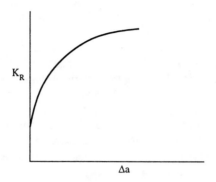

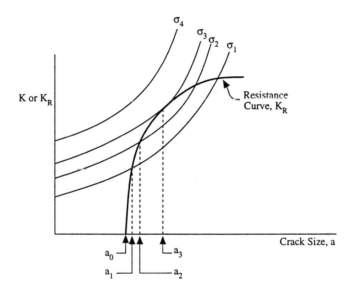

***Figure 3.11*** *(a) K-resistance curve as a function of crack size; (b) $K_R$ vs. crack size curve superimposed on the K vs. crack size trend for different stress levels to illustrate instability.*

### 3.5 Subcritical Crack Growth

Under the conditions of cyclic loading or sustained loading in the presence of hostile environment or both, cracks in metals can grow at K levels well below the $K_{IC}$ of the material. Such crack growth is known as subcritical crack growth and is paramount in determining the service life of components containing defects that are smaller than the critical size. In this section, we will briefly review the LEFM approach for characterizing the rates at which subcritical cracks grow under cyclic loading and/or the presence of hostile environment. We begin with a discussion of fatigue crack growth and then follow with a discussion of stress corrosion cracking (SCC) and corrosion fatigue crack growth.

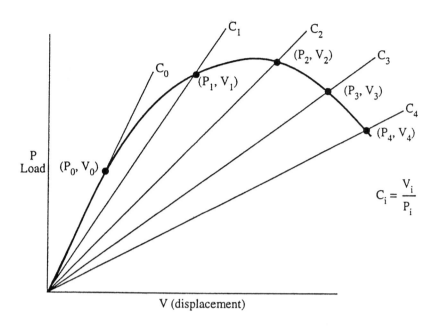

**Figure 3.12** *The load vs. displacement behavior during an R-curve test. The various points $(P_i, V_i)$ on the P-V diagram are used to calculate $a_{\it eff}$ and $K_p$ to characterize the R-curve.*

### 3.5.1 Fatigue Crack Growth

If we consider several identical cracked specimens and subject them to cyclic loading where the load (P) fluctuates between fixed maximum and minimum values, $P_{max}$ and $P_{min}$, respectively, at a specified frequency, $\nu$, and a specified loading waveform, the amount of crack extension during each cycle, da/dN, in the most general case is given by:

$$\frac{da}{dN} = f(\Delta P, R, \nu, waveform, a) \tag{3.32}$$

where $\Delta P = P_{max} - P_{min}$ and $R = P_{min} / P_{max}$. During early research on fatigue crack growth, the influence of each of the above parameters was studied empirically. The results conclusively showed that in benign environments such as room temperature air with low humidity levels (40% or less), the cyclic frequency, $\nu$, and the waveform do not significantly influence fatigue crack growth rates in metals or alloys. We will see later that in the presence of hostile environments, this is not the case. In this section, we will exclude the effects of environment which equation (3.32) simplifies considerably. In the early nineteen sixties [3.21, 3.22], Paris and co-workers suggested that the dependence of cyclic load and crack size on da/dN could be combined into a single parameter, $\Delta K$, where $\Delta K = K_{max} - K_{min}$ and $K_{max}$ and $K_{min}$ are the stress intensity parameters corresponding to $P_{max}$ and $P_{min}$, respectively. Thus under the conditions of constant amplitude loading in which $P_{max}$ and $P_{min}$ are held constant, da/dN = f ($\Delta K$).

Following the above proposal by Paris and co-workers [3.21, 3.22], several experimental studies were conducted and it was shown that the relationship between log (da/dN) and log ($\Delta K$) was unique for a constant R over a very wide range of crack growth rate behavior. A schematic of this relationship for different values of R is shown in Figure 3.13a, and in Figure 3.13b actual data are presented for a 300M steel [3.23]. The behavior over the wide range of da/dN can be divided into three regions. Region I is called the near-threshold region in which the slope of the log (da/dN) vs. log ($\Delta K$) behavior is high. As $\Delta K \to \Delta K_{th}$, or the threshold value of $\Delta K$, the crack grows at a diminishingly small rate. Ideally, $\Delta K_{th}$ is defined as the $\Delta K$ value for which da/dN becomes 0. Operationally, it is defined as the $\Delta K$ at which da/dN is $10^{-10}$ m/cycle. Region II, also known as the Paris region, is represented as a straight line on a log (da/dN) vs. log ($\Delta K$) plot. In this region, the fatigue crack growth rates can be described by:

$$\frac{da}{dN} = c\,(\Delta K)^{n_2} \qquad (3.33)$$

where c and $n_2$ are regression parameters. Region III is characterized by an increasing slope of the log (da/dN) vs. log ($\Delta K$) curve. In this region, the maximum value of K during a cycle approaches the critical K for fracture, $K_c$, for brittle materials. For ductile materials, stable crack growth becomes a factor and enhances the crack growth rate.

Several equations have been used to describe the wide range fatigue crack growth behavior. Since these equations rely heavily on regression parameters, most are successful in representing the wide range fatigue crack growth behavior. As an example, the following equation [3.24] is proposed:

$$\frac{1}{da/dN} = \frac{A_1}{(\Delta K)^{n_1}} + \frac{1}{c}\left[\frac{1}{(\Delta K)^{n_2}} - \frac{1}{((1-R)K_c)^{n_2}}\right] \qquad (3.34)$$

The above equation is called the 3-component model and in Region II it reduces to the simpler equation (3.33). The equation has five regression constants $A_1$, $n_1$, c, $n_2$, and $K_c$. The constants $A_1$ and $n_1$ represent the behavior in Region I, c and $n_2$ represent the behavior in Region II, and $K_c$ sets the $K_{max}$ value at which the crack growth becomes unstable. The constants $A_1$ and c are functions of R while $n_1$, $n_2$, and $K_c$, to a first approximation, can be considered as not dependent on R.

ASTM Standard E-647 [3.25] describes in detail the method for fatigue crack growth testing. Two types of tests are conducted to characterize the wide range fatigue crack growth behavior. The constant amplitude load test consists of applying cyclic loads in which the $P_{max}$ and $P_{min}$ values are fixed. The crack length is measured as a function of applied cycles. The a vs. N curve is differentiated at different points to determine da/dN and the $\Delta K$ value corresponding to the point of da/dN calculation is also obtained. Thus, each test yields several da/dN vs. $\Delta K$ points. This is called the $\Delta K$-increasing test because the value of $\Delta K$ continuously increases as the test progresses. This test is usually well suited for characterizing the crack growth rates in Regions II and III of the da/dN vs. $\Delta K$ relationship.

The second type of test is called the $\Delta K$-decreasing test in which the applied loading $\Delta P$ is periodically decreased while keeping R constant in a manner such that the following relationship is satisfied [3.26]:

$$\Delta K = \Delta K_0 \exp(-c_1 \Delta a) \qquad (3.35)$$

where $c_1$ is a constant in the range of 0.04 to 0.08mm$^{-1}$. As $\Delta K$ decreases, the plastic zone size ahead of the crack tip also decreases in size. Following the above equation ensures that $dr_p / r_p$, the fractional decrease in plastic zone size, remains constant [3.26]. This avoids retardation effects due to decrease in the plastic zone size as the fatigue crack advances.

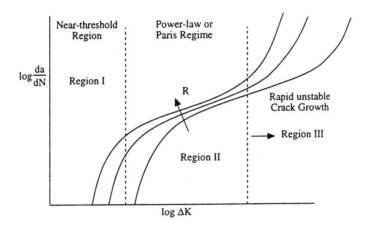

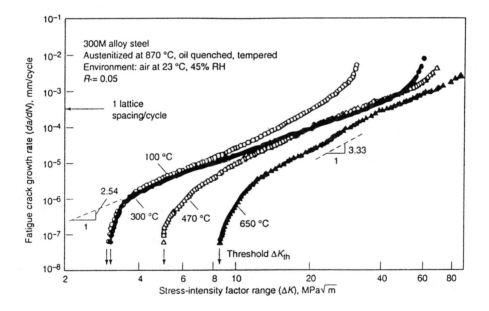

**Figure 3.13** *(a) A schematic of the wide range fatigue crack growth behavior; (b) fatigue crack growth rate data over a wide range for 300M alloy steel (Ref. 3.23).*

For da/dN to be uniquely correlated with $\Delta K$, it is important to maintain linear elastic conditions. In fatigue crack growth testing, linear elastic conditions are ensured by requiring that at all times:

$$W - a \geq \frac{4}{\pi} \left( \frac{K_{max}}{\sigma_0} \right)^2 \qquad (3.36)$$

Since K levels during fatigue are much lower than the $K_{IC}$ values, LEFM methodology for FCGR testing is applicable to a much broader class of materials than the $K_{IC}$ method. Also note that no requirements are imposed to specify restrictions on the thickness beyond the minimum thickness needed to prevent the specimen from buckling. It is widely believed, from empirical observations, that thickness is not a significant variable in determining the fatigue crack growth behavior in Regions I and II.

With the knowledge of the fatigue crack growth behavior of a material, fatigue lives of components made from the material can be determined. Several problems which illustrate this application are included as exercise problems in this chapter.

### 3.5.2 Stress Corrosion or Environment Induced Cracking

In several materials, subcritical crack growth below $K_{IC}$ can occur in the presence of sustained loading in certain environments. This mechanism of crack growth is known as the stress corrosion cracking (SCC). The presence of all three conditions namely, (i) environmentally sensitive material, (ii) sustained loading, and (iii) an aggressive environment, are requisites for SCC. This form of fracture, since it occurs at low K levels, is quite brittle due to lack of plastic deformation. The crack growth can be intergranular as well as transgranular. Several mechanisms and models have been proposed for SCC but they can be classified as anodic dissolution models and mechanical fracture models [3.27]. Among the anodic dissolution models, there is an active path intergranular SCC model in which anodic current density is set up due to compositional differences between the grain boundaries and in the interior of the grains resulting in transfer of material from the crack tip region due to an electrochemical reaction. It is also assumed that the crack tip consists of bare fresh metal which is much more active compared to the remainder of the crack surface which passivates and does not participate in the electrochemical reactions. The film rupture SCC model assumes that the stress at the crack tip acts to open the crack tip and rupture the protective film exposing bare metal which dissolves rapidly resulting in crack growth.

The mechanical fracture models are based on weakening of the atomic bonds in the crack tip region due to corrosion or due to atomic hydrogen entry into the metal in the crack tip region and subsequent rupture of the weakened ligaments. Within this class of models, there is a corrosion tunnel model which assumes that corrosion tunnels form at the crack tip until the remaining ligaments fracture. The hydrogen embrittlement mechanism has been proposed to be the operative mechanism in stress corrosion cracking in ferritic steels, nickel base alloys, titanium alloys, and aluminum alloys. Hydrogen can enter the lattice from both gaseous and aqueous phases. In the aqueous phase, the hydrogen ion reduction can often be the cathodic reaction where atomic hydrogen is formed on the surface which partially combines to form hydrogen gas, but some of it remains in the atomic form and enters the lattice through the crack tip causing it to embrittle. In the gaseous form, the mechanism is illustrated in Figure 3.14. In this model, the hydrogen embrittlement is depicted as a five-step sequential process [3.28] leading to an embrittled zone near the crack tip.

**Fracture Mechanics Approach to SCC** Fracture mechanics concepts have been applied to quantify stress corrosion cracking rates and to determine threshold levels of applied K below which SCC does not occur. Figure 3.15 shows a plot of da/dt vs. K behavior commonly observed for several material/environment systems. These trends can be characterized by testing several specimens under dead-weight loading conditions while the specimen is immersed in the test environment created by building an environment control chamber around the specimen. An alternate method of testing involves wedge opening a cracked specimen by cranking the torque on a bolt as shown in Figure 3.16. After apply-

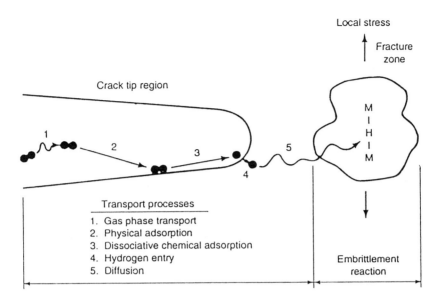

*Figure 3.14* Sequential process illustrating the mechanism of hydrogen embrittlement (Ref. 3.28).

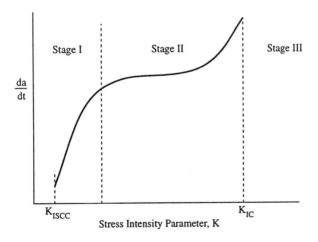

*Figure 3.15* Schematic of the relationship between time rate of crack growth and applied K for stress corrosion cracking.

ing sufficient torque on the bolt, the specimen is placed in an environment chamber and the crack size is periodically recorded. As the crack grows, the bearing load on the bolt decreases and the K value also decreases. Thus, the crack growth rate can be recorded at different K levels. One of the primary objectives of this testing is to measure a threshold value of K below which no stress corrosion cracking occurs, $K_{ISCC}$, as shown in Figure 3.15. This is measured in a bolt-loaded test by continuing the test until the K value decreases to $K_{ISCC}$ at which point the crack ceases to grow any further.

The threshold value of K for SCC measured by the constant load technique is often different from the threshold value of K for SCC obtained from the bolt-loaded specimens because the former technique measures the initiation threshold while the latter measures the crack arrest threshold. The constant load method of determining $K_{ISCC}$ has recently been standardized by ASTM [3.29]. The da/dt vs. K relationship can be used to predict lives of structures operating under conditions where fracture occurs by SCC. However, most design criteria specify that the maximum allowable K levels not exceed the $K_{ISCC}$ of the material/environment conditions. Thus, SCC can be avoided.

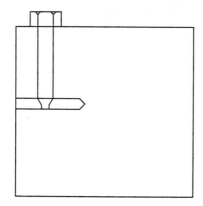

*Figure 3.16* Bolt-loaded test specimen used for stress corrosion cracking tests.

*3.5.3 Corrosion-Fatigue Crack Growth*

Corrosion-fatigue refers to the phenomenon of cracking in materials under the combined action of fatigue loading and a corrosive environment. Earlier in this section, we discussed fatigue crack growth in benign environments and showed that loading frequency and waveform do not influence the fatigue crack growth rate. In the presence of deleterious environments, both loading frequency and loading waveform influence the cyclic crack growth rate as shown in Figures 3.17 and 3.18, respectively [3.28]. In the case of cyclic frequency, the crack growth rate increases with decreasing frequency of loading at the same $\Delta K$. The effects of different waveform are complex and a systematic picture does not emerge from the data. In the presence of deleterious environment, load ratio takes on increased importance as seen in Figure 3.19 for a high strength low alloy (HSLA) steel. In benign environments, the difference between the fatigue crack growth rates for high- and low-load ratios is considerably less for this material in Region II of the crack growth behavior.

## 3.6 Limitations of LEFM

Linear elastic fracture mechanics is well suited for applications which involve high strength, low fracture toughness materials such as aluminum alloys used in aircraft structures. However, several structural materials used in the construction of chemical reactors, highways, bridges, and power plants

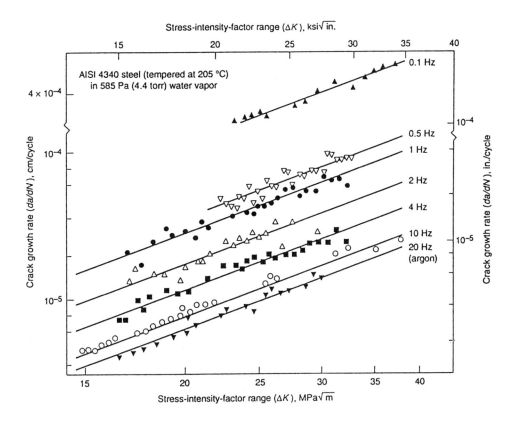

**Figure 3.17** Influence of cyclic frequency on the fatigue crack growth behavior of 4340 in dehumidified argon and in water vapor (Ref. 3.28).

do not fall in this category. Brittle fractures in these materials do not occur often and are preceded by ductile or stable crack growth. Fracture is also accompanied by significant plastic deformation. The latter condition violates the assumptions of LEFM as illustrated in the following example.

**Example Problem 3.1**

Suppose that a low alloy steel has a yield strength of 350 MPa and a $K_{IC}$ value of 110 MPa$\sqrt{m}$. Determine the smallest size compact specimen that will be needed to measure a valid $K_{IC}$. Also, estimate the load capacity of the machine needed to perform the test.

*Solution:*

The minimum size requirements, B and W, are given by equation (3.29).
Substituting $K_Q = K_{IC} = 110$ MPa$\sqrt{m}$ and $\sigma_0 = 350$ MPa, we get:

$$B = W - a = 0.246 m = 24.6 cm$$

Thus, the specimen must be 24.6 cm thick or 49.2 cm wide. Using the K expression from Table 3.1 for the CT specimen, we can calculate that for a/W = 0.5:

$$F(a/W) = \frac{K}{P} BW^{\frac{1}{2}} \approx 9.6$$

$$\text{or} \quad P = \frac{110(.246)(.492)^{\frac{1}{2}}}{9.6} \approx 2MN = 2000 kN$$

Thus, the machine must have a minimum load capacity of 2,000 kN which is approximately equal to 449,000 lbs.

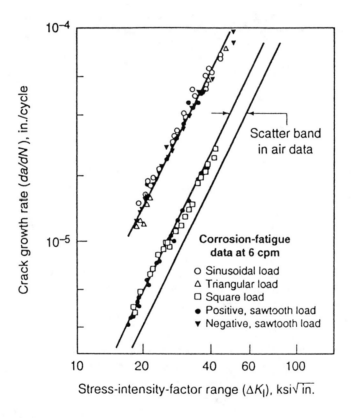

*Figure 3.18* *Effect of cyclic loading waveform on the corrosion-fatigue crack growth behavior of a 15Ni-5Cr-3Mo steel (Ref. 3.28).*

From the above example it is clear that testing such a large specimen is not practical. Note that the volume of the specimen will be 89,321 cm³ and it will weigh approximately 678 kg. The material and machining costs plus the handling and testing costs of such a test will be prohibitive. It also means that the minimum dimensions of the component will also have to be the same for the results to be applicable. These are very severe restrictions on the use of LEFM.

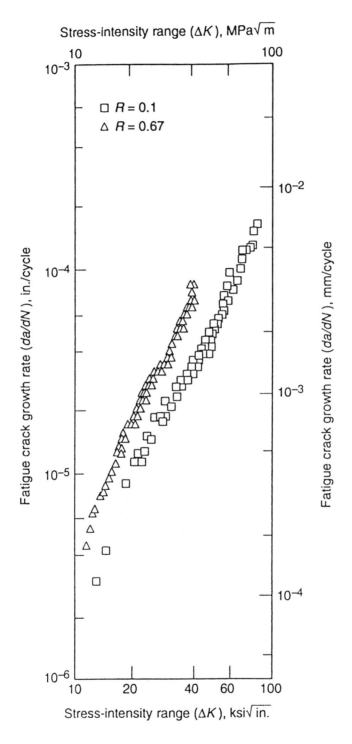

**Figure 3.19** *Effect of load ratio on the corrosion-fatigue crack growth rates of a HSLA-80 steel (Ref. 3.28).*

## Example Problem 3.2

A 5cm-wide compact specimen of the same material as in Example Problem 3.1 is to be used for fatigue crack growth testing. For the application being considered, it is desirable to obtain fatigue crack growth rate data for a load ratio of 0.5 and a $\Delta K$ level of 30 MPa$\sqrt{m}$. What is the minimum ligament size (W - a) for the test to be valid using LEFM concepts.

*Solution*:

The validity criterion in this case is given by equation (3.36). For $\Delta K = 30$ MPa$\sqrt{m}$ and R = 0.5, the $K_{max} = 60$ MPa$\sqrt{m}$. Substituting this into equation (3.36) we get:

$$W - a \geq \frac{4}{\pi} \left( \frac{60}{350} \right)^2 = 0.0374 m = 3.74 cm$$

Therefore, the maximum a/W for which valid data can be obtained = 2.26/5.0 = 0.452. Since fatigue crack growth testing can commence at a/W = 0.3 in CT specimens, the crack can be grown by 0.76cm before the data will no longer be considered as being valid by LEFM requirements. Nevertheless, it is possible to use a specimen 5.0cm wide to generate the fatigue crack growth data.

From Example 3.2 it is evident the LEFM has more applications in materials that have low strength and high fracture toughness under the conditions of cyclic loading than under monotomic loading. However, even for cyclic loading, the consideration of small cracks often limits the applicability of LEFM. Small cracks are important in estimating the fatigue crack growth life of components. For LEFM to apply, the crack size must exceed a certain length such as given by the right-hand side of equation (3.36). For cracks that do not meet this requirement, LEFM is not valid because the crack size becomes comparable to the plastic zone size. This limitation is illustrated by Example Problem 3.3.

## Example Problem 3.3

An internally pressurized cylinder from the same material as in Example Problems 3.1 and 3.2 is subjected to a maximum cyclic hoop stress of 250MPa. This component can have long axial cracks on the inside surface which run from one end of the cylinder to the other. The cylinder is inspected using a magnetic particle technique which can detect all cracks that are 0.5mm or more deep. Comment on whether LEFM may be used to evaluate the in-service inspection interval. Assume that the cylinder wall is 10mm thick.

*Solution:*

The crack size at inspection is much smaller than the wall thickness. Therefore, the stress intensity parameter can be approximated by the following expression:

$$K = 1.12 \, \sigma \sqrt{\pi a}$$

For LEFM to be valid, the crack size, a, must be greater than the right-hand side of equation (3.36) at all times. K for a stress level of 250MPa and a = 0.5mm is 11.09 MPa$\sqrt{m}$. Thus:

$$a \geq \frac{4}{\pi} \left( \frac{11.09}{350} \right)^2 = 0.00128 m = 1.28 mm$$

Since the inspection crack depth of 0.5mm is smaller than 1.28mm, LEFM cannot be applied to estimate the inspection interval of this component.

There are other situations besides small cracks for which LEFM is limited in describing the fatigue crack growth behavior. Several fatigue crack growth problems occur due to thermal-mechanical stresses or dominantly thermal stresses due to transient and steady-state temperature distributions in components. The materials used in thermal-mechanical fatigue applications are typically low strength and high fracture toughness materials. The stresses during these applications can easily exceed the yield strength of the material and can vary with time. LEFM is severely limited for such applications. Examples of such applications are jet engine components, steam headers, turbine casings, thick wall pipes, and nozzles in power plants.

There are limitations to the use of LEFM when sustained loads are applied at temperatures high enough to cause creep deformation at the crack tip. If creep deformation occurs, the stresses redistribute with time near the crack tip and $K$ no longer uniquely determines the crack tip stresses. In the case where the creep deformation is limited to a small region near the crack tip, $K$ characterizes the stresses beyond the creep zone and influences the stresses within the creep zone, but the relationship between stress and $K$ within the creep zone is not unique. Stresses in this region depend both on $K$ and time, $t$, and the creep properties of the material. When the creep deformation becomes extensive, $K$ is of no consequence in determining the crack tip stress fields. Thus, there are clear limitations to LEFM in analyzing the integrity of high temperature components. Time-dependent fracture mechanics is needed for analyzing such problems.

Time is also a factor when the loading is very rapid such that inertial effects become important or when crack speed is a consideration. The LEFM concepts presented in this chapter do not describe such situations. Dynamic fracture mechanics concepts are used in such instances.

LEFM concepts described in this chapter are dominantly for mode I loading. When mixed-mode loading involving combined modes I, II, or III or simply modes II and III by themselves are encountered, there are no widely accepted approaches for analyzing these problems. This remains a topic of current research in the field of fracture mechanics.

LEFM concepts are also not directly applicable to anisotropic and inhomogeneous materials such as composites. The concept of $K$ is based on the assumption that the cracked bodies are homogeneous and isotropic. Also, mode-mixity is quite often encountered in dealing with cracks in composite materials. Development of methods for predicting fracture in such materials also remains a topic of current research.

## 3.7 Summary

In this chapter, the linear elastic fracture mechanics (LEFM) and the mechanics framework on which it is based were briefly reviewed. This background is absolutely essential for understanding the later chapters of this book. Rather than attempting to be complete in our description of LEFM, which is the subject of other textbooks, this chapter has attempted to emphasize the assumptions that determine the limitations of LEFM. These limitations serve as the motivation for studying elastic-plastic and time-dependent fracture mechanics concepts described in detail in the subsequent chapters.

The chapter began with a critical summary of the basic concepts of the crack extension force and the stress intensity parameter for characterizing crack growth and fracture. The origin of the former approach is embedded in the principle of energy balance while the latter is a straightforward stress-based approach. Although the two approaches were developed independently and seem different, they are mathematically equivalent.

Crack tip plasticity was considered in this chapter because allowing for some plastic deformation is essential to the use of any theory for predicting crack growth and fracture in metals. It was shown that the stress intensity parameter, $K$, uniquely determines the shape and size of the crack tip plastic zone regardless of which plastic yield criterion is chosen, provided small-scale yielding can be ensured. Conditions for ensuring small-scale yielding are stated.

The framework of LEFM is applied to address fracture toughness under plane strain and plane stress conditions as well as crack growth under cyclic loading (fatigue), stress corrosion cracking, and corrosion-fatigue. It was shown that within the limitations of LEFM, the crack growth rates can be uniquely characterized by the stress intensity parameter. This makes LEFM a very powerful analytical tool for addressing crack growth and fracture in a variety of structural components.

The chapter ended with a discussion of the limitations of LEFM with respect to plasticity which accompanies fracture in low strength, high fracture toughness materials; with respect to describing the behavior of small cracks which is important during consideration of fatigue loading; and with respect to creep deformation for high temperature components, mixed-mode loading and anisotropic and inhomogeneous materials.

**3.8 References**

3.1  D. Broek, "Elementary Engineering Fracture Mechanics", Fourth Edition, Kluwer Academic Publishers, Dordrecht, 1991.

3.2  T.L. Anderson, "Fracture Mechanics-Fundamental and Applications", Second Edition, CRC Press, Boca Raton, 1995.

3.3  E.E. Gdoutos, "Fracture Mechanics- An Introduction", Kluwer Academic Publishers, Dordrecht, 1993.

3.4  A.A. Griffith, "The Phenomena of Rupture and Flow in Solids", Philosophical Transactions, Series A, Vol. 221, 1921, pp. 163-197.

3.5  E. Orowan, "Fracture and Strength of Solids", Reports on Progress in Physics, Vol. XII, 1948, pp. 185-232.

3.6  G.R. Irwin, "Fracture Dynamics", Fracturing of Metals, American Society for Metals, 1948, pp. 147-166.

3.7  G.R. Irwin, "Analysis of Stresses and Strains Near the End of a Crack Traversing a Plate", Journal of Applied Mechanics, Vol. 24, 1957, pp. 361-364.

3.8  H.M. Westergaard, "Bearing Pressures and Cracks", Journal of Applied Mechanics, Vol. 6, 1939, pp. 49-53.

3.9  M.L. Williams, "On the Stress Distribution at the Base of a Stationary Crack", Journal of Applied Mechanics, Vol. 24, 1957, pp. 109-114.

3.10  A. Talug and K. Reifsnider, "Analysis and Investigations of Small Flaws", ASTM STP 637, 1977, pp. 81-96.

3.11  H. Tada, P.C. Paris, and G.R. Irwin, Stress Analysis of Cracks Handbook, Del Research Co., St. Louis, 1985.

3.12  D.P. Rooke and D.J. Cartwright, Compendium of Stress Intensity Factors, Her Majesty's Stationery Office, London, 1976.

3.13 J.C. Newman and I.S. Raju, "Stress Intensity Factor Equations for Cracks in Three-Dimensional Finite Bodies", Fracture Mechanics: Fourteenth Symposium - Volume I: Theory and Analysis, ASTM STP 791, American Society for Testing and Materials, 1983, pp. I-238 - I-265.

3.14 G.R. Irwin, "Plastic Zone Near a Crack and Fracture Toughness", Proceeding of the 7th Sagamore Conference, pp. IV-63-70, 1960.

3.15 D.S. Dugdale, "Yielding in Steel Sheets Containing Slits", Journal of Mechanics and Physics of Solids, Vol. 8, 1960, pp. 100-106.

3.16 G.I. Barenblatt, "The Mathematical Theory of Equilibrium Cracks in Brittle Fracture", Advances in Applied Mechanics, Vol. VII, Academic Press, 1962, pp. 55-129.

3.17 A. Saxena and S.J. Hudak, Jr., "Review and Extension of Compliance Relationships for Common Crack Growth Specimens", International Journal of Fracture, Vol. 14, 1978, pp. 453-469.

3.18 E 399-90: "Standard Test Method for Plane-Strain Fracture Toughness of Metallic Materials", 1994, Annual Book of ASTM Standards, Vol. 03.01, 1994, pp. 407-437.

3.19 BS 5447:1974, "Methods of Testing for Plane Strain Fracture Toughness ($K_{IC}$) of Metallic Materials", British Standards Institution, London, 1974.

3.20 E 561-94, "Standard Practice for R-Curve Determination", 1994 Annual Book of ASTM Standard, Vol. 03.01, 1994, pp. 489-501.

3.21 P.C. Paris, M.P. Gomez, and W.P. Anderson, "A Rational Analytic Theory of Fatigue, The Trend in Engineering", Vol. 13, 1961, pp. 9-14.

3.22 P.C. Paris and F. Erdogan, "A Critical Analysis of Crack Propagation Laws", Journal of Basic Engineering, Vol. 85, 1960, pp. 528-534.

3.23 D.W. Cameron, "Fatigue Properties in Engineering", ASM Handbook, Vol. 19, Fatigue and Fracture, 1996, pp. 15-26.

3.24 A. Saxena, S.J. Hudak, Jr., and G.M. Jouris, "A Three Component Model for Representing Wide Range Fatigue Crack Growth Rate Data, Engineering Fracture Mechanics, Vol.12, 1979, pp. 103-115.

3.25 E-647-93: "Standard Test Method for Measurement of Fatigue Crack Growth Rates", 1994 Annual Book of ASTM Standards, Vol. 03.01, 1994, pp. 569-596.

3.26 A. Saxena, S.J. Hudak, K. Donald, and D. Schmidt, "Computer-Controlled Decreasing Stress Intensity Technique for Low Rate Fatigue Crack Growth Testing", Journal of Testing and Evaluation, Vol. 6, No. 3, 1978, pp. 167-174.

3.27 G.H. Koch, "Stress-Corrosion Cracking and Hydrogen Embrittlement", ASM Handbook, Vol. 19, Fatigue and Fracture, ASM International, Metals Park, Ohio, 1996, pp. 483-506.

3.28 P.S. Pao, "Mechanisms of Corrosion Fatigue", ASM Handbook, Vol. 19, Fatigue of Corrosion Fatigue", ASM Handbook, Vol. 19, Fatigue and Fracture, ASM International, Metals Park, OH, 1996, pp. 185-192.

3.29 E 1681-95, "Standard Test Method for Determining a Threshold Stress Intensity Factor for Environment-Assisted Cracking of Metallic Materials Under Constant Load", 1996 Annual Book of Standards, Vol. 03.01, American Society for Testing and Materials, Philadelphia, 1996, (in press).

## 3.9 Exercise Problems

3.1 Show that for fracture under fixed grips and under fixed load:

$$\mathcal{G} = \left(\frac{P^2}{2B}\right)\left(\frac{dC}{da}\right)$$

and $\mathcal{G}$ is always equal to the rate of change of strain energy with crack size.

3.2 For plane stress conditions, show that the relationship between the Griffith's crack extension force and the stress intensity parameter is given by:

$$\mathcal{G} = \frac{K^2}{E}$$

3.3 Using the Von Mises criterion, derive the equations for the shape of the crack tip plastic zone for plane stress and plane strain conditions, respectively.

3.4 Starting from equation (3.23), derive equation (3.24).

3.5 If the Airy's stress function $\psi = \text{Re}\underline{\underline{Z}} + y\,\text{Im}\underline{Z}$, show that:

$$\sigma_x = ReZ - yImZ'$$
$$\sigma_y = ReZ - yImZ$$
$$\tau_{xy} = -ReZ'$$

where $Z$ is a complex analytic function and:

$$\overline{\underline{Z}} = \frac{d\overline{Z}}{dz}\,,\ \underline{Z} = \frac{d\overline{Z}}{dz}\,,\ Z' = \frac{dZ}{dz}$$

$z = x + iy$ the complex variable and Re and Im refer to the real and imaginary parts of the complex function.

3.6 For a wedge loaded crack in a semi-infinite plate the $Z$ function is given by:

$$Z = \frac{Pa}{\pi(z-a)z}\sqrt{\frac{1-(b/a)^2}{1-(a/z)^2}}$$

show that:

$$K_A = \frac{P}{\sqrt{\pi a}}\sqrt{\frac{a+b}{a-b}}$$

$$K_B = \frac{P}{\sqrt{\pi a}}\sqrt{\frac{a-b}{a+b}}$$

3.7 On your own, derive the crack tip stress field equations for mode I using the following Westergaard's function:

$$Z = \frac{K_I}{\sqrt{2\pi z}}$$

where $K_I$ = mode I stress intensity parameter.

3.8 Derive the crack tip displacement equations for mode I under plane stress conditions.

3.9 Using the principle of superposition and the results of Problem 3.6, show that for a crack in a semi-infinite plate subjected to pressure, $p$, on the crack surfaces, the stress intensity parameter, $K$ is given by:

$$K = p\sqrt{\pi a}$$

where $a$ = half crack length.

3.10 Discuss the sources of nonlinearity in load-displacement diagrams during fracture toughness testing. What additional information besides the load-displacement diagram are needed to separate the nonlinearity due to various contributions?

3.11 What are crack growth resistance curves and how can they be used for predicting instability in components which sustain stable crack growth prior to complete fracture?

3.12 What is the role of fatigue precracking in fracture toughness testing? What limitations are placed in selecting load levels and loading frequencies during precracking for obtaining $K_{IC}$ measurements and why?

3.13 The FCGR behavior of 2024-T4 Al alloy is given by the following equation:

$$\frac{da}{dN} = 8.5 \times 10^{-10} (\Delta K)^4$$

where da/dN is expressed in m/cycle and $\Delta K$ in MPa$\sqrt{m}$. A tension arm which activates a relay switch in a circuit breaker is made from this alloy and is under uniform tensile stress of 120 MPa due to spring forces. This arm has a width of 75mm. When the relay is activated, the stress relaxes completely. The average frequency of activation of the relay is 20/day. During a routine inspection of the circuit breaker, a through-the-thickness edge crack which is 5mm in length is uncovered. For the purpose of establishing the subsequent inspection interval, you are asked to calculate the crack length as a function of elapsed time in days. The fracture toughness of this material is 33 MPa$\sqrt{m}$. Predict the remaining life of the component and comment on the applicability of LEFM for this problem.

3.14 A compact specimen of 304 stainless steel is being used to generate FCGR data at a load ratio of 0.5. The width of the specimen is 50mm and the thickness is 6.25mm. The range of applied fatigue load is 2 x 10³N. If the yield strength of stainless steel is 200 MPa, what is the maximum crack size for which the data can be correlated by $\Delta K$. If FCGR data are needed up to $\Delta K$ of 45 MPa$\sqrt{m}$, what size specimens will be needed to obtain valid data?

3.15 Describe the differences between a constant load and a constant deflection stress corrosion cracking test and the significance of the results from each.

# CHAPTER 4

# ANALYSIS OF CRACKS UNDER ELASTIC-PLASTIC CONDITIONS

## 4.1 Introduction

In the previous chapter, we considered the limitations of linear elastic fracture mechanics for testing materials in which fracture is accompanied by large-scale plastic deformation. These are the high fracture toughness, low yield strength materials used extensively in power generation, construction, and chemical industries. Because of the success in the use of LEFM on high strength, low toughness materials used in the aerospace applications, it is tempting to overcome the limitations of LEFM for extending it to predict fracture in components made from the so-called "ductile" materials.

In looking to extend the fracture mechanics concepts to conditions for which scale of plasticity is no longer a limitation, it is prudent to take advantage of lessons learned during the development of LEFM. Let us briefly recount the reasons why the stress intensity parameter, K, successfully characterizes fracture under dominantly elastic conditions:

- K uniquely characterizes the crack tip stress field in a region which is sufficiently large in comparison to the region in which microscopic scale damage develops. The latter region is called the process zone.

- K uniquely characterizes the size and shape of the crack tip plastic zone. In other words, two bodies subjected to the same K level will have identical plastic zone sizes and shapes regardless of how different they are in size, shape, crack size, and loading configuration. In order to meet these similitude requirements, it is also essential that both bodies satisfy the required conditions for K-dominance and the crack tip regions experience the same levels of constraint. The restrictions in the use of LEFM are all related to meeting these requirements.

- K, through its unique relationship with $\mathcal{G}$, also characterizes the rate of release of strain energy during crack extension. For fracture under fixed displacement conditions, the strain energy released is used for supplying the required energy for plastic deformation and also for the creation of new surfaces which accompany fracture.

In the quest for a new approach for characterizing fracture under elastic-plastic and fully-plastic conditions, it seems reasonable to look for a parameter which meets the same set of conditions. The J-integral, which comes closest to meeting several of these properties is discussed next.

## 4.2 Rice's J-Integral

In 1968, J.R. Rice [4.1, 4.2] published papers in which he discussed the potential of a path-independent integral, J, for characterizing fracture in nonlinear-elastic materials. This integral is identical in form to the static component of the energy-momentum tensor for characterizing generalized forces on dislocations and point defects introduced by Eshelby. G.P. Cherapnov [4.3] working independently in the former Soviet Union during the same period as Rice, also presented a formulation of an integral similar to Rice's J.

The definition of J is formulated for nonlinear-elastic materials. Recall from discussion in Chapter 2 that nonlinear elastic materials are similar to plastic materials when the loading is in only one direction. In the mathematical analyses which follow, monotonically loaded plastic materials are modeled as nonlinear elastic materials with the restriction that unloading is not permitted.

Rice recognized that for any nonlinear-elastic, planar, homogeneous, and isotropic body in a state of static equilibrium, a certain integral, designated J, along a closed path is always equal to 0. Let $\Phi$ be a closed contour bounding a region A occupied by the body as shown in Figure 4.1. Let $x_1$, $x_2$ be the fixed Cartesian coordinate system to which all quantities are referenced:

$$J_\phi = \oint_\phi \left( W n_1 - \mathbf{T} \cdot \frac{\partial \mathbf{u}}{\partial x_1} \right) ds \tag{4.1}$$

where W = elastic strain energy density given by:

$$W = \int_0^{\varepsilon_{ij}} \sigma_{ij} d\varepsilon_{ij} \tag{4.2}$$

$$\sigma_{ij} = \frac{\partial W}{\partial \varepsilon_{ij}} \tag{4.3}$$

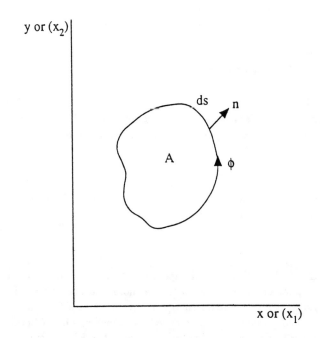

*Figure 4.1* *A counter-clockwise closed contour, $\phi$.*

Also, it follows that **T** = Traction vector defined according to the outward normal **n** to the contour $\Phi$, and **u** = displacement vector = $u_1 \mathbf{i} + u_2 \mathbf{j}$:

$$T_i = \sigma_{ij} n_j \tag{4.4}$$

Thus, equation (4.1) becomes:

$$J_\phi = \oint_\phi (Wn_1 - T_i \frac{\partial u_i}{\partial x_1}) \, ds$$

$$J_\phi = \oint_\phi (Wn_1 - (\sigma_{ij} n_j) \frac{\partial u_i}{\partial x_1}) \, ds \qquad (4.5)$$

using the Green's theorem (see any standard text book on advanced mathematics for statement and proof of the Green's theorem), we get:

$$J_\phi = \int_A (\frac{\partial W}{\partial x_1} - \frac{\partial}{\partial x_j}(\sigma_{ij} \frac{\partial u_i}{\partial x_1})) \, dx_1 \, dx_2 \qquad (4.6)$$

In the absence of body forces and assuming small deformations, we can write:

equilibrium equations:
$$\frac{\partial \sigma_{ij}}{\partial x_j} = 0 \qquad (4.7)$$

strain-displacements relations:
$$\varepsilon_{ij} = \varepsilon_{ji} = \frac{1}{2}(\frac{\partial u_i}{\partial x_j} + \frac{\partial u_j}{\partial x_i})$$

$$\frac{\partial W}{\partial x_1} = \frac{\partial W}{\partial \varepsilon_{ij}} \frac{\partial \varepsilon_{ij}}{\partial x_1} = \sigma_{ij} \frac{\partial \varepsilon_{ij}}{\partial x_1}$$

$$= \frac{1}{2} \sigma_{ij} [\frac{\partial}{\partial x_1}(\frac{\partial u_i}{\partial x_j} + \frac{\partial u_j}{\partial x_i})] = \sigma_{ij} \frac{\partial}{\partial x_j}(\frac{\partial u_i}{\partial x_1})$$

Applying equation (4.7) to the above equation, we get:

$$\frac{\partial W}{\partial x_1} = \frac{\partial}{\partial x_j}(\sigma_{ij} \frac{\partial u_i}{\partial x_i})$$

Substituting the above equation in equation (4.6), we get:

$$J_\phi = 0$$

Rice defined the J-integral for a cracked body as follows:

$$J = \int_\Gamma (Wn_1 - T_i \frac{\partial u_i}{\partial x_1}) \, ds \qquad (4.8)$$

where $\Gamma$ is a counter clockwise contour as shown in Figure 4.2 which begins at a point on the lower crack surface and ends on any point on the upper crack surface.

$$= \int_\Gamma W dy - T_i \frac{\partial u_i}{\partial x} ds \quad \text{(in the x,y coordinate system)}$$

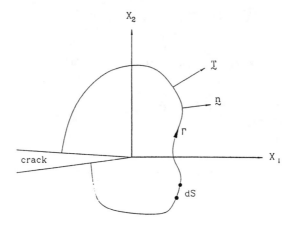

**Figure 4.2** *A 2-dimensional cracked body with a contour $\Gamma$ originating from the lower crack surface and going counterclockwise and terminating at the upper crack surface. The traction vector at any point on the contour is marked T.*

### 4.2.1 Path - Independence of J - Integral

It will now be shown that the value of J is independent of the path $\Gamma$ if the path originates at any point on the lower crack surface and goes counter clockwise and ends at any point on the upper crack surface. Figure 4.3 shows a closed path $\phi$ which begins and ends at point A on the crack surface but can be divided into four segments $\Gamma_1$, $\Gamma_2$, $\Gamma_3$, and $\Gamma_4$, as shown in Figure 4.3. Paths $\Gamma_2$ and $\Gamma_4$ are parallel to the crack surface which lies along the x-axis as shown. Since J along the closed path is 0, we can write the following:

$$J_{\Gamma_1} + J_{\Gamma_2} + J_{\Gamma_3} + J_{\Gamma_4} = 0 \tag{4.9}$$

Note that:

$$n_1 = \frac{dy}{ds} \text{ and } dy = 0 \text{ along } \Gamma_2 \text{ and } \Gamma_4$$

Also, since the crack surfaces are traction free, $T_i = 0$ along $\Gamma_2$ and $\Gamma_4$. This leads to $\Gamma_2 = \Gamma_4 = 0$. Thus, equation (4.9) becomes:

$$J_{\Gamma_1} = -J_{\Gamma_3}$$

and it follows the J calculated along any counter clockwise path which begins at the lower crack surface and ends at the upper crack surface is equal.

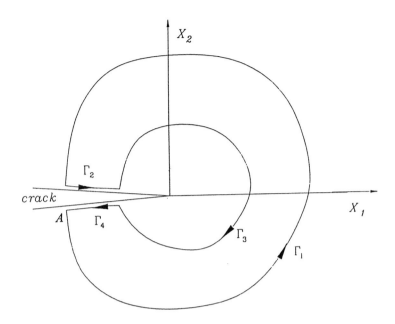

*Figure 4.3 A closed contour around a cracked tip with four distinct segments.*

In the context of fracture, the path-independence of J is an important property. It will be shown later in this chapter that J uniquely characterizes the crack tip stress fields. Thus, once we calculate the value of J around any admissible path $\Gamma$, its value can be used to estimate the crack tip stress and strain fields. The fact that J can be estimated from stress and displacement quantities far away from the crack tip is very important in numerical analysis for determining J because numerical solutions are often not accurate in the immediate vicinity of the crack tip but increase in accuracy as one moves away from the crack tip. The path-independence of J implies that J can also be measured at points remote from the crack tip. Methods of measuring J will be discussed in the next chapter.

*4.2.2 Relationship Between J and Potential Energy*

J can be shown to relate to the rate of change of potential energy with respect to change in crack size. This interpretation of J is useful in showing that under linear elastic conditions, $J \equiv \mathcal{G}$, the Griffith's crack extension force. Also, alternate methods for determining J for cracked bodies rely on the relationship between potential energy and J. This relationship is derived next.

Let us consider a two-dimensional cracked body with crack length, a, subjected to prescribed tractions and displacements along parts of the body as shown in Figure 4.4. The tractions and displacements are independent of the crack size. The thickness of the body is constant and without loss of generality, it can be assumed to be unity. The potential energy, U, is given by:

$$U = \int_A W(x,y)\,dA - \int_{S_T} T_i u_i\,ds \qquad (4.10)$$

where A = area of the body and $T_i$ and $u_i$ are the tractions and displacements, respectively, applied along the boundary $\Gamma$. Differentiating U with 'a' yields:

$$\frac{dU}{da} = \int_A \frac{dW}{da} dA - \int_\Gamma T_i \frac{du_i}{da} ds \tag{4.11}$$

The contour in the second integral on the right-hand side of the above equation has been extended to include the entire boundary, $\Gamma$, because $du_i / da = 0$ along $S_u$, the portion of the boundary along which the displacements are prescribed. Also, the crack surfaces and the boundary of the body, other than $S_T$ and $S_u$, are traction free.

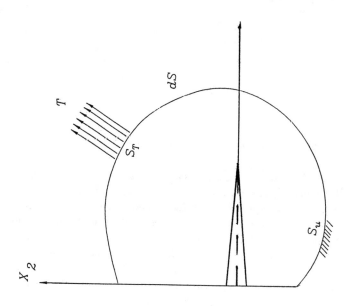

**Figure 4.4** *A planar cracked body with a boundary defined by $\Gamma$. $S_T$ and $S_u$ represent regions where the tractions and displacements along the boundary are defined.*

If we now move to a new coordinate system located at the crack tip, $x_1$, $x_2$, we have:

$$x_1 = x - a \quad \text{and} \quad x_2 = y$$

Hence:

$$\frac{d}{da} = \frac{\partial}{\partial a} + \frac{\partial}{\partial x_1} \frac{\partial x_1}{\partial a} = \frac{\partial}{\partial a} - \frac{\partial}{\partial x_1} \quad (\text{since } \partial x_1 = -\partial a)$$

Thus, (4.11) becomes:

$$\frac{dU}{da} = \int_A \left(\frac{\partial W}{\partial a} - \frac{\partial W}{\partial x_1}\right)dA - \int_\Gamma T_i \left(\frac{\partial u_i}{\partial a} - \frac{\partial u_i}{\partial x_1}\right) ds \qquad (4.12)$$

$$\frac{\partial W}{\partial a} = \frac{\partial W}{\partial \varepsilon_{ij}} \frac{\partial \varepsilon_{ij}}{\partial a} = \sigma_{ij} \frac{\partial \varepsilon_{ij}}{\partial a}$$

$$\int_A \frac{\partial W}{\partial a} dA = \int_A \sigma_{ij} \frac{\partial \varepsilon_{ij}}{\partial a} dA$$

Application of the Green's Theorem and recognizing the $T_i = \sigma_{ij} n_j$, we get:

$$\int_A \sigma_{ij} \frac{\partial \varepsilon_{ij}}{\partial a} dA = \int_\Gamma T_i \frac{\partial u_i}{\partial a} ds$$

Also from Green's Theorem:

$$\int_A \frac{\partial W}{\partial x_1} dA = \int_\Gamma W dx_2$$

or

$$\frac{dU}{da} = \int_\Gamma \left(T_i \frac{\partial u_i}{\partial a} ds - W dx_2 - \left(T_i \left(\frac{\partial u_i}{\partial a} - \frac{\partial u_i}{\partial x_1}\right)\right) ds\right.$$

or

$$-\frac{dU}{da} = \int_\Gamma W dx_2 - T_i \frac{\partial u_i}{\partial x_1} ds$$

$$= \int_\Gamma \left(W n_1 - T_i \frac{\partial u_i}{\partial x_1}\right) ds$$

Hence, $J = -dU/da$ for a body of unit thickness. For a body of thickness B:

$$J = -\frac{1}{B} \frac{dU}{da} \qquad (4.13)$$

Thus, J can be interpreted as being directly related to the rate of change of potential energy of the body with respect to crack length. This definition of J will be used in the next chapter to derive formulae for determining J. For linear elastic bodies, by comparing equation (4.13) and equation (3.6), it can be shown that $J \equiv \mathcal{G}$. Hence, if we define a fracture criterion based on J, it will be entirely consistent with the theory of linear elastic fracture mechanics considering that $\mathcal{G}$ is uniquely related to K, equation (3.15).

### 4.2.3 J - Integral for Blunt Notches

If we consider a body with a blunt notch as shown in Figure 4.5, J can be defined around the root of the notch by taking a path along the notch surface from A to B. Since the notch surface is traction free:

$$J = \int_{\Gamma_t} W dy \qquad (4.14)$$

where $\Gamma_t$ is a path along the surface of the notch. Since $dy = 0$ along the surfaces of the notch that are parallel to the x-axis, J represents the average strain energy density along the root of the notch from A to B. Thus, J may be used as a criterion for crack initiation for blunt notches.

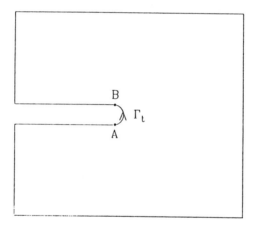

*Figure 4.5* Blunt notch body with a contour taken around the surface of the notch.

## 4.3 J-Integral, Crack Tip Stress Fields and Crack Tip Opening Displacement

The interest in J-integral as a fracture criterion spurred up considerably when relationships were found which uniquely relate J to the crack tip stress fields and to the crack tip opening displacement (CTOD).

### 4.3.1 Relationship Between J and Crack Tip Stress Fields

In 1968, in separate papers, Huchinson [4.4, 4.5] and Rice and Rosengren [4.6] showed that a unique relationship exists between the magnitude of J and the stress and strain fields at the crack tip for nonlinear elastic materials. These stress fields have come to be widely known as the Huchinson, Rice, and Rosengren (HRR) stress fields in the fracture mechanics literature. Let us now look at the derivation of these stress fields.

Since J is path independent, we can choose any convenient path around the crack and evaluate J without loss of generality. Let us then choose a circular path of radius r and its center located at the crack tip as shown in Figure 4.6. We use the following relationships:

$$y = r\sin\theta \text{ or } dy = r\cos\theta\, d\theta$$

$$ds = r\, d\theta$$

Thus, J can be written as:

$$J = r \int_{-\pi}^{\pi} \left[ W(r,\theta) \cos\theta - T_i \frac{\partial u_i}{\partial x_1} \right] d\theta$$

or

$$\frac{J}{r} = \int_{-\pi}^{\pi} \left[ W(r,\theta) \cos\theta - T_i \frac{\partial u_i}{\partial x_1} \right] d\theta \tag{4.15}$$

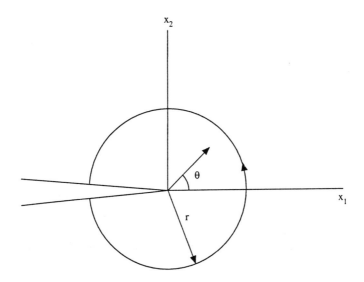

*Figure 4.6* A circular path of radius r around the crack tip with the crack tip at the center of the circular path.

All terms in the integrand are functions of the products of stress and strain. Therefore, one can write the following:

$$\sigma_{ij}\, \varepsilon_{ij}\, f_{ij}(\theta,m) \propto \left(\frac{J}{r}\right)$$

$$\sigma_{ij}\, \varepsilon_{ij} \propto \left(\frac{J}{r}\right) g_{ij}(\theta,m) \tag{4.16}$$

where $f_{ij}$ and $g_{ij}$ are functions of $\theta$ which correspond to the various stress and strain components. Let us assume the following nonlinear stress-strain relationship representing the plastic portion of equation 2.27, given in the uniaxial form as:

$$\frac{\varepsilon}{\varepsilon_o} = \alpha \left(\frac{\sigma}{\sigma_o}\right)^m \tag{4.17}$$

where $\varepsilon_o$ and $\sigma_o$ are the strain and stress, respectively, at the yield point and m is the plasticity exponent. Substituting equation (4.17) into equation (4.16) yields the following relationship for the crack tip stress and strain fields:

$$\sigma_{ij} = \sigma_o \left(\frac{J}{\alpha \sigma_o \varepsilon_o I_m r}\right)^{\frac{1}{1+m}} \hat{\sigma}_{ij}(\theta, m) \tag{4.18a}$$

$$\varepsilon_{ij} = \alpha \varepsilon_o \left(\frac{J}{\alpha \sigma_o \varepsilon_o I_m r}\right)^{\frac{m}{1+m}} \hat{\varepsilon}_{ij}(\theta, m) \tag{4.18b}$$

where all quantities except $I_m$ and the angular functions have been defined previously. The values of $I_m$ for different values of m are shown in Figure 4.7 for plane stress and plane strain conditions, respectively. The following equations also provide the approximate values of $I_m$:

for plane strain: $\quad I_m = 6.568 - .4744m + .0404m^2 - .001262m^3 \tag{4.19a}$

for plane stress: $\quad I_m = 4.546 - .2827m + .0175m^2 - .45816 \times 10^{-4}m^3 \tag{4.19b}$

The angular functions $\hat{\sigma}_{ij}$ and $\hat{\varepsilon}_{ij}$ are listed in Appendix I in the back of the book for various values of m and $\theta$. Equations (4.18) can be considered as being analogous to the crack tip stress fields derived for linear elastic conditions. By substituting $m = 1$ and $J = K^2/E$, the linear dependence of crack tip stress and strain on K and the inverse square root dependence on r can be readily obtained.

### 4.3.2 Relationship Between J and CTOD

A simple relationship between J and the crack opening displacement (CTOD) can be derived by considering the Dugdale [4.7] model of a strip deformation zone at the crack tip as shown in Figure 4.8. If we take an integration path which follows the boundary of the strip deformation zone ABC at the crack tip along which $dy = 0$, the J integral simplifies to:

$$J = \int_a^{a+c} \sigma_o \frac{\partial}{\partial x}(u_2^+ - u_2^-) \, dx$$

where $u_2^+$ and $u_2^-$ are the displacements in the y-direction corresponding to upper and lower crack surfaces. Hence:

$$J = \sigma_o(u_2^+ - u_2^-)]_a^{a+c} = \sigma_o \delta_t \qquad (4.20)$$

where $\delta_t$ = crack tip opening displacement. Thus, equation (4.20) relates J directly to the crack tip opening displacement.

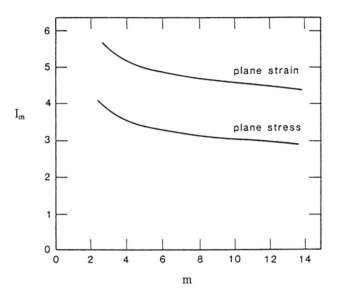

**Figure 4.7** *Variation in the values of $I_m$ with m for plane stress and plane strain conditions.*

Since the above derivation is for the special case assumed by the Dugdale model, in other words, the limiting case of a perfect plastic material ($m \to \infty$), a more general relationship between CTOD and J is needed. Figure 4.9 shows the various displacement components near the crack tip region along with a somewhat arbitrary definition of CTOD. In this case, CTOD is defined by Tracy [4.8] as the crack opening at the intercepts of two symmetric 45° lines from the deformed crack tip and the crack profile. From the crack tip fields, the displacement component $u_2$ along the crack profile can be given as follows:

$$u_2 = \alpha \varepsilon_o \left( \frac{J}{\alpha \varepsilon_o \sigma_o I_m} \right)^{\frac{m}{m+1}} r^{\frac{1}{1+m}} \hat{u}(\pi) \qquad (4.21)$$

The above equation is derived from equation (4.18b) and the use of strain-displacement relationships. Using these relationships, Shih [4.9] obtained the following equation:

$$\delta_t = d_m \frac{J}{\sigma_o} \qquad (4.22)$$

where $d_m$ is a constant which varies with the value of m. The values of $d_m$ for several values of m were obtained by Shih from finite element analysis. These are shown in Figure 4.10 for plane stress and plane strain. The value of $d_m$ also varies slightly with $\sigma_o/E$, but it varies much more strongly with m. Equation (4.22) clearly establishes the general relationship between the J-integral and CTOD. One can see that for plane stress and $m \to \infty$, the value of $d_m$ approaches 1 which is the same as predicted by equation (4.20) for the Dugdale model.

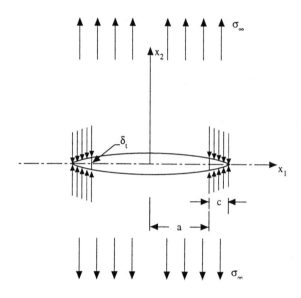

*Figure 4.8* *The Dugdale model for deformation at the crack tip in a thin sheet in an elastic perfectly plastic material.*

## 4.4 J-Integral as a Fracture Parameter and its Limitations

In the previous sections we have shown that the path-independent J-integral can also be interpreted as the rate of release of potential energy for nonlinear elastic materials; it is identical to $\mathcal{G}$ in the linear elastic regime; it can be related uniquely to the amplitude of the crack tip stress fields and also to the crack tip opening displacements. Comparing these properties of J to the reasons why K successfully characterized fracture under dominantly elastic conditions, J appears to be an attractive fracture parameter. However, one should be reminded again of some of the implicit assumptions which determine the conditions for which J retains the above properties. Specifically, nonlinear elasticity or deformation theory of plasticity applies only to elastic-plastic materials for monotonic loading. In other words, no unloading must occur. The second limitation is due to the small deformation theory which is used in proving path-independence of J, in deriving the relationship between J and potential energy and also in deriving the relationships between J and the crack tip stress fields and the CTOD. In ductile materials, fracture can occur in a process zone in which the deformations can no longer be considered small. That is, the strains can exceed 10 percent. Therefore, this limitation of J must be addressed before accepting J as a fracture parameter.

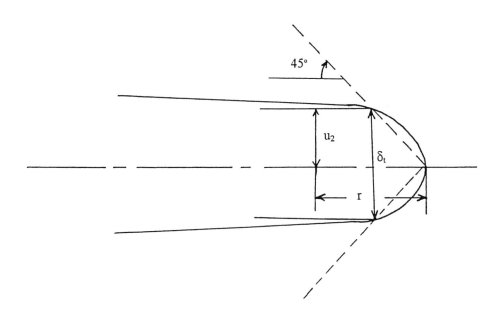

**Figure 4.9** *Definition of crack tip opening displacement (CTOD), $\delta_t$.*

### 4.4.1 $J_{Ic}$ and J - $\Delta a$ Curves

J.A. Begley and J.D. Landes in 1972 [4.10,4.11] were the first to experimentally demonstrate the potential of J as a fracture parameter. In the subsequent extensions of their early work, they hypothesized a relationship between the applied J and the amount of ductile crack extension, $\Delta a$, as shown in Figure 4.11. They also proposed a physical ductile tearing process during different stages of fracture ranging from the initially sharp crack prior to loading to extensive stable crack extension also shown in Figure 4.11. When crack extension occurs, some material which is initially in front of the crack tip lies in the wake region which is traction free. Thus, this material element is unloaded as it goes from the front to behind the crack tip with crack advance, a condition not permitted by the framework of the deformation theory of plasticity. In order to avoid this restriction, Begley and Landes proposed to use J as a fracture criterion only to specify the onset of ductile tearing, point 3 in Figure 4.11. They defined this point as the $J_{Ic}$, or the critical J in mode I at the onset of ductile tearing. Figure 4.12 reproduces their test results showing that the J is indeed independent of specimen geometry within the normal scatter observed during fracture testing. In these landmark experiments, several bend-type specimens and center-crack tension (CCT) specimens were tested, each with a different crack size. Figure 4.12 shows the plot of J vs. the applied load-line displacement. Arrows indicate the points on each of those plots where ductile tearing was noted to have initiated. The critical J values at crack initiation for both geometries vary between 0.14 to 0.185 MJ/m². Considering that $K_{IC}$ values can vary by 15 percent and the errors are squared when the same data are represented in terms of J, this was considered to be a very encouraging first result. As we will discuss in Chapter 6, a precise definition of $J_{Ic}$ can also reduce some of the scatter apparent in these results. At the time these first results were reported, no precise definition of $J_{Ic}$ was available. Also, the method of estimating J used in these early studies was not as accurate as the methods available today.

Operationally, $J_{Ic}$ is defined by the intersection of the crack blunting line which is approximated by the straight line described by $J = 2\sigma_0 \Delta a$, and the line which defines the J-$\Delta a$ curve, Figure 4.11. This construction is necessary to define the point of crack initiation because its physical detection in

94  *Nonlinear Fracture Mechanics for Engineers*

practice with a high consistency is difficult as was evidenced in the first set of results shown in Figure 4.12. An argument can be made that since the slope of the tearing portion of the J-Δa curve can be in error due to J not being strictly valid in that region, the consistency in the measurement $J_{Ic}$ will most likely suffer from this shortcoming. Further, we have also not resolved the issue of use of J in the presence of finite deformations in the process zone near the crack tip. In the discussion that follows, we address these issues by bringing in additional arguments and also by pointing to additional analyses which permit us to use J as a fracture criterion, despite these shortcomings.

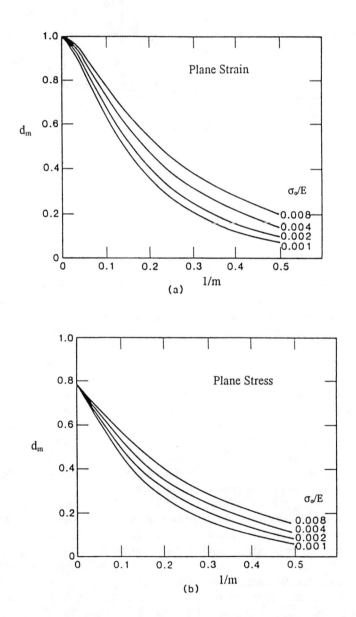

**Figure 4.10** *Dependence of $d_m$ on 1/m for (a) plane stress and (b) plane strain conditions.*

Analysis of Cracks Under Elastic-Plastic Conditions    95

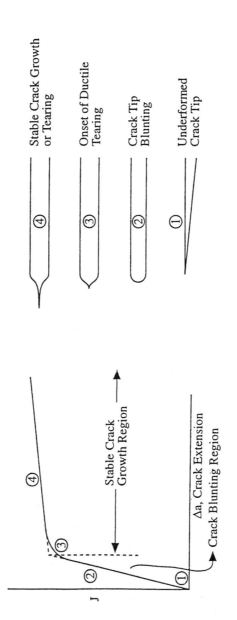

**Figure 4.11** Schematic representation of crack tip events during stable, ductile tearing (Ref. 4.11).

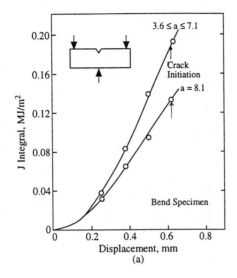

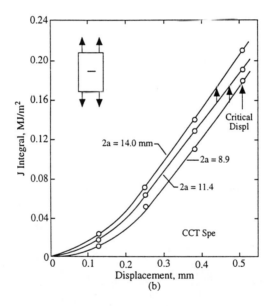

***Figure 4.12*** *The first experimental results of Landes and Begley demonstrating the independence of $J_{Ic}$ from specimen geometry (Ref. 4.11).*

### 4.4.2 Influence of Geometry and Deformation on J-Dominance

In order to achieve J-dominance, it is necessary that the crack tip conditions are the same for different geometries and that they are controlled by the magnitude of the applied J. The extent of the process zone in which the large deformation conditions exist can be expected to extend one CTOD (or $\delta_t$) distance beyond the crack tip. This zone of large deformation, also known as the zone of intense deformation, is surrounded by a larger zone in which J-controlled conditions apply. Thus, we impose the condition that, in order for J to be a valid fracture parameter, all pertinent length parameters in the test specimen such as the crack size, a, ligament size, W-a, where W is the planar width of the specimen, and the thickness, B, all exceed several times of $\delta_t$. Thus, we can require that to ensure J-dominated conditions:

$$a, W - a, B \geq c \left(\frac{J}{\sigma}\right) \tag{4.23}$$

where the value of c is on the order of 20, which will be considered further in Chapter 6.

McClintock [4.12] in his early work on plasticity aspects of fracture has pointed out that for perfectly plastic materials, the slip line fields for cracked specimens of different geometries can be vastly different. For example, Figure 4.13 shows these fields for three specimen geometries including an edge crack subjected to pure bending (SENB) and the center crack tension (CCT) specimen and the double edge notch (DEN) specimen. He, thus, argued that the crack tip stress and strain fields must be geometry-dependent. We then argue that strain hardening is required to have a clear J-dominated region and even in the presence of strain hardening, the extent of the region of J-dominance will depend on the specimen geometry. McMecking and Parks [4.13] performed an inelastic finite element analysis of the center crack tension (CCT) and the single edge notch bend (SENB) specimens allowing for consideration of finite deformation to explore the conditions which must be applied to ensure J dominance. For the CCT specimen with an a/W value of 0.5, m = 10, and $\varepsilon_0$ = .0033, the results of their computational analysis are replotted in Figure 4.14. This figure plots the normalized value of the stress, $\sigma_{yy}/\sigma_o$, as a function of normalized distance from the crack tip, $r\sigma_o/J$, for specimens of several different remaining ligament sizes, (W-a). The stress is expected to be independent of the ligament size if J-dominated conditions prevail. The solid line plots the results for the case of small-scale yielding. It is seen that in order for $\sigma_{yy}$ to be independent of the ligament size, the value of W-a for CCT specimens must satisfy the following condition:

$$b = W - a \geq 200 \left(\frac{J}{\sigma_0}\right) \tag{4.24}$$

For bend specimens, the J-dominated conditions were realized for:

$$b = W - a \geq 30 \left(\frac{J}{\sigma_0}\right) \tag{4.25}$$

The above results were confirmed by an extensive analysis by Shih and German [4.14] involving more values of m and specimen geometries. These results are shown in Figures 4.15 and 4.16 for m values of 3 and 10.

From the above discussion, it can be concluded that J-dominance can be assured in test specimens and also in components provided the conditions in equations (4.24) and (4.25) can be satisfied. This is extremely valuable in selecting adequate specimen sizes for fracture testing and for specifying the validity requirements for qualifying data, as will be discussed in Chapter 6.

### 4.4.3 Hutchinson-Paris Condition for J-Dominated Crack Growth

In the previous discussion, we have already built a rigorous rationale for use of J as a fracture parameter at the onset of ductile crack initiation. However, as we discussed earlier, it is necessary to also ensure some degree of J-dominated stable crack extension so that the crack initiation point such as shown in Figure 4.11 can be consistently measured. Hutchinson and Paris [4.15] in their analysis showed that for limited amounts of crack extension, J continues to characterize stable crack extension. The amount of crack extension for which J characterizes the crack growth will depend on geometry, size, and the strain hardening exponent, m. This analysis is described in the discussion which follows.

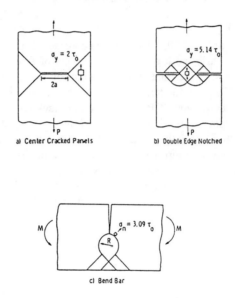

*Figure 4.13* Slip-lines for perfectly plastic materials for specimens of three geometries including (a) center crack tension (CCT), (b) double edge notch (DEN) under tension loading, and (c) single edge notch under pure bending (SENB) (Ref. 4.11).

The derivation of Hutchinson and Paris relies on the fact that even beyond initiation of crack growth, several metals sustain only small amounts of crack extension in comparison to the other pertinent length dimensions which characterize the mechanics of the problem. If crack growth is controlled by J, it is both necessary as well as sufficient that nearly proportional plastic deformation occur everywhere except in the small region near of the crack tip. In other words, the increment in strain must be uniquely related to the increment in J in the dominant portion of the body. Under these conditions, the deformation and incremental plasticity theories yield identical results. Figure 4.17 shows a schematic of the deformation zones ahead of the crack tip.

For power-law hardening materials, we can rewrite equation 4.18b as follows:

$$\varepsilon_{ij} = K_m \left(\frac{J}{r}\right)^{\left(\frac{m}{m+1}\right)} \hat{\varepsilon}_{ij}(\theta) \qquad (4.26)$$

where $K_m$ consists of several terms from equation (4.18b). We can differentiate equation (4.26) and determine the increment in strain with simultaneous increase in J and the crack length. The latter is due to each point ahead of the crack tip moving closer to the crack tip and changing the value of r. Let us assume that the crack lies along the x-axis and is assumed to advance an amount da in the x-direction.

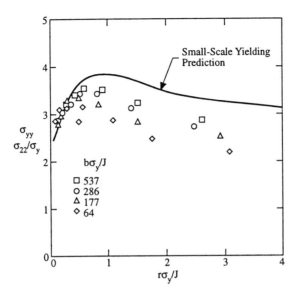

*Figure 4.14* Normal stress distribution ahead of the crack tip for a CCT specimen with a/W = 0.5, m = 10, and $\varepsilon_o$ = 0.0033. The solid curve is a prediction from small-scale yielding (Ref. 4.13).

The strain field of equation (4.26) continues to hold, even in in the presence of crack growth, with r and θ centered at the current crack tip location. In other words, we are examining the strain increment in an element of material which was originally at r + da and is now only a distance r away from the current crack tip:

$$d\varepsilon_{ij} = \frac{m}{m+1} K_m (\frac{1}{r})^{\frac{m}{1+m}} (J)^{-\frac{1}{m+1}} dJ \hat{\varepsilon}_{ij}(\theta) - K_m J^{\frac{m}{m+1}} da \frac{\partial}{\partial x} (r^{-\frac{m}{m+1}} \hat{\varepsilon}_{ij}(\theta))$$

using $\quad \frac{\partial}{\partial x} = \cos\theta \frac{\partial}{\partial r} - \frac{\sin\theta}{r} \frac{\partial}{\partial \theta} \quad$ we get:

$$d\varepsilon_{ij} = K_m (\frac{J}{r})^{\frac{m}{m+1}} \{ \frac{m}{m+1} \frac{dJ}{J} \hat{\varepsilon}_{ij}(\theta) + \frac{da}{r} \hat{\beta}_{ij}(\theta) \} \tag{4.27}$$

where

$$\hat{\beta}_{ij}(\theta) = \frac{m}{m+1} \cos\theta \, \hat{\varepsilon}_{ij}(\theta) + \sin\theta \frac{\partial}{\partial \theta} \hat{\varepsilon}_{ij}(\theta)$$

or

$$\frac{d\varepsilon_{ij}}{\varepsilon_{ij}} = \frac{m}{m+1} \frac{dJ}{J} + \frac{da}{r} \frac{\hat{\beta}_{ij}(\theta)}{\hat{\varepsilon}_{ij}(\theta)} \tag{4.28}$$

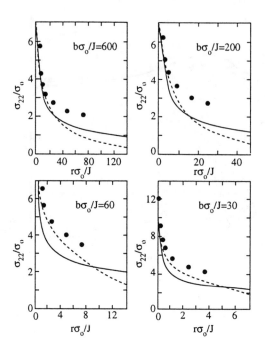

*Figure 4.15* Variation in normal stress ahead of the crack tip for SENB (----) and CCT (___) specimens from small-scale-yielding to fully-plastic behavior for a/W = 0.75, m = 3, along with the HRR field shown as (...) (Ref. 4.14).

The first term in equation (4.28) is very clearly the proportional term which implies that the increment in strain is directly proportional to the increment in J, dJ. The second term is the nonproportional term. Therefore, for J-controlled crack growth to occur, the first term must overwhelm the second. Since the magnitudes of m/(m+1) and all the angular terms are on the order of unity, this condition translates into the following requirement:

$$\frac{dJ}{J} >> \frac{da}{r} \tag{4.29}$$

$$\frac{1}{J}\frac{dJ}{da} >> \frac{1}{r} \tag{4.30}$$

The physical meaning of the term dJ/da is shown in Figure 4.18. If along the $J_R$ curve, the magnitude of J/(dJ/da) remains small in comparison to the region R where J fields dominate, J-controlled crack growth will occur. For fully-yielded specimens, R = W-a, the uncracked ligament size. Hence, equation (4.30) can be written as:

$$\frac{J}{(\frac{dJ}{da})} << R = W - a$$

or

$$\frac{W-a}{J}\frac{dJ}{da} = \omega \gg 1$$

The requirement of $\omega \gg 1$ is in addition to the requirement of equation (4.23) which is needed to insure that CTOD is still small in comparison to the ligament size, thus, the small deformation theory holds everywhere except in the small process zone near the crack tip.

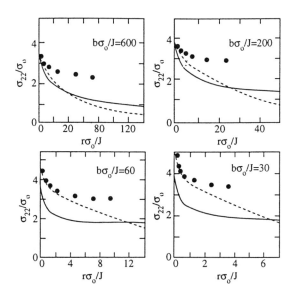

*Figure 4.16* Same as Figure 4.15 except m = 10 (Ref. 4.14).

The extent of J-controlled crack growth also depends on dJ/da. In other words, larger amounts of crack growth can be characterized by J in a specimen of given size if the slope of the $J_R$ curve is high. The subscript R on J refers to crack growth resistance as opposed to the applied value of J which does not have the subscript.

An additional question that remains unanswered is the minimum value of $\omega$ which will guarantee J-controlled crack growth. The value depends on the plasticity exponent m and the geometry of the specimen. Shih et al. [4.16] performed nonlinear finite element analysis assuming deformation theory of plasticity which allowed for different amounts of crack growth in a compact specimen with a W = 200mm and a = B = 100mm. They computed J along several contours which pass through the body at various distances from the crack tip. The results are shown in Figure 4.18. The preservation of path-independence is an indication of the J-controlled crack growth conditions. It is seen from these results that path-independence is preserved up to $\Delta a = 4$mm. The material used in this study is an ASTM grade A533 steel and the value of $\omega = 40$. For primarily bend geometries, analysis of Shih et al. [4.16] reported that J-controlled conditions exist for $\Delta a \leq 0.06$ (W-a) and for $\omega \geq 10$. In Chapter 6, we will explore the experimental implications of these conditions during the measurement of $J_{Ic}$.

102  *Nonlinear Fracture Mechanics for Engineers*

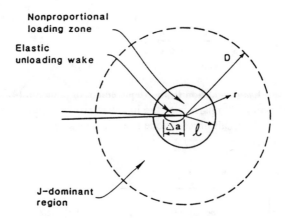

*Figure 4.17* Schematic representation of crack tip deformation zones for a growing crack (Ref. 4.15).

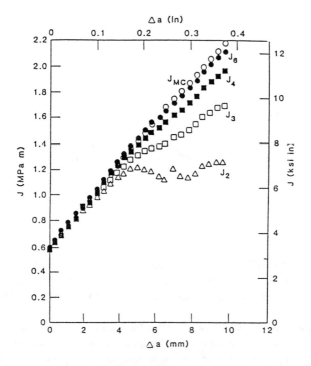

*Figure 4.18* J-resistance curves for different contours on a A533B, 4T compact specimen. The higher the number, the further removed is the contour from the crack tip. $J_{mc}$ is the value of J calculated from the load vs. load-line relationship discussed in detail in Chapter 5.

## 4.5 Summary

In this chapter, the J-integral first introduced by Rice was defined and its analytical basis for applicability as a fracture parameter for ductile materials was examined in detail. It was shown that

J-integral which is defined with the assumptions associated with the deformation theory of plasticity, can still be used as a fracture parameter for limited amounts of crack extension because it characterizes the amplitude of the crack tip stress fields. It was shown that, provided certain size requirements are met and the amount of crack extension is limited, the deformation fields ahead of the crack tip are essentially controlled by J even in the presence of crack growth. These findings pave the way for J to be applied as a characterizing parameter for fracture in ductile materials for which LEFM is unsuitable, provided certain restrictions are followed. These restrictions are discussed in detail in this chapter, and their validity is explored through analytical results, finite element analysis results, and also by experimental observations.

## 4.6 References

4.1 J.R. Rice, "A Path-Independent Integral and the Approximate Analysis of Strain Concentration by Notches and Cracks", Journal of Applied Mechanics, Trans. ASME, Vol. 35, 1968, pp. 379-386.

4.2 J.R. Rice, "Mathematical Analysis in the Mechanics of Fracture", Fracture - An Advanced Treatise, Vol II, H. Liebowitz, editor, Academic Press, N.Y., 1968, pp. 191-308.

4.3 G.P. Cherapnov, "Crack Propagation in Continuous Media", Applied Mathematics and Mechanics, (trans.P.M.M.), Vol. 31, 1967, pp. 476-488.

4.4 J.W. Hutchinson, "Singular Behavior at the End of a Tensile Crack in a Hardening Material", Journal of Mechanics and Physics of Solids, Vol. 16, 1968, pp. 13-131.

4.5 J.W. Hutchinson, "Plastic Stress and Strain Fields at a Crack Tip", Journal of Mechanics and Physics of Solids, Vol. 16, 1968, pp. 337-347.

4.6 J.R. Rice and G.F. Rosengren, "Plane Strain Deformation Near a Crack Tip in a Power-Law Hardening Material", Journal of Mechanics and Physics of Solids, Vol. 16, pp. 1-12.

4.7 D.S. Dugdale, "Yielding of Steel Sheets Containing Slits", Journal of Mechanics and Physics of Solids", Vol.8, 1960, pp. 100-104.

4.8 D.M. Tracy, "Finite Element Solutions for Crack Tip Behavior in Small-Scale Yielding", Journal of Engineering Materials and Technology, Vol. 98, 1976, pp. 146-151.

4.9 C.F. Shih, "Relationship Between Crack Tip Opening Displacement for Stationary and Extending Cracks", Journal of Mechanics and Physics of Solids , Vol. 29, 1981, pp. 305-326.

4.10 J.A. Begley and J.D. Landes, "The J-Integral as a Fracture Criterion", Fracture Toughness, ASTM STP 514, American Society for Testing and Materials, 1972, pp. 1-23.

4.11 J.D. Landes and J.A. Begley, "The Effect of Specimen Geometry on $J_{IC}$", Fracture Toughness, ASTM 514, American Society for Testing and Materials, 1972, pp. 24-39.

4.12 F.A. McClintock, "Plasticity Aspects of Fracture", Fracture - An Advanced Treatise, Vol. 3, H. Liebowitz, editor, Academic Press, New York, 1971, pp. 47-255.

4.13 R.M. McMeeking and D.M. Parks, "On Criterion for J-Dominance of Crack Tip Fields in Large-Scale Yielding", Elastic-Plastic Fracture, ASTM STP 668, American Society for Testing and Materials, 1979, pp. 175-194.

4.14 C.F. Shih and M.D. German, "Requirements for a One Parameter Characterization of Crack Tip Stress Fields in Large-Scale Yielding", Elastic-Plastic Fracture, ASTM STP 668, American Society for Testing and Materials, 1979, pp. 175-194.

4.15 J.W. Hutchinson and P.C. Paris, "Stability Analysis of J-Controlled Crack Growth", Elastic-Plastic Fracture, ASTM STP 668, American Society for Testing and Materials, 1979, pp. 37-64.

4.16 C.F. Shih, H.G. Delorenzi, and W.R. Andrews, "Studies on Crack Initiation and Stable Crack Growth", Elastic-Plastic Fracture, ASTM STP 668, American Society for Testing and Materials, 1979, pp. 65-120.

## 4.7 Exercise Problems

4.1 What limitations of deformation theory of plasticity influence the validity of J as a fracture criterion? What arguments are used to justify using it as a fracture criterion inspite of these limitations?

4.2 What do you understand by the following terms?
a. Strain energy release rate
b. Elastic-plastic and fully-plastic cracked bodies
c. HRR singularity

4.3 Show the following relationship between cartesian and polar coordinates in two dimensions:

$$\frac{\partial}{\partial x} = \cos\theta \frac{\partial}{\partial r} - \frac{\sin\theta}{r} \frac{\partial}{\partial \theta}$$

4.4 Prove the J-Integral for linear elastic cracked bodies is equal to the Griffith's crack extension force, $\mathcal{G}$.

4.5 Knowing that near the crack tip:

$$\sigma_{ij}\, \varepsilon_{ij} = \left(\frac{J}{I_m r}\right) g_{ij}(\theta)$$

derive equations for the various stress and strain components for a material obeying the Ramberg-Osgood relationship.

4.6 What is the significance of J-integral being path-independent?

4.7 Derive the Hutchinson-Paris condition for J-controlled crack growth on your own showing all the steps.

4.8 Discuss the significance of the relationship between J-integral and the crack tip opening displacement (CTOD).

4.9 What is $J_{Ic}$ and how does it differ from $K_{Ic}$? If the $J_{Ic}$ of a steel is 35 kJ/m², estimate the $K_{IC}$ of the material. Comment on whether this estimate if $K_{Ic}$ is conservative, optimistic or realistic.

4.10 Why does the minimum uncracked ligament size required to insure J-dominance depend on specimen geometry? What is the implication of this on the suitability of the center crack tension (CCT) specimen as the preferred geometry for fracture toughness testing.

4.11 Starting from equation (4.18b), derive the equation for how displacements in the y-direction vary near the crack tip.

4.12 Starting from equation (4.21) and using the definition of CTOD given by Tracy, show that the CTOD is given by the equation (4.22).

# CHAPTER 5

# METHODS OF ESTIMATING J-INTEGRAL

In the previous chapter, J-integral was defined and shown to characterize the process of ductile tearing under elastic-plastic and fully-plastic conditions, provided certain conditions were met. The usefulness of a fracture parameter in engineering applications is dependent upon how readily it can be calculated for cracked components of different geometries and loading. Further, its magnitude should be easily measurable in test specimens used for characterizing crack growth resistance. In this chapter, we will be discussing methods commonly used for calculating and measuring the values of J-integral for a variety of geometries including both specimens and components.

## 5.1 Analytical Solutions

The definition of J as a path-independent integral given by equation (4.8) can be used to determine its value, if the stress and displacements are known everywhere along the chosen path. The following definition of J is most useful for this purpose:

$$J = \int_\Gamma W dy - T_i \frac{\partial u_i}{\partial x} ds \qquad (5.1)$$

where all symbols are as defined and explained in Chapter 4. Equation (5.1) is most often used for determining J in conjunction with nonlinear finite element analyses of cracked bodies in which the stress and displacement distributions have been obtained numerically. There are some configurations for which J can be easily determined by inspection [5.1.2]. Although these configurations are not of much practical value, they serve the purpose of illustrating the use of equation (5.1).

**Example Problem 5.1**
Determine J for a semi-infinite notch in an infinite strip of height 2h which is subjected to uniform displacement, $u_0$, by clamping forces along its upper and lower surfaces, as shown Figure 5.1.

*Solution:*
We choose a counter-clockwise path shown by dashed lines in Figure 5.1 which goes along the boundary of the notched body. The various segments of the path are numbered for convenience. Along segments 2 and 4, dy = 0 and also:

$$\frac{\partial u_i}{\partial x} = 0$$

because the displacement u is uniform. Hence, these segments do not contribute to the value of J. Along segments 1 and 5, which are located far away from the crack tip at $x = -\infty$, all components of stress as well as $\partial u_i/\partial x = 0$. Therefore, these segments also do not contribute to the value of J. Thus, the sole contribution to J comes from segment 3 for which $\partial u_i/\partial x = 0$ and J is given by the following equation:

$$J = \int_{-h}^{h} W(x=\infty)\,dy \qquad (1)$$

or

$$J = 2h\,W_\infty$$

$$W_\infty = \int_0^{\varepsilon_2} \sigma_2\,d\varepsilon_2$$

because all other components of stress are 0. Hence, $W_\infty = \sigma_0\,\varepsilon_2$ for a perfectly plastic material and $\varepsilon_2 = 2u/2h$ where $u$ = applied remote displacement. Hence:

$$J = \frac{2h\sigma_0 u}{h} = 2\sigma_0 u \qquad (2)$$

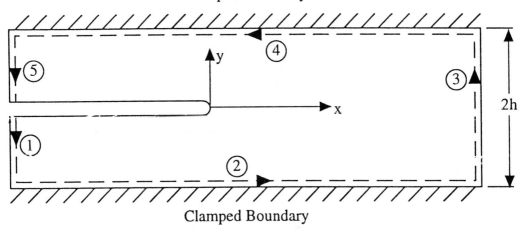

*Figure 5.1* Semi-infinite notch in an infinite strip of height 2h subjected to uniform displacements along its lower and upper surfaces.

### Example 5.2

Determine J for a semi-infinite notch in an infinite strip of height 2h which is subjected to pure bending moment, M per unit thickness, of its beam like arms as shown in Figure 5.2.

*Solution:*

We again choose a path along the boundaries of the cracked body as shown in Figure 5.2. Along segments 2 and 4 of the contour, $dy = 0$ and $T_i = 0$. Hence, these segments do not contribute to the J-integral. Along segment 3 there are no stresses, hence, both W and $T_i$ vanish making the contribution to J along that segment 0. Therefore, the only contribution to J-integral is from segments 5 and 1. Along these segments, $dy = -ds$ and $T_y = 0$ and $T_x = -\sigma_{xx}$ because we assume pure bending. Therefore, J is given by the following equations:

$$J = \int_h^{-h} (W - \sigma_{xx} \frac{\partial u_x}{\partial x}) dy$$

$$= \int_h^{-h} (W - \sigma_{xx} \varepsilon_{xx}) dy$$

$$= 2 \int_0^h (\sigma_{ij} \varepsilon_{ij} - W) dy$$

Note that since $\sigma_{xx}$ and $\varepsilon_{xx}$ are the only nonzero components of stress and strain in a state of pure bending $\sigma_{ij} \varepsilon_{ij} = \sigma_{xx} \varepsilon_{xx}$. Also, the term $(\sigma_{ij} \varepsilon_{ij} - W)$ is the complementary energy density, $\Omega$, given by:

$$\Omega = \int_0^{\sigma_{ij}} \varepsilon_{ij} d\sigma_{ij}$$

Thus, $J = 2\Omega h$

In this example, J is related to the complementary energy density while as in the previous example J was directly related to the strain energy density.

In Chapter 4, we derived the relationship between CTOD, $\delta_t$, and J, equation 4.20, using the Dugdale formulation of the plastic zone size in small-scale yielding as follows:

$$J = \sigma_0 \delta_t$$

using the symbols used in Figure 4.8 to designate various quantities, we write the results from a derivation given by Muskhelishvilli [5.2] as follows:

$$\frac{c}{a+c} = 2\sin^2[\frac{\pi}{4} \frac{\sigma^\infty}{\sigma_0}] \tag{5.2}$$

The above equation can be rewritten for $\sigma^\infty/\sigma_0 << 1$ as:

$$\frac{c}{a} = \sec[\frac{\pi}{2} \frac{\sigma^\infty}{\sigma_0}] - 1 \tag{5.3}$$

110  *Nonlinear Fracture Mechanics for Engineers*

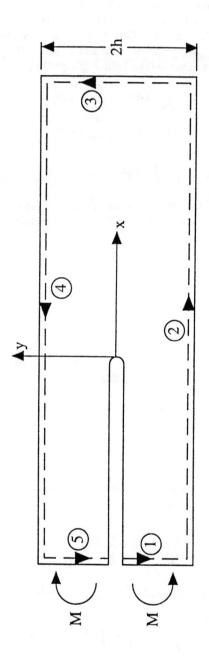

*Figure 5.2  Semi-infinite notch in an infinite strip of height 2h subjected to pure bending with a bending moment M per unit thickness.*

The stress and displacement distributions near the crack tip are as follows: $\sigma_{yy} = \sigma_0$ for a < |x| < (a+c), $\sigma_{yy} = 0$ for |x| < a, $v_y = 0$ for |x| ≥ (a+c). The solution leads to an expression for CTOD as follows:

$$\delta_t = v_y(a,0^+) - v_y(a,0^-) = \frac{8}{\pi} \frac{\sigma_0}{E} a \ln \left[\sec\left(\frac{\pi}{2} \frac{\sigma^\infty}{\sigma_0}\right)\right] \tag{5.4}$$

Substituting equation (5.4) into the equation for J, equation (4.20), we get:

$$J = \frac{8}{\pi} \frac{\sigma_0^2}{E} a \ln \left[\sec\left(\frac{\pi}{2} \frac{\sigma^\infty}{\sigma_0}\right)\right] \tag{5.5}$$

The above equation holds only for a finite crack in an infinite plate when $\sigma^\infty/\sigma_0 << 1$.

In the discussion above, we have seen how J-integral can be obtained using its definition as a path-independent integral. The property of path-independence permits us to select the contour in regions where the stresses and displacements are accurately known, avoiding the complexities of stresses in the immediate crack tip region. In the subsequent sections, we consider other methods of determining J for more realistic configurations.

## 5.2 Determination of J in Test Specimens

In test specimens, we have the benefit of being able to measure the applied load and also the load-point displacement. We thus make use of the relationship between J and the potential energy, equation 4.13, to determine the values of J. This class of methods of determining J is divided into purely experimental methods and those which combine experimentally measured quantities with an analytical formula. The latter are referred as the semi-empirical methods and are the ones most widely used for determining J in the test specimens. Therefore, the bulk of our discussion will focus on these methods.

### 5.2.1 Experimental Method of Determining J

This method of determining J is analogous to the compliance method of determining K and requires several specimens. The method was originally developed by J.A. Begley and J.D. Landes [5.3] and was used in their original work in which they proposed J as a fracture criterion.

In the experimental method, load-displacement curves are generated for specimens of unit thickness with different initial crack lengths. Each specimen is loaded to displacement levels that can be sustained whithout crack extension. The area under the load-displacement (P-V) curve Figure 5.3a to specified values of displacements, $V_1$, $V_2$, $V_3$, ---- $V_i$, are obtained graphically to determine corresponding values of the potential energy, $U_1$, $U_2$, $U_3$,----$U_i$ as schematically shown in Figure 5.3a. For prescribed displacement conditions, the area under the load-displacement curve is U(V). This relationship will be derived in Section 5.2.2. Thus, the load-displacement curves can be divided into segments associated with each level of displacement and the areas $A_1$, $A_2$, $A_3$,----$A_i$ of each segment is identified with the corresponding values of $U_i$. For each fixed value of displacement, $U_i$ can be plotted as a function of the crack length, $a_i$, as shown schematically in Figure 5.3b. The slope of U-a curve, dU/da, is then measured at various points to yield the values of J. If specimen thickness is other than unity, dU/da must be divided by the thickness, B, in order to obtain the value of J. The resulting J values can be plotted as a function of the applied displacement for various crack sizes, as shown in Fig 5.3c.

As we can see from the above discussion, this procedure is quite tedious and requires several specimens and the resultant J-V curves are also material and specimen geometry dependent. In other words, if the material changes or the specimen geometry changes, a new set of curves must be devel-

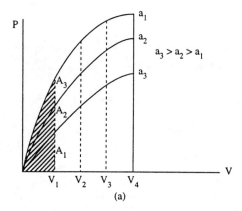

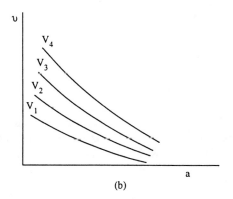

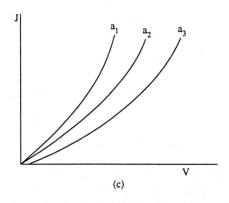

**Figure 5.3** Schematic showing the empirical method of determining J from load-displacement diagrams.

oped. Therefore, it is essential to develop simpler and more generally applicable alternatives that are described below.

### 5.2.2 Semi-Empirical Methods of Determining J

In this section, we will derive expressions for estimating J using the load-displacement record from a single specimen. The crack size is also specified and will be considered fixed. Recall the definition of potential energy, U, equation (4.10):

$$U(a) = \int_A W(x,y)\,dA - \int_{S_T} T_i u_i\,ds$$

Under the conditions of fixed displacement, the second term on the right-hand side of the above equation will be 0. Therefore:

$$U(a) = \int_A W(x,y)\,dA \tag{5.6}$$

From equation (5.6), it can be deduced that U(a) is the integral of the strain energy density over the planar area of the body. In other words, it is the strain energy of the body given by the area under the load-displacement record shown in Figure 5.4. Therefore, the potential energy difference, for a pair of specimens of unit thickness with crack sizes a and $a+\Delta a$ is given by the area between the two load-displacement records as shown in Figure 5.4. $-\Delta U$ can be estimated by the following equation:

$$-\Delta U = -\int_0^V \left(\frac{\partial P}{\partial a}\right)_V \Delta a\,dV \tag{5.7}$$

Hence:

$$J = \lim_{\Delta a \to 0} = -\frac{\Delta U}{\Delta a} = -\int_0^V \left(\frac{\partial P}{\partial a}\right)_V dV \tag{5.8a}$$

For a body of thickness, B, J is given by:

$$J = -\frac{1}{B}\int_0^V \left(\frac{\partial P}{\partial a}\right)_V dV \tag{5.8b}$$

Let us now turn to fixed load conditions. U(a) for this condition is given by:

$$U(a) = \int_A W(x,y)\,dA - PV \tag{5.9}$$

where P is the fixed load. -U(a), for this example, is given by the area shown in Figure 5.5 and $-\Delta U$ for a pair of specimens with crack sizes a and $a + \Delta a$ is as shown by the shaded area in Figure 5.5. In this case:

114  *Nonlinear Fracture Mechanics for Engineers*

$$-\Delta U = \int_0^P \left(\frac{\partial V}{\partial a}\right)_P dP$$

Therefore:

$$J = \lim_{\Delta a \to o} -\frac{\Delta U}{\Delta a} = -\int_0^P \left(\frac{\partial V}{\partial a}\right)_P dP \qquad (5.10a)$$

for a specimen with a thickness, B:

$$J = \frac{1}{B} \int_0^P \left(\frac{\partial V}{\partial a}\right)_P dP \qquad (5.10b)$$

Equations (5.8b) and (5.10b) can be used to estimate J from the load-displacement record of a single specimen as will be illustrated in the following examples.

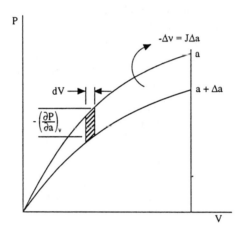

*Figure 5.4* Load-displacement diagrams of two cracked bodies that are identical in all ways except one has an incrementally longer crack length compared to the other. Both are loaded to the same displacement V.

**J for a Deep Edge Crack Specimen Subject to Pure Bending** The geometry and loading configuration of this specimen is shown in Figure 5.6. The simple derivation that follows was first shown by Rice in 1971 [5.4] and became the starting point for several similar analyses for various fracture mechanics specimens. Rice recognized that for configurations with a single characteristic length dimension dominating the deformation behavior such as the uncracked ligament, W - a = b, for a deeply cracked bend specimen, the relative additional angle of rotation due to the presence of the crack, $\theta_c$, can be estimated from a dimensional analysis. If M is the applied bending moment which also determines $\theta_c$, then:

$$\theta_c = f\left(\frac{M}{Bb^2}\right) \qquad (5.11)$$

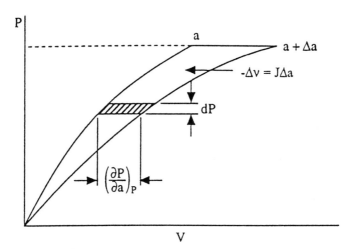

*Figure 5.5* Load-displacement diagram of two cracked bodies that are identical in all ways except one has an incrementally longer crack length compared to the other. Both are loaded to fixed load P.

The only other parameters which can enter the relationship are material parameters which are constant such as the exponent, m, which is dimensionless and the flow strength with dimensions of stress having the same units as $M/Bb^2$. We can make the argument of f to be nondimensional by introducing the flow stress term in the denominator. However, since it is a constant, it is unnecessary. If we experimentally obtain a M - θ plot as shown in Figure 5.7, we can write the following relationship:

$$\theta = \theta_{\text{no crack}} + \theta_c \tag{5.12}$$

$$\frac{\partial \theta}{\partial a} = \frac{\partial \theta_c}{\partial a}$$

because $\theta_{\text{no crack}}$ does not vary with a:

$$J = \frac{1}{B} \int_0^M \left(\frac{\partial \theta_c}{\partial a}\right)_M dM$$

$$\left(\frac{\partial \theta_c}{\partial a}\right)_M = \left(\frac{\partial \theta}{\partial a}\right)_M = -\left(\frac{\partial \theta}{\partial b}\right)_M$$

or

$$J = -\frac{1}{B} \int_0^M \left(-\frac{2M}{Bb^3}\right) f' \left(\frac{M}{Bb^2}\right) dM$$

where

$$f' = df / d\left(\frac{M}{Bb^2}\right)$$

116  Nonlinear Fracture Mechanics for Engineers

$$d\theta_c = f'\left(\frac{M}{Bb^2}\right) \cdot \frac{dM}{Bb^2}$$

Hence:

$$J = \frac{1}{B} \int_0^{\theta_c} \frac{2M}{b} d\theta_c = \frac{2}{Bb} \int_0^{\theta_c} M d\theta_c \tag{5.14}$$

where $\int_0^{\theta_c} M d\theta_c$ is the area under the M-$\theta_c$ curve as shown in Figure 5.7. In implementing equation (5.14), the full angle $\theta$ is usually measured. However, for a deeply cracked specimen, $\theta_{no\ crack}$ is small and we can usually assume that $\theta_c \approx \theta$. If the specimen is a 3-point bend configuration of length L, with P being the point load and the displacement of the load-point being designated V, we can write:

$$M = PL/2 \quad \text{and} \quad V \approx \left(\frac{\theta_c}{2} L\right)$$

Hence:

$$d\theta_c = (2/L) dV$$

and

$$\int_0^\theta M d\theta = \int_0^V P dV$$

Therefore:

$$J = \frac{2}{Bb} \int_0^V P dV \tag{5.15}$$

To obtain accurate estimates of J from equation (5.15) for 3-point bend specimens, a/W for the specimen must be $\geq 0.5$. Also note in this case, the displacement due to no crack is small in comparison to the displacement due to the crack; therefore, the distinction between displacement V and the additional displacement due to crack, $V_c$, is unnecessary.

**Merkle-Corten Analysis of a Compact Specimen**  Figure 5.8 shows a standard compact specimen used frequently for fracture toughness and crack growth testing. For very deeply cracked compact specimens (a/W $\geq 0.6$), the bending component of the load dominates and equation (5.15) provides a reasonable estimate of the J-intergral. Merkle and Corten [5.5] provided a more general expression which may be used to determine J over a wider range of crack sizes. Because of the importance of this specimen geometry in fracture mechanics testing, it is reasonable to present this analysis in detail.

Consider a fully-yielded compact specimen of unit thickness from a material with yield strength, $\sigma_0$. The load at which the specimen is fully yielded is designated $P_0$. Figure 5.9a shows the stress distribution in one half of the specimen at load $P_0$ for a nonhardening material. For a combination of bend and tension loading, the internal stress will be more biased in the negative direction to maintain a force equilibrium in the y-direction. If we consider $\alpha$ = ratio of the internal stress block equilibriating the applied load to the internal stress field in the remaining ligament, b, we have:

$$P_0 = \sigma_0 \alpha b \tag{5.16}$$

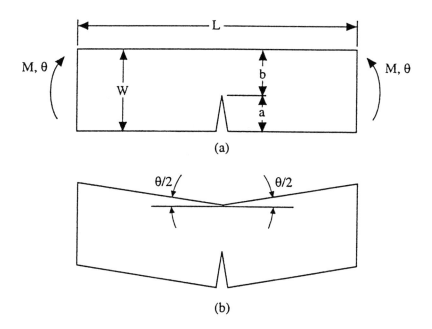

**Figure 5.6** Single edge notch specimen subject to pure bending: (a) undeformed specimen, (b) deformed specimen.

Equating the plastic moment, $M_0$, to the movement of the applied load $P_0$ about the centroid of the net section, we get:

$$M_o = P_o(a+b/2) = \sigma_o \frac{b^2}{4}(1-\alpha^2) \tag{5.17}$$

Equations (5.16) and (5.17) can be combined to eliminate $P_o$ and yield:

$$\alpha^2 + 2(\frac{2a}{b} + 1)\alpha - 1 = 0 \tag{5.18}$$

The quadratic equation (5.18) can be solved for its positive root to give a value of $\alpha$ as follows:

$$\alpha = [(\frac{2a}{b})^2 + 2(\frac{2a}{b}) + 2]^{\frac{1}{2}} - (\frac{2a}{b} + 1) \tag{5.19}$$

Equations (5.16) and (5.17) provide only a lower bound plastic limit analysis of the net section which is based on equilibrium ignoring the state of stress triaxiality at the crack tip. However, since J from this analysis is based on areas under the load-displacement diagrams which include effects due to crack tip triaxiality, these effects are automatically included in the estimates of J.

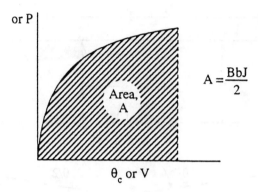

**Figure 5.7** *The relationship between the area under the load-displacement diagram (of the moment-angle diagram) and J.*

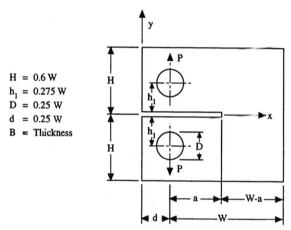

**Figure 5.8** *The geometry of a standard compact specimen.*

If we assume a rigid rotation about the neutral axis as shown in Figure 5.9b, the relationship between the rotation angle, $\theta_p$, and the load-line displacement, $V_p$, is given by:

$$\theta_p = \frac{V_p}{a + (1 + \alpha)\frac{b}{2}}$$

or

$$V_p = \frac{W}{2}\left(2 - (1 - \alpha)\frac{b}{W}\right)\theta_p \tag{5.20}$$

The plastic angle, $\theta_p$, is only a function of the ratio of the applied load to the limit load. Hence:

$$\theta_p = \hat{g}\ (P/P_0) \tag{5.21}$$

or the inverse relationship is:

$$P = P_0 \, g \, (\theta_p) \tag{5.22}$$

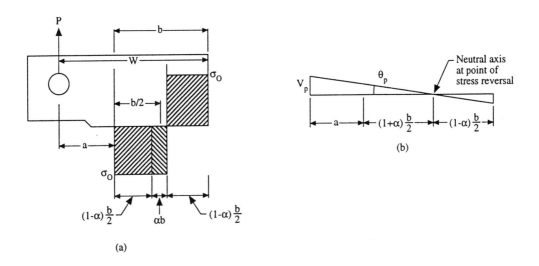

*Figure 5.9* (a) Fully yielded CT specimen; (b) displacement diagram of the deformed CT specimen.

Next, J is obtained as follows:

$$J = \int_o^P (\frac{\partial V}{\partial a})_P \, dP = -\int_o^P (\frac{\partial V}{\partial b})_P \, dP$$

$$V = V_e + V_p$$

where $V_e$ = elastic part of the deflection
$V_p$ = plastic part of the deflection

Also, no distinction is made between deflection due to crack and total deflection, V. Thus:

$$J = - [\int_o^P (\frac{\partial V_e}{\partial b})_P \, dP + \int_o^P (\frac{\partial V_p}{\partial b})_P \, dP \,] \tag{5.23}$$

120    *Nonlinear Fracture Mechanics for Engineers*

The first term on the right-hand side of equation (5.23) is equal to $\mathcal{G}$, the Griffith's crack extension force.
Thus, from equations (5.23) and (5.8b) applied to plastic part of deflection:

$$J = \mathcal{G} \int_o^P (\frac{\partial V_p}{\partial b})_P \, dP = \mathcal{G} + \int_o^{V_p}(\frac{\partial P}{\partial a})_{V_p} \, dV_p \qquad (5.24)$$

By differentiating equation (5.23b), we get:

$$(\frac{\partial P}{\partial b})_{V_p} = g \frac{\partial P_0}{\partial b} + P_0 \frac{\partial g}{\partial b} \qquad (5.25)$$

Combining equations (5.20) and (5.21), we get:

$$V_p = \frac{W}{2} [\, 2 - (1-\alpha)\frac{b}{W}\,] \, \hat{g} \, (\frac{P}{P_o}) \equiv F(\frac{P}{P_o}, \frac{b}{W}) \qquad (5.26)$$

$$(\frac{\partial V_p}{\partial b})_P = (\frac{\partial V_p}{W \partial(b/W)})_P = \frac{1}{W}[\,\frac{1}{W}\frac{\partial F}{\partial(b/W)} - \frac{1}{W}\frac{\partial F}{\partial(P/P_0)} \cdot \frac{P}{P_o^2}\frac{dP_o}{d(b/W)}\,]$$

or

$$-(\frac{\partial V_p}{\partial b})_P = \frac{1}{W}\frac{\partial F}{\partial(P/P_o)}\frac{P}{P_o^2}\frac{dP_o}{d(b/W)} - \frac{1}{W}\frac{\partial F}{\partial(b/W)}$$

or

$$-(\frac{\partial V_p}{\partial b})_P = \frac{1}{P_o}\frac{dP_o}{db}\frac{P}{P_o}\frac{\partial F}{\partial(P/P_o)} - \frac{1}{W}\frac{\partial F}{\partial(b/W)} \qquad (5.27)$$

$$(\frac{\partial V_p}{\partial P})_a = \frac{\partial F}{\partial(P/P_o)}\frac{1}{P_o} \qquad (5.28)$$

Substituting equation (5.28) into (5.27), we get:

$$-(\frac{\partial V_p}{\partial b})_P = \frac{1}{P_o}\frac{dP_o}{db} P (\frac{\partial V_p}{\partial P})_a - \frac{1}{W}\frac{\partial F}{\partial(b/W)}$$

or

$$J = -\int_o^P(\frac{\partial V_p}{\partial b})_P \, dP = \frac{1}{P_o}\frac{dP_o}{db}\int_o^{V_p} P dV_P - \frac{1}{W}\int_o^P(\frac{\partial F}{\partial b/W}) dP \qquad (5.29)$$

Differentiating equation (5.16), we get:

$$\frac{dP_o}{db} = \sigma_o \left( \alpha + b \frac{d\alpha}{db} \right)$$

From equation (5.19), we can obtain:

$$\frac{d\alpha}{db} = \frac{1}{b} \frac{(1 + 2\alpha - \alpha^2)\alpha}{1 + \alpha^2}$$

hence:

$$\frac{1}{P_o} \frac{dP_o}{db} = \frac{2}{b} \frac{1 + \alpha}{1 + \alpha^2} \tag{5.30}$$

We can also show that:

$$-\frac{1}{W} \frac{\partial F}{\partial (b/W)} = \frac{1 - 2\alpha - \alpha^2}{2(1 + \alpha^2)} \hat{g} = \frac{2}{b} \frac{(1 - 2\alpha - \alpha^2)}{(1 + \alpha^2)^2} \alpha V_p$$

Hence, the plastic part J, $J_p$, is given by:

$$J_p = \frac{2}{b} \frac{1 + \alpha}{1 + \alpha^2} \int_o^{V_p} P \, dV_p + \frac{2\alpha}{b} \frac{1 - 2\alpha - \alpha^2}{(1 + \alpha^2)^2} \int_o^P V_p \, dP \tag{5.31}$$

and the value of J by:

$$J = J_\beta + J_p \tag{5.32}$$

The first term on the right-hand side of equation (5.31) contains a term which is the area under the load displacement curve or the strain energy per unit thickness, and the other term contains the complementary energy per unit thickness. Equation (5.32) can be expressed as:

$$J = J_\beta + \frac{\eta}{b} \int_o^{V_p} dV_p + \frac{\eta_c}{b} \int_o^P V_p \, dP \tag{5.33}$$

where $\eta = 2(1+\alpha)/(1+\alpha^2)$ and $\eta_c = 2\alpha(1-2\alpha-\alpha^2)/(1+\alpha^2)^2$. The values of $\alpha$, $\eta$, and $\eta_c$ are listed in Table 5.1 for various values of a/W. From Table 5.1 it is evident that as a/W → 1, equation (5.33) approaches Rice's formula for pure bending, equation (5.15), as expected. Also, $\eta_c$ is small compared to $\eta$ and the complementary strain energy is also usually much smaller than the strain energy term. Hence, the second term in equation (5.31) may not contribute very significantly to the value of J. A much simpler expression has been developed by Ernst et.al [5.6] to estimate $J_p$ in compact specimens of thickness, B. This expression is given by:

$$J_p = \frac{A_p}{Bb}(2 + 0.522\frac{b}{W}) \tag{5.34}$$

where $A_p$ is the plastic area under the load-displacement curve. The total J can be calculated as the sum of the elastic and plastic parts.

*Table 5.1 Values of $\alpha$, $\eta$, and $\eta_c$ for Several Values of a/W in Equation (5.33)*

| a/W | $\alpha$ | $\eta$ | $\eta_c$ |
|-----|----------|--------|----------|
| 0   | 0.414    | 2.414  | 0        |
| 0.1 | 0.357    | 2.407  | 0.089    |
| 0.2 | 0.303    | 2.387  | 0.154    |
| 0.4 | 0.205    | 2.313  | 0.207    |
| 0.5 | 0.162    | 2.265  | 0.200    |
| 0.6 | 0.123    | 2.213  | 0.176    |
| 0.7 | 0.088    | 2.159  | 0.141    |
| 0.8 | 0.055    | 2.104  | 0.097    |
| 1.0 | 0        | 2.00   | 0        |

**J for Center Crack Tension Geometry** Figure 5.10 shows the center crack tension (CCT) specimen and the load-plastic displacement curves for estimating J. The expression for $J_p$ is given by [5.34]:

$$J_p = \frac{A_p}{Bb} \tag{5.35}$$

where $A_p$ is defined in Figure 5.10b. The derivation of this relationship is left as a homework exercise (see Exercise Problem 5.4).

## 5.3 J for Growing Cracks

The equations for estimating J discussed so far were derived for constant crack length in which the non-linearity between the load-deflection behavior is due to plastic deformation only. Now let us consider plots of moment, M, vs. the plastic part of the rotation angle, $\theta_p$, for specimens with different constant crack lengths, $a_o$, $a_1$, $a_2$ --- etc. as shown in Figure 5.11 for a deeply cracked specimen in bending. Next, we superimpose on this family of curves, the M vs. $\theta_p$ diagram of a specimen in which the initial crack length was $a_o$, but the crack grew to $a_2$ during loading. The onset of crack growth was at point 0 and crack length increased to $a_2$ at C with A as the intermediate point when the crack length was $a_1$. Rewriting equation (5.11) in the inverse form:

$$M = Bb^2 H(\theta_p) \tag{5.36}$$

where, as before, B = specimen thickness and b = uncracked ligament of the specimen:

$$J_p = \frac{2}{Bb}\int_o^{\theta_p} 2Bb^2 H(\theta_p) d\theta_p = \int_o^{\theta_p} 2bH(\theta_p) d\theta_p \tag{5.37}$$

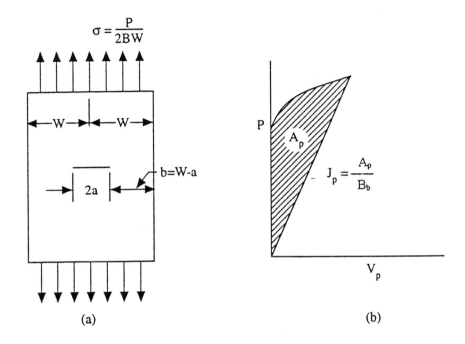

**Figure 5.10** (a) Geometry of a center crack tension (CCT) specimen and (b) relationship between $J_p$ and the area under the load-displacement curve.

For a growing crack:

$$dJ_p = 2bH(\theta_p)\, d\theta_p + 2db \int_o^{\theta_p} H(\theta_p)\, d\theta_p$$

$$= 2bH(\theta_p)\, d\theta_p - 2da \int_o^{\theta_p} H(\theta_p)\, d\theta_p$$

Hence:

$$J_p = \int_o^{\theta_p} 2bH(\theta_p)\, d\theta_p - \int_{a_o}^{a} \frac{J_p}{b}\, da \qquad (5.38)$$

We can use the above equation to estimate the value of $J_p$ for growing cracks provided the crack size information is available for the entire $M$ - $\theta_p$ diagram. Let us illustrate the procedure by estimating the correct value of $J_p$ for points B and C in Figure 5.11. The correct value of $_pJ$ at point A from equation (5.38) is:

$$J_{pA} = J_{po'} - \int_{a_o}^{a_1} \frac{J_{po'}}{b}\, da$$

$$J_{po'} = J_{po} + \frac{2}{Bb} \int_{\theta_{po}}^{\theta_{po'}} M\, d\theta_p$$

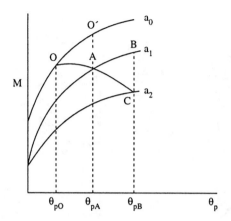

**Figure 5.11** *M-$\theta_p$ diagram for several crack sizes superimposed on one for a growing crack.*

If the crack length increment between points O, A, and C is infinitesimal, then the value of $J_{pA}$ can be written as:

$$J_{pA} = J_{po} + \frac{2A_{oo'}}{Bb} - \frac{J_{po'}}{b}(a_1 - a_o)$$

$$= [J_{po} + \frac{2A_{oo'}}{Bb}][1 - \frac{a_1 - a_o}{b}] \simeq [J_{po} + \frac{2A_{oA}}{Bb_o}][1 - \frac{a_1 - a_o}{b_o}]$$

The area $A_{oo'}$, which is the area under curve oo', can be approximated by the area under the curve OA if $(a_1 - a_o)$ is small. Similarly, the $J_p$ value for point C will be:

$$J_{pC} \simeq [J_{pA} + \frac{2A_{AC}}{Bb_1}][1 - \frac{a_2 - a_1}{b_1}]$$

at any point along the M - $\theta_p$ curve:

$$J_{pi+1} \simeq [J_{pi} + \frac{2A_{i,i+1}}{Bb_i}][1 - \frac{a_{i+1} - a_i}{b_i}] \qquad (5.39)$$

Ernst et.al. [5.6] have used the same procedure to derive the expression for the compact type (CT) specimen as follows:

$$J_{pi+1} = [J_{pi} + \frac{\eta A_{i,i+1}}{Bb_i}][1 - \gamma_i \frac{a_{i+1} - a_i}{b_i})] \qquad (5.40)$$

where $A_{i,i+1}$ is the area under the load-displacement diagram between the successive points considered and $\gamma$ at any point is given by:

$$\gamma = 1 + 0.76 \frac{b}{W} \qquad (5.41)$$

and η is as given earlier by equation (5.33). The derivation of the above equation is left as a homework exercise (See Problem 5.10).

## 5.4 Numerically Obtained J-Solutions

In this section we will discuss a method for estimating J for elastic, elastic-plastic, and fully plastic conditions for several crack configurations that are important in applications. The estimation of J in these situations is not dependent on the measurement of the load-line displacement. Instead, the displacement is computed from the deformation properties of the material, $\alpha$, $\sigma_0$, $\varepsilon_0$, and m. However, prior to presenting these results, it is important to develop a framework which specifies these equations in approximately the same form.

Under linear-elastic conditions, the J-integral in a body with crack length, a, $J_e$, crack mouth displacement, $\delta$, and the load-line displacement, V, can be related to applied load, P, by equations that take the following form:

$$\frac{J_e}{\sigma_0 \varepsilon_0 a} = [\frac{P}{P_0}]^2 F_{1e} (a/W) \tag{5.42}$$

where $P_0$ = limit-load as defined before and $F_{1e}$ is a function of the dimensionless crack size a/W where W = width of the cracked body. Similarly, the form of equations for the crack mouth displacement, $\delta_e$, and the load-line displacement, $V_e$, are as follows:

$$\frac{\delta_e}{\varepsilon_0 a} = [P/P_o] F_{2e} (a/W) \tag{5.43}$$

$$\frac{V_e}{\varepsilon_0 a} = [P/P_o] F_{3e} (a/W) \tag{5.44}$$

where $F_{2e}$ and $F_{3e}$ are functions of only a/W. Now consider an incompressible fully plastic material which obeys the Ramberg-Osgood equation. Ilyushin [5.7] noted that solution of the boundary value problems for such materials involving a single load or displacement parameter which increases monotonically and maintains proportional loading, has the following properties. The field quantities increase in direct proportion to the applied stress or displacement raised to power m, the Ramberg-Osgood exponent. Thus, for an applied remote stress, $\sigma$, the stress, strain, and displacements are given by [5.8]:

$$\frac{\sigma_{ij}}{\sigma_0} = (\sigma/\sigma_0)^m F_{1p} (r,m) \tag{5.45a}$$

$$\frac{\varepsilon_{ij}}{\varepsilon_0} = \alpha (\sigma/\sigma_0)^m F_{2p} (r,m) \tag{5.45b}$$

$$\frac{u_i}{\varepsilon_0 l} = \alpha (\sigma/\sigma_0)^m F_{3p} (r,m) \tag{5.45c}$$

where $l$ is some significant length parameter and $F_{1p}$, $F_{2p}$, and $F_{3p}$ are dimensionless functions of spatial position, $r$. This functional dependence directly implies that quantities such as J-integral, $J_p$, crack mouth displacement, and the load-line displacement can be given by:

$$\frac{J_p}{\alpha \varepsilon_0 \sigma_0 a} = [P/P_0]^{m+1} f_1(a/W,m) \qquad (5.46a)$$

$$\frac{\delta_p}{\alpha \varepsilon_0 a} = [P/P_0]^m f_2(a/W,m) \qquad (5.46b)$$

$$\frac{V_p}{\alpha \varepsilon_0 a} = [P/P_0]^m f_3(a/W,m) \qquad (5.46c)$$

The applied load in this case appears explicitly in the above equations. The functions $f_1$, $f_2$, and $f_3$ depend only on a/W and m and are independent of the applied load. This implies complete separability of the load and crack size dependence of the J, $\delta$, and V.

### 5.4.1 Elastic-Plastic Estimation Procedure

Fully plastic crack solutions and analyses apply only to conditions where the body has yielded everywhere and the plastic strains are much larger than the elastic strains. Most engineering situations may be elastic-plastic in nature. Therefore, it is necessary to develop an estimation scheme for J and the displacement quantities for elastic-plastic bodies. Interpolation schemes for this purpose have been developed by Shih and Hutchinson [5.9] which take the following form:

$$J = J_e(a_e) + J_p(a,m) \qquad (5.47a)$$

$$\delta = \delta_e(a_e) + \delta_p(a,m) \qquad (5.47b)$$

$$V = V_e(a_e) + V_p(a,m) \qquad (5.47c)$$

where $J_e(a_e)$, $\delta_e(a_e)$, and $V_e(a_e)$ are the elastic values of J, $\delta$, and V corresponding to the crack length adjusted for plasticity, $a_e$; this is the Irwin corrected crack length given by [5.10]:

$$a_e = a + \phi \, r_y \qquad (5.48)$$

where

$$r_y = \frac{1}{\beta \pi} \left(\frac{m-1}{m+1}\right) \left(\frac{K}{\sigma_0}\right)^2$$

$$\phi = \frac{1}{1 + (P/P_0)^2} \qquad (5.49)$$

$\beta = 2$ for plane stress and for plane strain, $\beta = 6$. The above equations are approximate and have been shown to provide sufficiently accurate estimates of J-integral and other quantities under elastic-plastic conditions for engineering applications. The accuracy of equations (5.47) are expected to be better for materials with a high m value. In these equations, the elastic parts of J, $\delta$, and V are readily available in handbooks [5.11]. By comparison to elastic solutions, the plastic solutions are available for fewer configurations. The plastic solutions for some important configurations are given in the following sections.

*5.4.2 Fully-Plastic J-Solutions for Test Specimens*

Expressions for estimating the plastic components of J, $\delta$, and V are listed for the standard compact type (CT) specimen shown in Figure 5.8, for center-crack-tension (CCT) specimen, Figure 5.10a, for single-edge-notch bend (SENB) specimen, Figure 5.6, for single edge notch tension (SENT) specimen, Figure 5.12, and for the double edge notch (DEN) specimen, Figure 5.13. These results are adapted from the work of Kumar, German, and Shih [5.11].

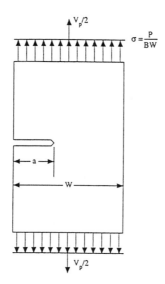

**Figure 5.12** *Geometry of a SENT specimen.*

CT Specimen

$$J_p = \alpha \sigma_0 \varepsilon_0 (W-a) h_1 (a/W, m) (P/P_0)^{m+1} \quad (5.50a)$$

$$\delta_p = \alpha \varepsilon_0 a h_2 (a/W, m) (P/P_0)^m \quad (5.50b)$$

$$V_p = \alpha \varepsilon_0 a h_3 (a/W, m) (P/P_0)^m \quad (5.50c)$$

where $h_1$, $h_2$, and $h_3$ are functions of a/W and m which are obtained from finite element analyses. Note that $\alpha$ in these equations refers to the constant in the Ramberg-Osgood equation and is different from the ratio in equation (5.16). That ratio is represented by $\eta_1$ in subsequent equations. These values are listed in Table 5.2 for plane stress and plane strain conditions. P is the applied load per unit thickness:

$$P_0 = 1.455 \eta_1 (W-a) \sigma_0 \quad \text{for plane strain} \tag{5.51}$$

and

$$P_0 = 1.071 \eta_1 (W-a) \sigma_0 \quad \text{for plane stress} \tag{5.52}$$

$$\eta_1 = [(\frac{2a}{W-a})^2 + 2 (\frac{2a}{W-a}) + 2]^{\frac{1}{2}} - [(\frac{2a}{W-a}) + 1] \tag{5.53}$$

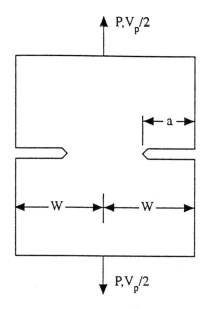

*Figure 5.13* Geometry of a DEN specimen.

*CCT Specimen*

$$J_p = \alpha \varepsilon_0 \sigma_0 a(1-a/W) h_1 (a/W,m) (P/P_0)^{m+1} \tag{5.54}$$

*Table 5.2* h Functions for Standard Compact Type Specimens (a) Plane Strain and (b) Plane Stress
*(from Kumar, German, and Shih, EPRI Report NP 1931, 1981)*

| a/W | m= | 1 | 2 | 3 | 5 | 7 | 10 | 13 | 16 | 20 |
|---|---|---|---|---|---|---|---|---|---|---|
| 1/4 | $h_1$ | 2.23 | 2.05 | 1.78 | 1.48 | 1.33 | 1.26 | 1.25 | 1.32 | 1.57 |
|  | $h_2$ | 17.9 | 12.5 | 11.7 | 10.8 | 10.5 | 10.7 | 11.5 | 12.6 | 14.6 |
|  | $h_3$ | 9.85 | 8.51 | 8.17 | 7.77 | 7.71 | 7.92 | 8.52 | 9.31 | 10.9 |
| 3/8 | $h_1$ | 2.15 | 1.72 | 1.39 | 0.970 | 0.693 | 0.443 | 0.276 | 0.176 | 0.098 |
|  | $h_2$ | 12.6 | 8.18 | 6.52 | 4.32 | 2.97 | 1.79 | 1.10 | 0.686 | 0.370 |
|  | $h_3$ | 7.94 | 5.76 | 4.64 | 3.10 | 2.14 | 1.29 | 0.793 | 0.494 | 0.266 |
| 1/2 | $h_1$ | 1.94 | 1.51 | 1.24 | 0.919 | 0.685 | 0.461 | 0.314 | 0.216 | 0.132 |
|  | $h_2$ | 9.33 | 5.85 | 4.30 | 2.75 | 1.91 | 1.20 | 0.788 | 0.530 | 0.317 |
|  | $h_3$ | 6.41 | 4.27 | 3.16 | 2.02 | 1.41 | 0.888 | 0.585 | 0.393 | 0.236 |
| 5/8 | $h_1$ | 1.76 | 1.45 | 1.24 | 0.974 | 0.752 | 0.602 | 0.459 | 0.347 | 0.248 |
|  | $h_2$ | 7.61 | 4.57 | 3.42 | 2.36 | 1.81 | 1.32 | 0.983 | 0.749 | 0.485 |
|  | $h_3$ | 5.52 | 3.43 | 2.58 | 1.79 | 1.37 | 1.00 | 0.746 | 0.568 | 0.368 |
| 3/4 | $h_1$ | 1.71 | 1.42 | 1.26 | 1.033 | 0.864 | 0.717 | 0.575 | 0.448 | 0.345 |
|  | $h_2$ | 6.37 | 3.95 | 3.18 | 2.34 | 1.88 | 1.44 | 1.12 | 0.887 | 0.665 |
|  | $h_3$ | 4.86 | 3.05 | 2.46 | 1.81 | 1.45 | 1.11 | 0.869 | 0.686 | 0.514 |
| 1 | $h_1$ | 1.57 | 1.45 | 1.35 | 1.18 | 1.08 | 0.950 | 0.850 | 0.730 | 0.630 |
|  | $h_2$ | 5.39 | 3.74 | 3.09 | 2.43 | 2.12 | 1.80 | 1.57 | 1.33 | 1.14 |
|  | $h_3$ | 4.31 | 2.99 | 2.47 | 1.95 | 1.79 | 1.44 | 1.26 | 1.07 | 0.909 |

a

| a/W | m= | 1 | 2 | 3 | 5 | 7 | 10 | 13 | 16 | 20 |
|---|---|---|---|---|---|---|---|---|---|---|
| 1/4 | $h_1$ | 1.61 | 1.46 | 1.28 | 1.06 | 0.903 | 0.729 | 0.601 | 0.511 | 0.395 |
|  | $h_2$ | 17.6 | 12.0 | 10.7 | 8.74 | 7.32 | 5.74 | 4.63 | 3.75 | 2.92 |
|  | $h_3$ | 9.67 | 8.00 | 7.21 | 5.94 | 5.00 | 3.95 | 3.19 | 2.59 | 2.023 |
| 3/8 | $h_1$ | 1.55 | 1.25 | 1.05 | 0.801 | 0.647 | 0.484 | 0.377 | 0.284 | 0.220 |
|  | $h_2$ | 12.4 | 8.20 | 6.54 | 4.56 | 3.45 | 2.44 | 1.83 | 1.36 | 1.02 |
|  | $h_3$ | 7.80 | 5.73 | 4.62 | 3.25 | 2.48 | 1.77 | 1.33 | 0.990 | 0.746 |
| 1/2 | $h_1$ | 1.40 | 1.08 | 0.901 | 0.686 | 0.558 | 0.436 | 0.356 | 0.298 | 0.238 |
|  | $h_2$ | 9.16 | 5.67 | 4.21 | 2.80 | 2.12 | 1.57 | 1.25 | 1.03 | 0.814 |
|  | $h_3$ | 6.29 | 4.15 | 3.11 | 2.09 | 1.59 | 1.18 | 0.938 | 0.774 | 0.614 |
| 5/8 | $h_1$ | 1.27 | 1.03 | 0.875 | 0.695 | 0.593 | 0.494 | 0.423 | 0.370 | 0.310 |
|  | $h_2$ | 7.47 | 4.48 | 3.35 | 2.37 | 1.92 | 1.54 | 1.29 | 1.12 | 0.928 |
|  | $h_3$ | 5.42 | 3.38 | 2.54 | 1.80 | 1.47 | 1.18 | 0.988 | 0.853 | 0.710 |
| 3/4 | $h_1$ | 1.23 | 0.977 | 0.833 | 0.683 | 0.598 | 0.506 | 0.431 | 0.373 | 0.314 |
|  | $h_2$ | 6.25 | 3.78 | 2.89 | 2.14 | 1.78 | 1.44 | 1.20 | 1.03 | 0.857 |
|  | $h_3$ | 4.77 | 2.92 | 2.24 | 1.66 | 1.38 | 1.12 | 0.936 | 0.800 | 0.666 |
| 1 | $h_1$ | 1.13 | 1.01 | 0.775 | 0.680 | 0.650 | 0.620 | 0.490 | 0.470 | 0.420 |
|  | $h_2$ | 5.29 | 3.54 | 2.41 | 1.91 | 1.73 | 1.59 | 1.23 | 1.17 | 1.03 |
|  | $h_3$ | 4.23 | 2.83 | 1.93 | 1.52 | 1.39 | 1.27 | 0.985 | 0.933 | 0.824 |

b

The equations for plastic parts of δ and V are identical in form to equations 5.50 b and 5.50c for the CT specimens and are, therefore, not repeated:

$$P_0 = \frac{4(W-a)\sigma_0}{\sqrt{3}} \quad \text{for plane strain} \tag{5.55}$$

and

$$P_0 = 2(W-a)\sigma_0 \quad \text{for plane stress} \tag{5.56}$$

The functions $h_1$, $h_2$, and $h_3$ for various values of a/W and m are given in Table 5.3 for plane strain and plane stress.

*SENB Specimen*

The forms of the equations for the plastic parts of J, δ, and V are identical in form to the ones for CT specimens given in equations (5.50) and are, therefore, not repeated here. The function $h_1$, $h_2$, and $h_3$ are given in Table 5.4. The following expressions for the limit-load, $P_0$, are, however, different from those of the CT specimen.

$$P_0 = \frac{1.455\sigma_0 (W-a)^2}{L} \quad \text{for plane strain} \tag{5.57}$$

$$P_0 = \frac{1.071\sigma_0 (W-a)^2}{L} \quad \text{for plane stress} \tag{5.58}$$

where L = length of the specimen (Figure 5.6).

*SENT Specimen*

The form of the equation for J for this case is identical to that for the CCT specimen, equation (5.54) and those for the plastic part of δ and V are identical in form to the rest of the specimens described previously. The $h_1$, $h_2$, and $h_3$ functions are listed in Table 5.5 for plane stress and plane strain conditions. The limit-load $P_0$ are given by the following equations:

$$P_0 = 1.455\eta_2 (W-a)\sigma_0 \quad \text{for plane strain}$$

$$P_0 = 1.072\eta_2 (W-a)\sigma_0 \quad \text{for plane stress}$$

$$\eta_2 = [1+(\frac{a}{W-a})]^{\frac{1}{2}} - \frac{a}{W-a}$$

**Table 5.3** h Functions for CCT Specimens (a) Plane Strain and (b) Plane Stress (from Kumar, German, and Shih, EPRI Report NP 1931, 1981)

| a/W | m= | 1 | 2 | 3 | 5 | 7 | 10 | 13 | 16 | 20 |
|---|---|---|---|---|---|---|---|---|---|---|
| 1/8 | $h_1$ | 2.80 | 3.61 | 4.06 | 4.35 | 4.33 | 4.02 | 3.56 | 3.06 | 2.46 |
|  | $h_2$ | 3.05 | 3.62 | 3.91 | 4.06 | 3.93 | 3.54 | 3.07 | 2.60 | 2.06 |
|  | $h_3$ | 0.303 | 0.574 | 0.840 | 1.30 | 1.63 | 1.95 | 2.03 | 1.96 | 1.77 |
| 1/4 | $h_1$ | 2.54 | 3.01 | 3.21 | 3.29 | 3.18 | 2.92 | 2.63 | 2.34 | 2.03 |
|  | $h_2$ | 2.68 | 2.99 | 3.01 | 2.85 | 2.61 | 2.30 | 1.97 | 1.71 | 1.45 |
|  | $h_3$ | 0.536 | 0.911 | 1.22 | 1.64 | 1.84 | 1.85 | 1.80 | 1.64 | 1.43 |
| 3/8 | $h_1$ | 2.34 | 2.62 | 2.65 | 2.51 | 2.28 | 1.97 | 1.71 | 1.46 | 1.19 |
|  | $h_2$ | 2.35 | 2.39 | 2.23 | 1.88 | 1.58 | 1.28 | 1.07 | 0.890 | 0.715 |
|  | $h_3$ | 0.699 | 1.06 | 1.28 | 1.44 | 1.40 | 1.23 | 1.05 | 0.888 | 0.719 |
| 1/2 | $h_1$ | 2.21 | 2.29 | 2.20 | 1.97 | 1.76 | 1.52 | 1.32 | 1.16 | 0.978 |
|  | $h_2$ | 2.03 | 1.86 | 1.60 | 1.23 | 1.00 | 0.799 | 0.664 | 0.564 | 0.466 |
|  | $h_3$ | 0.803 | 1.07 | 1.16 | 1.10 | 0.968 | 0.796 | 0.665 | 0.565 | 0.469 |
| 5/8 | $h_1$ | 2.12 | 1.96 | 1.76 | 1.43 | 1.17 | 0.863 | 0.628 | 0.458 | 0.300 |
|  | $h_2$ | 1.71 | 1.32 | 1.04 | 0.707 | 0.524 | 0.358 | 0.250 | 0.178 | 0.114 |
|  | $h_3$ | 0.844 | 0.937 | 0.879 | 0.701 | 0.522 | 0.361 | 0.251 | 0.178 | 0.115 |
| 3/4 | $h_1$ | 2.07 | 1.73 | 1.47 | 1.11 | 0.895 | 0.642 | 0.461 | 0.337 | 0.216 |
|  | $h_2$ | 1.35 | 0.857 | 0.596 | 0.361 | 0.254 | 0.167 | 0.114 | 0.081 | 0.0511 |
|  | $h_3$ | 0.805 | 0.700 | 0.555 | 0.359 | 0.254 | 0.168 | 0.114 | 0.081 | 0.0516 |
| 7/8 | $h_1$ | 2.08 | 1.64 | 1.40 | 1.14 | 0.987 | 0.814 | 0.688 | 0.573 | 0.461 |
|  | $h_2$ | 0.889 | 0.428 | 0.287 | 0.181 | 0.139 | 0.105 | 0.084 | 0.068 | 0.0533 |
|  | $h_3$ | 0.632 | 0.400 | 0.291 | 0.182 | 0.140 | 0.106 | 0.084 | 0.068 | 0.0535 |

a

| a/W | m= | 1 | 2 | 3 | 5 | 7 | 10 | 13 | 16 | 20 |
|---|---|---|---|---|---|---|---|---|---|---|
| 1/8 | $h_1$ | 2.80 | 3.57 | 4.01 | 4.47 | 4.65 | 4.62 | 4.41 | 4.13 | 3.72 |
|  | $h_2$ | 3.53 | 4.09 | 4.43 | 4.74 | 4.79 | 4.63 | 4.33 | 4.00 | 3.55 |
|  | $h_3$ | 0.350 | 0.661 | 0.997 | 1.55 | 2.05 | 2.56 | 2.83 | 2.95 | 2.92 |
| 1/4 | $h_1$ | 2.54 | 2.97 | 3.14 | 3.20 | 3.11 | 2.86 | 2.65 | 2.47 | 2.20 |
|  | $h_2$ | 3.10 | 3.29 | 3.30 | 3.15 | 2.93 | 2.56 | 2.29 | 2.08 | 1.81 |
|  | $h_3$ | 0.619 | 1.01 | 1.35 | 1.83 | 2.08 | 2.19 | 2.12 | 2.01 | 1.79 |
| 3/8 | $h_1$ | 2.34 | 2.53 | 2.52 | 2.35 | 2.17 | 1.95 | 1.77 | 1.61 | 1.43 |
|  | $h_2$ | 2.71 | 2.62 | 2.41 | 2.03 | 1.75 | 1.47 | 1.28 | 1.13 | 0.988 |
|  | $h_3$ | 0.807 | 1.20 | 1.43 | 1.59 | 1.57 | 1.43 | 1.27 | 1.13 | 0.994 |
| 1/2 | $h_1$ | 2.21 | 2.20 | 2.06 | 1.81 | 1.63 | 1.43 | 1.30 | 1.17 | 1.00 |
|  | $h_2$ | 2.34 | 2.01 | 1.70 | 1.30 | 1.07 | 0.871 | 0.757 | 0.666 | 0.557 |
|  | $h_3$ | 0.927 | 1.19 | 1.26 | 1.18 | 1.04 | 0.867 | 0.758 | 0.668 | 0.560 |
| 5/8 | $h_1$ | 2.12 | 1.91 | 1.69 | 1.41 | 1.22 | 1.01 | 0.853 | 0.712 | 0.573 |
|  | $h_2$ | 1.97 | 1.46 | 1.13 | 0.785 | 0.617 | 0.474 | 0.383 | 0.313 | 0.256 |
|  | $h_3$ | 0.975 | 1.05 | 0.970 | 0.763 | 0.620 | 0.478 | 0.386 | 0.318 | 0.273 |
| 3/4 | $h_1$ | 2.07 | 1.71 | 1.46 | 1.21 | 1.08 | 0.867 | 0.745 | 0.646 | 0.532 |
|  | $h_2$ | 1.55 | 0.970 | 0.685 | 0.452 | 0.361 | 0.262 | 0.216 | 0.183 | 0.148 |
|  | $h_3$ | 0.929 | 0.802 | 0.642 | 0.450 | 0.361 | 0.263 | 0.216 | 0.183 | 0.149 |
| 7/8 | $h_1$ | 2.08 | 1.57 | 1.31 | 1.08 | 0.972 | 0.862 | 0.778 | 0.715 | 0.630 |
|  | $h_2$ | 1.03 | 0.485 | 0.310 | 0.196 | 0.157 | 0.127 | 0.109 | 0.097 | 0.0842 |
|  | $h_3$ | 0.730 | 0.452 | 0.313 | 0.198 | 0.157 | 0.127 | 0.109 | 0.097 | 0.0842 |

b

**Table 5.4** h Functions for SENB Specimens (a) Plane Strain and (b) Plane Stress (from Kumar, German, and Shih, EPRI Report NP 1931, 1981)

| a/W | m= | 1 | 2 | 3 | 5 | 7 | 10 | 13 | 16 | 20 |
|---|---|---|---|---|---|---|---|---|---|---|
| 1/8 | $h_1$ | 0.936 | 0.869 | 0.805 | 0.687 | 0.580 | 0.437 | 0.329 | 0.245 | 0.165 |
|  | $h_2$ | 6.97 | 6.77 | 6.29 | 5.29 | 4.38 | 3.24 | 2.40 | 1.78 | 1.19 |
|  | $h_3$ | 3.00 | 22.1 | 20.0 | 15.0 | 11.7 | 8.39 | 6.14 | 4.54 | 3.01 |
| 1/4 | $h_1$ | 1.20 | 1.034 | 0.930 | 0.762 | 0.633 | 0.523 | 0.396 | 0.303 | 0.215 |
|  | $h_2$ | 5.80 | 4.67 | 4.01 | 3.08 | 2.45 | 1.93 | 1.45 | 1.09 | 0.758 |
|  | $h_3$ | 4.08 | 9.72 | 8.36 | 5.86 | 4.47 | 3.42 | 2.54 | 1.90 | 1.32 |
| 3/8 | $h_1$ | 1.33 | 1.15 | 1.02 | 0.084 | 0.695 | 0.556 | 0.442 | 0.360 | 0.265 |
|  | $h_2$ | 5.18 | 3.93 | 3.20 | 2.38 | 1.93 | 1.47 | 1.15 | 0.928 | 0.684 |
|  | $h_3$ | 4.51 | 6.01 | 5.03 | 3.74 | 3.02 | 2.30 | 1.80 | 1.45 | 1.07 |
| 1/2 | $h_1$ | 1.41 | 1.09 | 0.922 | 0.675 | 0.495 | 0.331 | 0.211 | 0.135 | 0.074 |
|  | $h_2$ | 4.87 | 3.28 | 2.53 | 1.69 | 1.19 | 0.773 | 0.480 | 0.304 | 0.165 |
|  | $h_3$ | 4.69 | 4.33 | 3.49 | 2.35 | 1.66 | 1.08 | 0.669 | 0.424 | 0.230 |
| 5/8 | $h_1$ | 1.46 | 1.07 | 0.896 | 0.631 | 0.436 | 0.255 | 0.142 | 0.084 | 0.041 |
|  | $h_2$ | 4.64 | 2.86 | 2.16 | 1.37 | 0.907 | 0.518 | 0.287 | 0.166 | 0.080 |
|  | $h_3$ | 4.71 | 3.49 | 2.70 | 1.72 | 1.14 | 0.652 | 0.361 | 0.209 | 0.102 |
| 3/4 | $h_1$ | 1.48 | 1.15 | 0.974 | 0.693 | 0.500 | 0.348 | 0.223 | 0.140 | 0.074 |
|  | $h_2$ | 4.47 | 2.75 | 2.10 | 1.36 | 0.936 | 0.618 | 0.388 | 0.239 | 0.127 |
|  | $h_3$ | 4.49 | 3.14 | 2.40 | 1.56 | 1.07 | 0.704 | 0.441 | 0.272 | 0.144 |
| 7/8 | $h_1$ | 1.50 | 1.35 | 1.20 | 1.02 | 0.855 | 0.690 | 0.551 | 0.440 | 0.321 |
|  | $h_2$ | 4.36 | 2.90 | 2.31 | 1.70 | 1.33 | 1.00 | 0.782 | 0.613 | 0.459 |
|  | $h_3$ | 4.15 | 3.08 | 2.45 | 1.81 | 1.41 | 1.06 | 0.828 | 0.649 | 0.486 |

a

| a/W | m= | 1 | 2 | 3 | 5 | 7 | 10 | 13 | 16 | 20 |
|---|---|---|---|---|---|---|---|---|---|---|
| 1/8 | $h_1$ | 0.676 | 0.600 | 0.548 | 0.459 | 0.383 | 0.297 | 0.238 | 0.192 | 0.148 |
|  | $h_2$ | 6.84 | 6.30 | 5.66 | 4.53 | 3.64 | 2.72 | 2.12 | 1.67 | 1.26 |
|  | $h_3$ | 2.95 | 20.1 | 14.6 | 12.2 | 9.12 | 6.75 | 5.20 | 4.09 | 3.07 |
| 1/4 | $h_1$ | 0.869 | 0.731 | 0.629 | 0.479 | 0.370 | 0.246 | 0.174 | 0.117 | 0.059 |
|  | $h_2$ | 5.69 | 4.50 | 3.68 | 2.61 | 1.95 | 1.29 | 0.897 | 0.603 | 0.307 |
|  | $h_3$ | 4.01 | 8.81 | 7.19 | 4.73 | 3.39 | 2.20 | 1.52 | 1.01 | 0.508 |
| 3/8 | $h_1$ | 0.963 | 0.797 | 0.680 | 0.527 | 0.418 | 0.307 | 0.232 | 0.174 | 0.105 |
|  | $h_2$ | 5.09 | 3.73 | 2.93 | 2.07 | 1.58 | 1.13 | 0.841 | 0.626 | 0.381 |
|  | $h_3$ | 4.42 | 5.53 | 4.48 | 3.17 | 2.41 | 1.73 | 1.28 | 0.948 | 0.575 |
| 1/2 | $h_1$ | 1.02 | 0.767 | 0.621 | 0.453 | 0.324 | 0.202 | 0.128 | 0.0813 | 0.029 |
|  | $h_2$ | 4.77 | 3.12 | 2.32 | 1.55 | 1.08 | 0.655 | 0.410 | 0.259 | 0.097 |
|  | $h_3$ | 4.60 | 4.09 | 3.09 | 2.08 | 1.44 | 0.874 | 0.545 | 0.344 | 0.129 |
| 5/8 | $h_1$ | 1.05 | 0.786 | 0.649 | 0.494 | 0.357 | 0.235 | 0.173 | 0.105 | 0.047 |
|  | $h_2$ | 4.55 | 2.83 | 2.12 | 1.46 | 1.02 | 0.656 | 0.472 | 0.286 | 0.130 |
|  | $h_3$ | 4.62 | 3.43 | 2.60 | 1.79 | 1.26 | 0.803 | 0.577 | 0.349 | 0.158 |
| 3/4 | $h_1$ | 1.07 | 0.786 | 0.643 | 0.474 | 0.343 | 0.230 | 0.167 | 0.110 | 0.044 |
|  | $h_2$ | 4.39 | 2.66 | 1.97 | 1.33 | 0.928 | 0.601 | 0.427 | 0.280 | 0.114 |
|  | $h_3$ | 4.39 | 3.01 | 2.24 | 1.51 | 1.05 | 0.680 | 0.483 | 0.316 | 0.129 |
| 7/8 | $h_1$ | 1.086 | 0.928 | 0.810 | 6.46 | 0.538 | 0.423 | 0.332 | 0.242 | 0.205 |
|  | $h_2$ | 4.28 | 2.76 | 2.16 | 1.56 | 1.23 | 0.922 | 0.702 | 0.561 | 0.428 |
|  | $h_3$ | 4.07 | 2.93 | 2.29 | 1.65 | 1.30 | 0.975 | 0.742 | 0.592 | 0.452 |

b

**Table 5.5** *h Functions for SENT Specimens for (a) Plane Strain Conditions and (b) Plane Stress Conditions (from Kumar, German, and Shih, EPRI Report NP 1931, 1981)*

### a

| a/W | | m=1 | 2 | 3 | 5 | 7 | 10 | 13 | 16 | 20 |
|---|---|---|---|---|---|---|---|---|---|---|
| 1/8 | $h_1$ | 4.95 | 6.93 | 8.57 | 11.5 | 13.5 | 16.1 | 18.1 | 19.9 | 21.2 |
|     | $h_2$ | 5.25 | 6.47 | 7.56 | 9.46 | 11.1 | 12.9 | 14.4 | 15.7 | 16.8 |
|     | $h_3$ | 26.6 | 25.8 | 25.2 | 24.2 | 23.6 | 23.2 | 23.2 | 23.5 | 23.7 |
| 1/4 | $h_1$ | 4.34 | 4.77 | 4.64 | 3.82 | 3.06 | 2.17 | 1.55 | 1.11 | 0.712 |
|     | $h_2$ | 4.76 | 4.56 | 4.28 | 3.39 | 2.64 | 1.81 | 1.25 | 0.875 | 0.552 |
|     | $h_3$ | 10.3 | 7.64 | 5.87 | 3.70 | 2.48 | 1.50 | 0.970 | 0.654 | 0.404 |
| 3/8 | $h_1$ | 3.88 | 3.25 | 2.63 | 1.68 | 1.06 | 0.539 | 0.276 | 0.142 | 0.0595 |
|     | $h_2$ | 4.54 | 3.49 | 2.67 | 1.57 | 0.946 | 0.458 | 0.229 | 0.116 | 0.048 |
|     | $h_3$ | 5.14 | 2.99 | 1.90 | 0.923 | 0.515 | 0.240 | 0.119 | 0.060 | 0.0246 |
| 1/2 | $h_1$ | 3.40 | 2.30 | 1.69 | 0.928 | 0.514 | 0.213 | 0.0902 | 0.0385 | 0.0119 |
|     | $h_2$ | 4.45 | 2.77 | 1.89 | 0.954 | 0.507 | 0.204 | 0.0854 | 0.0356 | 0.0110 |
|     | $h_3$ | 3.15 | 1.54 | 0.912 | 0.417 | 0.215 | 0.085 | 0.0358 | 0.0147 | 0.0045 |
| 5/8 | $h_1$ | 2.86 | 1.80 | 1.30 | 0.697 | 0.378 | 0.153 | 0.0625 | 0.0256 | 0.0078 |
|     | $h_2$ | 4.37 | 2.44 | 1.62 | 0.080 | 0.423 | 0.167 | 0.0671 | 0.0272 | 0.00823 |
|     | $h_3$ | 2.31 | 1.08 | 0.681 | 0.329 | 0.171 | 0.067 | 0.0268 | 0.0108 | 0.0033 |
| 3/4 | $h_1$ | 2.34 | 1.61 | 1.25 | 0.769 | 0.477 | 0.233 | 0.116 | 0.059 | 0.0215 |
|     | $h_2$ | 4.32 | 2.52 | 1.79 | 1.03 | 0.619 | 0.296 | 0.146 | 0.0735 | 0.0267 |
|     | $h_3$ | 2.02 | 1.10 | 0.765 | 0.435 | 0.262 | 0.125 | 0.0617 | 0.0312 | 0.0113 |
|     | $h_1$ | 1.91 | 1.57 | 1.37 | 1.10 | 0.925 | 0.702 | | | |
|     | $h_2$ | 4.29 | 2.75 | 2.14 | 1.55 | 1.23 | 0.921 | From Kumar, German and Shih, | | |
|     | $h_3$ | 2.01 | 1.27 | 0.988 | 0.713 | 0.564 | 0.424 | EPRI Report NP 1931, 1981 | | |

### b

| a/W | | m=1 | 2 | 3 | 5 | 7 | 10 | 13 | 16 | 20 |
|---|---|---|---|---|---|---|---|---|---|---|
| 1/8 | $h_1$ | 3.58 | 4.55 | 5.06 | 5.30 | 4.96 | 4.14 | 3.29 | 2.60 | 1.92 |
|     | $h_2$ | 5.15 | 5.43 | 6.05 | 6.01 | 5.47 | 4.46 | 3.48 | 2.74 | 2.02 |
|     | $h_3$ | 26.1 | 21.6 | 18.0 | 12.7 | 9.24 | 5.98 | 3.94 | 2.72 | 2.0 |
| 1/4 | $h_1$ | 3.14 | 3.26 | 2.92 | 2.12 | 1.53 | 0.960 | 0.615 | 0.400 | 0.230 |
|     | $h_2$ | a/W | 4.30 | 3.70 | 2.53 | 1.76 | 1.05 | 0.656 | 0.419 | 0.237 |
|     | $h_3$ | 10.1 | 6.49 | 4.36 | 2.19 | 1.24 | 0.630 | 0.362 | 0.224 | 0.123 |
| 3/8 | $h_1$ | 2.81 | 2.37 | 1.94 | 1.37 | 1.01 | 0.677 | 0.474 | 0.342 | 0.226 |
|     | $h_2$ | 4.47 | 3.43 | 2.63 | 1.69 | 1.18 | 0.762 | 0.524 | 0.372 | 0.244 |
|     | $h_3$ | 5.05 | 2.65 | 1.60 | 0.812 | 0.525 | 0.328 | 0.223 | 0.157 | 0.102 |
| 1/2 | $h_1$ | 2.46 | 1.67 | 1.25 | 0.776 | 0.510 | 0.286 | 0.164 | 0.0956 | 0.0469 |
|     | $h_2$ | 4.37 | 2.73 | 1.91 | 1.09 | 0.694 | 0.380 | 0.216 | 0.124 | 0.0607 |
|     | $h_3$ | 3.10 | 1.43 | 0.871 | 0.461 | 0.286 | 0.155 | 0.088 | 0.0506 | 0.0247 |
| 5\8 | $h_1$ | 2.07 | 1.41 | 1.105 | 0.755 | 0.551 | 0.363 | 0.248 | 0.172 | 0.107 |
|     | $h_2$ | 4.30 | 2.55 | 1.84 | 1.16 | 0.816 | 0.523 | 0.353 | 0.242 | 0.150 |
|     | $h_3$ | 2.27 | 1.13 | 0.771 | 0.478 | 0.336 | 0.215 | 0.146 | 0.100 | 0.0616 |
| 3/4 | $h_1$ | 1.70 | 1.14 | 0.910 | 0.624 | 0.447 | 0.280 | 0.181 | 0.118 | 0.0670 |
|     | $h_2$ | 4.24 | 2.47 | 1.81 | 1.15 | 0.798 | 0.490 | 0.314 | 0.203 | 0.115 |
|     | $h_3$ | 1.98 | 1.09 | 0.784 | 0.494 | 0.344 | 0.211 | 0.136 | 0.0581 | 0.0496 |
|     | $h_1$ | 1.38 | 1.11 | 0.962 | 0.792 | 0.677 | 0.574 | | | |
|     | $h_2$ | 4.22 | 2.68 | 2.08 | 1.54 | 1.27 | 1.04 | From Kumar, German and Shih | | |
|     | $h_3$ | 1.97 | 1.25 | 0.969 | 0.716 | 0.591 | 0.483 | EPRI Report NP 1931, 1981 | | |

*DEN Specimens*

The expression for J is identical in form to the CCT specimen, equation (5.54a), and is, therefore, not repeated here. The expressions for the plastic parts of $\delta$ and V are given below:

$$\delta_p = \alpha \varepsilon_0 (W-a) h_2 (a/W, m) (P/P_0)^m \tag{5.59a}$$

$$V_p = \alpha \varepsilon_0 (W-a) h_3 (a/W, m) (P/P_0)^m \tag{5.59b}$$

The values of $h_1$, $h_2$, and $h_3$ for plane stress and plane strain are given in Table 5.6. The limit-load, $P_0$, for unit thickness is given by the following equations:

$$P_0 = (0.72 + 1.82(1-a/W))\sigma_0 W \quad \text{for plane strain} \tag{5.60}$$

and

$$P_0 = \frac{4}{\sqrt{3}} (W-a)\sigma_0 \quad \text{for plane stress} \tag{5.61}$$

### 5.4.3 J-Solutions for Cracks in Infinite Bodies

He and Hutchinson [5.12] have derived expressions for J for fully-plastic bodies containing cracks in the center or on the edge of semi-infinite cracked bodies. These solutions are invaluable in applications and can be used to extend the J expressions for several configurations described in the previous section. For an edge crack in an infinite body subjected to uniform tension, $\sigma$, the plastic part of J, $J_p$, is given by the following expressions for plane strain:

$$J_p = 1.21\pi\sqrt{m} \, \frac{\varepsilon_0 a}{\sigma_0^m} \left(\frac{\sqrt{3}}{2}\sigma\right)^{m+1} \tag{5.62}$$

and

$$J_p = 1.21\pi\sqrt{m} \, \frac{\alpha\varepsilon_0 a}{\sigma_0^m} \sigma^{m+1} \quad \text{for plane stress} \tag{5.63}$$

The expressions for estimating J for cracks inside the plate are the same as equations (5.57) and (5.58) except the 1.21 factor is replaced with unity.

**Example Problem 5.3**

Using the fully-plastic J expressions for DEN specimens and equation (5.57), estimate the value of $h_1$ for the DEN specimen for the condition $a/W \to 0$ for plane strain.

*Solution:*

We can rewrite the expression for $J_p$ for DEN specimen for plane strain as follows:

$$J_p = \alpha\sigma_0\varepsilon_0 \, a(1-a/W) h_1 \left[\frac{\sigma(2W)}{(0.72+1.82(1-a/W))\sigma_0 W}\right]^{m+1}$$

**Table 5.6** h Functions for DEN Specimens for (a) Plane Strain Conditions and (b) Plane Stress Conditions (from Kumar, German, and Shih, EPRI Report NP 1931, 1981)

| a/W | m= | 1 | 2 | 3 | 5 | 7 | 10 | 13 | 16 | 20 |
|---|---|---|---|---|---|---|---|---|---|---|
| 1/8 | $h_1$ | 4.576 | 6.176 | 7.736 | 9.04 | 10.8 | 12.88 | 14.88 | 16.64 | 19.52 |
|  | $h_2$ | 0.732 | 0.852 | 0.961 | 1.14 | 1.29 | 1.50 | 1.70 | 1.94 | 2.17 |
|  | $h_3$ | 0.063 | 0.126 | 0.200 | 0.372 | 0.571 | 0.911 | 1.30 | 1.74 | 2.29 |
| 1/4 | $h_1$ | 4.4 | 5.28 | 5.52 | 6.6 | 7.00 | 7.28 | 7.44 | 7.56 | 7.68 |
|  | $h_2$ | 1.56 | 1.63 | 1.70 | 1.78 | 1.80 | 1.81 | 1.79 | 1.78 | 1.76 |
|  | $h_3$ | 0.267 | 0.479 | 0.698 | 1.11 | 1.47 | 1.92 | 2.25 | 2.49 | 2.73 |
| 3/8 | $h_1$ | 4.299 | 4.886 | 5.126 | 5.126 | 4.913 | 4.486 | 3.978 | 3.524 | 2.99 |
|  | $h_2$ | 2.51 | 2.41 | 2.35 | 2.15 | 1.94 | 1.68 | 1.44 | 1.25 | 1.05 |
|  | $h_3$ | 0.637 | 1.05 | 1.40 | 1.87 | 2.11 | 2.20 | 2.09 | 1.92 | 1.67 |
| 1/2 | $h_1$ | 4.44 | 4.86 | 4.96 | 4.86 | 4.64 | 2.24 | 3.82 | 3.20 | 3.05 |
|  | $h_2$ | 3.73 | 3.40 | 3.15 | 2.71 | 2.37 | 2.01 | 1.72 | 1.40 | 1.38 |
|  | $h_3$ | 1.26 | 1.92 | 2.37 | 2.79 | 2.85 | 2.68 | 2.40 | 1.99 | 1.94 |
| 5/8 | $h_1$ | 5.056 | 5.408 | 5.52 | 5.472 | 5.248 | 4.80 | 4.064 | 3.776 | 3.632 |
|  | $h_2$ | 5.57 | 4.76 | 4.23 | 3.46 | 2.97 | 2.48 | 2.02 | 1.82 | 1.66 |
|  | $h_3$ | 2.36 | 3.29 | 3.74 | 3.90 | 3.68 | 3.23 | 2.66 | 2.40 | 2.19 |
| 3/4 | $h_1$ | 6.985 | 8.265 | 9.557 | 11.250 | 12.610 | 14.53 | 15.862 | 15.063 | 23.194 |
|  | $h_2$ | 9.10 | 7.76 | 7.14 | 6.64 | 6.83 | 7.48 | 7.79 | 7.14 | 11.10 |
|  | $h_3$ | 4.73 | 6.26 | 7.03 | 7.63 | 8.14 | 9.04 | 9.40 | 8.58 | 13.5 |
| 7/8 | $h_1$ | 16.23 | 28.34 | 44.577 | 89.61 | 160.02 | 389.76 | 888.11 | 1794.5 | 4366.3 |
|  | $h_2$ | 20.1 | 19.4 | 22.7 | 36.1 | 58.9 | 133.0 | 294.0 | 585.0 | 1400.0 |
|  | $h_3$ | 12.7 | 18.2 | 24.1 | 40.4 | 65.8 | 149.0 | 327.0 | 650.0 | 1560.0 |

a

| a/W | m= | 1 | 2 | 3 | 5 | 7 | 10 | 13 | 16 | 20 |
|---|---|---|---|---|---|---|---|---|---|---|
| 1/8 | $h_1$ | 4.664 | 6.6 | 8.16 | 10.96 | 13.68 | 17.92 | 22.72 | 28.32 | 36.96 |
|  | $h_2$ | 0.853 | 1.05 | 1.23 | 1.55 | 1.87 | 2.38 | 2.96 | 3.65 | 4.70 |
|  | $h_3$ | 0.0729 | 0.159 | 0.26 | 0.504 | 0.821 | 1.41 | 2.18 | 3.16 | 4.73 |
| 1/4 | $h_1$ | 4.04 | 4.92 | 5.44 | 5.92 | 6.16 | 6.32 | 6.36 | 6.36 | 6.36 |
|  | $h_2$ | 1.73 | 1.82 | 1.89 | 1.92 | 1.91 | 1.85 | 1.80 | 1.75 | 1.70 |
|  | $h_3$ | 0.296 | 0.537 | 0.770 | 1.17 | 1.49 | 1.82 | 2.02 | 2.12 | 2.20 |
| 3/8 | $h_1$ | 3.354 | 3.791 | 3.818 | 3.58 | 3.311 | 2.910 | 2.59 | 2.331 | 1.800 |
|  | $h_2$ | 2.59 | 2.39 | 2.22 | 1.86 | 1.59 | 1.28 | 1.07 | 0.922 | 0.709 |
|  | $h_3$ | 0.658 | 1.04 | 1.30 | 1.52 | 1.55 | 1.41 | 1.23 | 1.07 | 0.830 |
| 1/2 | $h_1$ | 2.96 | 2.94 | 2.76 | 2.34 | 2.02 | 1.69 | 1.464 | 1.25 | .416 |
|  | $h_2$ | 3.51 | 2.82 | 2.34 | 1.67 | 1.28 | 0.944 | 0.762 | 0.630 | 0.232 |
|  | $h_3$ | 1.18 | 1.58 | 1.69 | 1.56 | 1.32 | 1.01 | 0.809 | 0.662 | 0.266 |
| 5\8 | $h_1$ | 2.544 | 2.32 | 2.064 | 1.664 | 1.411 | 1.179 | 1.038 | .7456 | .03232 |
|  | $h_2$ | 4.56 | 3.15 | 2.32 | 1.45 | 1.06 | 0.790 | 0.657 | 0.473 | 0.0277 |
|  | $h_3$ | 1.93 | 2.14 | 1.95 | 1.44 | 1.09 | 0.809 | 0.665 | 0.487 | 0.0317 |
| 3/4 | $h_1$ | 2.199 | 1.906 | 1.626 | 1.305 | 1.112 | .934 | .840 | .396 |  |
|  | $h_2$ | 5.90 | 3.37 | 2.22 | 1.30 | 0.966 | 0.741 | 0.636 | 0.312 |  |
|  | $h_3$ | 3.06 | 2.67 | 2.06 | 1.31 | 0.978 | 0.747 | 0.638 | 0.318 |  |
| 7/8 | $h_1$ | 1.931 | 1.634 | 1.394 | 1.119 | .966 | .891 | .759 | .818 | .642 |
|  | $h_2$ | 8.02 | 3.51 | 2.14 | 1.27 | 0.971 | 0.775 | 0.663 | 0.596 | 0.535 |
|  | $h_3$ | 5.07 | 3.18 | 2.16 | 1.30 | 0.980 | 0.779 | 0.665 | 0.597 | 0.538 |

b

$$\lim a/W \to 0, J_p = \frac{\alpha \varepsilon_0}{\sigma_0^m} a h_1 \left[\frac{\sigma}{1.27}\right]^{m+1}$$

The above expression must be the same as equation (5.62). Thus, equating the two, we get:

$$1.21 \pi \sqrt{m} \left(\frac{\sqrt{3}}{2}\right)^{m+1} = h_1(a/W \to 0, m) \left(\frac{1}{1.27}\right)^{m+1}$$

or

$$h_1(a/W \to 0, m) = 1.21 \pi \sqrt{m} (1.0998)^{m+1}$$

This approach can be used to extend the J expressions for SENB, SENT, and the CCT specimens to small crack sizes.

### Example Problem 5.4

A compact specimen which is 50mm wide and 25mm thick and has a crack size of 25mm is made from a 2.25Cr - 1Mo steel and is being tested at 538°C. The Ramberg-Osgood constants for this material at 538°C are as follows:

$\sigma_0$ = 140 Mpa
$\alpha$ = 2.2
$\varepsilon_0$ = 8 x 10$^{-4}$
m = 5.0
E = 175 x 10$^3$ Mpa
$\nu$ = 0.3

If the applied load is 12.5 kN, estimate the value of J for plane stress and plane strain conditions. Explain the large difference between the two values. Is this also a problem in characterizing J-resistance curves from compact type specimens?

*Solution:*

$$J = J_e(a_e) + J_p$$

$a_e$ = effective crack length = $a + \phi \, r_y$

$$\phi = \frac{1}{1 + (P/P_0)^2} \; ; \; r_y = \frac{1}{\beta \pi} \left(\frac{m-1}{m+1}\right) \left(\frac{K}{\sigma_0}\right)^2$$

$$K = \frac{P}{BW^{1/2}} \frac{(2+a/W)}{(1-a/W)^{3/2}} [0.886 + 4.64(a/W) - 13.32(a/W)^2 + 14.72(a/W)^3 - 5.6(a/W)^4]$$

B = 0.025m     W = 0.050m
K = 21.6 Mpa√m

Let us calculate J for plane stress conditions first:

$\beta = 2$, hence $r_y = 0.00395$
$P_0 = 1.071\, \eta_1\, (W-a)\, \sigma_0\, B$
$\eta_1 = 0.1623$ from equation (5.53) for $a/W = 0.5$
$P_0 = 0.0152\, MN = 15.2\, kN$

Thus:
$$\phi = \frac{1}{1+(\frac{12.5}{15.2})^2} = 0.596$$

Thus:
$$a_e = .025 + .596 \times .00395 = .0273$$

$$J_e(a_e) = \frac{[K(a_e)]^2}{E} = \frac{(25.1)^2}{175 \times 10^3} = 3.6 \times 10^{-3}\, MJ/m^2 = 3.6\, kJ/m^2$$

$$J_p = \alpha \sigma \varepsilon_0\, (W-a)\, h_1\, (0.5, 5)\, (P/P_0)^{m+1}$$

$$= .00616 \times .686 \times (\frac{12.5}{15.2})^6 = 1.3 \times 10^{-3}\, MJ/m^2 = 1.3\, kJ/m^2$$

$J = 3.6 + 1.3 = 4.9\, kJ/m^2$

For plane strain, J is calculated as follows:
$r_y = .00132$ because $\beta = 6$
$P_0 = 1.455\, \eta_1\, (W-a)\, \sigma_0\, B = 0.0206\, MN = 20.6\, kN$

$$\phi = \frac{1}{1+(\frac{12.5}{20.6})^2} = 0.731$$

$$a_e = .025 + .731(.00132) = .02596\, m$$

$$J_e(a_e) = (1-v^2)\frac{[K(a_e)]^2}{E} = \frac{(22.59)^2}{175 \times 10^3}(.91) = 2.65 \times 10^{-3}\, MJ/m^2 = 2.65\, kJ/m^2$$

$$J_p = .00616 \times 0.919\, (\frac{12.5}{20.6})^6 = 2.82 \times 10^{-4} = .282\, kJ$$

Thus:
$$J = 2.65 + .281 = 2.931\, kJ$$

138    Nonlinear Fracture Mechanics for Engineers

The difference between the J values for plane stress and plane strain is very significant. This is a problem in estimating J in components under elastic-plastic loading if it is not very evident whether plane stress or plane strain should be assumed. However, it is not significant in test specimens for which measured values of deflections are used in estimating J. The correct state-of-stress is already reflected in the deflection response to the applied load.

*5.4.4 J-Solutions for Cracked Cylinders*

The performance of pressure vessel and pipes is often evaluated in the presence of cracks which can have axial or circumferential orientations. Kumar, German, and Shih [5.11] have provided J solutions for a few cases. Due to their practical importance, we will list these J solutions in their entirety.

**Internally Pressurized Cylinder with an Internal Axial Crack** The configuration in question is shown in Figure 5.14. The J under fully-plastic conditions and the opening at the crack mouth are given by the following equations:

$$J_p = \alpha \sigma_0 \varepsilon_0 (1-a/W) \, a h_1 \, (a/W, m, W/R_i)(p/p_0)^{m+1} \tag{5.64}$$

In this case, $R_i$ the inside radius of the cylinder, $W = R_0 - R_i$ = wall thickness of the pipe, $R_0$ = outside radius of the pipe, $p_0$ = limit pressure given by:

$$p_0 = \frac{2}{\sqrt{3}} \frac{(W-a)\sigma_0}{R_i + a} \tag{5.65}$$

where a = crack depth:

$$\delta_p = \alpha \sigma_0 \varepsilon_0 a h_2 \, (a/W, m, W/R_i)(p/p_0)^m \tag{5.66}$$

$h_1$ and $h_2$ are dimensionless functions given in Table 5.7 for various values of m and also for values of $W/R_i$ of 0.05, 0.1, and 0.2.

**Cylinder with an Internal Circumferential Crack Under Remote Tension** This configuration is shown in Figure 5.15. This $J_p$, $\delta_p$, $V_p$, and $\delta_t$ are given by the following equations [5.11].

$$J_p = \alpha \sigma_0 \varepsilon_0 (1-a/W) \, a h_1 \, (a/W, m, W/R_i)(P/P_0)^{m+1} \tag{5.67}$$

$$\delta_p = \alpha \varepsilon_0 \, a h_2 \, (a/W, m, W/R_i)(P/P_0)^{m+1} \tag{5.68}$$

$$V_p = \alpha \varepsilon_0 \, a h_3 \, (a/W, m, W/R_i)(P/P_0)^m \tag{5.69}$$

$$\delta_t = \alpha \varepsilon_0 \, a h_4 \, (a/W, m, W/R_i)(P/P_0)^m \tag{5.70}$$

where P = applied remote load and

$$P_0 = \frac{2}{\sqrt{3}} \sigma_0 \pi (R_0^2 - (R_i + a)^2) \tag{5.71}$$

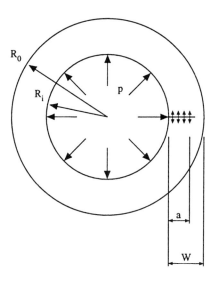

*Figure 5.14* Internally pressurized cylinder with an axial internal crack.

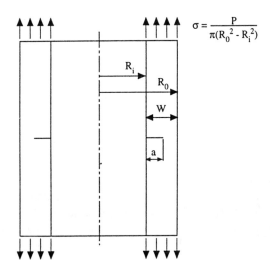

*Figure 5.15* Cylinder with an internal circumferential crack under remote tension.

*Table 5.7  h Functions for Internally Pressurized Cylinder with an Axial Crack on the Inside Surface for Wall Thickness to Inside Radius Ratio of (a) 0.2, (b) 0.1, and (c) 0.05 (from Kumar, German, and Shih, EPRI Report NP 1931, 1981)*

| a/W | m= | 1 | 2 | 3 | 5 | 7 | 10 |
|---|---|---|---|---|---|---|---|
| 1/8 | $h_1$ | 6.32 | 7.93 | 9.32 | 11.5 | 13.12 | 14.94 |
|  | $h_2$ | 5.83 | 7.01 | 7.96 | 9.49 | 10.67 | 11.96 |
| 1/4 | $h_1$ | 7.00 | 8.34 | 9.03 | 9.59 | 9.71 | 9.45 |
|  | $h_2$ | 5.92 | 8.72 | 7.07 | 7.26 | 7.14 | 6.71 |
| 1/2 | $h_1$ | 9.79 | 10.37 | 9.07 | 5.61 | 3.52 | 2.11 |
|  | $h_2$ | 7.05 | 6.97 | 6.01 | 3.70 | 2.28 | 1.25 |
| 3/4 | $h_1$ | 11.00 | 5.54 | 2.84 | 1.24 | 0.83 | 0.493 |
|  | $h_2$ | 7.35 | 3.86 | 1.86 | 0.556 | 0.261 | 0.129 |

a

| a/W | m= | 1 | 2 | 3 | 5 | 7 | 10 |
|---|---|---|---|---|---|---|---|
| 1/8 | $h_1$ | 5.22 | 6.64 | 7.59 | 8.76 | 9.34 | 9.55 |
|  | $h_2$ | 5.31 | 6.25 | 6.88 | 7.65 | 8.02 | 8.09 |
| 1/4 | $h_1$ | 6.16 | 7.49 | 7.96 | 8.08 | 7.78 | 6.98 |
|  | $h_2$ | 5.56 | 6.31 | 6.52 | 6.40 | 6.01 | 5.27 |
| 1/2 | $h_1$ | 10.5 | 11.6 | 10.7 | 6.47 | 3.95 | 2.27 |
|  | $h_2$ | 7.48 | 7.72 | 7.01 | 4.29 | 2.58 | 1.37 |
| 3/4 | $h_1$ | 16.1 | 8.19 | 3.87 | 1.46 | 1.05 | 0.787 |
|  | $h_2$ | 9.57 | 5.40 | 2.57 | 0.706 | 0.370 | 0.232 |

b

| a/W | m= | 1 | 2 | 3 | 5 | 7 | 10 |
|---|---|---|---|---|---|---|---|
| 1/8 | $h_1$ | 4.50 | 5.79 | 6.62 | 7.65 | 8.07 | 7.75 |
|  | $h_2$ | 4.96 | 5.71 | 6.20 | 6.82 | 7.02 | 6.66 |
| 1/4 | $h_1$ | 5.57 | 6.91 | 7.37 | 7.47 | 7.21 | 6.53 |
|  | $h_2$ | 5.29 | 5.98 | 6.16 | 6.01 | 5.63 | 4.93 |
| 1/2 | $h_1$ | 10.8 | 12.8 | 12.8 | 8.16 | 4.88 | 2.62 |
|  | $h_2$ | 7.66 | 8.33 | 8.13 | 5.33 | 3.20 | 1.65 |
| 3/4 | $h_1$ | 23.1 | 13.1 | 5.87 | 1.90 | 1.23 | 0.883 |
|  | $h_2$ | 12.1 | 7.88 | 3.84 | 1.01 | 0.454 | 0.240 |

c

The $h_1$, $h_2$, $h_3$, and $h_4$ functions are listed in Table 5.8 for several values of $W/R_i$, m, and a/W. These solutions can also be extended to a/W → 0 using the procedure used in Example Problem 5.3. It is left as an exercise problem for the student to derive that result (See Exercise Problem 5.13). J solutions for several other configurations can be found in References 5.13.

### 5.4.5 Reference Stress Method

The reference stress method is a way to use the abundantly available K solutions to obtain an approximate estimate of J-integral. This method, which was first developed by Ainsworth [5.14], defines a reference stress in the following way:

$$\sigma_{ref} = (P/P_o)\, \sigma_0 \tag{5.72}$$

*Table 5.8 h Functions for a Circumferential Crack on the Inside Surface of a Cylinder Wall in Uniform Tension With a Thickness to Inside Radius Ratio of (a) 0.2, (b) 0.1, and (c) 0.05 (from Kumar, German, and Shih, EPRI Report NP 1931, 1981)*

| a/W | m= | 1 | 2 | 3 | 5 | 7 | 10 |
|---|---|---|---|---|---|---|---|
| 1/8 | $h_1$ | 3.78 | 5.00 | 5.94 | 7.54 | 8.99 | 11.1 |
|  | $h_2$ | 4.56 | 5.55 | 6.37 | 7.79 | 9.10 | 11.0 |
|  | $h_3$ | 0.369 | 0.700 | 1.07 | 1.96 | 3.04 | 4.94 |
| 1/4 | $h_1$ | 3.88 | 4.95 | 5.64 | 6.49 | 6.94 | 7.22 |
|  | $h_2$ | 4.40 | 5.12 | 5.57 | 6.07 | 6.28 | 6.30 |
|  | $h_3$ | 0.673 | 1.25 | 1.79 | 2.79 | 3.61 | 4.52 |
| 1/2 | $h_1$ | 4.40 | 4.78 | 4.59 | 3.79 | 3.07 | 2.34 |
|  | $h_2$ | 4.36 | 4.30 | 3.91 | 3.00 | 2.26 | 1.55 |
|  | $h_3$ | 1.33 | 1.93 | 2.21 | 2.23 | 1.94 | 1.46 |
| 3/4 | $h_1$ | 4.12 | 3.03 | 2.23 | 1.546 | 1.30 | 1.11 |
|  | $h_2$ | 3.46 | 2.19 | 1.36 | 0.638 | 0.436 | 0.325 |
|  | $h_3$ | 1.54 | 1.39 | 1.04 | 0.686 | 0.508 | 0.366 |

a

| a/W | m= | 1 | 2 | 3 | 5 | 7 | 10 |
|---|---|---|---|---|---|---|---|
| 1/8 | $h_1$ | 4.00 | 5.13 | 6.09 | 7.69 | 9.09 | 11.1 |
|  | $h_2$ | 4.71 | 5.63 | 6.45 | 7.85 | 9.09 | 10.9 |
|  | $h_3$ | 0.548 | 0.733 | 1.13 | 2.07 | 3.16 | 5.07 |
| 1/4 | $h_1$ | 4.17 | 5.35 | 6.09 | 6.93 | 7.30 | 7.41 |
|  | $h_2$ | 4.58 | 5.36 | 5.84 | 6.31 | 6.44 | 6.31 |
|  | $h_3$ | 0.757 | 1.35 | 1.93 | 2.96 | 3.78 | 4.60 |
| 1/2 | $h_1$ | 5.40 | 5.90 | 5.63 | 4.51 | 3.49 | 2.47 |
|  | $h_2$ | 4.99 | 5.01 | 4.59 | 3.48 | 2.56 | 1.67 |
|  | $h_3$ | 1.555 | 2.26 | 2.59 | 2.57 | 2.18 | 1.56 |
| 3/4 | $h_1$ | 5.18 | 3.78 | 2.57 | 1.59 | 1.31 | 1.10 |
|  | $h_2$ | 4.22 | 2.79 | 1.67 | 0.725 | 0.48 | 0.300 |
|  | $h_3$ | 1.86 | 1.73 | 1.26 | 0.775 | 0.561 | 0.360 |

b

| a/W | m= | 1 | 2 | 3 | 5 | 7 | 10 |
|---|---|---|---|---|---|---|---|
| 1/8 | $h_1$ | 4.04 | 5.23 | 6.22 | 7.82 | 9.19 | 11.1 |
|  | $h_2$ | 4.82 | 5.69 | 6.52 | 7.90 | 9.11 | 10.8 |
|  | $h_3$ | 0.680 | 0.759 | 1.17 | 2.13 | 3.23 | 5.12 |
| 1/4 | $h_1$ | 4.38 | 5.68 | 6.45 | 7.29 | 7.62 | 7.65 |
|  | $h_2$ | 4.71 | 5.56 | 6.05 | 6.51 | 6.59 | 6.39 |
|  | $h_3$ | 0.818 | 1.43 | 2.03 | 3.10 | 3.91 | 4.69 |
| 1/2 | $h_1$ | 6.55 | 7.17 | 6.89 | 5.46 | 4.13 | 2.77 |
|  | $h_2$ | 5.67 | 5.77 | 5.36 | 4.08 | 2.97 | 1.88 |
|  | $h_3$ | 1.80 | 2.59 | 2.99 | 2.98 | 2.50 | 1.74 |
| 3/4 | $h_1$ | 6.64 | 4.87 | 3.08 | 1.68 | 1.30 | 1.07 |
|  | $h_2$ | 5.18 | 3.57 | 2.07 | 0.808 | 0.472 | 0.316 |
|  | $h_3$ | 2.36 | 2.18 | 1.53 | 0.772 | 0.494 | 0.330 |

c

Defining a reference strain, $\varepsilon_{ref}$, as the uniaxial strain corresponding to an applied stress, $\sigma_{ref}$, we can write:

$$\varepsilon_{ref} = \alpha \varepsilon_o \left(\frac{\sigma_{ref}}{\sigma_o}\right)^m \tag{5.73}$$

If, as an example, we use the expression for $J_p$ for the CT specimen, Equation 5.50a:

$$J_p = \sigma_{ref} \varepsilon_{ref} (W-a) h_1 (a/W, m) \tag{5.74}$$

If $h_1 (a/W, m)$ is a weak function of m, it may be assumed that $h_1 (a/W, m) \approx h_1 (a/W, 1)$. The function $h_1 (a/W, 1)$ can be obtained from the K-calibration expression (see Exercise Problem 5.16). Thus, the K solutions can be used to obtain J solutions if an expression to estimate the limit load for the cracked body can be derived. These solutions are considered to have reasonable accuracy for engineering purposes and are particularly useful in obtaining J for semi-elliptical cracks under complex loading. Equation (5.74) also provides better estimates of J for materials whose stress-strain characteristics are not accurately represented by the Ramberg-Osgood relationship. This is particularly useful if the strains are in the vicinity of the yield point.

## 5.5 Summary

This chapter discusses the methods of determining J-integral for test specimens as well as for several crack configurations of practical interest. We began by describing an example of how J can be obtained analytically for simple configurations in which stresses can be easily obtained along a contour. Experimental methods of determining J were discussed next. In test specimens, J can be determined by using formulae that are based upon the applied load, load-line displacement, the crack size, and the relevant length dimensions of the specimen. A method of estimating J for specimens in which crack growth has occurred was also discussed.

Numerical solutions for several test specimen configurations and crack models resembling actual components were also discussed. Examples are provided for further understanding of the use and limitations of the various methods of determining J.

## 5.6 References

5.1  J.R. Rice, "A Path-Independent Integral and the Approximate Analysis of Strain Concentration as Notches and Cracks", Journal of Applied Mechanics, Trans. ASME, Vol. 35, 1968, pp. 379-386.

5.2  N.I. Muskhelishvili, "Some Basic Problems of Mathematical Theory of Elasticity", Noordhoff, Groninger, The Netherlands, 1963.

5.3  J.A. Begley and J.D. Landes, "The J-Integral as a Fracture Criterion", Fracture Toughness, Part II, ASTM STP 514, American Society for Testing and Materials, Philadelphia, Pa, 1972, pp. 1-20.

5.4  J.R. Rice, P.C. Paris, and J.G. Merkle, "Some Further Results of J-Integral Analysis and Estimates", Progress in Flaw Growth and Fracture Testing, ASTM STP 536, American Society for Testing and Materials, Philadelphia, Pa., 1973, pp. 231-245.

5.5  J.G. Merkle and H.T. Corten, "A J-Integral Analysis of Compact Specimen, Considering Axial Forces and Bending Effects", Journal of Pressure Vessel Technology, Vol. 96, 1974, pp. 286-292.

5.6  H.A. Ernst, P.C. Paris, and J.D. Landes, "Estimations of J-Integral and Tearing Modulus T from a Single Specimen Test Record", Fracture Mechanics: Thirteenth Conference, ASTM STP 743, American Society for Testing and Materials, 1981, pp. 476-502.

5.7  A.A. Ilyushin, "The Theory of Small Elastic-Plastic Deformations", Prikadnaia Matematika i Mekhanika, PMM Vol. 10, 1946, pp. 347-356.

5.8  N.L. Goldman and J.W. Hutchinson, "Fully Plastic Crack Problems: The Center-Cracked Strip Under Plane Strain", International Journal of Solids and Structures, Vol. 11, 1975, pp. 575-591.

5.9  C.F. Shih and J.W. Hutchinson, "Fully Plastic Solutions and Large Scale Yielding Estimates for Plane Stress Crack Problems", Transactions of ASME, Journal of Engineering Materials and Technology, Series H, Vol. 98, 1976, pp. 289-295.

5.10  V. Kumar and C.F. Shih, "Fully Plastic Crack Solutions, Estimation Scheme and Stability Analyses for Compact Specimen", in Fracture Mechanics; Twelfth Conference, ASTM STP 700, American Society for Testing and Materials, 1980, pp. 406-438.

5.11  V.Kumar, M.D. German, and C.F. Shih, " An Engineering Approach for Elastic-Plastic Fracture Analysis", EPRI Report NP-1931, Electric Power Research Institute, Palo Alto, CA, July 1981.

5.12  M.Y. He and J.W. Hutchinson, "Fully-Plastic J-Solutions for Cracks in Infinite Solids", Elastic-Plastic Fracture, Second Symposium, Vol. I - Inelastic Crack Analysis, ASTM STP 803, American Society for Testing and Materials, 1983, pp. I291-I305.

5.13  V. Kumar and M.D. German, "Elastic-Plastic Fracture Analysis of Through-Wall and Surface Flaws in Cylinders", EPRI Report NP-5596, Electric Power Research Institute, Palo Alto, CA, 1988.

5.14  R.A. Ainsworth, "The Assessment of Defects in Structures of Strain Hardening Materials", Engineering Fracture Mechanics, Vol. 19, 1984, pp.633-640.

## 5.7 Exercise Problems

5.1  Describe in words the methods of determining J and comment on their relative advantages and disadvantages.

5.2  Show on your own that:

$$J = - \int_0^V (\frac{\partial P}{\partial a})_V \, dV = \int_0^P (\frac{\partial V}{\partial a})_P \, dP$$

144  Nonlinear Fracture Mechanics for Engineers

5.3 Show that for a center crack tension specimen the plastic part of J is given by:

$$J_p = \frac{PV_p}{2BW(1-a/W)} \left(\frac{m-1}{m+1}\right)$$

*Hint*: From dimensional analysis:

$$V_p = bf\left(\frac{P}{Bb}\right)$$

and for power-law hardening material, it can be assumed that $V_p = CP^m$.

5.4 Show that the plastic part of J for a center crack tension specimen is given by the following relationship:

$$J_p = \frac{A_p}{B(W-a)}$$

where $A_p$ is as shown in Figure 5.10b.

5.5 Derive on your own the Rice's formula for calculating J in a single edge notch specimen subjected to pure bending.

5.6 Schematically show why J estimates should be modified for growing cracks.

5.7 Describe the method for experimentally measuring J.

5.8 Suppose you are testing a double edge notch specimen in the laboratory. Derive an expression for estimating J which utilizes the measured load vs. load-line deflection diagram obtained during the test.

5.9 How would you modify the expression derived in Problem 5.8 for growing cracks? Explain your derivation.

5.10 Derive equation (5.41) in the chapter for compact specimens.

5.11 Calculate the value of J for an internally pressurized cylinder of diameter 50cm and a wall thickness of 2.5cm. Assume that the cylinder has an axial crack which is 0.5cm deep and it contains steam at a pressure of 4.5MPa. Assume the following parameters for the tensile properties of the pipe material:

$E$ = 175 x 10³ Mpa
$\sigma_0$ = 138 Mpa
$m$ = 5.0
$\varepsilon_0$ = 7.88 x 10⁻⁴
$\alpha$ = 1.27

5.12 Show that equations (5.34) and (5.50) provide consistent estimates of J for compact specimens. Discuss the differences in the results from the two approaches.

5.13 Derive an expression extending the J-solution for a cylinder containing a very small internal circumferential crack under uniform remote tension such that $a/W \to 0$, where a is the crack depth and W the wall thickness of the cylinder.

5.14 Do the same as for Problem 5.13 except for an internally pressurized cylinder containing an axial crack.

5.15 Rice's formula for estimating J for a deeply cracked single edge specimen subject to bending, Equation (5.15), is strictly valid only for estimating the plastic part of J. For estimating the total J, one can write an expression of the type shown in Equation (5.32). Further, $\mathcal{J}_e$ may be written in the form:

$$\mathcal{J}_e = \frac{\eta_e}{b} \int_o^{V_e} P\, dV$$

Show that $\eta_e \simeq 2$ for a/W ≥ .45 for this specimen and the total J can therefore be estimated by:

$$J = \frac{2A}{Bb}$$

where A = total area under the load-displacement diagram.

5.16 Derive an expression for estimating $h_1$ (a/W, 1) for estimating $J_p$ via the reference stress approach in Equation (5.50a) for a CT specimen utilizing the K-calibration function given in Example Problem 5.4. Compare your results with the values reported in Table 5.2.

CHAPTER 6

# CRACK GROWTH RESISTANCE CURVES

Fracture in brittle materials occurs by simple separation of the specimen (or component) into two or more pieces. At the microscopic level, a type of brittle fracture is known as cleavage fracture which is characterized by the crack propagation along preferred crystallographic planes. For example, in ferritic steels which have a body-centered cubic structure, cleavage fracture occurs by crack propagation along the (100) planes. A second type of brittle fracture occurs by microvoid coalescence and is common among high strength aluminum alloys which have a face-centered cubic structure. The common characteristics among all brittle fractures are low fracture toughness and a sudden unstable fracture which can be catastrophic in structural components. A single fracture toughness value is sufficient for characterizing both types of brittle fractures.

Ductile fractures, on the other hand, are accompanied by significant amounts of crack tip plasticity. Also, instability is preceded by stable crack growth, which is defined as the condition when the fracture can be easily stabilized by either decreasing or holding constant the value of applied J. Instability during ductile fracture occurs by the onset of cleavage fracture after some stable crack growth, or by plastic instability (or collapse), or by rapid rate of increase in the applied value of the crack tip parameter (K or J). An example of the latter under dominantly linear elastic conditions is when dK/da exceeds $dK_R/da$ as discussed previously in Chapter 3. Due to these complex circumstances under which fracture occurs, several approaches have evolved for characterizing the various aspects of ductile and brittle fractures.

In this chapter, we will be focusing on characterizing brittle and ductile fractures under elastic-plastic and fully-plastic loading conditions. Brittle fracture under linear-elastic conditions is completely specified by a single fracture toughness value given by the critical stress intensity parameter, K. When the thickness of the specimen is sufficient to satisfy plane strain conditions, the critical value of K is known as $K_{IC}$, as described earlier in Chapter 3. Under a more general state of stress, the critical value of K is called $K_C$ and an additional restriction is placed on its use, in that its use is limited to structures which have the same thickness as the specimen used to determine the fracture toughness. For ductile fractures under linear elastic conditions, the crack growth resistance curves are represented as $K_R$ as a function of crack extension, $\Delta a$. This approach was described in Chapter 3 and will, therefore, not be repeated here. The $J_e$, $J_{Ic}$, and $J_R$-curves discussed in the next section complement the linear-elastic parameters, $K_{IC}$, $K_C$, and $K_R$-curves described in the context of linear-elastic fracture. In addition to these parameters, we will also discuss the crack tip opening displacement (CTOD) concept.

## 6.1 Fracture Parameters Under Elastic-Plastic Loading

Based on the discussions in Chapter 4, the J-integral can be used to characterize stable fracture and the onset of instability in some cases, under elastic-plastic conditions. Figure. 6.1 shows a typical relationship between the applied value of J and the amount of stable crack extension. This relationship implies ductile fracture and is known as the J-resistance or the $J_R$-curve. Such curves are usually developed for plane strain conditions by selecting specimens of sufficient thickness and also by side-grooving the specimens along the crack plane to induce plane strain conditions uniformly across the thickness of the specimen. The full $J_R$-curve is a material property and is indicative of the materials toughness. A single value, $J_{Ic}$, is often used for convenience to represent the toughness of the material. This value is the critical value of J at the onset of stable crack growth (or ductile tearing) and is obtained from the $J_R$-curve following a procedure described in later sections of this chapter.

Sometimes instability can develop under elastic-plastic conditions without significant stable crack extension. This condition is similar to the development of instability at $K_{IC}$ (or $K_C$) under linear-elastic

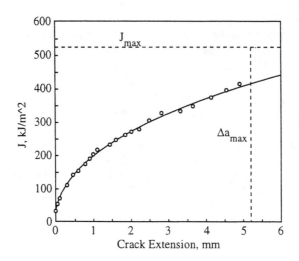

***Figure 6.1*** *Typical J-resistance ($J_R$ vs. $\Delta a$) curve.*

conditions, except the presence of elastic-plastic conditions necessitate the use of the J-integral approach. This critical value of J is referred to as $J_c$ and can be a material property, independent of specimen geometry. However, $J_c$ can vary with thickness just as $K_C$ does in the linear-elastic regime.

The point on the $J_R$-curve, $J_u$, at which instability occurs following significant amounts of stable crack extension is not a material property and can vary with specimen geometry size and the loading configuration. Nevertheless, $J_u$ can be useful in defining the limits of ductile fracture behavior and its value can be predicted using the instability theory which is described in the next chapter.

Crack tip opening displacement (CTOD), $\delta_t$, is an alternative to J-integral for representing fracture toughness and crack growth resistance of materials under elastic-plastic loading. CTOD is defined as the displacement normal to the crack plane due to elastic and plastic deformation at variously defined locations near the original (prior to an application of load) crack tip. If the definition of CTOD is chosen as in Chapter 4 (Section 4.3.2), a direct relationship between CTOD and J-integral exists. However, due to difficulties in measuring CTOD, according to the above definition, only operational measures of CTOD can be used. Just as with J, several CTOD parameters are defined to describe the various stages and events during ductile and brittle fracture. It is, therefore, useful to define these critical CTOD values before proceeding to describe the methods of their determination.

$\delta_{Ic}$ is a value of CTOD at the onset of stable crack extension, thus it corresponds to $J_{Ic}$ in the J-integral approach. $\delta_c$ is the CTOD value corresponding to unstable crack extension prior to significant crack extension. This value corresponds to $J_c$ in the J-integral approach. $\delta_u$ is the CTOD at instability subsequent to substantial stable crack growth and corresponds with the $J_u$. It is specimen size, geometry, and loading configuration dependent. $\delta_m$ is the CTOD value which corresponds to the first attainment of a maximum load plateau for fully plastic behavior. This value is also specimen size and geometry dependent.

In the subsequent section, a general experimental procedure is described which can be used to obtain fracture toughness and/or crack growth resistance by all applicable methods. Even though several fracture toughness and stable crack growth measurement standards are available [6.1 - 6.8], efforts are underway

to unify the procedures into a single test method [6.9]. Subsequently, the test record can be analyzed to obtain the desired fracture and crack growth parameters.

## 6.2 Experimental Methods for Determining Stable Crack Growth and Fracture

### *6.2.1 Overall Test Method*

The objective of the procedure is to apply load to a fatigue precracked, standard specimen of one of several geometries to obtain one or both of the following responses: (a) unstable crack extension, including significant "pop-in", resulting in instability in the specimen; (b) stable crack extension or ductile tearing. As mentioned earlier, fracture instability without significant stable crack extension results in a single fracture toughness value determined at the point of instability. Stable tearing results in a crack growth resistance curve, also called the R-curve (or $J_R$-curve) from which the fracture parameters can be determined. The load and load-point displacements are continuously measured during the test. For developing the R-curve, it is also necessary to obtain periodic measurements of crack length. The load, load-point displacement, and crack size data along with specimen geometry and size information are used to obtain the R-curve. Prior to discussing the details of data analysis, some additional experimental details about the method are provided.

### *6.2.2 Test Specimen Geometries and Preparation*

Three specimens geometries shown in Figures 6.2 to 6.4 are used frequently for fracture toughness and stable crack growth resistance testing. These geometries are the SENB, CT, and the Disk Shaped Compact Specimen, DCT. All specimen dimensions are proportioned to the width, W. The crack plane orientation and loading directions are chosen based on the application of the data. W/B ratios, (where B = thickness) other than 2 may be used. For example, for SENB specimens $1 \leq W/B \leq 4$ and for CT and DCT specimens $2 \leq W/B \leq 4$ may be utilized [6.9]. All specimens must be fatigue precracked to generate reproducible results. The maximum stress intensity parameter, $K_{max}$, during precracking should not exceed $0.0002E$ MPa$\sqrt{m}$ ($0.001 E$ Ksi$\sqrt{in}$) or 70% of the final load achieved during testing. If higher loads are needed to quickly initiate the crack, it must be ensured that the last 50% of the fatigue crack extension is carried out meeting the above requirements. The length of the fatigue precrack from the machined notch must not be less than 5% of the total crack size, $a_0$, and also not less than 1.3mm (0.05in). Side grooves on the order of 0.1B deep on both surfaces of the specimen are often used to ensure a straight crack front during testing. It is highly recommended to machine the side-grooves after precracking. The thickness of the specimen between the notches of the side groove is designated $B_N$.

### *6.2.3 Loading Apparatus and Displacement Gauges*

The detailed loading fixtures and clevis designs are given in various ASTM standards for specimens of different geometries. Figure 6.5 shows a schematic of the fixture used for 3-point bend load and Figure 6.6 shows the clevis for the CT specimens.

Figure 6.7 shows the schematic of a displacement gauge used for monitoring the load-point displacement. Several alternative designs are available for displacement gauges. Some utilize strain gauges configured in a full Wheatstone Bridge as transducers while others are based on linear voltage differential transformers (LVDT) as the transducers. The displacement measuring system must be capable of reliably detecting changes in displacements on the order of $2.5 \times 10^{-3}$ mm ($10^{-4}$ in) or better to be useful in these tests.

## 6.2.4 Crack Length Measurement

The initial and final crack lengths can be measured directly on the fracture surface after completing the test. Several techniques are used to highlight the final crack length. The initial crack length is the end of the fatigue crack and is easily discerned as can be seen in Figure 6.8. To facilitate the measurement of

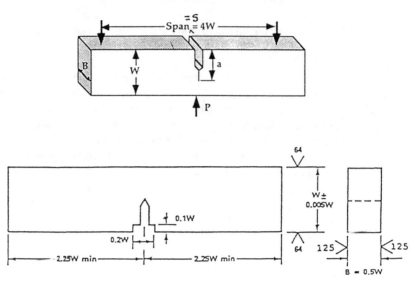

*Figure 6.2 Single edge notch bend (SENB) specimen.*

the final crack size, after stable crack growth, the test is often interrupted prior to fracture. It is subsequently fractured under fatigue loading so that the boundary between final crack size and the fatigue crack is clearly marked on the fracture surface. Thus, the final crack size can be accurately measured. In yet another technique, the specimen is briefly heated by a flame to oxidize the precrack and the stable crack region and subsequently fractured. An example of the results from this technique are shown in Figure 6.8. In this technique the fracture surface is destroyed and is not available for fracto- graphic studies. Steel specimens can often be cooled in liquid nitrogen and then fractured quickly to generate a cleavage fracture which is quite different in appearance from the stable crack growth region. Thus, the amount of stable crack growth can be easily measured.

Two techniques are used to measure crack size at intermediate points between the initial and final crack lengths. The first is the heat tinting technique already described earlier. However, to generate the R-curve, this technique can be used only with multiple specimens with each specimen loaded to different amounts of stable crack extension. Thus, the J and crack increment, $\Delta a$, can be determined for each specimen and plotted to obtain the R-curve. The other technique which permits measurement of the crack size at several points during each test is called the unloading compliance technique [6.10]. This technique is briefly described. For details about implementing the unloading compliance technique for the various specimen geometries, one should refer to the appropriate ASTM Standards [6.3, 6.9]. Briefly, in this method, the specimen is periodically, partially unloaded and the slope of the load-displacement curve during unloading is calculated. Since the unloading is elastic, the instantaneous value of the compliance, $C_i$, can be obtained and the crack length at that instant can be ascertained. By strategically selecting the unloading points, the entire R-curve can be obtained from a single specimen. The method of estimating crack length using the unloading compliance is illustrated using the example of a CT specimen. Figure 6.9 shows a load-displacement diagram with several unloadings. The various unloadings are labeled 1, 2, 3, 4, ---i, etc.

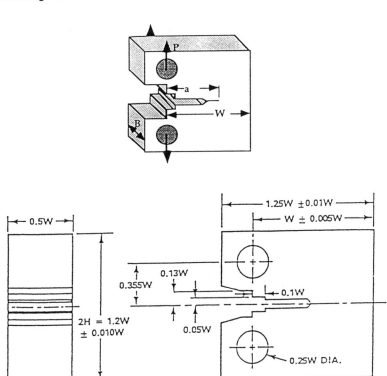

Figure 6.3 *Compact type, CT, specimen geometry.*

The compliances corresponding to those points are $C_1$, $C_2$, $C_3$, --- $C_i$, etc. The relationship between compliance and crack size for CT specimens can be given as follows [6.11]:

$$C_i = \frac{1}{E'B_e} \left(\frac{W+a}{W-a_i}\right)^2 [2.163 + 12.21\,(a_i/W) - 20.065\,(a_i/W)^2$$

$$- 0.9925\,(a_i/W)^3 + 20.609\,(a_i/W)^4 - 9.9314\,(a_i/W)^5] \tag{6.1}$$

Where:

$$E' = \frac{E}{1-v^2} \tag{6.2}$$

$$B_e = B - \left(\frac{B-B_N}{B}\right)^2 \tag{6.3}$$

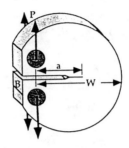

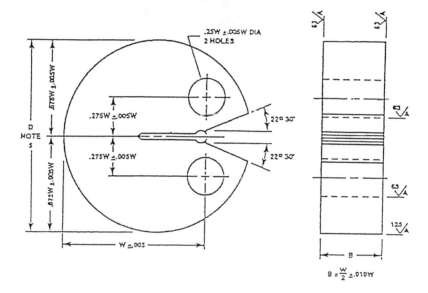

*Figure 6.4* Disk-shaped compact specimen.

The inverse relationship between crack size and compliance is as follows [6.11]:

$$a_i /W = [1.000196 - 4.06319\mu + 11.242\mu^2 - 106.043\mu^3 + 464.335\mu^4 - 650.677\mu^5 ] \qquad (6.4)$$

where:

$$\mu = \frac{1}{[B_e E' C_i]^{1/2} + 1} \qquad (6.5)$$

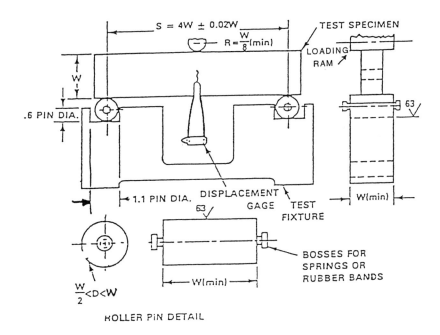

*Figure 6.5 Schematic of the fixture used for 3-point bend load.*

For large plastic displacements, it may be necessary to use rotation correction on the measured compliances. The details of the method are given in the standard [6.9]. Due to several experimental difficulties, frequently the measured compliance at the initial crack size differs slightly from the calculated value from equation (6.1). Appropriate adjustments to the value of the elastic modulus are made to force the initially predicted crack size to agree with measured value from the crack surface. Using the adjusted elastic modulus, the crack size is predicted for the remainder of the unloadings from equation (6.4). The predicted final crack size can be compared with the measured final crack size from the fracture surface of the tested specimen to ensure the accuracy of crack length measurements at the intermediate points.

Attempts have been made to use an electric potential drop method for measurement of crack size [6.12]. In this technique, the electrical resistance change in the specimen is measured as the crack length increases. Thus, by continuously measuring the change in electrical resistance during the fracture test, the crack size at any instant can be obtained. This technique has not been adopted in any of the current fracture test standards but has been used by researchers. The problems in the technique stem from not being able to separate change in electrical resistance due to crack tip plasticity from the change in resistance due to crack extension. Therefore, this technique is more suitable for measuring subcritical cracking such as during fatigue crack growth and creep crack growth in which the amounts of crack extension during the test are larger than the typical fracture tests. Also, by comparison to fracture toughness tests, the scale of crack tip plasticity is much smaller during subcritical cracking which makes this technique more appropriate for those applications.

### 6.2.5 Final Loading of the Specimen and Post-test Measurements

Precracked specimens are loaded under displacement gauge or machine crosshead or actuator displacement control. The rate of loading should be such that the maximum load in the test is achieved in 0.1 to 10 minutes. The time to perform an unload/reload sequence should be kept to a minimum and

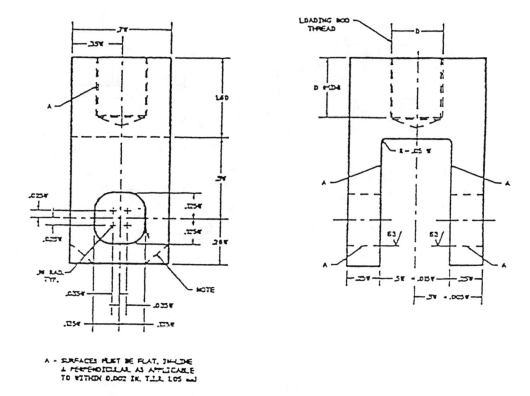

*Figure 6.6* Loading clevises for a compact specimen.

should not exceed 10 minutes. Loading of the specimen should be continued until either a sudden fracture occurs or sufficient stable crack growth data have been collected. The unloading during an unloading compliance measurement should not exceed approximately 50% of the current load.

After completing the test, the specimen is fractured (if it did not fracture during the test) in one of three ways: directly in liquid $N_2$, by fatigue loading, or at ambient temperature after heat tinting. Final crack size and crack size at the end of precracking are measured at nine points along the thickness of the specimen as shown in Figure 6.10. The average crack sizes are determined for use in the subsequent data analysis.

### 6.2.6 Data Analysis and Qualification

There are three types of responses during fracture testing, Figure 6.11. The first consists of unstable fracture without significant prior stable crack growth. The second is unstable fracture following stable crack growth and the third type consists of stable crack growth with no instability. From the first two types of fracture responses, the fracture toughness at the instability point can be obtained. The resistance curves can be obtained from the stable crack growth portions of the second and third type of tests. We will first discuss the procedure for obtaining the fracture toughness at instability and then consider the procedure for obtaining the resistance curves. As an example, we will consider analysis of data from the compact type specimens. Similar procedures can be followed for other geometries. These are described in detail in the ASTM standard [6.9].

**Determination of Fracture Instability Toughness** We will consider determining fracture instability toughness using the J-integral and the CTOD concepts. If the load-displacement response is linear-elastic, fracture toughness can be represented by $K_{Ic}$, as discussed earlier in Chapter 3.

J is calculated at the instability point using formulae derived in Chapter 5. This provisional value of toughness is termed $J_Q$. For CT specimens, if the amount of stable crack extension preceding instability is negligible, J is given by:

$$J_Q = (1-v^2)\frac{K^2}{E} + J_p \tag{6.6}$$

where $J_p$ is given by equation (5.34). In order to calculate $J_p$, the plastic portion of the area under the load vs. load-line displacement curve is needed. The area $A_p$ is shown in Figure 6.13. Crack extension, $\Delta a_p$, is considered to be negligible if it meets the following condition:

$$\Delta a_p \leq 0.2mm + J_Q/M\sigma_Y \tag{6.7}$$

where M = 2 and $\sigma_Y = \frac{1}{2}(\sigma_{YS} + \sigma_{un})$, the average of the 0.2% yield strength and the ultimate tensile strength. If $J_Q$ meets the crack extension condition of equation (6.7) and also meets the conditions:

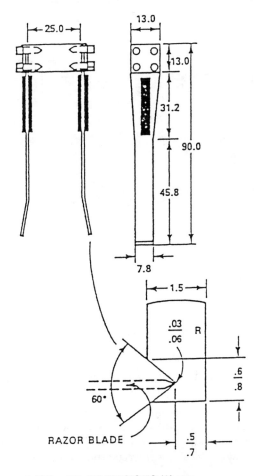

NOTE – ALL DIMENSIONS IN mm

**Figure 6.7** *Schematic drawing of a clip-on displacement gauge (Ref. 6.3).*

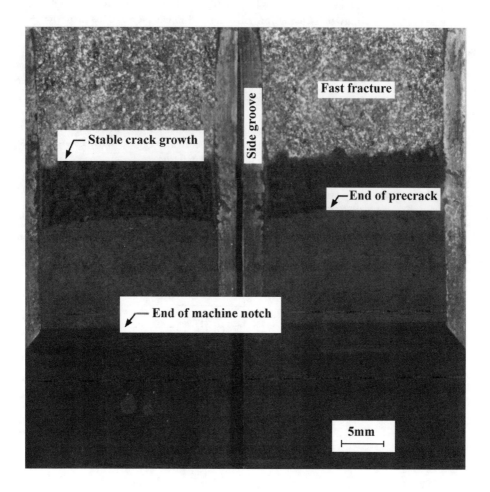

*Figure 6.8* Fractured specimen showing the end of the pre-crack and the rest of the fracture surface.

$$B, W-a_0 \geq 200 \, J_Q/\sigma_Y \qquad (6.8)$$

then $J_Q = J_c$. As mentioned before, $J_c$ is independent of the in-plane dimensions of the specimen and also the specimen geometry but may depend on thickness of the specimen.

If equation (6.7) is not satisfied, the $J_Q$ value at instability is labeled as $J_u$. The $J_u$ value is not independent of specimen size or geometry; therefore, no size criterion is applied in qualifying $J_Q$ as $J_u$. It is merely an instability toughness value which can be used to indicate that sudden fracture can occur in the material. Beyond conveying the above characteristics of the material, $J_u$ should not be used for

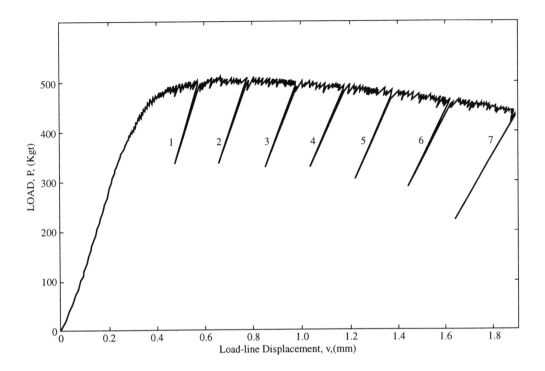

*Figure 6.9* Load-displacement record showing the unloadings during a fracture test.

any other purpose except for qualitative comparison of fracture toughness behavior between different materials, if similar geometry and specimen's size are used for fracture toughness testing of each material. The CTOD during any point in a fracture test is calculated using the following equation [6.7, 6.9]:

$$\delta_i = \frac{K_i^2 (1-\upsilon^2)}{2\sigma_{YS}E} + \frac{[r_{pi} (W-a_i) + \Delta a_i] V_{pi}}{[r_{pi} (W-a_i) + a_i + z]} \qquad (6.9)$$

where

$a_0$ = original crack length
$a_i$ = instantaneous crack length
$\Delta a_i$ = $a_i - a_0$
$V_{pi}$ = plastic displacement at the instant 'i'
$\sigma_{YS}$ = 0.2% yield strength of the material
z = distance of knife edge measurement point from the load-line (if displacement is measured along
  the load-line, z = 0)
$r_{pi}$ = plastic rotation factor given by:

158  *Nonlinear Fracture Mechanics for Engineers*

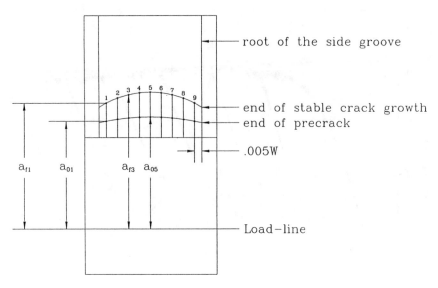

precrack length, $a_0 = .125[.5(a_{01} + a_{09}) + a_{02} + \ldots a_{08}]$
final crack length, $a_f = .125[.5(a_{f1} + a_{f9}) + a_{f2} + \ldots a_{f8}]$

*Figure 6.10* Schematic illustration of average crack size measurement.

$$r_{pi} = 0.4(1 + \alpha_i) \qquad (6.10)$$

$$\alpha_i = 2\left[\left(\frac{a_i}{W-a_i}\right)^2 + \left(\frac{a_i}{W-a_i}\right) + \frac{1}{2}\right]^{\frac{1}{2}} - 2\left[\frac{a_i}{W-a_i} + \frac{1}{2}\right] \qquad (6.11)$$

Figure 6.13 shows the geometric relationships between $V_{pi}$ and $\delta_i$ for an SENB specimen in which $r_{pi} = 0.44$ and $z =$ distance between the knife edge and the specimen front edge. The derivation of equation [6.9] for this case is straightforward. Using equations (6.9) to (6.11), we can easily determine the $\delta_c$ and $\delta_i$ for a test in which the load-displacement diagram terminates with an instability. $\delta_c$ represents the fracture toughness value which is independent of the in-plane dimensions and geometry of the specimen (but could be dependent on thickness) and is valid only if the following two conditions are met:

$$B, W-a_0 \geq 300\delta_Q \qquad (6.12a)$$

$$\Delta a_p < 0.2 + 0.7\delta_Q \qquad (6.12b)$$

where $\delta_Q$ is the CTOD value corresponding to the instability point. If either of these conditions in (6.12 a and b) are not met, $\delta_Q = \delta_u$ and is not a valid measurement of $\delta_c$.

Crack Growth Resistance Curves 159

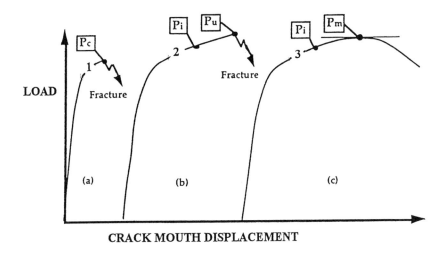

*Figure 6.11* Three types of load-displacement diagrams encountered during fracture toughness testing.

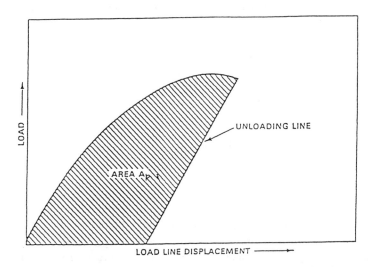

*Figure 6.12* Schematic illustrations of the method to determine the plastic area, $A_P$, in a load-displacement diagram.

**Determination of the Crack Growth Resistance Curves** Crack growth resistance is estimated using the J-integral approach or the CTOD approach. The resultant curves from the two approaches are known as the J-Resistance (or $J_R$) or $\delta$-Resistance ($\delta_R$) curves. As mentioned before, these curves are also referred to as resistance curves or R-curves for stable crack growth. We will begin with the determination of the $J_R$ curve and then we will discuss the determination of the $\delta_R$ curve.

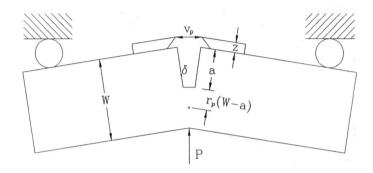

*Figure 6.13* Schematic relationship between CTOD and the measured displacement.

As mentioned earlier in Figure 6.1, the $J_R$ curve consists of a plot of the J vs. crack extension in the region of J-controlled crack growth and can be used to derive the value of $J_{Ic}$ which corresponds to the onset of stable crack growth. To ensure J-controlled crack growth, the measurement capacity of the specimen is governed by the following equations. The maximum J-integral capacity of the specimen is given by the smaller of:

$$J_{max} = (W-a)\, \sigma_Y/20 \tag{6.13a}$$

$$J_{max} = B\, \sigma_Y/20 \tag{6.13b}$$

Similarly, the maximum crack extension capacity of the specimen is given by:

$$\Delta a_{max} = 0.25\,(W-a_0) \tag{6.14}$$

The J-integral value as a function of crack extension, $\Delta a$, is plotted as shown in Figure 6.14. The $J_{max}$ and $\Delta a_{max}$ lines are also plotted and if any data lie outside of the range bounded by $J_{max}$ and $\Delta a_{max}$ lines, it must be excluded from further consideration. The value of J-integral at various points is calculated using equation (5.40) for CT specimens and $\Delta a$ is estimated from the compliance method as described earlier in this chapter. Having obtained the $J_R$ curve, we will now turn to the method for estimating $J_{Ic}$.

To estimate $J_{Ic}$, an equation of the following type is fitted through a select group of J vs. $\Delta a$ points on the $J_R$ curve:

$$\ln J = \ln C_1 + C_2 \ln (\Delta a) \tag{6.15}$$

where $\Delta a$ is in mm and $C_1$, $C_2$ are regression constants obtained from linear regression of the ln J vs.

ln$\Delta$a data. The data that are included in the regression are those that lie within a certain interval determined

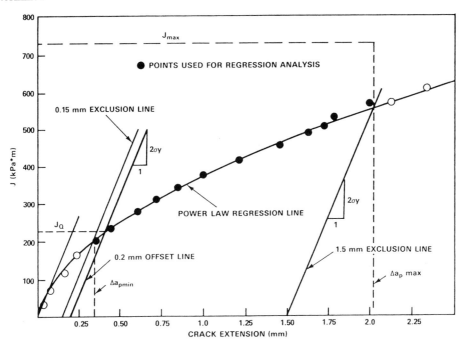

**Figure 6.14** *The J vs. $\Delta a$ relationship with the $J_{max}$ and $\Delta a_{max}$ lines plotted along with the other data exclusion lines.*

in the following way. The first step is to plot what is termed the construction line defined by the following equation:

$$J = 2\sigma_Y \Delta a \tag{6.16}$$

Straight lines parallel to the construction line are plotted at $\Delta a$ values of 0.15mm, 0.2mm, 0.5mm, and 1.5mm as shown in Figure 6.14. Only data between the 0.15mm and 1.5mm exclusion lines that also fall below the $J_{limit} = (W - a_0)\,\sigma_Y/15$ are used for regression analysis. In order to ensure that the available data are uniformly distributed in the interval, we should be sure that at least one point lies between the 0.15 and 0.5mm offset lines and at least one other point lies between the 0.5mm and 1.5mm lines. The $J_{Ic}$ is given by the J value at the intersection of the fitted $J_R$ curve and the 0.2mm offset line. A very detailed procedure for determining $J_{Ic}$ is given in the relevant ASTM standards [6.2, 6.9]. If $J_Q$ is the candidate $J_{Ic}$ value determined in the above fashion, it qualifies for being called $J_{Ic}$ if:

$$B, W - a_0 \geq 25\frac{J_Q}{\sigma_Y} \tag{6.17}$$

and the slope of the $J_R$ curve obtained by regression at $\Delta a_Q$ is less than $\sigma_Y$. The latter condition is imposed to account for problems encountered with materials with very high strain hardening characteristics, such as stainless steel, in which the $J_R$ curve is very steep and the power-law representation of equation (6.15) is not a valid approach.

The CTOD resistance curve, $\delta_R$-curve, can be determined by using equation (6.9) to estimate $\delta$ for various amounts of crack extension as shown in Figure 6.15. The $\delta_{max}$ is given by:

$$\delta_{max} = \frac{W - a_0}{20} \tag{6.18}$$

and $\Delta a_{max}$ is the same as given by equation (6.14). To estimate the value of $\delta_{Ic}$, the procedure followed is very similar to the one used for $J_{Ic}$ determination. The selected $\delta$ vs. $\Delta a$ data are used to derive the following regression line:

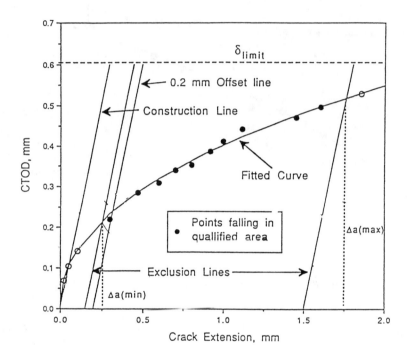

*Figure 6.15* CTOD ($\delta$) vs. $\Delta a$ curve similar to the one shown in Figure 6.14 for the J-integral approach.

$$\ln\delta = \ln C_1 + C_2 \ln(\Delta a) \tag{6.19}$$

The data are selected between the exclusion lines drawn at offsets of 0.15mm and 1.5mm crack extensions and the $\delta_{limit}$ line given by $\delta_{limit} = (W - a_0)/15$. The $\delta_{Ic}$ is given by the value of $\delta$ at the intersection between the $\delta_R$ curve derived from regression and the 0.2mm offset line similar to the way in which $J_{Ic}$ was determined.

### 6.3 Special Considerations for Weldments

Weld joints typically contain macroscopic heterogeniety in mechanical properties such as yield strength and strain hardening characteristics in the weld metal/base metal interface region. The heterogeniety is caused by differences in chemistry and/or microstructures between the base metal and the weld metal as well as by the thermal cycles during the welding process. Figure 6.16 shows the various microstructures produced during submerged arc welding (SMAW) process used to weld Cr-Mo steels. As one goes across the fusion line from the weld metal side to the microstructure representing the original base metal, several different microstructures in the heat-affected zone (HAZ) of the base metal are encountered, as labeled schematically in Figure 6.16.

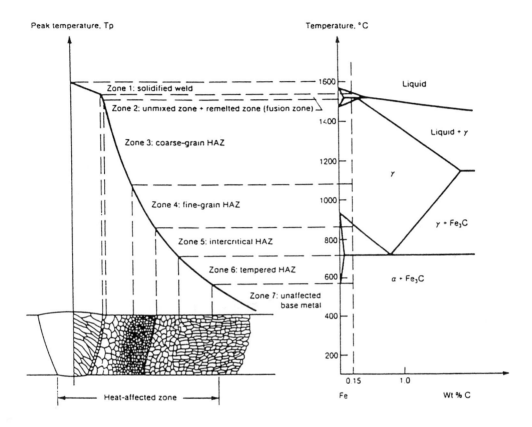

*Figure 6.16* *Different regions of the weldment characterized by microstructural differences. (From R. Viswanathan, "Damage Mechanisms and Life Assessment of High-Temperature Components," ASM International, 1989. Reproduced with permission.)*

These differences in microstructures destroy the symmetry of the stress and strain fields for bodies in which the cracks are located in the interface region, even if the loading and specimen geometry are symmetric with respect to the crack such as seen in Figure 6.17. Notably, the development of plasticity can be quite asymmetric for such cases and the presence of combined mode I and mode II is common even when the loading is normal to the crack plane.

Since the path-independence of J-integral is dependent on the assumption of homogeneous material properties, the relationship between far-field measurements of J from load-deflection diagrams and the crack tip stress fields must be reconsidered. Similarly, the relationship between load-line deflection and the crack tip opening displacement such as given by equation (6.9) must also be reconsidered. Therefore, there is a need to examine the applicability of the J-integral and CTOD approaches for weldments.

The analytically simplest case is that of a crack located at a perfect bi-material interface with remote tensile loading applied normal to the plane of the crack. For weldments of similar materials one can assume that the elastic properties of the two materials are identical but the plastic properties are not. Numerical solutions to this problem have been derived by Shih, Asaro, and O'Dowd [6.13 - 6.15] for elastic, small-scale yielding and large scale yielding, respectively.

Although weldment problems do not typically involve significant mismatch in elastic properties (elastic modulus, E, and the Poisson's ration, ν), it is useful to briefly review some of the basic differences between crack tip fields for a homogeneous material and one with an elastic mismatch across the interface, with the crack located along the interface.

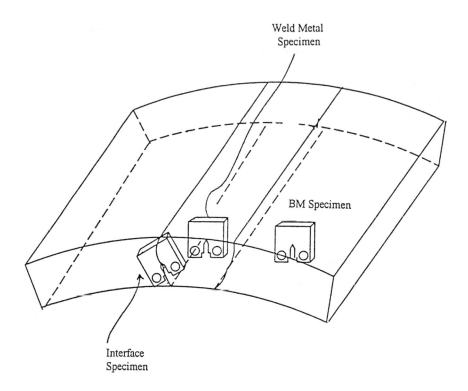

*Figure 6.17* Fracture specimens oriented to sample fracture toughness of the various regions of a weldment.

This problem has been addressed since the 1950s when Williams [6.16] applied the Airy stress function approach to this problem. Subsequently, complete solutions were developed by Cherapanov [6.17], England [6.18], Erdogan [6.19], and Rice and Sih [6.20]. In the engineering treatment of this mathematically complex subject, it is sufficient to present the significant conclusions of these analyses. These are discussed below.

The stress fields of interfacial cracks are inherently mixed mode regardless of the nature of far-field loading. In other words, even if the only far-field load is normal to the crack plane, components of mode I and mode II are present in the crack tip vicinity. Separate tensile and shear modes cannot be unambiguously defined since both tensile and shear stresses exist at the interface. One exception to this is the double edge notch tension (DENT) geometry where the remote loading is, in fact, pure mode I.

The stress and displacement singularities have an unbounded oscillatory nature in a region very near to the crack tip. The region of dominance of this oscillatory stress field is so small that it is not of any practical consequence for fracture [6.21]. In fact, it has been shown that if the slightest plastic deformation occurs near the crack tip, the oscillatory field disappears. From the small scale yielding analysis [6.14], it is clear that the stress and strain (or displacement) fields on the length scales comparable to the size of the dominant plastic zone are governed by the lower yield strength or the less strain hardening of the two materials (higher value of the Ramberg-Osgood exponent m). The yield strength or the plastic deformation properties of the stronger materials have limited influence on the near tip fields. The singular fields well within the plastic zone are governed by the strain-hardening behavior of the more plastically compliant (lower strain hardening) material. The asymptotic behavior of the lower strain hardening material is similar to a material with identical plastic properties bonded to a rigid substrate.

For opening dominated load states, the fields that develop near bimaterial interfaces are more intense than those in homogeneous material for the same value of J or the remote load. From a dimensional analysis, Shih et al. [6.15] have proposed a plausible form for the asymptotic crack tip stress field deep into the plastic zone (i.e., $r \to 0$) as:

$$\sigma_{ij} = \sigma_0 \left(\frac{J}{\alpha \sigma_0 \varepsilon_0 r}\right)^{\frac{1}{1+m}} h_{ij}\left(\theta, \xi, \frac{r}{J/\sigma_0}, m\right) \qquad (6.20)$$

where all constants pertain to the weaker materials and $h_{ij}$ is a dimensionless function with a weak dependence on $r/(J/\sigma_0)$, $\theta$ is the angle, and $\xi$ is a function of the plastic phase angle which depends, among other things, on the mode mixity. The value of $\xi$ at a given distance r is given by the following equation taken from reference [6.16]:

$$\tan \xi = \left(\frac{\tau_{r\theta}(\theta = 0)}{\sigma_{\theta\theta}(\theta = 0)}\right)_r \qquad (6.21)$$

Nakagaki et al. [6.17] have analytically shown that for the special configuration considered above, J-integral is path independent even across the material interface (see Exercise Problem 6.12) provided proportional loading prevails throughout the body. Also, some empirical formulae for determining J from applied load-displacement records are also valid under these assumptions. None of the attributes discussed above are valid if the crack does not lie along the sharp interface. The semi-empirical formulae for determining J are valid only if the remote loading is normal to the crack plane.

The interface toughness represents the materials resistance to fracture and is an experimentally determined quantity. The toughness may not be a single material parameter, but may be a function of the amount of mode I and mode II components along the interface which depends on the differences in properties between the two materials. Since the fracture toughness can vary from location to location in the weldment, it is necessary that a judicious choice be made about notch placement to sample the toughness characteristics of the region where the weldment is most likely to fail during service. Some configurations of notch placement are shown in Figure 6.17.

## Example Problem 6.1

Figure 6.9 shows an actual load-displacement diagram of a 3-point bend specimen. The objective is to determine the $J_R$-curve and $J_{Ic}$. This specimen is the size of a normal Charpy specimen with a thickness = 10mm, W (width) = 10mm, and S (span) which is the distance between the two outer loading points = 50.8mm. The test was performed on a plane-sided specimen (i.e., no side-grooves). The load-displacement curve contains periodic unloadings as shown in the figure. The specimen is made from a plain carbon steel with the following properties. The 0.2% yield strength,

$\sigma_{YS} = 380$ MPa and the ultimate tensile strength, $\sigma_{UTS} = 620$ MPa. The precrack length and the final crack length measurements along nine, approximately equally spaced points along the specimen thickness yield the measurements listed in Table 6.1.

*Solution:*

The first step in obtaining the $J_R$-curve is to check the validity of a test based on crack front straightness requirement. The average precrack length, $a_0$, and the final crack length, $a_p$, are deter-

*Table 6.1 Crack Length Measurements From the Fracture Surface of the Specimen Used in Example 6.1*

| Measurement Point | Precrack Length (mm) | Final Crack Length (mm) | Crack Extension (mm) |
|---|---|---|---|
| .005W | 4.92 | 5.15 | 0.23 |
| 0.125W | 5.16 | 6.18 | 1.02 |
| 0.25W | 5.36 | 6.28 | 0.92 |
| 0.375W | 5.36 | 6.27 | 0.91 |
| 0.500W | 5.36 | 6.41 | 1.05 |
| 0.625W | 5.36 | 6.37 | 1.01 |
| 0.75W | 5.36 | 6.29 | 0.93 |
| 0.875W | 5.12 | 6.21 | 1.09 |
| 0.995W | 4.91 | 5.16 | 0.25 |

mined as follows [6.3]: average of the two near surface measurements are combined with the remaining seven measurements and an overall average is calculated. From this procedure, $a_0 = 5.25$ and $a_p = 6.15$mm. The maximum permitted variation according to the ASTM standard in the measured value of the precrack length along the nine points is $\pm 7\%$ from the average, $a_0$. Thus, all measurements must lie between 4.88 mm and 5.62 mm. All precrack length measurements along the specimen thickness satisfy this requirement. Next, the measured crack extension, $\Delta a_p$, at either of the two near surface points must not differ from the measurement at the center by more than $\pm 0.02W$, or in this case by $\pm .2$mm. Thus, these measurements must fall in the range of $1.05 \pm 0.21$mm which, as we can see, is not met in this test. A conclusion from this observation is that the subsequent specimens must be side-grooved. Side-grooves with depth equal to 10% of the specimen thickness on each side of the specimen may lead to the desired result. Despite failing to meet this requirement, we will continue to process the data for determining the $J_R$-curve and the $J_{Ic}$ point to illustrate the method. It should also be mentioned that crack front straightness requirement is necessary because the J-integral, as calculated from the methods discussed in Chapter 5, is for 2-dimensional configurations in which the crack length is assumed to be identical along the specimen thickness. Therefore, this requirement must not be taken lightly.

The next step is to determine the crack size at all the intermediate points for which the unloading compliance values were measured. Including the initial and final measurements, compliance values can be obtained from the 8 points on the load-deflection diagram as numbered on Figure 6.9. These compliance values are listed in the table below (Table 6.2).

The compliance expression for a 3-point bend specimen based on the load-point deflection is given by [6.18]:

$$C = \frac{2.1S^2}{W^2 EB} \left(\frac{a/W}{1-a/W}\right)^2 \qquad (1)$$

where C = compliance and all other terms are as explained before. The compliance $C_0$ corresponds to the initial crack size $a_0$. Since we have measured $C_0$ and $a_0$, we can solve for E, the elastic modulus. This value of E is termed as the measured value, $E_m$. Thus:

**Table 6.2** *Compliance and Area Under the Curve Measurements at the Various Points on the Load-Displacement Diagram Shown in Figure 6.9*

| Measurement Print | Load (N) | Measured Compliance (mm/N) | Predicted Crack Size (mm) | (a/W) | $\Delta A^1_i$ (Joules) |
|---|---|---|---|---|---|
| 1 | 3188.2 | 3.41 x 10$^{-5}$ | 5.25 | .525 | 0 |
| 2 | 4905 | 3.597 x 10$^{-5}$ | 5.32 | .532 | .597 |
| 3 | 4959 | 3.94 x 10$^{-5}$ | 5.43 | .543 | .459 |
| 4 | 4905 | 3.995 x 10$^{-5}$ | 5.446 | .545 | .459 |
| 5 | 4742.8 | 4.812 x 10$^{-5}$ | 5.667 | .567 | .456 |
| 6 | 4526.5 | 4.995 x 10$^{-5}$ | 5.72 | .572 | .503 |
| 7 | 4093.9 | 5.722 x 10$^{-5}$ | 5.89 | .589 | .579 |
| 8 | 3553.2 | 6.703 x 10$^{-5}$ | 6.076 | .608 | .590 |

1. $\Delta A_i$ is the additional plastic area under the load-displacement diagram between the current and the previous point.

$$E_m = \frac{2.1S^2}{W^2 C_0 B} \left(\frac{a_0/W}{1-a_0/W}\right)^2 \qquad (1a)$$

Substituting for the various terms we get $E_m$ = 194 x 10$^3$ MPa. The published value of E for carbon steels is approximately 207 x 10$^3$ MPa. Since the measured value of E is within 10% of the published value, the test is valid. We will subsequently use $E_m$ to obtain crack sizes at the other unloading points from the following equation which was obtained by rearranging equation (1):

$$\left(\frac{a/W}{1-a/W}\right)^2 = \frac{CW^2 E_m B}{2.1S^2} = 35.8 \times 10^6 C$$

or

$$\frac{a/W}{(1-a/W)} = 5.983 \times 10^3 \sqrt{C}$$

or

$$a/W = \frac{\sqrt{C}}{1.671 \times 10^{-4} + \sqrt{C}} \qquad (2)$$

Note that the compliance value must be substituted in m/N instead of mm/N as given in Table 6.2 to be consistent in units. Substituting the measured values of compliance in equation (2), we obtain the crack sizes as listed in Table 6.2. As another check, we compare the predicted crack extension of 6.076 - 5.25 = 0.826mm with the measured crack extension on the fracture surface of $(a_p - a_0)$ = 0.9mm. These values compare quite favorably. Thus, we are ensured that the unloading compliance technique has yielded accurate measurements of crack size at various instances during the test.

We now proceed to calculate the J-integral values at the various points on the load-displacement diagram. The formula for estimating the instantaneous value of J, $J_i$, at various points is given by:

$$J_i = J_{ei} + J_{pi} \qquad (3)$$

$$J_{ei} = \frac{K_i^2 (1 - v^2)}{E}$$

where

$$K_i = \frac{P_i S}{B W^{1.5}} f(a_i/W) \qquad (4)$$

$$f(a_i/W) = \frac{3 \alpha^{.5}}{2(1 + 2\alpha)(1 - \alpha)^{1.5}} (1.99 - \alpha(1 - \alpha))(2.15 - 3.93\alpha + 2.7\alpha^2)$$

$$\alpha = a_i/W$$

$$J_{pi} = J_{p(i-1)} + \frac{2}{b_{i-1}} \frac{A_i - A_{i-1}}{B} [1 - \frac{a_i - a_{i-1}}{b_{i-1}}] \qquad (5)$$

In equation (5), $A_i$ is plastic area corresponding to point 'i' and $A_{i-1}$ is the same area corresponding to point i -1. Therefore, $A_i - A_{i-1} = \Delta A_i$. Substituting $\Delta a_i$ into equation (5) and calculating $K_i$ and then substituting $J_{pi}$ and $J_{ei}$ into equation (3), we get the quantities listed in Table 6.3.

The $J_R$ vs. $\Delta a$ is plotted in Figure 6.18. We now turn to qualification of data and the determination of $J_{Ic}$. The maximum J-integral capacity of the specimen is given in equations 6.13a and b. Since in this specimen B = W, the condition in equation (13a) is controlling. Thus:

$$J_{max} = ((.01 - .00525)(680 + 320)/2) \times 10^3 / (20) = 118.75 KJ/m^2$$

The $J_{max}$ line is drawn in Figure 6.18. Only the last two data points are disqualified based on this condition. The next qualification condition is given by equation (6.14) for the maximum crack extension capability. This condition results in:

$$\Delta a_{max} = 0.25(10 - 5.25) = 1.0625 mm$$

*Table 6.3 Calculation of J-Integral Values*

| Point | $a_i$ | $b_i$ | $\Delta a_i$ | $K_i$ | $J_{pl}$ | $J_i$ | $\Delta a$ |
|---|---|---|---|---|---|---|---|
| | (mm) | (mm) | (mm) | MPa$\sqrt{m}$ | KJ/m² | KJ/m² | (mm) |
| 1 | 5.25 | 4.75 | 0 | 37.9 | 0 | 6.314 | 0 |
| 2 | 5.32 | 4.68 | .07 | 59.1 | 25.12 | 40.47 | .07 |
| 3 | 5.43 | 4.57 | .11 | 61.1 | 44.7 | 61.11 | .18 |
| 4 | 5.446 | 4.554 | .016 | 60.65 | 64.8 | 80.97 | .196 |
| 5 | 5.667 | 4.533 | .221 | 61.6 | 83.94 | 100.62 | .417 |
| 6 | 5.72 | 4.28 | .053 | 59.5 | 107.16 | 122.7 | .47 |
| 7 | 5.89 | 4.11 | 0.17 | 56.3 | 134.2 | 148.1 | .64 |
| 8 | 6.076 | 3.924 | 0.186 | 51.5 | 162.8 | 174.45 | .826 |

**Notes:** $\Delta a_i = a_i - a_{i-1}$

KJ/m² = 1000 j/m²

$E = 207 \times 10^3$ MPa and $\nu = 0.3$

$\Delta a = \sum_{i=1}^{i} \Delta a_i$

All data obtained in the test are qualified on this basis. Thus, the $J_R$-curve is defined by all points that lie in the box between the $J_{max}$ and $\Delta a_{max}$ lines.

We next turn to the determination of $J_{Ic}$. Another exclusion line given by $J_{limit} = (W - a_0) \sigma_Y /15 = 158$ kJ/m² is drawn on Figure 6.18. All data fall below this line. Straight lines parallel to the construction line at $\Delta a$ values of 0.15mm, 0.2mm, and 0.5mm are also drawn. Only data that lie between 0.15mm and 1.5mm offset lines and also below the $J_{limit}$ line are used for $J_{Ic}$ calculation. These include all but the first four data points and the last data point. Since two data points lie between the 0.15mm and 0.5mm offset line and one more lies between 0.5mm and 1.5mm offset lines, the data are suitable for proceeding to calculate $J_Q$, the tentative value of $J_{Ic}$. Conducting a regression analysis to fit equation (6.15) provides the value of $C_1 = 205.06$, $C_2 = 0.75548$ where $\Delta a$ is in mm and J in kJ/m². The 0.2mm offset line intersects the fitted curve at J of 78.75 kJ/m² which is termed $J_Q$. $J_Q = J_{Ic}$, if the condition in equation (6.17) is met. The value of 25 $J_Q/\sigma_Y$ = .00393m = 3.93mm. Since B = 10mm and W - $a_0$ = 4.75mm, $J_{Ic} = J_Q$. Thus, $J_{Ic}$ = 78.75 kJ/m². The equivalent K value, $K_J$ = 133.8 MPa$\sqrt{m}$. This number can be used as a conservative estimate for $K_{Ic}$. However, to measure a valid $K_{Ic}$, a specimen with minimum W-$a_0$ and B of 0.3m is needed. Hence, W will be approximately 0.62m which is 62 time the size of the specimen tested in this example. This clearly illustrates the advantages of the elastic-plastic fracture toughness test methods. The $J_R$-curve consists of the plot shown in Figure 6.18 with points within the box defined by the 0.15mm offset line and the $J_{max}$ line considered to be valid according to the ASTM test method. From measurements provided in this test, the CTOD R-curve cannot

170  Nonlinear Fracture Mechanics for Engineers

be obtained because the deflection was measured only at the load-point. If the deflection had also been measured on the front-face, we could have also represented the toughness using the CTOD approach.

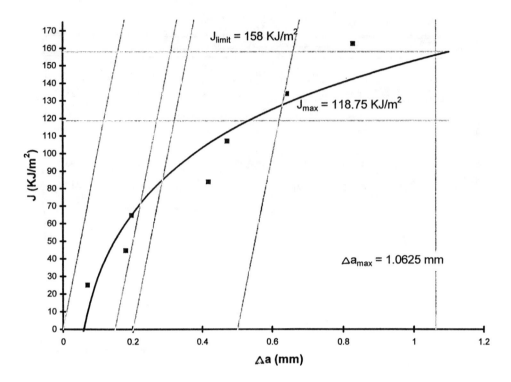

*Figure 6.18* $J_R$-curve from the test with the load-displacement record shown in Figure 6.9.

## 6.4 Summary

In this chapter, methods of determining the fracture toughness using the J-integral and CTOD approaches were discussed in detail. Several measures of fracture toughness were defined. The onset of ductile tearing is defined by $J_{Ic}$ and $\delta_{Ic}$ in the two approaches. The crack growth resistance is characterized by the $J_R$-curve and the $\delta_R$-curve, respectively. The details of the experimental method for determining the $J_{Ic}$, $\delta_{Ic}$, $J_R$, and $\delta_R$ curves based on testing deeply cracked bend and compact type specimens were described. Special considerations for testing weldments were also discussed. An example using a charpy size specimen in 3-point bend loading configuration for characterizing fracture toughness of a carbon steel illustrated the use of the ASTM standard method for determining the various fracture toughness parameters.

## 6.5 References

6.1  "Standard Test Method for Plane-Strain Fracture Toughness of Metallic Materials", ASTM Standard E-399-90, Annual Book of ASTM Standards, Vol. 03.01, 1992, pp. 506-536.

6.2  "Standard Test Method for $J_{Ic}$, a Measure of Fracture Toughness", ASTM Standard E-813-89, Annual Book of ASTM Standards, Vol. 03.01, 1992, pp. 732-746.

6.3 "Standard Test Method for Determining J-R Curves", ASTM Standard E-1152-87, Annual Book of ASTM Standards, Vol. 03.01, 1992, pp. 847-857.

6.4 "Standard Test Method for Crack-Tip Opening Displacement (CTOD) Testing", ASTM Standard E-1290-89, Annual Book of ASTM Standards, Vol. 03.01, pp. 946-961.

6.5 M.G. Dawes, "Elastic-Plastic Fracture Toughness Based on COD and J-Integral", Elastic-Plastic Fracture ASTM STP 668, American Society for Testing and Materials, 1979, pp. 307-333.

6.6 "Standard Practice for R-Curve Determination", ASTM Standard E-561-92a, Annual Book of ASTM Standards, Vol. 03.01, 1992, pp. 597-608.

6.7 "Methods for Crack Opening Displacement (COD) Testing", BS 5762, British Standards Institution, London, 1979.

6.8 "Methods of Testing for Plane Strain Fracture Toughness ($K_{Ic}$) of Metallic Materials", BS 5447, British Standards Institute, 1974.

6.9 "Standard Method for Measurement of Fracture Toughness", Draft of Proposed ASTM Standard, September 1994. Committee E-08 on Fatigue and Fracture, ASTM, Philadelphia.

6.10 W.R. Andrews, G.A. Clarke, P.C. Paris, and D.W. Schmidt, "Single Specimen Tests for $J_{Ic}$ Determination", Mechanics of Crack Growth, ASTM STP 590, ASTM, 1976, pp. 27-42.

6.11 A. Saxena and S.J. Hudak, Jr., "Review an Extension of Compliance Expressions for Common Crack Growth Specimen", International Journal of Fracture, Vol. 14, October 1978, pp. 453-468.

6.12 M.G. Vassilaros and E.M. Hackett, "J-Integral R-Curve Testing of High Strength Steels Utilizing the Direct-Current Potential Drop Method", Fracture Mechanics, Fifteenth Symposium, ASTM STP 833, 1984, pp. 535-552.

6.13 C.F. Shih and R.J. Asaro, "The Elastic-Plastic Analysis of Cracks in Bimaterial Interfaces: Part I - Small Scale Yielding", Journal of Applied Mechanics, Vol. 55, 1988, pp. 299-316.

6.14 C.F. Shih and R.J. Asaro, "The Elastic-Plastic Analysis of Cracks in Bimaterial Interfaces: Part II - Structure of Small Crack Yielding", Journal of Applied Mechanics, Vol. 56, 1989, pp. 763-779.

6.15 C.F. Shih, R.J. Asaro, and N.P. O'Dowd, "The Elastic-Plastic Analysis of Cracks in Bimaterial Interfaces: Part III - Large-Scale Yielding", Journal of Applied Mechanics, Vol. 58, 1991, pp. 450-463.

6.16 C.F. Shih, "Cracks on Bimaterial Interfaces: Elasticity and Plasticity Aspects", Materials Science and Engineering, Vol. A143, 1991, pp. 77-90.

6.17 M. Nakagaki, C.W. Marschall, and F.W. Brust, "Elastic-Plastic Fracture Mechanics Evaluations of Stainless Steel Tungsten/Inert Gas Welds", Nonlinear Fracture Mechanics: Volume II - Elastic-Plastic Fracture, ASTM STP 995, 1989, pp. 214-243.

172  *Nonlinear Fracture Mechanics for Engineers*

6.18  P.C. Paris, H. Tada, and G.R. Irwin, "Stress Intensity of Cracks Handbook", Second Edition, Paris Productions, St. Louis, Mo., 1985.

## 6.6 Exercise Problems

6.1  What are the advantages of the J-integral approach to fracture toughness testing over the plane strain fracture toughness measurement for ductile materials? Illustrate your arguments with a numerical example.

6.2  We measure fracture toughness of the material in the J-integral approach by parameters such as $J_{Ic}$, $J_R$-curve, and $J_u$. Why are these various measures of fracture toughness necessary?

6.3  If you are asked to determine both the $J_R$-curve and the $\delta_R$-curve using SENB specimen, make a list of measurements that you must make during the test with a justification for each.

6.4  The $J_R$-curve of a steel with a 0.2% yield strength of 380 MPa and ultimate tensile strength is 620 MPa is represented by the following equation:

$$\ln J = 5.32 + 0.75 \ln(\Delta a)$$

where J is in KJ/m$^2$ and $\Delta a$ is mm. You are asked to experimentally determine the $J_R$-curve of a steel that is likely to be similar because the yield and ultimate strengths of this unknown material are approximately the same and the microstructure also has considerable similarity. In addition to the $J_{Ic}$, you are also asked to measure a valid $J_R$-curve for crack extensions up to 2mm. What size compact specimen will you select to enable you to make these measurements from a single specimen. Justify your reccommendations in detail.

6.5  (a) What is the rationale for the construction line given by the equation $J = 2\sigma_Y \Delta a$ in relationship to the measurement of $J_{Ic}$ ?

(b) For $J_Q$ to be valid $J_{Ic}$, equation (6.17) requires that the following condition be met:

$$B, W - a_0 \geq 25 J_Q / \sigma_Y$$

What is the significance of this condition, in other words, why is it needed?

6.6  Using Figure 6.13 and other necessary information, derive equation (6.9), the relationship between $\delta_i$, $K_i$, and $V_{pi}$ .

6.7  Estimate the value of J in Example Problem 6.1 without applying the correction for growing cracks. Is it necessary to consider the complexity of the growing crack J formula in this case? If yes, at what amount of crack extension does it become important?

6.8  The following J vs. $\Delta a$ data for a steel $\sigma_{YS}$ = 400 MPa and $\sigma_{UTS}$ = 550 MPa is obtained using a side grooved CT specimen with B = 25mm, $B_N$ = 20mm, W = 50mm, and $a_0$ = 25mm.

Calculate the $J_{Ic}$ of the material and the $J_R$-curve. Comment on the validity of $J_{Ic}$ and the $J_R$-curve data. You may assume that the crack-front straightness requirements are met by the specimen.

| J (kJ/m²) | Δa (mm) |
|---|---|
| 105 | 0.32 |
| 175 | 0.39 |
| 189 | 0.85 |
| 226 | 1.20 |
| 255 | 1.65 |
| 305 | 2.00 |

6.9 The load-displacement diagram during a fracture test shows significant nonlinear behavior prior to sudden fracture. When the fracture surface of the specimen is examined, no stable crack extension beyond the crack opening stretch is measured. The J-integral measured at fracture is 100 KJ/m². The 0.2% yield strength of the materials is 400 MPa and the ultimate tensile strength is 600 MPa. The specimen was a charpy size specimen with a W = B = 10, and $a_0$ of 4mm. What fracture toughness information can you extract out of this test and what is its engineering significance?

6.10 The cross-section of a longitudinal seam welded pipe is shown in the following figure along with dimensions. It is found that cracks during service grow preferentially along the fusion line between the weld metal and the base metal. You are asked to choose the geometry, size, and orientation of a suitable specimen for fracture toughness testing. What will the specimen be? Give explanations for your choices.

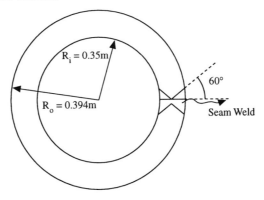

*Figure 6.19* Cross-section of a longitudinal seam-welded pipe.

174  *Nonlinear Fracture Mechanics for Engineers*

6.11 The relationship between crack size and elastic compliance of a compact specimen for a measurement location on the front-face is given by the following equation:

$$C_i = \frac{1}{EB}\left(\frac{W+a_i}{W-a_i}\right)^2 \left(1 + \frac{0.25}{a_i/W}\right) f(a_i/W)$$

$$f(a_i/W) = [1.614 + 12.68(a_i/W) - 14.23(a_i/W)^2 - 16.61(a_i/W)^3 + 35.05(a_i/W)^4 - 14.49(a_i/W)^5]$$

for a/W values of 0.45, 0.55, and 0.60, estimate the relative change in compliance dC/C for a relative change in crack size da/a.

6.12 Show that for a crack located at the interface of an ideal bimaterial, the J-integral is path-independent.

# CHAPTER 7

# INSTABILITY, DYNAMIC FRACTURE, AND CRACK ARREST

A question that design engineers often face is, at what load is the structure likely to fail? In the previous chapter, we concentrated on the methods used for characterizing the fracture resistance of materials. However, fracture resistance of the material is only one side of the equation dealing with this question. The other side of the equation, which is equally important in determining when instability (or catastrophic fracture) will occur, is dependent on the size, geometry, and loading characteristics of the structure. For example, consider a cracked body of a ductile material subjected to constant deflection along the load-line. If the imposed deflection is sufficiently large to result in a high enough load to cause fracture initiation, the crack will grow. However, the crack growth will most likely cease after some crack extension because the load will decrease as the crack length becomes larger and the body becomes more compliant. On the other hand, other loading configuration and material characteristics could lead to a catastrophic fracture. In this chapter, we will discuss methods that are used to predict fracture stability and instability in structures.

When very high loading rates are applied, a fraction of the applied energy is converted into kinetic energy of the crack which is therefore not available for fracture propagation. In addition, the fracture resistance itself can be sensitive to the rate of loading and there can be complications due to reflected stress waves. Thus, there are several additional considerations involved in predicting fracture under very high loading rates. In Section 7.2, we will be concerned with methods for predicting fracture under very high loading rates or dynamic conditions. The next topic in this chapter will deal with crack arrest, which is concerned with stabilizing a crack which is propagating in an unstable manner. The final topic in this chapter will deal with experimental methods in dynamic fracture and crack arrest.

## 7.1 Fracture Instability

We begin the discussion on predicting fracture instability by precisely defining what is meant by stable and unstable fracture. Stable fracture is formally defined as the condition of crack growth during which the applied value of J (or K for linear elastic conditions) equals the resistance of the material to fracture, $J_R$ ($J = J_R$). Thus, in order to continue the fracture process (to grow the crack further), the applied J must increase. This is illustrated in Figure 7.1 in which we have plotted a $J_R$ versus $\Delta a$ curve along with a family of curves which show how applied J increases with crack extension for different load levels, $P_1$, $P_2$, $P_3$, etc., note that $P_1 > P_2 > P_3 > P_4$. If we follow the curve for load $P_1$ up until point A, $J \geq J_R$; but beyond point A, $J < J_R$. Thus, if the load is increased to $P_1$ gradually and then held constant, the crack will grow by an amount equal to $\Delta a_A$ and then stop. Similarly, if the load is gradually increased to $P_2$, the crack will grow to point B and then stop growing. Both of these examples represent stable crack growth.

Unstable fracture occurs when not only $J \geq J_R$ but also $\partial J/\partial a \geq \partial J_R/\partial a$, thus the fracture continues to propagate spontaneously until complete separation occurs. This condition is illustrated in Figure 7.1 for load $P_3$. If the load is increased gradually to $P_3$, the fracture will be stable until Point C beyond which it will become unstable. At load $P_4$, the fracture will be unstable from the beginning and will not be preceded by any stable crack growth such as in the previous case. This example serves as an introduction for the tearing instability theory of Paris et al. [7.1-7.4].

### 7.1.1 Tearing Modulus
Paris and co-workers [7.1, 7.2] defined a tearing modulus, T, as follows:

$$T = \frac{E}{\sigma_0^2} \left(\frac{\partial J}{\partial a}\right) \tag{7.1}$$

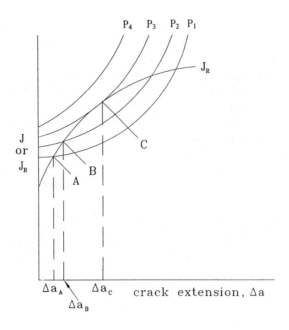

*Figure 7.1* The applied $J$ and $J_R$ vs. crack extension, $\Delta a$, curves.

The above equation yields the value of the applied value of T. Since J is dependent upon variables other than a, the partial derivative is appropriate. The resistance to tearing is given by:

$$T_R = \frac{E}{\sigma_0^2}\left(\frac{dJ_R}{da}\right) \tag{7.2}$$

We write $dJ_R/da$ as total derivative because ideally $J_R$ is a function of $\Delta a$ only. If $J > J_R$ but $T < T_R$, the crack growth is expected to be stable. On the other hand, if $J > J_R$ and $T \geq T_R$, unstable crack extension will occur.

To predict instability, we need expressions for determining T and $T_R$. Since J for a component can be written as J(P, a, m), $(\partial J/\partial a)$ can be obtained by differentiation and an expression for estimating the value of T can be obtained. $T_R$ can be obtained by finding the slope of the $J_R$ curve. Figure 7.2 shows a plot of T vs. $T_R$ for a Ni-Cr-Mo-V rotor steel obtained from several tests [7.1]. The 45° line representing $T = T_R$ is the partition between stable and unstable fracture. The tests from which data are presented in Figure 7.2 were performed on bend specimens of the same size at two different temperatures. The data support the validity of the tearing modulus theory. Since the specimens were of the same geometry and size and were loaded similarly, we find an average value of $T_R$ at instability to be 36. However, this value should not be regarded as a material constant because it can differ from one geometry to another.

Next, we turn to an example in which the tearing modulus theory is used to clearly illustrate how machine compliance, $C_M$, can influence the point of instability during fracture. Load-control conditions can be regarded as a machine with infinite compliance or zero stiffness, and load-line displacement control conditions can be regarded as infinite stiffness or zero compliance. An in-between condition can be modeled as spring in series with the specimen, Figure 7.3, with the remote displacement being fixed.

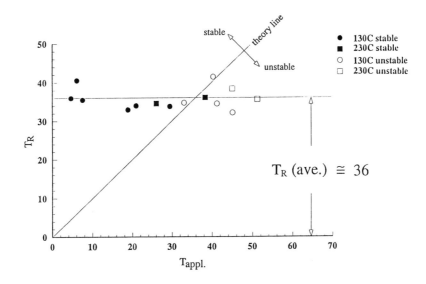

*Figure 7.2* $T_R$ vs. $T$ for Ni-Cr-Mo-V steel in 3-point bending tests (Ref. 7.1).

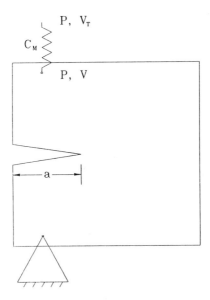

*Figure 7.3* Schematic representation of a cracked specimen in series with a spring representing the stiffness of the loading machine.

In the case of a component, the spring represents the stiffness of the overall structure of which the cracked sub-component is a part. P and V represent the load and displacement, respectively, of the cracked specimen and P, $V_T$ represent the load and deflection, respectively, of the overall system. Since the spring and specimen are in series they are subjected to the same load. We can write the following relationships:

178  *Nonlinear Fracture Mechanics for Engineers*

$$V_T = C_M P + V \tag{7.3}$$

$$J = J(P, a) \tag{7.4}$$

Since the material is given, we need not mention the dependence of J on m.

$$dJ = \left(\frac{\partial J}{\partial a}\right)_P da + \left(\frac{\partial J}{\partial P}\right)_a dP \tag{7.5}$$

Under the conditions of fixed total deflection $V_T$:

$$dV_T = C_M dP + \left(\frac{\partial V}{\partial a}\right)_P da + \left(\frac{\partial V}{\partial P}\right)_a dP = 0$$

or

$$dP = -da \left(\frac{\partial V}{\partial a}\right)_P / \left(C_M + \left(\frac{\partial V}{\partial P}\right)_a\right) \tag{7.6}$$

Combining equations (7.5) and (7.6), we get:

$$\left(\frac{\partial J}{\partial a}\right)_{V_T} = \left(\frac{\partial J}{\partial a}\right)_P - \frac{\left(\frac{\partial J}{\partial P}\right)_a \left(\frac{\partial V}{\partial a}\right)_P}{C_M + \left(\frac{\partial V}{\partial P}\right)_a} \tag{7.7}$$

or

$$T = \frac{E}{\sigma_o^2} \left[\left(\frac{\partial J}{\partial a}\right)_P - \left(\frac{\partial J}{\partial P}\right)_a \left(\frac{\partial V}{\partial a}\right)_P \left(\frac{1}{C_M + \left(\frac{\partial V}{\partial P}\right)_a}\right)\right] \tag{7.8}$$

In this example, the deflection in all linkages between those fixed points minus the specimen deflections, V, is included in estimating machine compliance, $C_M$. Since machine deflection is elastic, $C_M$ is a constant.

For the load control condition, $C_M = \infty$, thus:

$$T = \frac{E}{\sigma_o^2} \left(\frac{\partial J}{\partial a}\right)_P \tag{7.9}$$

which is an expected result showing that the load controlled condition is the worst for unstable fracture. More importantly, equation (7.8) very explicitly shows how machine stiffness (in the case of specimens) and system stiffness (in the case of components) can influence the fracture instability point.

**J-T Diagram** The J-T diagram shown in Figure 7.4 shows an alternate way of predicting instability using the tearing modulus theory. In this approach, the J vs. T relationship for both the applied values and the material resistance are plotted. Instability occurs when the applied J-T curve intersects the $J_R$-$T_R$ curve.

Figure 7.5 shows an experimental setup for the verification of the tearing modulus theory. In this setup, the position of the rollers can be varied to continuously change the value of $C_M$, or compliance of the machine [7.3]. Thus, the instability point in terms of the value of T can also be varied. The plot of T vs. $T_R$ shown in Figure 7.6 is for several steels. Considering that sometimes instability can also result from the onset of cleavage fracture, which was discussed in Chapter 6 and is not described by tearing modulus theory, the data show strong support for the validity of the tearing modulus theory. In addition to cleavage, constraint at the crack tip also varies with geometry and can cause the slope in the $J_R$-curve to be different. Since $T_R$ depends on this slope, the predictive ability of the theory is also influenced by this factor. The issue of constraint and cleavage fracture will be addressed in Chapter 8, in detail.

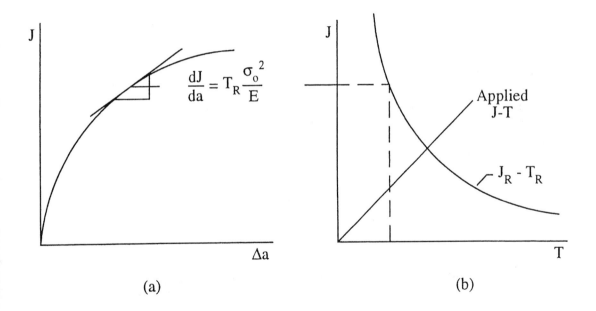

*Figure 7.4 The J-T diagram for predictive instability.*

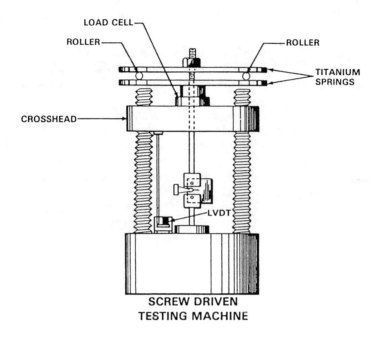

*Figure 7.5* Schematic of a test system to vary machine compliance (Ref. 7.3).

**Expression for Estimating T in a Deeply Cracked 3-Point Bend Specimen** The Rice's formula, equation (5.15), can be used to estimate J in a deeply cracked 3-point bend specimen. For such a geometry and loading, the following closed form expression for estimating T can be derived. For details of the derivation, the reader should consult references 7.1 and 7.5.

$$T = \frac{E}{\sigma_o^2} \left[ -\frac{J}{b} + \frac{4P^2}{Bb^2} \left( \frac{C}{1 + C \left( \frac{\partial P}{\partial V} \right)_a} \right) \right] \tag{7.10}$$

where C is the combined compliance of the machine and the uncracked specimen given below:

$$C = C_M + (V_{no\ crack}) / P \tag{7.11}$$

$$V_{no\ crack} = \frac{P}{4BE} \left( \frac{L}{W} \right)^3$$

L = specimen span, and W = specimen width

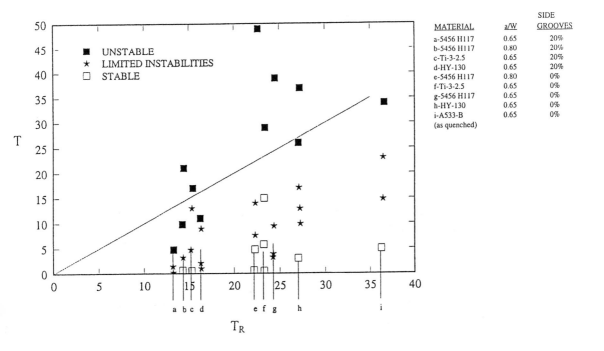

*Figure 7.6* $T$ vs. $T_R$ *plot for steels demonstrating the validity of the Tearing Modulus Theory (Ref. 7.3).*

### Example Problem 7.1
Equation (7.10) provides an expression for determining the tearing modulus, T, for the condition of remote constant deflection. Simplify the equation for the conditions of (a) dead loading and (b) under fixed load-line displacement.

*Solution:*
Under dead loading conditions, $C_M = \infty$, thus C in equation (7.11) is also $\infty$. Substituting $C \to \infty$ in equation (7.10) as a limit, we get:

$$T = \frac{E}{\sigma_0^2} \left(\frac{\partial J}{\partial a}\right)_{V_T} = \frac{E}{\sigma_0^2} \left[-\frac{J}{b} + \frac{4P^2}{Bb^2}\left(\frac{\partial V}{\partial P}\right)_a\right]$$

Under fixed load-line displacement conditions, $C_M = 0$, therefore:

$$C = \frac{(L/W)^3}{4BE}$$

Substituting for C in equation (7.10) yields:

$$T = \frac{E}{\sigma_0^2}\left[-\frac{J}{b} + \frac{4P^2}{Bb^2}\left(\frac{1}{\frac{4BE}{(L/W)^3} + \left(\frac{\partial P}{\partial V}\right)_a}\right)\right]$$

If we also assume that the material is elastic-perfectly plastic, $(\partial P/\partial V)_a$ is 0 for fully plastic conditions. Under dead loading, it implies that instability will occur as soon as fully plastic conditions are attained. Under fixed load-line displacement conditions, the expression for T reduces to:

$$T = \frac{E}{\sigma_0^2}\left[-\frac{J}{b} + \left(\frac{P}{Bb}\right)^2 \frac{(L/W)^3}{E}\right]$$

Since $J_R$ depends only on crack extension $\Delta a$, $T_R$ is relatively straightforward to obtain. One way is to fit a polynomial to represent the $J_R$ vs. $\Delta a$ curve and then differentiate it to get the value of $T_R$. Thus, $T_R$ and $J_R$ can be obtained for different values of $\Delta a$ to plot on a J-T diagram. The applied J vs. T relationship is not as straightforward to obtain. If the initial crack size is known, we can choose points on the $J_R$ curve and calculate the value of T corresponding to those points. Since the applied J is the same as $J_R$, the applied J vs. T relationship can be obtained. The point of instability is given by the intersection of the $J_R$ vs. $T_R$ and the J vs. T plots as explained before. This procedure is explained in more detail in the following example.

**Example Problem 7.2**

The $J_R$ vs. $\Delta a$ curve for a structural aluminum alloy is given by the following equation:

$$J_R = 57.62 + 115.39\,\Delta a - 9.876\,\Delta a^2$$

where $J_R$ is in kJ/m² and $\Delta a$ is in mm. A three-point bend specimen of this material has the following dimensions; B = 25 mm, W = 50 mm, S = 200 mm, and a = 25 mm. If we assume that the material is nonhardening and its deformation properties are given by E = 70 Gpa, $\nu$ = 0.3, $\sigma_0$ = 350 Mpa, $\varepsilon_0$ = 5 x 10⁻³, $\alpha$ = 1.62. The specimen is loaded under the conditions of fixed load-point displacement rate of 0.25 mm/minute. Estimate the load-displacement curve and the crack size as a fuction of time up to the point of instability. You may assume the specimen becomes fully plastic prior to any stable crack extension.

*Solution:*

Since the specimen thickness is 25 mm, we asume that plane strain conditions dominate the specimen behavior. If we were dealing with a thin sheet, plane stress would have been a more appropriate assumption. If we rewrite the expression for $J_R$ such that $J_R$ is in kJ/m² and a is in meters, we get:

$$J_R = 57.62 + 115.39 \times 10^3 (a - a_0) + 9.876 \times 10^6 (a - a_0)^2 \qquad (1)$$

By differentiating the above equation, we can get:

$$T_R = \frac{E}{\sigma_0^2} \frac{dJ_R}{da} = \frac{70 \times 10^3}{(350 \times 10^3)^2} [115.39 \times 10^3 - 19.752(a-a_0) \times 10^6]$$

substituting $a_0 = .025$ m (or 25 mm), we get:

$$T_R = 348.112 - 11.287 \times 10^3 a \quad (2)$$

The values of $J_R$ and $T_R$ from equations (1) and (2) can be obtained for various values of a as shown in Table 7.1:

**Table 7.1** *Values of J, $J_R$, T, $T_R$, P, and V as a Function of a for the 3-Point Bend Specimen in Example Problem 7.2*

| a (m) | J=$J_R$ (kJ/m²) | $T_R$ | P(kN) | V(m) | T |
|---|---|---|---|---|---|
| 0.025 | 57.62 | 65.93 | 39.78 | 4.61 x 10⁻⁴ | 1.34 |
| 0.0255 | 112.85 | 60.29 | 38.21 | 9.21 x 10⁻⁴ | 2.68 |
| .0260 | 163.134 | 54.65 | 36.66 | 1.36 x 10⁻³ | 3.95 |
| 0.265 | 208.48 | 49.00 | 35.15 | 1.774 x 10⁻³ | 5.16 |
| .0270 | 248.9 | 43.36 | 33.67 | 2.164 x 10⁻³ | 6.297 |
| .028 | 314.9 | 32.07 | 30.81 | 2.862 x 10⁻³ | 8.328 |
| .029 | 361.16 | 20.79 | 28.07 | 3.44 x 10⁻³ | 10.01 |
| .030 | 387.67 | 9.50 | 25.46 | 3.876 x 10⁻³ | 11.28 |

The applied value of $J = J_R$ during the stable crack growth portion of the test. The applied value of T is calculated as follows. For the nonhardening material, the Rice's formula for estimating J for a fully plastic body is given by:

$$J \simeq \frac{2PV}{Bb} \quad (3)$$

We assume that J is dominated by its plastic part, and note that $da = -db$. Thus, T is given by:

$$T = \frac{E}{\sigma_0^2} \frac{\partial J}{\partial a} = \frac{E}{\sigma_0^2} \frac{2PV}{Bb^2} \quad (4)$$

Since the crack extension occurs under fully plastic conditions, P = limit-load, $P_0$, at all times during the stable portion of the test. $P_0$ for this configuration is given by:

$$P_0 = \frac{1.455 \sigma_0 b^2}{L} B \tag{5}$$

Substituting $P_0$ for P in equation (3) yields:

$$V = \frac{L}{2.91 \sigma_0 b} J \tag{6}$$

Substituting $P_0$ for P in equation (4) yields:

$$T = \frac{E}{\sigma_0} \frac{2.91}{L} V = \frac{E}{\sigma_0^2} \frac{J}{b} \tag{7}$$

From equations (5), (6), and (7), the values of P, V, and T for the test can be estimated for various crack sizes. These values are also tabulated in Table 7.1. From the results of these calculations, failure is predicted between crack lengths of 0.029 to 0.030 meters. The approximate load-displacement values (neglecting elastic deflections) are also given in Table 7.1 up to the point of instability. In the above approach, several convenient assumptions are made such as nonhardening material and neglecting elastic deformation. In Excerise Problems 7.4 and 7.5 the reader is asked to solve this problem without these assumptions.

### 7.1.2 Failure Assessment Diagrams (FAD)

In very ductile materials, failure may be caused by plastic collapse as opposed to brittle fracture or J-based instability criterion. Two criterion failure assessment diagram (FAD) was introduced [7.6, 7.7] to describe the interaction between fracture and plastic collapse. We will first describe the concept as it is used for small-scale yielding conditions and then extend it to the J-based analysis. The plasticity corrected value of K, $K_{eff}$, based on the Dugdale strip plastic zone model which was derived in Chapter 5. Equation (5.5) for a center crack panel is as follows:

$$K_{eff} = \sigma_0 \sqrt{\pi a} \left[ \frac{8}{\pi^2} \ln \sec \left( \frac{\pi \sigma}{2 \sigma_o} \right) \right]^{\frac{1}{2}} \tag{7.12}$$

Equation (7.12) can be modified for real structures by replacing $\sigma_o$ with a plastic collapse stress, $\sigma_c$. The value of $\sigma_c$ depends on $\sigma_o$ and the flaw size relative to the width of the structure. For example, for a center crack panel containing a half crack size, a, half width = W, $\sigma_c$ will be given by:

$$\sigma_c = \sigma_o (1 - a/W) \tag{7.13}$$

Next, the K ratio, $K_r$, and the stress ratio, $S_r$, are defined as follows:

$$K_r = \frac{K}{K_{eff}}, \quad S_r = \frac{\sigma}{\sigma_c}$$

We can write $K_r$ as:

$$K_r = S_r \left[ \frac{8}{\pi^2} \ln \sec \left( \frac{\pi}{2} S_r \right) \right]^{-\frac{1}{2}} \tag{7.14}$$

Equation (7.14) defines FAD as shown in Figure 7.7. This curve represents the locus of predicted failure points. If the toughness is large, the structure fails by collapse and, if it is low, it fails by brittle fracture. For an in-between situation, equation (7.14) defines an interaction criterion. This modification expresses the driving force in a dimensionless form, eliminates the dependence on crack length, and, to first approximation, also removes the geometry dependence of the strip yield model as shown in Exercise Problem 7.10. To use FAD, one can compute $K_r$ as $K / K_{IC}$ and $S_r$ as $\sigma/\sigma_c$, where $\sigma_c$ is the collapse stress. If the point falls below the curve in Figure 7.6, the structure is considered safe and if it falls above the line, it would be considered unsafe. This fracture asessment method has come to be known as the R6 approach [7.7].

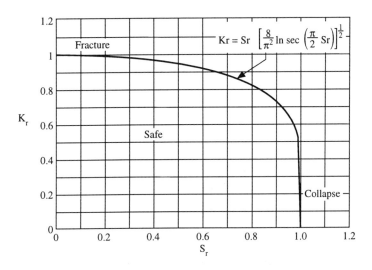

*Figure 7.7 Failure assessment diagram (FAD) derived from the Dugdale's strip yield model.*

## Example Problem 7.3

A tension plate in a bridge must support a uniform stress of 300 Mpa. The plate is 100mm wide and has been found to contain two cracks, one on each edge of the plate along a common plane normal to the loading direction. Both cracks are 10mm long. If the $K_{IC}$ of the material is 62.5 Mpa $\sqrt{m}$ and the yield strength is 440 Mpa. Using the FAD diagram, you are asked to determine whether the panel will fail.

*Solution:*

The K expression for the above geometry is represented by the double edge notch (DEN) panel under uniform tension. The following K-calibration expression can be found in standard handbooks for this geometry:

$$K = \sigma\sqrt{\pi a}\, (\frac{1}{1-a/W})^{\frac{1}{2}}\, [\, 1.122 - .561\, (a/W)$$

$$- .205\, (a/W)^2 + .471\, (a/W)^3 - 0.19\, (a/W)^4\, ] \qquad (1)$$

where σ = remote applied stress, a = crack size, and W = half width of the panel. Substituting for W = 50mm = .05m, a = 10mm = .01m, and σ = 300MPa, the applied K = 59.7. Thus, $K_r$ = K/$K_{IC}$ = 0.95, $S_r = \sigma/\sigma_e$, where $\sigma_e$ is calculated from equation (7.13), $\sigma_e$ = 440 (1 - 0.2) = 352, and $S_r$ = 0.852. Plotting the point $S_r$ = 0.852 and $K_r$ = .952 in Figure 7.6 shows that the structure is in the unsafe region by the interaction criterion. According to the plastic collapse criterion or the fracture mechanics criterion separately, the structure would be predicted as being safe.

The above approach yields the lower bound, conservative fracture predictions and its application is limited to small-scale yielding conditions. Some conservatism can be reduced by accounting for strain hardening and the limitation of small-scale yielding can be removed by going to a J-based approach as suggested by Bloom [7.8] and by Shih et al. [7.9]. This is accomplished by defining the J-ratio, $J_r$, and stress ratio, $S_r$, as follows:

$$J_r = \frac{J_e}{J_{eff} + J_p} = K_r^2 \qquad (7.15)$$

or

$$S_r = \frac{P}{P_o} \qquad (7.15a)$$

where $J_{eff}$ = plastic zone corrected value of J corresponding to the crack size, $a_e = a + \phi r_y$. The relationship defining the FAD can be obtained by the approach described as follows:

$$J_e(a) = \tilde{J}_e\, (P/P_o)^2 = \tilde{J}_e\, S_r^2$$

$$J_p(a,m) = \tilde{J}_p\, (P/P_o)^{m+1} = \tilde{J}_p\, S_r^{m+1}$$

where $\tilde{J}_e$ and $\tilde{J}_p$ are simply $J_e(P_0/P)^2$ and $J_p(P_0/P)^{m+1}$, respectively.

Thus:

$$J_r = \frac{S_r^2}{H_e S_r^2 + H_m S_r^{m+1}} \tag{7.16}$$

where:

$$H_e = \tilde{J}_e(a_e) / \tilde{J}_e(a) \tag{7.17a}$$

$$H_m = \tilde{J}_p(a,m) / \tilde{J}_e(a) \tag{7.17b}$$

The functions $H_e$ and $H_m$ can be obtained from the K-calibration functions and the $h$ factors described in Chapter 5. To simplify the calculations, $H_e$ may be assumed to be approximately one, which amounts to neglecting in elevation in J due to the plastic zone correction term. For structures approaching the limit-load, this contribution is small. Figures 7.8a and 7.8b show the computed FAD diagrams by Anderson for center crack panels for m = 5, 10 and 20 for an a/W = 0.5, and for various a/W values for m = 10, respectively [7.11]. For comparison the FAD from the strip model (or the R6 approach) are also plotted in these figures. This approach can also be expressed in terms of the CTOD crack driving force and material resistance curves [7.11].

## 7.2 Fracture Under Dynamic Conditions

This section deals with understanding fracture behavior when the crack is traveling at a high speed or the load is being applied at a high rate, or both. Several effects which influence fracture analyses become important when these conditions occur. For example, inertia effects become important in estimating the crack driving force and also in the materials resistance to tearing. This is particularly important when a structure must be designed to stabilize unstable fractures in high impact applications such as armor for ballistic protection, high energy piping systems in power plants designed to resist fracture under accident conditions, design of aircraft, bridge, and automobile structures for accident conditions, etc. In this section, we will discuss the most significant concepts of fracture mechanics that have evolved to deal with this class of problems. This topic has been treated in more detail in the book by Kanninen and Popelar [7.5] and by Freund [7.12] to which the readers are referred for additional background, especially in the area of mechanics of dynamic fracture.

To proceed further, it is convenient to divide dynamic fracture problems into two categories. The first deals with rapidly growing cracks in bodies subjected to constant or slowly varying loads while the second category of problems deals with the behavior of stationary cracks which are subjected to rapidly varying loads. Both are important from a practical standpoint as metioned earlier.

### 7.2.1 Early Analysis of Speeds of Propagating Cracks

In linear-elastic bodies, if the crack extension force is given by $\mathcal{G}$ and the resistance of the mateiral is given by $\mathcal{G}_R$, the difference $\mathcal{G} - \mathcal{G}_R$ is the energy available for driving crack speeds. The early approaches were simplistic in that they ignored the effects of inertia in the estimation $\mathcal{G}$ and also assumed that $\mathcal{G}_R$ is independent of crack size and crack speed. We know that neither assumption is true in the strictest sense

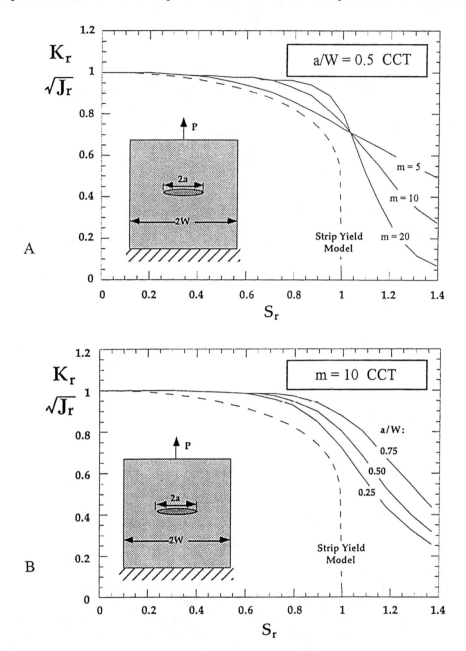

**Figure 7.8** (a) FAD for center crack panel with a/W = 0.5 and n = 5, 10, and 30. (b) FAD for center crack panel for n = 10 and varying a/W. From Kumar et al. (Ref.7.10).

and we will consider those effects in a later section. Nevertheless, if we proceed with these assumptions of Mott [7.13] in analyses developed in the late 1940s, interesting results are obtained which contribute significantly to the understanding of the problem.

**Mott, Robert, and Wells Analysis** The assumptions in this quasi-static analysis which ignores all inertia effects include (1) applied stress is constant and (2) crack speed, $V_s$, is small in comparison to the velocity of sound in the body, $c_0$. The analysis was carried out for a configuration identical to the one assumed by Griffith, that is, an infinite plate containing a crack of length 2a subjected to remote uniform tension, σ. Mott estimated the kinetic energy, $E_k$, as follows [7.13]:

$$E_k = \frac{1}{2} \rho V_s^2 \iint (u_x^2 + u_y^2) \, dx \, dy \tag{7.18}$$

where $u_x$ and $u_y$ are the x and y components of displacements in the crack tip region and ρ = density of the material. He further recognized from dimensional analysis that displacements must be proportional to the product of crack size and the remote strain σ/E. Applying it to equation (7.18), one can show that:

$$E_k = \frac{1}{2} k \rho a^2 V_s^2 (\sigma/E)^2 \tag{7.19}$$

where k = proportionality constant. Next, we can rewrite Griffith's original equation also including the kinetic energy term as follows:

$$\mathcal{G}(t) = \frac{dF}{dA} - \frac{dU}{dA} - \frac{dE_k}{dA} \tag{7.20}$$

where $\mathcal{G}(t)$ = instantaneous value of the Griffith's crack extension force, F = external work, U = strain energy, and A = crack area. If no external work is done and we assume unit thickness so that dA = d(2a), we can write the following equation:

$$\mathcal{G}(t) = \frac{1}{2}\frac{d}{da}\left[\frac{\pi\sigma^2 a^2}{E} - \frac{1}{2} k\rho a^2 V_s^2 \left(\frac{\sigma}{E}\right)^2\right] = 2\gamma$$

where γ = work of fracture which includes the work of plastic deformation accompanying crack extension plus the surface energy. From the above equation if we assume $dV_s/da = 0$, it can be easily shown that:

$$V_s = \left(\frac{2\pi}{k}\right)^{\frac{1}{2}} \left(\frac{E}{\rho}\right)^{\frac{1}{2}} \left(1 - \frac{a_0}{a}\right)^{\frac{1}{2}} \tag{7.21}$$

where $a_0 = 2E\gamma/\sigma^2\pi$.

Mott's derivation had the unrealistic assumption that $dV_s/da = 0$, which was addressed in subsequent analyses as described later. Robert and Wells [7.14] determined the kinetic energy of the body numerically which allowed them to estimate the value of $(2\pi/k)^{1/2}$ as 0.38. In equation (7.21), $V_s = 0$ when $a = a_0$, the critical crack size for fracture instability. Recognizing that $(E/\rho)^{1/2}$ is velocity of sound in the media (or the longitudinal wave velocity in a one-dimensional rod), Equation (7.21) can be written as:

$$V_s = 0.38\, c_o \left(1 - \frac{a_o}{a}\right)^{\frac{1}{2}} \qquad (7.22)$$

Thus, the limiting speed, $V_{\infty}$, of the crack is $0.38\, c_o$ and it occurs for $a \gg a_o$. The measured limiting velocities of cracks have been found to be between 0.2 to 0.4 of $c_0$ [7.15] which is in agreement with the above analytical estimate. Since the predicted limiting velocity of crack propagation is on the order of $c_0$, a condition which is not in agreement with the original assumption of the analysis that $V_s \ll c_0$, the good agreement with the experimental results can be fortuitous.

Stroh [7.15] suggested an alternate value for limiting crack velocity, $V_{\infty}$. He argued that the limiting velocity must be independent of the surface energy of the material. Thus, it must be the same for $\gamma = 0$ where $\gamma$ = work of fracture, as mentioned earlier. This is equivalent to a disturbance moving along a free surface which moves at the Rayleigh velocity (the surface wave speed, $c_R$). For a Poison's ratio of 0.3, $c_R = 0.57 c_o$, which is a bit higher than the experimentally measured values of limiting crack speeds mentioned before.

In 1960, Berry [7.17] and Dulaney and Brace [7.18] simultaneously published their modifications of the Mott analysis arriving at essentially the same results. The latter of the two analyses is more straightforward and is, therefore, presented here. They recognized the problem in Mott's assumption of setting $dV_s/da = 0$ to arrive at Equation (7.21). If no external work is supplied, the conservation of energy requires that:

$$\frac{1}{2} k \rho^2 a^2 V_s^2 (\sigma/E)^2 - \frac{\pi \sigma^2 a^2}{E} + 4\gamma a = \text{constant} \qquad (7.23)$$

From Griffith's analysis:

$$\gamma = \frac{\pi \sigma^2 a_o}{2E}$$

If we substitute for $\gamma$ in the energy conservation equation and also substitute the initial condition $V_s = 0$ at $a = a_o$, the value of the constant on the right-hand side of equation (7.23) can be derived as $\pi \sigma^2 a_o^2/E$. It can then be readily derived that:

$$V_s = \left(\frac{2\pi}{k}\right)^{\frac{1}{2}} \left(\frac{E}{\rho}\right)^{\frac{1}{2}} \left(1 - \frac{a_o}{a}\right) \qquad (7.24)$$

Note the slight difference between equations (7.24) and (7.21). Freund [7.19, 7.20] performed numerical calculations which suggest that $c_R$ is a more reasonable estimate of the limiting crack velocity than the Robert and Well's estimate. Therefore, the most accurate and current estimate for crack speed is given by:

$$V_s = c_R \left(1 - \frac{a_o}{a}\right) \qquad (7.24a)$$

Table 7.2 lists the approximate value of elastic wave speeds for common engineering materials [7.5, 7.21]. In this table, $c_R$ is the Rayleigh wave speed (shear wave speed at the surface), $c_0$ is the speed of sound in the medium and is also equal to longitudinal wave speed in a one dimensional bar, $c_1$ is the longitudinal wave speed, and $c_2$ is the shear wave speed in the medium. All the wave speeds are relevant in dynamic fracture as will become apparent in the subsequent discussion.

*Table 7.2 Approximate Values of Elastic Constants and Elastic Wave Speeds for Common Engineering Materials (adapted from Refs. 7.5, 7.21)*

|  | Steel | Cu | Al | Glass | Rubber |
|---|---|---|---|---|---|
| E (Gpa) | 210 | 120 | 70 | 70 | 20 |
| $\rho$ (kg/m³) | 7800 | 8900 | 2700 | 2500 | 900 |
| $\nu$ | 0.29 | 0.34 | 0.34 | 0.25 | 0.5 |
| $c_R$ (m/sec) | 2980 | 2120 | 2920 | 3080 | 26 |
| $c_0$ (m/sec) | 5190 | 3670 | 5090 | 5300 | 46 |
| $c_1$ (m/sec) | 5940 | 4560 | 6320 | 5800 | 1040 |
| $c_2$ (m/sec) | 3220 | 2250 | 3100 | 3350 | 27 |

*7.2.2 Inertial Effects in Dynamic Fracture*

In this section, we will discuss results of dynamic fracture analyses which include the inertial effects in the calculation of the crack driving force and also on the crack growth resistance. Recall that these were ignored in all analyses presented in the previous section. Let us condiser a crack in the $x_1$, $x_2$ plane with the $x_2$ axis normal to the crack plane and the crack extension occuring in the $x_1$ direction. Combining equations of motion, the strain-displacement relationships, and the constitutive equations for linear elasticity, the following equation must be satisfied by the displacement field:

$$G\, u_{i,jj} + (\lambda + G)\, u_{j,ij} = \rho \ddot{u}_i \tag{7.25}$$

where $G$ is the shear modulus and $\lambda$ is given by:

$$\lambda = \frac{2G\nu}{(1-2\nu)} \tag{7.26}$$

and each dot denotes a partial derivative with time. Assuming an infinite body subjected to uniform tension, $\sigma$, and a crack moving with a velocity, V, the dynamic stress intensity factor as a function of time, K(t), is given by:

$$K(t) = \alpha(V_s)\, K(0) \tag{7.27}$$

where $\alpha$ is a function of crack velocity $V_s$ and $K(0)$ is the stress intensity parameter under static conditions ($V_s = 0$). The function $\alpha(0) = 1$ when $V_s = 0$ and it decreases to 0 as the crack approaches the Rayleigh speed. An approximate expression for $\alpha$ with an accuracy of 5% is given by Rose [7.22] as follows:

$$\alpha(V_s) = (1 - \frac{V_s}{c_R})(1 - hV_s)^{\frac{1}{2}} \quad (7.28)$$

where

$$h \approx \frac{2}{c_1}(\frac{c_2}{c_R})^2 [1 - (\frac{c_2}{c_1})]^2 \quad (7.29)$$

$c_1, c_2, c_R$ are longitudinal, transverse, and Rayleigh (surface) wave velocities as defined earlier. Freund [7.23] estimated the instantaneous strain energy release rate, $\mathcal{G}(t)$, for the same problem as follows:

$$\mathcal{G}(t) = g(V_2) \mathcal{G}(0) \quad (7.30)$$

where

$$g(V_s) = 1 - \frac{V_s}{c_R} \quad (7.31)$$

and $\mathcal{G}(0)$ is the strain energy release rate for $V_s = 0$. Combining equations (7.31) and (7.27) yields the relationship between the strain energy release rate and the stress intensity parameter given by equation (7.32):

$$\mathcal{G} = \frac{1+\nu}{E}(\frac{V_s}{c_2})^2 (1 - (\frac{V_s}{c_1})^2)^{\frac{1}{2}} \frac{K^2}{D(V_s)} \quad (7.32)$$

where

$$D(V_s) = 4[1 - (\frac{V_s}{c_1})^2]^{\frac{1}{2}} [1 - (\frac{V_s}{c_2})^2]^{\frac{1}{2}} - [2 - (\frac{V_s}{c_2})^2]^2 \quad (7.33)$$

$$\text{As } V_s \to 0, D(V_s) \to \frac{4}{\kappa+1}(\frac{V_s}{c_2})^2 \quad (7.34)$$

where

$$\frac{c_1^2}{c_2^2} = \frac{\kappa+1}{\kappa-1}$$

Thus, for $V_s = 0$, equation (7.32) reduces to its static counterpart. When the crack velocity, $V_s$, is finite, the relationship between $\mathcal{G}$ and K is dependent on it. In small-scale yielding, the relationship between crack tip stress and K is given by:

$$\sigma_{ij} = \frac{K(t)}{\sqrt{2\pi r}} f_{ij}(\theta, V_s) \tag{7.35}$$

The function $f_{ij}$ reduces to the quasi-static case for $V_s = 0$. The displacement functions $\mu$ and $\mu$ also become a function of crack speed. Thus, the assumption in Mott's analysis for kinetic energy calculation in equation (7.18) that the displacement components are independent of crack speed was an idealization. The detailed derivations of equation (7.35) are available in references [7.5] and [7.24]. These are useful derivations for students with primary interests in applied mechanics, but somewhat peripheral to this more physical phenomena and practically oriented text. The governing equation for Mode I crack propagation at high speed is given by:

$$K(t) = K_{ID}(V_s) \tag{7.36}$$

where $K_{ID}$ is the dynamic fracture toughness which varies with crack speed. We have already defined the relationship between the instantaneous K and the crack speed. Figure 7.9 shows a typical relationship between $K_{ID}$ value and crack speed for a structural steel at 40°C [7.25]. Other similar results can be found in the work by Rosakis et al. [7.26]. This trend is represented by the following relationship:

$$K_{ID} = \frac{K_{IA}}{1 - (V_s/V_l)^{m_1}} \tag{7.37}$$

where $m_1$, $V_l$, and $K_{IA}$ are regression constants.

$V_l$ can be interpreted as some limiting velocity such as $c_R$ or $0.57\ c_0$ as discussed earlier. However, if sufficient data are available, $V_l$ can simply be derived from regression of the data.

In finite bodies, the stress and the stress intensity factors are influenced by the reflected wave as shown schematically in Figure 7.10. In order to estimate dynamic stress intensity parameters for such situations, (some experimental techniques for determining dynamic stress intensity parameters will be described later in this section) one has to resort to numerical analyses or to experiments. For example, a comparison of dynamic stress intensity parameters and static stress intensity parameters in polymers and steel are given in Figure 7.11 [7.5]. The major influence causing the difference between the static and dynamic cases is due to the reflected stress waves as pointed out by Kanninen [7.27] by computations on the double cantilever beam (DCB) specimens. The relative distributions of kinetic energy, strain energy, and fracture energy in a DCB specimen analyzed in the above study as a function of crack extension is shown in Figure 7.12 [7.5]. The kinetic energy rises up to a $\Delta a$ value of approx-imately 33mm with the subsequent decrease due to reflected stress waves from the specimen boundary. Similar behavior has also been observed for compact specimens [7.28].

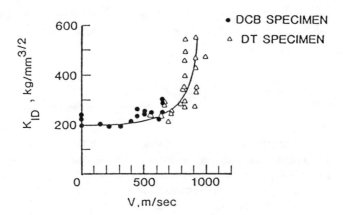

*Figure 7.9* Dynamic fracture toughness values for a ship steel obtained from DCB specimens and from wide-plate tests at 40° C (Ref. 7.24).

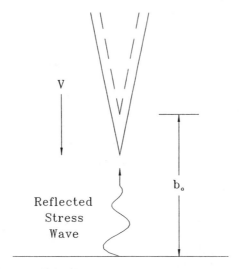

*Figure 7.10* Propagating cracks encountering reflected stress waves from the specimen surface.

Figure 7.13 shows a comparison between a dynamic calculation performed by Popelar and Kanninen [7.29] and experiments of Kaltoff et al. [7.30] in which K was measured by the method of optical caustics in DCB specimens of Araldite B. The specimens were suddenly wedge-loaded to K levels sufficient to drive the crack at high speeds, and then the cracks were monitored until arrest. In this case K is plotted as a function of time clearly showing the oscillation in the stress intensity parameter as a result of reflected waves following crack arrest. Since the calculation included the material's visoelastic response, damping of the waves with time was also captured in the predicted behavior, as it is also included in the measured K values.

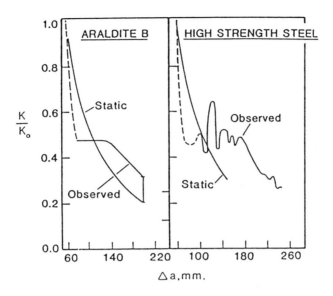

*Figure 7.11* Stress intensity parameters for fast moving cracks in DCB specimens by method of caustics in (a) Araldite B and (b) in high strength steel (Ref. 7.5).

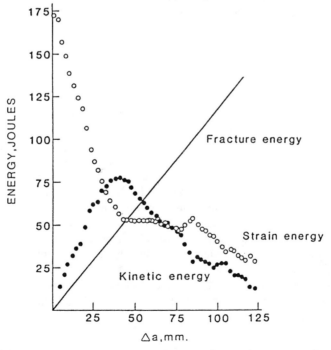

*Figure 7.12* Distribution of various components of energy during dynamic crack propagation in DCB specimen experiments (Ref. 7.5).

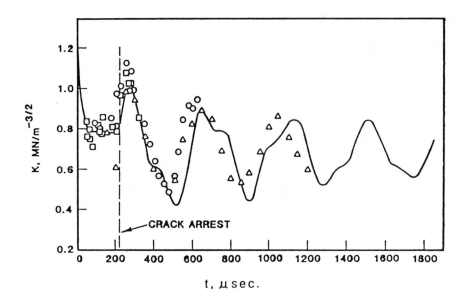

*Figure 7.13* Stress intensity factor as a function of time following crack arrest. The data points are measurements made using the technique of optical caustics and the solid line represents calculations (Ref. 7.35).

### 7.2.3 Rapid Loading of a Stationary Crack

There are several practical situations which involve rapid laoding of structures. For example, gun blast can produce loading in a period of a millisecond ($\sim 10^{-3}$ seconds). As we have also seen in the previous section, discrete stress waves produced by reflection from the free surfaces of the specimen or the body can influence the crack tip stress fields. Figure 7.14 shows a typical load-time response of a rapidly loaded body. The nominal load is shown by the dashed line and the oscillating load due to stress waves is shown as the solid line. The oscillations reduce in amplitude with time due to damping effects. Since the stress waves interact with the crack tip stress fields, the relationship between the externally applied load and K or J becomes questionable. Thus, it is important to know when dynamic effects must be considered in the determination of J-integral or the stress intensity factor, K. Since it is obvious that consideration of dynamic effects are not easy, the reverse question if very important. Specifically, up to what loading rates are the quasi-static formulae for estimating K and J applicable? Recent work of Nakamura and Shih [7.31-7.32] and others [7.33] has shown that the quasi-static expressions can be used in the majority of practical applications. The salient features of their analyses are presented in this section.

A mathematical expression for energy release rate in terms of crack tip fields for 2-d problems involving dynamic conditions is as follows:

$$J = \lim_{\Gamma \to 0} \int_\Gamma ((U + T) n_1 - \sigma_{ij} n_j \frac{u_i}{\partial x_1}] \, ds \qquad (7.38)$$

where $\Gamma$ is a counter clockwise contour enclosing the crack tip but very close to the crack tip:

$$W = \int_{-\infty}^{t} \sigma_{ij} \frac{\partial^2 u_i}{\partial t' \partial x_j} \, dt' \qquad (7.39)$$

$$T = \frac{1}{2} \rho \frac{\partial u_i}{\partial t} \frac{\partial u_i}{\partial t} \qquad (7.39a)$$

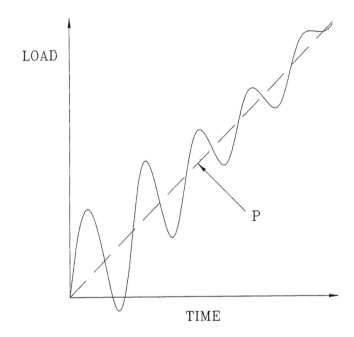

**Figure 7.14** *Load-time response of a rapidly loaded body.*

Note that W and T are stress-work and kinetic energy densities, and t' is a dummy integration variable. Note also that J in equation (7.38) is not path-independent and is calculated in the vicinity of the crack-tip. The path-dependence of J in equation (7.38) can be explained by the following argument. If we choose two distinct contours around the crack tip a finite distance from the crack tip and the stress wave passes through one but not the other, the J values are expected to be different.

The integral in equation (7.38) can be evaluated numerically from finite element solutions of cracked bodies considering the dynamic effects. Nakamura et al. [7.31 - 7.32] performed such analysis for the three-point bend specimen frequently used in dynamic fracture toughness measurements. They also introduced a transition time, $\tau$, to provide an estimate of the time beyond application of the loading at which a J-dominated field is established in the crack-tip region and the deep crack J estimation formula (equation 5.15) is applicable. They proposed that an estimate of transition time can be determined from the time history of the relative magnitudes of the total kinetic energy and the total deformation energy of the specimen. In particular, $\tau$, is defined as the time when both of these energies are equal. Thus for $t \gg \tau$, deformation energy dominates and for $t \ll \tau$, kinetic energy dominates. These components cannot be measured, but Nakamura et al. [7.32] proposed the following approximate expression for calculating the ratio of the kinetic energy, $E_k$, and the deformation energy, U:

$$\frac{E_k}{U} = (S\frac{W\dot{V}(t)}{c_o V(t)}) \tag{7.40}$$

where $c_o$ = velocity of sound, V(t) is the load-point displacement, $\dot{V}(t)$ is the time rate of load-point displacement:

$$S = (\frac{LBEC}{2W})^{\frac{1}{2}} \tag{7.41}$$

where E = elastic modulus, L = length, B = thickness, W = width, and C = elastic compliance of the specimen. In the above analysis, inertia effects are considered but the effect of discrete waves is ignored. Therefore, this equation is not valid until elastic waves make several passes through the width of the specimen and their effects begin to damp out. This limitation is not significant since such wave effects are not important at the transition time as discussed later in this section. The model also assumes that plastic deformation is not important prior to transition time. A dimensionless coefficient, D, is introduced as follows:

$$D = \frac{t\dot{V}(t)}{V(t)}\Big|_\tau \tag{7.42}$$

Then, if $\tau$ is defined as the time when $E_k/U = 1$, we get:

$$\tau = \frac{DSW}{c_o} \tag{7.43}$$

Note that $V(t) \propto t^\gamma$, $D = \gamma$. If the applied load-point displacement varies linearly with time, D = 1. Figure 7.15 shows a comparison of predicted values of $E_k/U$ from equation (7.40) with those calculated from finite element analysis. The value of $\tau$ is also indicated in the figure. The close agreement between the finite element calculation with the estimate from equation (7.40) clearly establishes the validity of the simple equation. It is found that from equation (7.40), $\tau c_1/W \sim 28$ (or $\tau c_o/W \sim 24$) in dimensionless

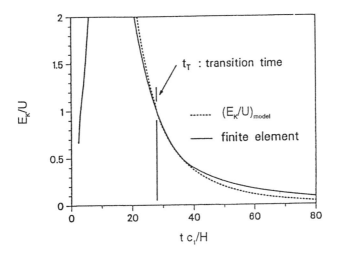

**Figure 7.15** *Ratio of the total kinetic energy ($E_k$) to the total deformation energy (U) from Equation 7.40 and that from the finite element calculations of Nakamura et al. (Refs. 7.30, 7.31).*

units in this example. Recall from earlier discussion that $c_1$ is the longitudinal wave velocity in an unbounded medium, while $c_o$ = speed of sound which is the longitudinal wave velocity in a one dimensional bar. The value of $\tau$ from finite element calculations yielded $\tau\, c_1/W \approx 27$. Since $W/c_1$ represents the time for the longitudinal wave to travel to the specimen surface, we can see that discrete stress wave effects are not very important at these time scales because the stress wave has already bounced back and forth 27 times and has most likely suffered significant attenuation.

For a deeply cracked 3-point bend specimen (SENB), Nakamura et al. [7.32] evaluated the validity of the following equation for estimating J for rapid loading, $J_{dc}$:

$$J_{dc}(t^*) = \frac{2}{B(W-a)} \int_0^{V(t^*)} p(t)\, dV(t) \qquad (7.44)$$

where $t^*$ represents any arbitrary time. The results of these calculations are shown in Figure 7.16. The value of $J_{dc}$ estimated from equation (7.44), is normalized with an average value J, $J_{ave}$, estimated from the calculation. In this calculation, 3-d effects were also included, thus, the calculated values of J varied across the thickness of the specimen. For comparison with $J_{dc}$, a single value, $J_{ave}$, was obtained by averaging J values across the thickness. The transition time is also indicated in the figure. In the short time, $t < \tau$, the $J_{dc}$ computed from equation (7.44) was considerably higher than $J_{ave}$. However, for $t > 2\tau$, the ratio of $J_{dc}$ and $J_{ave}$ is essentially constant. The constant value is somewhat greater than 1 which is attributed to the 3-d effects included in the calculation and not due to dynamic effects. Recall, equation (7.44) is a 2-d equation. As shown in Figure 7.17, the J at the specimen edge is significantly lower than at the mid-point leading to $J_{ave}$ values that are smaller than the mid-point values. Nakamura et al. [7.32] recommended the use of equation (7.44) for $t > 2\tau$. Thus, if fracture in the specimen occurs after $t > 2\tau$, this equation can be used to estimate its fracture toughness.

**Example Problem 7.4**
A 3-point bend geometry AISI 4340 steel specimen with dimensions W = 50mm, B = 25mm, L = 200mm, and a/W = 0.5 is loaded rapidly. Calculate the transition, $\tau$, for this test. Assume that the load increases linearly in time.

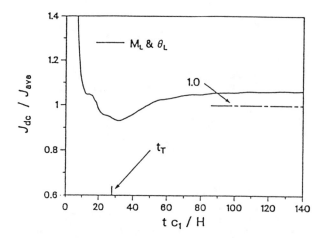

**Figure 7.16** *Comparison of $J_{dc}$ from Equation (7.44) with the through-thickness average, $J_{ave}$, as function of time for 3-point bend specimens (Ref. 7.31).*

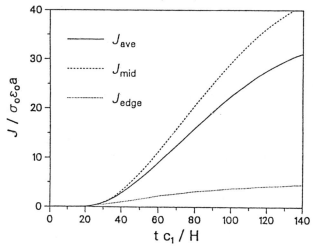

**Figure 7.17** *Normalized values of J calculated by 3-d finite element analysis for a 3-point bend specimen as a function of time at the edge and the mid-point of the specimen. A through-the-thickness average value is also shown for comparison (Ref. 7.31).*

*Solution:*
We will use Equation (7.43) to estimate the value of $\tau$. Since the loading is linear, D = 1. From Table 7.1, $c_o$ = 5190 m/sec and the value of S is given in Equation (7.41):

$$S = \left(\frac{LBEC}{2W}\right)^{\frac{1}{2}}$$

Rearranging equation (1) of Example Problem 6.1, we can write the following:

$$BEC = \frac{2.1L^2}{W^2}\left(\frac{a/W}{1-a/W}\right)^2$$

For a/W = 0.5, BEC = 2.1 = 33.6. Thus:

$$S = \left(\frac{33.6}{2}\left(\frac{L}{W}\right)\right)^{\frac{1}{2}} = 8.197$$

Substituting for S, W, and $c_o$ into Equation (7.43), we get $\tau = 379.5$ μ sec. For equation (7.44) to be valid for estimating J for dynamic conditions, the fracture must initiate at times on the order of a millisecond after application of the load.

Thus, in fracture tests conducted under rapidly loading conditions, we must also measure the time to crack initiation in addition to the other measurements performed for the quasi-static case. These techniques will be discussed in Section 7.4. The transition time estimated in Example Problem 7.3 for steel is typically smaller than the fracture initiation time for 3-point bend specimens of ductile materials in which fracture initiation is accompanied by significant plasticity. Fracture initiation times in circumferentially notched specimens can be on the order of 35 μ sec [7.33]. Thus, to interpret data from such tests, finite element analysis considering dynamic effects are needed.

## 7.3 Crack Arrest

Kanninen and Popelar [7.5] state, "If initiation of crack propagation cannot be precluded and the consequences of fracture are sufficiently large, then a crack arrest strategy is mandated as a second line of defense against a catastrophic rupture." This statement sums up the need for considering and designing applications for crack arrest involving high pressure gas and steam lines, nuclear pressure vessels, and in aircraft structures as well as other applications. In this section, we will analyze crack arrest conditions and from that infer a methodology that can be used to design for crack arrest. The foremost question that can be asked in the context of crack arrest is at what ciritcal K or J level will an unstable crack moving at a high speed come to an arrest? In this section, we will attempt to address this question. Most research to date on crack arrest has focused on linear-elastic conditions, therefore, the bulk of the discussion in this section will also pertain to those conditions. Since crack arrest occurs under decreasing J or K conditions in large components, the majority of the situations involve linear elastic conditions.

An interesting set of results was obtained by Hahn et al. [7.34] using DCB specimen schematically shown in Figure 7.18. In this specimen of height 2h, width, W, and thickness, B, the machined notch tip is blunted with an arbitrary radius. The load is applied by driving a wedge at the crack mouth until the crack begins to grow. By increasing the crack tip radius, the specimen can be loaded to a K-level which exceeds $K_{IC}$ by a higher amount prior to initiation of brittle fracture. Once the crack initiates, it accelerates due to the excess strain energy released beyond the value required to trigger brittle fracture. The crack growth is measured as a function of time by use of timing wires which rupture with the crack as shown in Figure 7.18.

The K-calibration function for these specimens has been derived by several researchers [7.35-7.37]. For W - a > 2h, the K-calibration is given by [7.35]:

$$\frac{KBh^{\frac{1}{2}}}{2\sqrt{3}P} = \left(\frac{a}{h} + 0.64\right) \tag{7.45}$$

where P is the applied load.

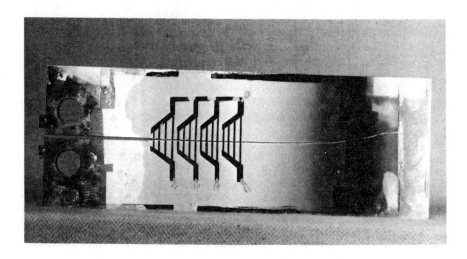

*Figure 7.18 A picture of a DCB specimen commonly used in crack arrest studies.*

The results of the experiments of Hahn et al. [7.34] are shown in Figure 7.19 in which the measured crack vs. time behavior shows essentially a constant crack growth rate. Also, due to a constant applied deflection at the loading point, first the crack jumps and is then arrested. The crack jump length was dependent on the radius of the machine notch as expected. Figure 7.19 shows the relationship between crack speed and crack growth predicted by three different analyses and also the experimental data. The analysis methods are discussed in detail. The first analysis method, termed the quasi-static analysis, assumed that the fracture energy is given by $K_{IC}$ and the excess energy went into the kinetic energy of the crack. Thus, the speed of the crack can be estimated. The approach used for estimating the crack speed was similar to one given by equations (7.18 and 7.19) for semi-infinite body. As would be expected, a bell-shaped curve is predicted. The predicted behavior varies considerably from the experiment in the trend and also in the crack jump length. Therefore, this analysis and the assumptions that go with it must not be reasonable. The next analysis assumed that the critical $\mathcal{G}$ for a crack growing at high speed is the one at the time of the arrest. Thus, the predicted length of crack jump matches exactly with the observed value, but the crack speed vs. crack extension curve still varies considerably from the observed trend. Also, the highest predicted crack speed exceeds the limiting crack velocity as shown in Figure 7.19. Again, we must conclude that the assumptions in the analysis are not reasonable. A dynamic approach is able to predict the observed trend very accurately even though the fracture energy is considered to be constant and given by $K_{IC}$. In this approach, inertia effects are included in the estimation of K, equation (7.27). Thus, in the final analysis, static analyses are not suitable for addressing cracks growing at high speeds and dynamic analyses are required.

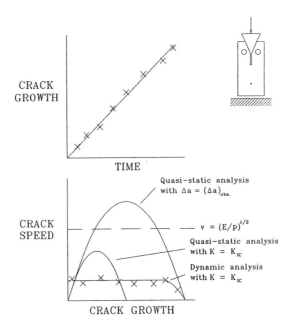

*Figure 7.19* Comparison of static and dynamic analyses of crack propagation and arrest in DCB test specimen, Hahn et al. (Ref. 7.34), Kanninen and Popelar (Ref. 7.5).

Further experiments conducted by Kalthoff et al. [7.30] and by Kobayashi et al. [7.28] reinforced the need for dynamic analysis and were quite helpful in finalizing approaches used for predicting crack arrest. Kalthoff's experiments (briefly described earlier in the chapter) were more extensive and are described here in detail. This study was conducted on several DCB specimens of Araldite in which the methods of caustics (see Section 7.4) was used to determine the stress intensity parameter at various instances during the test. The crack position as a function of time was also measured. The results are shown in Figure 7.20 which also shows the actual crack arrest points and the statically interpreted arrest points. The difference between the two appears to increase with crack jump length reinforcing the point previously made about the inadequacy of the static analysis. Also, the K value at crack arrest, $K_{Ia}$, seems to be a constant among all the tests. The observed trends of K vs. crack extension also vary significantly from those predicted by the static analysis. Near the arrest point, 'ringing' effects due to discrete waves are also very evident since the measured values of K bounce above and below the arrest point, as explained earlier in Figure 7.13.

From the previous discussion it is clear that equation (7.36) is most appropriate for predicting crack growth at high speeds. From equations (7.27) and (7.28) it follows that K(t) at the arrest point is essentially equal to K(0), the statically calculated value of K. Similarly, from equation (7.37) it follows that $K_{ID}(0) = K_{IA}$. Thus, crack arrest occurs when statically calculated value of K becomes equal to $K_{IA}$ which is a material property derived from regression of critical K vs. crack speed data, equation (7.37).

A second crack arrest criterion can be derived from the results of Figure 7.20 which shows that crack arrest occurs when the applied K reaches a very definite value equal to $K_{Ia}$. If we can assume that the applied K at crack arrest must be the same as the statically calculated K, then, according to this criterion, crack arrest occurs when $K(0) = K_{Ia}$. It is clear that the two crack arrest approaches are only equal if $K_{Ia} = K_{IA}$ and also the statically and dynamically calculated values of K at crack arrest are equal. The former

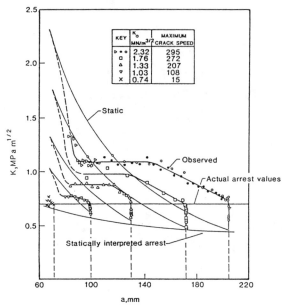

**Figure 7.20** *Comparison of calculated dynamic stress intensity factors and values observed by methods of caustics (Ref. 7.29).*

condition is essentially true because both $K_{Ia}$ and $K_{IA}$ are the same material properties, albiet derived somewhat differently. Although $K_{Ia}$ is measured direclty while $K_{IA}$ is inferred from regression of critical K vs. crack speed data, they both describe the same material behavior. The second condition that dynamically and statically calculated values of K be essentially the same is somewhat tricky. In an infinite medium where there are no intervening effects due to reflected waves, this is true. Recall that equation (7.28) was derived for an infinite medium. In a finite medium it is true only if arrest occurs prior to the first-reflected wave reaching the crack tip. For example, in Figure 7.20 this would be the case for short crack jump tests while it may not be true for the long crack jump tests. However, in Figure 7.20, crack arrest in all specimens seemed to have occurred at the same K, regardless of speed and crack jump distance, provided dynamically calculated (or measured) K values are used to interpret the data. If static K values are used to interpret the data, it would appear that $K_{IA}$ at crack arrest is dependent on crack speed. Therefore, the need to estimate dynamic stress intensities is once again reinforced.

**Example Problem 7.5**

An infinite plate of 4340 steel containing a center crack that is 20cm long is subjected to a time varying uniform tensile stress in a direction normal to the crack plane. This stress, σ, increases steadily in value with time until the crack becomes unstable and then the value of stress remains constant. Further, dynamic fracture experiments have yielded the following data:

$K_{Ia}$ = 55 Mpa $\sqrt{m}$
$K_{ID}$ (for crack velocity of 1000 m/sec) = 165 Mpa $\sqrt{m}$

If the $K_{IC}$ for this material is 60 Mpa $\sqrt{m}$, estimate the crack velocity as a function of time.

*Solution:*

We use equation (7.36) as the governing equation for the problem. The left-hand side of equation (7.36) can be given by equation (7.27) as:

$$K(t) = \alpha(V_s)\, \sigma\sqrt{\pi a(t)} \qquad (1)$$

Recall that since we are dealing with infinite medium, reflected stress waves are not significant. The value of $\alpha$ ($V_s$) is given by equations (7.28) and (7.29). From Table 7.2, using the values of $c_1$, $c_2$, $c_R$ for steel, we get:

$$h = 8.24 \times 10^{-5} \text{ (m/s)}^{-1}$$

$$\alpha(V_s) = (1 - \frac{V_s}{2980})(1 - 8.24 \times 10^{-5})^{\frac{1}{2}}$$

Thus:

$$K(t) = (1 - \frac{V_s}{2980})(1 - 8.24 \times 10^{-5})^{\frac{1}{2}} \sigma\sqrt{\pi a(t)} \qquad (2)$$

Next, we will determine the constants used in equation (7.37). We use $K_{IA} = K_{Ia} = 55$ Mpa $\sqrt{m}$. For $V_I$ we use the estimate that $V_I = c_R$ from Table 7.1, $c_R = 2980$ m/sec. Substituting for $V_I$, $K_{IA}$, and $K_{ID} = 165$ Mpa $\sqrt{m}$ for $V_S = 1100$ m/sec into equation (7.37), we get:

$$m_1 = \frac{\ln(.667)}{\ln(1000/2980)} = .3708$$

Note that if we had chosen $V_I = .38\ C_0$, the Robert and Wells approximation, the value of $m_1$ would be somewhat different. Also, if $K_{ID}$ had been measured for one more speed, the value of $V_I$ could have been determined empirically and we would not have to rely on assumptions. The stress at the initiation of fracture corresponds to $K = K_{IC}$, for a = 10cm; this stress = 107 Mpa. Now returning to equation (7.36), we find the following relationship for this application:

$$(1 - \frac{V_s}{2980})(1 - 8.24 \times 10^{-5} V_s)^{\frac{1}{2}} 107\sqrt{\pi a} = \frac{55}{(1 - \frac{V_s}{2980})^{.3708}}$$

or

$$(1 - \frac{V_s}{2980})(1 - 8.24 \times 10^{-5} V_s)^{\frac{1}{2}} (1 - (\frac{V_s}{2980})^{.3708}) = \frac{55}{107\sqrt{\pi a}} \qquad (2)$$

If we designate the left-hand side of equation (2) as $F(V_s)$ and the right-hand side as $F_1$ (a), Table 7.3 gives the values of F and $F_1$ as a function of their arguments:

**Table 7.3** Values of $F$ and $F_1$

| $V_s$ (m/sec) | $F(V_s)$ | a(m) | $F_1(a)$ |
|---|---|---|---|
| 0 | 1 | 0.1 | 0.917 |
| 20 | .837 | 0.12 | 0.837 |
| 100 | .6891 | 0.2 | 0.648 |
| 200 | .5854 | 0.3 | 0.5294 |
| 300 | .5090 | 0.4 | 0.4585 |
| 400 | .4470 | 0.5 | 0.4101 |
| 500 | .3945 | 0.6 | 0.3744 |
| 600 | .3488 | 0.7 | 0.34662 |
| 700 | .3086 | 0.8 | 0.32423 |
| 800 | .2728 | 0.9 | 0.3057 |
| 900 | .2408 | 1.0 | 0.2900 |
| 1000 | .2119 | 1.5 | 0.2368 |
| 1500 | .1045 | 2.0 | 0.2050 |
| 2980 | 0 | 5.0 | 0.1297 |

In the following figures, we first plot $V_s$ as a function of $F(V_s)$ and a as a function of $F_1(a)$. Since along the fracture path, $F(V_s) = F_1(a)$, we can find the combination of $V_s$ and a which yield that condition and make a second plot. The latter plot is the answer to the problem.

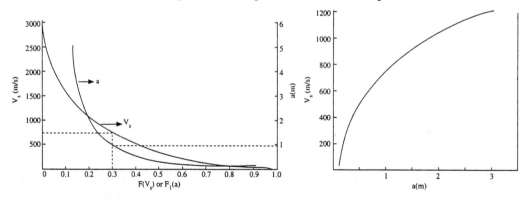

## 7.4 Test Methods for Dynamic Fracture and Crack Arrest

As discussed in the earlier sections, the fracture response of materials may be quite different when the load is applied rapidly or the crack velocity is high. Therefore, it is important to measure the materials resistance under rapidly applied loading and also the crack arrest toughness. This section deals with these topics.

*7.4.1 Dynamic Fracture Testing*

The special requirements for rapid-load plane-strain fracture toughness testing are given in Appendix A7 of the ASTM Standard E 399-90 [7.40]. The primary difference between this procedure and the static case is that the rate of loading exceeds the conventional rate of 2.75 Mpa$\sqrt{m}$/sec. These high rates of loading are achieved by servohydraulic machines and typically involve attaining maximum loads in the matter of a few milliseconds. The load vs. time, displacement vs. time, and the load-displacement curves are recorded during the test. These records are not only necessary to ensure that ringing is not present during the test, but provide an accurate measurement of test time. Ringing is not permitted up to the time the critical K, $K_Q$, is reached. The time to reach $K_Q$ is always reported along with test temperature and the $K_{IC}$ values. The qualification requirements for $K_{IC}$ are very similar to those for the quasi-static cases as discussed earlier in Chapter 3. The reader is referred to reference (7.38) for additional details of the test. This method has been extended by Joyce and Hackett for determining dynamic $J_R$ curves [7.39]. For details, consult this reference.

*7.4.2 Crack Arrest Testing*

To measure the critical K at crack arrest, $K_{Ia}$, it is necessary for the crack to first grow in an unstable manner and subsequently come to arrest. There are two primary approaches to achieve arrest conditions. The first which is more common involves obtaining crack extension under constant displacement conditions such that the crack driving force actually decreases once the crack initiates causing the crack to eventually arrest. The other approach is to use a large specimen under controlled temperature gradient. As the crack extends into increasing temperature regions in steels, the fracture resistance rises sharply causing the crack to arrest. The point of crack arrest can be used to compute $K_{Ia}$ at the temperature of that point. This method is obviously more cumbersome and requires sophisticated controlled heating and cooling devices. The more common approach of decreasing crack driving force is adopted in the standard ASTM test method for determining, $K_{ia}$, E1221-88 [7.40] and is briefly described below.

The loading apparatus used in $K_{Ia}$ testing is shown in Figure 7.21. The test specimens are very often side-grooved and the starter notch is often placed in a brittle weld-bead to get the crack to initiate rapidly. The wedge shown in Figure 7.22 is subjected to a load P which applies Mode I loading on the crack tip. The wedge load P and the crack mouth displacement, V, are recorded. The displacement V can be used to calculate the stress intensity parameter K. However, there are some uncertainties in the estimate of K using this method. For example, if crack tip plasticity occurs, the displacement associated with this plasticity does not contribute to the K value. This is illustrated in Figure 7.22 where the specimen is loaded to values of loads corresponding to $P_1$, $P_2$, $P_3$ and subsequently unloaded assuming that the crack size remains constant. The permanent residual displacement at zero load, $R_1$, $R_2$, $R_3$ are values of displacement corresponding to the crack tip plasticity and do not contribute to the K value. The standard test method requires that the specimen be loaded and unloaded from increasingly higher loads until fracture initiates such as at $P_4$ in Figure 7.22. The initiation of rapid fracture occurred at load $P_4$ in this test. In a constant displacement test, the crack tip plasticity decreases with crack extension. Hence, given sufficient time, the plastic displacement will convert into elastic displacement and contribute to the value of K. However, it is not clear whether sufficient time is allowed for this conversion to occur during a crack arrest test. Therefore, the ASTM Standard recommends that half of the plastic offset be used in calculating the value of K. For a given displacement $\delta$, the value of K can be calculated using the following expression for standard compact specimens in which the displacement gage is located at a distance of 0.25W from the load-line:

$$K = \frac{E\delta f(a/W) (BB_N)^{\frac{1}{2}}}{W^{\frac{1}{2}}} \qquad (7.46)$$

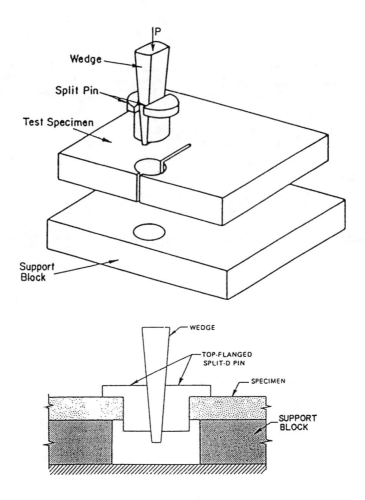

*Figure 7.21 Schematic and sectional views of the wedge and split-pin assembly, test specimen, and support block used in crack arrest testing (Ref. 7.38).*

where B, $B_N$ are the nominal thickness and the thickness in the notch plane, respectively:

$$f(a/W) = \frac{2.24(1.72 - 0.9\ (a/W) + (a/W)^2\ )\ (1 - a/W)^{\frac{1}{2}}}{(9.85 - 0.17(a/W) + 11\ (a/W)^2\ )} \quad (7.47)$$

$\delta$ = displacement at the crack mouth appropriately corrected for plasticity

The requirements for obtaining a valid $K_{Ia}$ measurement are as follows. If the crack length at crack arrrest is given by $a_a$ and the initial crack size by $a_0$, the following conditions must be met:

$$W - a_a \geq 0.15W \tag{7.48a}$$

$$W - a_a \geq 1.25 \left(\frac{K_a}{\sigma_{Yd}}\right)^2 \tag{7.48b}$$

$$B \geq \left(\frac{K_a}{\sigma_{Yd}}\right)^2 \tag{7.48c}$$

$$a_a - a_0 \geq \frac{1}{2\pi} \left(\frac{K_a}{\sigma_{Ys}}\right)^2 \tag{7.48d}$$

$\sigma_{Yd}$ = assumed dynamic yield strength which is typically 205 Mpa (30 ksi) higher than the 0.2% yield strength, $\sigma_{Ys}$. If all of the above conditions are met, $K_a = K_{ia}$.

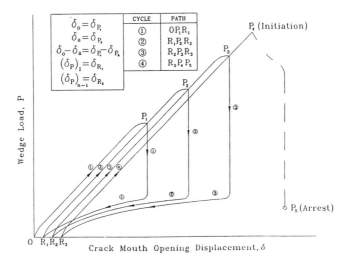

**Figure 7.22** *Wedge load vs. crack mouth opening displacement test record for a specimen tested using cyclic loading technique in which rapid crack growth and arrest occured during the fourth loading (Ref. 7.38).*

### 7.4.3 Method of Optical Caustics for Determining K and the J-Integral During Crack Arrest Testing

The method for determining K during crack arrest testing described in the previous section does not account for crack speed. Even if crack speed data were available, equation (7.46) will not be able to make use of this information. If crack velocity is measured during crack arrest testing, finite element analysis can be used to account for dynamic effects in estimating K. This method is considered an indirect method of obtaining K and relies on anlaytical tools not commonly available to experimentalists. The technique of optical caustics has been developed over the past few years to overcome this shortcoming. In the subsequent discussion, the fundamentals of this technique will be described and equations for estimating K and J for rapidly growing cracks are provided.

The method of caustics has been developed as a successful tool for direct measurement of stress intensity parameter K and the J-integral [7.41 - 7.46]. The caustic is a bright band of light surrounding a shadow zone around the crack tip caused by the reflection of the incident light by the distorted specimen surface. Figure 7.23 shows a schematic of an experimental set up for photographing caustics and Figure 7.24 shows a set of parallel light rays normally incident on an initially planar, reflective specimen which has been deformed by loading [7.45]. The deformed shape of the specimen surface is such that the virtual extension of the reflected light rays form an envelope in space as shown in Figure 7.24. This surface is called the caustic surface and is the locus of points of maximum luminosity. Its intersection with a plane located a distance $Z_0$ behind the specimen is called the casutic curve which bounds a dark region called the shadow spot. The size and shape of the caustic is determined by the characteristics of the strain field in the neighborhood of the crack tip and $Z_0$ which represents the distance between the reflective surface and the virtual image plane where the camera is focused to photograph the caustic, Figure. 7.24.

Suppose that the light reflected from a point $(x_1, x_2)$ on the specimen intersects the plane at $Z_0$ at a point $(X_1, X_2)$, where $(x_1, x_2)$ is a coordinate system on the specimen centered at the crack tip with $x_1$-axis along the length of the crack and $(X_1, X_2)$ is a system translated by a distance $Z_0$ behind the specimen on the image plane. The $(X_1, X_2)$ are given by the mapping function [7.41, 7.46]:

$$X_\alpha = x_\alpha - 2Z_0 \frac{\partial f(x_1, x_2)}{\partial x_\alpha} \qquad \alpha = 1, 2 \qquad (7.49)$$

$f = -u_3 (x_1, x_2)$, the out-of-plane displacement of the specimen

The caustic curve will exist if and only if the Jacobian determinant of the mapping vanishes [7.46]:

$$\det \left[ \frac{\partial X_\alpha}{\partial x_p} \right] = 0 \qquad (7.50)$$

The locus of points on the specimen satisfying equation (7.50) is called the initial curve. All points inside and outside the initial curve map outside the caustic curve, thus, leaving a dark spot in the region inside of the intersection between the caustic surface and the image plane as shown in Figure 7.24.

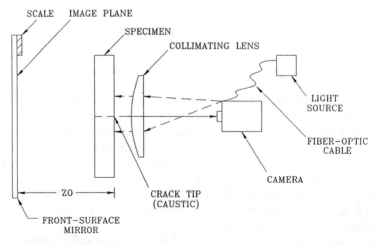

**Figure 7.23** *Schematic arrangement of apparatus to photograph the caustic formed during loading of a compact type specimen (Ref. 7.45).*

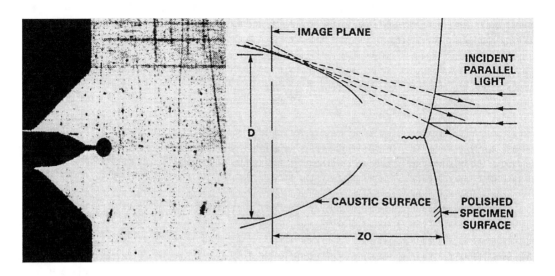

*Figure 7.24* Illustration of the formation of a caustic and and actual caustic in an experiment (Ref. 7.45).

The Mode I, plane stress out-of-plane displacement field is given by:

$$u_3 = \frac{-\nu BK}{E\sqrt{2\pi r}} \cos \frac{\theta}{2} \tag{7.51}$$

where E = elastic modulus, $\nu$ = Poisson's ratio, B = specimen thickness, and $r$ and $\theta$ are polar coordinates. Plane stress is chosen because it is most appropriate for a free surface. By substituting equation (7.51) into equations (7.49) and (7.50), it can be shown that the caustic is an epicycloid and K is related to the caustic diameter, D, which is measured as the width of the caustic in the $x_2$ direction by the following relationship [7.44]:

$$K = \frac{ED^{\frac{5}{2}}}{10.7 \, Z_0 \nu B} \tag{7.52}$$

The initial curve is circular with its radius, $r_0$, given by [7.44]:

$$r_0 = 0.316D = \left[ \frac{3B\nu KZ_0}{2\sqrt{2\pi}E} \right]^{\frac{2}{5}} \tag{7.53}$$

For the above equations to hold, the initial curve must completely engulf the plastic zone because equation (7.51) is valid only for the elastic region away from the plastic zone. If on the other hand the specimen is dominantly plastic, the HRR displacement fields may be used determine $u_3$ as follows:

$$u_3 = \frac{\alpha \sigma_0 B}{2E} \left[ \frac{JE}{\alpha \sigma_0^2 I_m r} \right]^{\frac{m}{m+1}} [E_{rr}(\theta,m) + (E_{\theta\theta}(\theta,m)] \tag{7.54}$$

212  *Nonlinear Fracture Mechanics for Engineers*

where $\alpha$, $\sigma_0$, and m are the Ramberg-Osgood constants as described earlier in Chapters 2 and 4. $I_m$ is a numerical factor also described in equation (4.19), $E_{rr}$ and $E_{\theta\theta}$ are dimensionless angular functions of $\theta$ and m. Substituting equation (7.54) into equations (7.49) and (7.50) yields the following result [7.49]:

$$J = S_m \frac{\alpha \sigma_0^2}{E} \left[ \frac{E}{\alpha \sigma_0 Z_0 B} \right]^{m+\frac{1}{m}} D^{3m+2/m} \tag{7.55}$$

The factor $S_m$ is listed in Table 7.4 and D is the maximum diameter of the caustic. For a nonhardening material, the relationship between J and the caustic diameter is [7.46]:

$$J = \frac{\sigma_0 D^3}{13.5 Z_0 B} \tag{7.56}$$

The initial curve for the case of extensive plasticity is not circular and its shape depends on the hardening exponent. The point on the initial curve that maps to the maximum value of $X_2$ on the caustic is at angle $\theta_{max}$ from the $x_1$ axis. If this distance is $r_0$ and is used to represent the initial curve size, the following values of $r_0$ are estimated:

$r_0 = 0.385$ D for m = 9 and
$r_0 = 0.40$ D for m = 50

The above relationships are based on HRR fields. Numerical analysis can be used to estimate more accurate relationships utilizing the full-field solutions. For more information, the reader is referred to other references [7.46]. Figure 7.25 shows the relationships between experimentally measured J and the

*Table 7.4 Values of $S_m$ (7.46)*

| Hardening Exponent m | $S_m$ |
|---|---|
| 1 | .0277 |
| 2 | .0513 |
| 3 | .0611 |
| 4 | .0660 |
| 5 | .0687 |
| 6 | .0701 |
| 7 | .0710 |
| 8 | .0715 |
| 9 | .0718 |
| 10 | .0719 |
| 15 | .0719 |
| 20 | .0717 |
| 25 | .0714 |

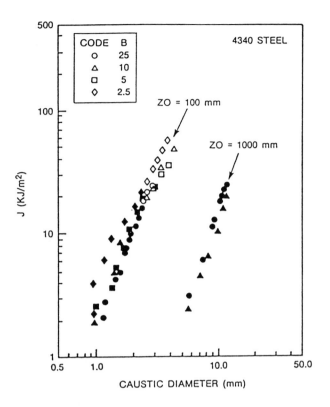

*Figure 7.25 Relationship between applied J and the caustic diameter for AISI 4340 steel compact type specimens (50 mm wide) and of thickness (mm), B, as a result of loading. All data were taken prior to crack initiation. Open symbols are from regions where the load-deflection plot was no longer linear (Ref. 7.45).*

caustic diameter for CT specimens of different thicknesses and two values of $Z_0$ [7.45]. These relationships are for AISI 4340 steel from which 50mm wide specimens of different thicknesses were tested while recording the load-displacement diagram to estimate J and simultaneously measuring the caustic diameter. The tests ranged from small-scale yielding condition to fully plastic condition. Such calibrations can be used to determine the value of J from dynamic tests. Figure 7.26 shows the caustic diameter at J levels for two different image planes. The total applied values of J are labeled on the figure and the value of $J_{pl}$ is indicated in the parantheses. Thus, these pictures were taken during dominantly elastic-to-elastic plastic loading.

The above method can be very effectively used to measure the value of J for a rapidly growing crack and will account for the dynamic effects. Thus, arrest fracture toughness values obtained from experiments which make use of the caustic method are more reliable.

**7.5 Summary**

In this chapter, we defined stable fracture as a condition of crack growth during which the applied J is always equal to the resistance to fracture, $J_R$. Unstable fracture was said to occur when $J > J_R$ and also $(\partial J/\partial a) \geq (dJ_R/da)$. Using these principles, a tearing instability theory was developed to predict the condi-

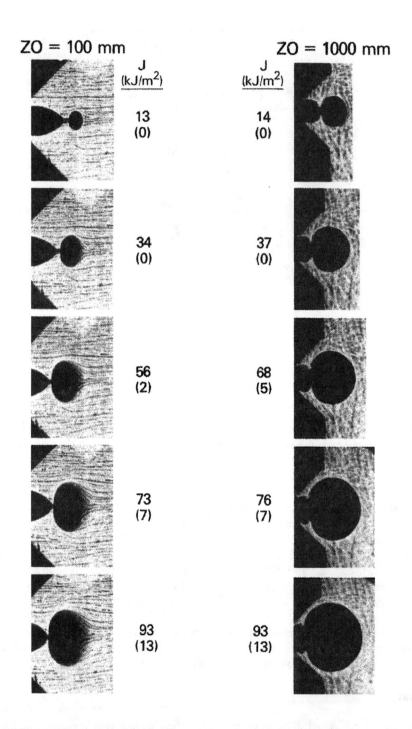

*Figure 7.26* Series of optical caustics for 5% nickel steel taken at two image planes. The values of J corresponding to each photograph are listed along with the value of plastic part of J in the parentheses (Ref. 7.45).

tions under which fast fracture occurs. The influence of loading configuration, specimen geometry, and size, as well as material characteristics on fracture stability (or instability) can be understood with this theory.

Dynamic fracture theories were developed to address problems of predicting fracture under rapid loading and also to understand the fracture behavior of cracks propagating at high speeds. When cracks propagate at high speeds, it was observed that part of the energy released goes into kinetic energy for driving crack speed. Also, inertial effects are important in estimating the correct crack driving force. Analytical methods to account for these considerations are described in this chapter. Similarly, for rapid loading reflected stress waves become an important consideration. However, it is shown that if fracture occurs after elapse of a few milliseconds from the time of application of the load, the reflected stress waves will have already attenuated significantly during that time period. Therefore, even though reflected stress waves are a concern during fracture at high loading rates, the majority of the situations can be addressed by ignoring their effects.

Test methods used for measuring dynamic fracture toughness, crack arrest fracture toughness, dynamic stress intensity parameter, and dynamic J-integral values using the method of optical caustics were also described in this chapter.

## 7.6 References

7.1 P.C. Paris, H. Tada, A. Zahoor, and H.A. Ernst, "The Theory of Instability of the Tearing Mode of Elastic-Plastic Crack Growth", in Elastic-Plastic Fracture, ASTM STP 668, American Society for Testing and Materials, Philadelphia, Pa., 1979, pp. 5-36.

7.2 H.A. Ernst, "Further Developments in the Tearing Instability Theory", in Elastic-Plastic Fracture-Second Symposium - Vol. II, ASTM STP 803, American Society for Testing and Materials, 1983, pp. 133-155.

7.3 J.A. Joyce and M.G. Vassilaros, "An Experimental Evaluation of Tearing Instability Using the Compact Specimen", in Fracture Mechanics: Thirteenth Conference, ASTM STP 743, American Society for Testing and Materials, 1981, pp. 525-542.

7.4 H.A. Ernst, P.C. Paris, and J.D. Landes, "Estimations of J-Integral and Tearing Modulus T from a single Specimen Test Record", Fracture Mechanics: Thirteenth Conference, ASTM STP 743, American Society for Testing and Materials, 1981, pp. 476-502.

7.5 M.F. Kanninen and C.F. Popelar, in Advanced Fracture Mecahnics, Oxford University Press, New York, 1985.

7.6 A.R. Dowling and C.H.A. Townley, "The Effects of Defects on Structural Failure: A Two Criteria Approach", International Journal of Pressure Vessels and Piping, Vol. 3, 1975, pp. 77-137.

7.7 R.P. Harrison, K. Loosemore, and I. Milne, "Assessments of the Integrity of Structures Containing Defects", CDGB Report R/H.RG, CEGB, U.K. 1976.

7.8 J.M. Bloom, "Prediction of Ductile Tearing Using a Proposed Strain Hardening Failure Assessment Diagram", International Journal of Fracture, Vol. 6, 1980, pp. R73-R77.

7.9   C.F. Shih, M.D. German, and V. Kumar, "An Engineering Approach for Examining Crack Growth and Stability in Flawed Structures", International Journal of Pressure Vessels and Piping, Vol. 9, 1981, pp. 159-196.

7.10  V. Kumar, M.D. German, and C.F. Shih, "An Engineering Approach to Elastic-Plastic Fracture Analysis", EPRI Report NP-1931, Electric Power Research Institute, Palo Alto, CA, 1981.

7.11  T.L. Anderson, R.H. Leggett, and S.J. Garwood, "The Use of CTOD Methods in Fitness for Purpose Analysis", The Crack Tip Opening Displacement in Elastic-Plastic Fracture Mechanics, Springer-Verlag, Berlin, 1986, pp. 281-313.

7.12  L.B. Freund, "Dynamic Fracture Mechanics", Cambridge University Press, Cambridge, U.K. 1990.

7.13  N.F. Mott, "Fracture of Metals: Theoretical Considerations", Engineering, Vol. 165, 1948, pp. 16-18.

7.14  D.K. Roberts and A.A. Wells, "The Velocity of Brittle Fracture", Engineering, Vol. 178, 1954, pp. 820-821.

7.15  J.I. Bluhm, "Fracture Arrest", in Fracture - Vol. V, H. Liebowitz (ed.), Academic Press, NY, 1969, pp. 1-69.

7.16  A.N. Stroh, "A Theory of Fracture of Metals", Advances in Physics, Vol. 6, 1957, pp.418-465.

7.17  E.N. Dulaney and W.F. Brace, "Velocity Behavior of a Growing Crack", Journal of Applied Physics, Vol. 31, pp. 2233-2236.

7.18  J.P. Berry, "Some Kinetic considerations of the Griffith Criterion for Fracture", Journal of Mechanics and Physics of Solids", Vol. 8, 1980, pp. 194-216.

7.19  L.B. Freund, "Crack Propagation in Elastic Solid Subjected to General Loading - I. Constant Rate of Extension", Journal of Mechanics and Physics of Solids, Vol. 20, 1972, pp. 129-140.

7.20  L.B. Freund, "Crack Propagation in an Elastic Solid Subjected to General Loading - II. Non-Uniform Rate Extension", Journal of Mechanics and Physics of Solids, Vol. 20, 1972, pp. 141-152.

7.21  H. Kohsky, "Stress Waves in Solids", Dover, New York, 1963.

7.22  L.R.F. Rose, "An Approximate (Wiener-Hopf) Kernel for Dynamic Crack Problems in Linear Elasticity and Viscoelasticity", Proceeding, Royal Society of London, Vol. A - 329, 1976, pp. 497-521.

7.23  L.B. Freund, "Energy Flux into the Tip of an Extending Crack in an Elastic Solid", Journal of Elasticity, Vol. 2, 1972, pp. 341-349.

7.24  T.L. Anderson, "Fracture Mechanics - Fundamentals and Application", CRC Press, Second Edition, 1995.

7.25 T. Kanazawa and S. Machids, "Fracture Dynamics Analysis of Fast Fracture and Crack Arrest Experiments", Fracture Tolerance Evaluation, Kanazawa et al. editors, Toyoprint, Japan, 1982.

7.26 A.J. Rosakis and L.B. Freund, "Optical Measurement of Plastic Strain Concentration at a Crack Tip in a Ductle Steel Plate", Journal of Engineering Materials Technology, Vo. 10A, 1982, pp. 115-120.

7.27 M.F. Kanninen, "An Analysis of Dynamic Crack Propagation and Arrest for a Material Having a Crack Speed Dependent Fracture Toughness", Prospects of Fracture Mechanics, G.C. Sih et al. (eds.), Noordhoff, Leyden, The Netherlands, 1974, pp. 251-266.

7.28 A.S. Kobayashi, K. Seo, J. Jou, and Y. Urabe, "A Dynamic Analysis of Modified Compact-Tension Specimens Using Homolite-100 and Polycarbonate Plates", Experimental Mechanics, Vol. 20, 1980, pp. 73-79.

7.29 C.H. Popelar and M.F. Kanninen, "A Dynamic Viscoelastic Analysis of Crack Propagation and Crack Arrest in a Double Cantilver Beam Test Specimen", Crack Arrest Methodology and Applications, G.T. Hahn and M.F. Kanninen (editors), ASTM STP 711, American Society for Testing and Materials, 1980, pp. 3-21.

7.30 J.F. Kalthoff, J. Bienart, and S. Winkler, "Measurements of Dynamic Stress Intensity Factors for Fast Running and Arresting Cracks in Double-Cantilever-Beam Specimens", Fast Fracture and Crack Arrest, ASTM STP 627, American Society for Testing and Materials, 1977, pp. 161-176.

7.31 T. Nakamura, C.F. Shih, and L.B. Freund, "Analysis of a Dynamically Loaded Three-Point-Bend Ductile Fracture Specimen", Engineering Fracture Mechanics, Vol. 25, 1986, pp. 323-329.

7.32 T. Nakamura, C.F. Shih, and L.B. Freund, "Three-Dimensional Transient Analysis of a Dynamically Loaded Three-Point-Bend Ductile Fracture Specimen", Nonlinear Fracture Mechanics: Volume I - Time-Dependent Fracture, ASTM STP 995, American Society for Testing and Materials, 1989, pp. 217-241.

7.33 E.M. Hackett, J.A. Joyce, and C.F. Shih, "Measurements of Dynamic Fracture Toughness of Ductile Materials", Nonlinear Fracture Mechanics: Volume I - Time-Dependent Fracture, ASTM STP 995, American Society for Testing and Materials, 1989, pp. 274-297.

7.34 G.T. Hahn, R.G. Hoagland, M.F. Kanninen, and A.R. Rosenfield, "A Preliminary Study of Fast Fracture and Arrest in a DCB Test Specimen", Dynamic Crack Propagation, G.C. Sih (ed.), Noordhoff, Leyden, 1973, pp. 649-662.

7.35 M.F. Kanninen, "A Dynamic Analysis of Unstable Crack Propagation and Arrest in the DCB Test Specimen", International Journal of Fracture, Vol. 10, 1974, pp. 415-430.

7.36 J.E. Srawley and B. Gross, "Stress Intensity Factors for Crack-Line-Loaded Edge Crack Specimens", Materials Research and Standards, Vol. 7, 1967, pp. 155-162.

7.37 W.B. Fitcher, "The Stress Intensity Factor for a Double Cantilever Beam", International Journal of Fracture, Vol. 22, 1983, pp. 133-143.

7.38 Standard Test Method for Plane-Strain Fracture Toughness of Metallic Materials, ASTM Standard E-399-90, Annual Book of ASTM of Standards, 1992, Vol. 03.01, pp. 506-536.

7.39 J.A. Joyce and E.M. Hackett, "An Advanced Procedure for J-R Curve Testing Using a Drop Tower", in Nonlinear Fracture Mechanics: Vol. I, Time-Dependent Fracture, ASTM STP 995, American Society for Testing and Materials, 1989, pp. 298-317.

7.40 Standard Test Method for Determining Plane-Strain Crack-Arrest Fracture Toughness, $K_{Ia}$, of Ferritic Steels, ASTM Standard E 1221-88, Annual Book of ASTM Standards, 1992, Vol. 03.01, pp. 879-894.

7.41 P. Mannogg, "Anwendung der Schattenpotik zur 2 Untersuchung des Zereibvorgangs von Platten", Ph.D. dissertation, University of Freiburg, Freiburg, Germany, 1964 (in german).

7.42 A.J. Rosakis, "Experimental Determination of the Fracture Initiation and Dynamic Crack Propagation Resistance of Structural Steels by the Optical Method of Caustics", Ph.D. thesis, Brown University, Providence, RI, 1982.

7.43 A.J. Rosakis and L.B. Freund, "Optical Measurement of the Plastic Strain Concentration of a Crack Tip in a Ductile Steel Plate", Engineering Materials and Technology, Vol. 104, April 1982, pp. 115-120.

7.44 A.J. Rosakis, C.C. Ma, and L.B. Freund, "Analysis of Optical Shadow Spot Method for a Tensile Crack in a Power-Law Hardening Material", Journal of Applied Mechanics, Vol. 105, 1983, pp. 777-782.

7.45 R.W. Judy, Jr. and R.J. Sanford, "Correlation of Optical Caustics with Fracture Behavior of High-Strength Steels", in Nonlinear Fracture Mechanics, Vol. I, Time Dependent Fracture, ASTM STP 995, American Society of Testing and Materials, 1989, pp. 340-357.

7.46 A.T. Zehnder, A.J. Rosakis, and R. Narashiman, "Measurement of the J-Integral with Caustics: An Experimental and Numerical Investigation", in Nonlinear Fracture Mechanics: Volume I, Time-Dependent Fracture, ASTM STP 995, American Society for Testing and Materials, 1989, pp. 318-339.

## 7.7 Exercise Problems

7.1 Equation (7.10) provides an equation for calculating the applied value of T in a deeply cracked 3-point bend specimen. Derive this expression on your own.

7.2 For the condition of dead-weight loading, derive an expression for estimating the tearing modulus, T, for a standard CT specimen. Do the same for the condition of fixed remote deflection.

7.3 Calculate the amount of stable crack extension and maximum load at instability for the specimen in Example Problem 7.2 if the specimen is loaded under dead-weight conditions.

7.4 Repeat Example 7.2 without the assumption that the material is nonhardening. Assume that m = 10 and the other constants are the same as in the example problem. Compare the load-displacement curves for the two cases.

7.5 Repeat Example Problem 7.2 without neglecting elastic deformation and without assuming that the specimen is fully plastic prior to onset of ductile crack growth. Compare the predicted load-displacement plots for this case with that calculated in the example problem.

7.6 What are the conditions which favor instability governed by plastic collapse as opposed to a tearing modulus approach? Describe what approach you will use to ensure that a proper fracture criterion is always used.

7.7 If the structural alloy in Example Problem 7.2 has a $J_{Ic}$ value of 80 KJ/m², estimate the additional plastic displacement sustained by the specimen from the onset of crack initiation to instability.

7.8 If the starting crack size in Example Problem 7.2 is selected as 10 mm and we can no longer use Rice's formula for estimating J, how will you go about predicting instability? You need only to describe your approach in words.

7.9 For a center crack tension specimen, derive the expression for failure assessment diagram (FAD) for m = 20 and compare it with the FAD derived for small scale yielding condition.

7.10 Draw the failure assessment diagram for a DEN specimen assuming that the width is infinite and compare it with the diagram shown in Figure 7.6 for the center crack tension (CCT) specimen. Recall that is was argued in Section 7.1.2 that the diagram for CCT specimen is to a first approximation independent of the specimen geometry. Repeat the problem for a 3-point bend specimen assuming that the specimen always maintains small-scale yielding conditions.

7.11 What additional effects did Mott, Robert, and Wells (MRW) analysis consider which were not considered in Griffith's fracture analysis? What were the simplifying assumptions in the MRW analysis which makes one suspect the predictions from the analysis about dynamic fracture behavior?

7.12 What modifications to the MRW analysis were proposed by Berry and by Dulaney and Brace? What is the principle significance(s) of the proposed modifications?

7.13 Derive the relationship between stress intensity parameter, K, and the crack extension force, $\Box$, for a crack moving at high speed.

7.14 The governing equation for dynamic fracture is given by:

$$K(t) = K_{ID}(V_a)$$

In your own words, explain what the various terms in the above equation mean and what considerations are important in determining the terms on the two sides of the above equation.

7.15 In dynamic tests involving rapid loading, why is it important to measure the time for crack initiation?

7.16 Discuss the conditions of crack arrest in a high pressure natural gas line assuming that a longitudinal through-the-thickness crack develops and becomes unstable. Explain what material data and anlysis will have to be performed to demonstrate when crack arrest will occur.

7.17 Calculate crack speed as a function of crack size in Example Problem 7.4 for an initial crack size of 50 cm. Compare your answer with that in the Example Problem and comment on the differences and similarities in the trends.

7.18 The yield strength of 5% Ni steel is 1000 Mpa and we can assume that for practical purposes it is essentially a nonhardening material. Using the photographs of caustics shown in Figure 7.26, evaluate the accuracies of equations (7.52) and (7.57). Assume that the thickness of the specimens is 23mm and use the measured values of J provided with the photographs.

# CHAPTER 8

# CONSTRAINT EFFECTS AND MICROSCOPIC ASPECTS OF FRACTURE

In Chapters 3 to 7, the primary emphasis of our discussion was on identifying a single parameter which uniquely characterizes the crack tip stress fields independent of the specimen size and geometry. This discussion led us to characterize stable and unstable fracture in terms of K if the small-scale-yielding conditions were met, and by the J-integral if an elastic-plastic or fully plastic condition prevailed. In reality, as also briefly mentioned in Chapter 3, both K and J characterize the amplitudes of the first and only singular term in a series expansion consisting of several higher order terms that describe the stress field ahead of the crack tip. The contribution of the second order terms depends on the distance from the crack tip, **r**, the geometry and size of the specimen, and the extent of crack tip plasticity. As one approaches the crack tip, **r** →0, the singular term dominates the magnitude of the stress and the contribution from the other terms can be neglected. However, if one moves further from the crack tip, that is no longer the case.

The damage leading to fracture in almost all instances develops ahead of the crack tip. Imagine a situation in which the damage develops or at least begins to develop in a region where the contribution to the local stress from the higher order terms is significant. Since the higher order terms are geometry- and size-dependent, would it then be possible to characterize fracture and/or stable crack growth by a single parameter? Would one then expect a single parameter characterization of fracture and stable crack growth behavior to be independent of specimen size or geometry? The answer to both questions is an obvious no.

In this chapter, we will first describe a mechanics frame-work for including the effects of the second order terms in the description of the crack tip stress fields and then proceed to describe how these concepts can be used to understand several fracture related phenomena that are of considerable engineering significance. Included among them is the excessive scatter observed in the fracture data during cleavage fracture, during the ductile-brittle transition, and also the effects of geometry and size beyond the $J_{Ic}$ point during ductile crack growth.

## 8.1 Higher Order Terms of Asymptotic Series

In considering the difference between the actual stress fields at the crack tip and the magnitude predicted by the singular term, it is best to separate the discussion of purely elastic behavior, referred to as the elastic T-stress in the literature, and that of the elastic-plastic and fully plastic behavior referred to as the J-Q approach.

### 8.1.1 Elastic T-Stress

As mentioned earlier in Chapter 3, Williams [8.1] was the first to derive an expression for the full-field crack tip stress distribution ahead of the crack tip in a cracked elastic solid. He proposed the stress-field in the vicinity of the crack to have the following form:

$$\sigma_{ij}(r,\theta) = Ar^{\frac{-1}{2}} f_{ij}(\theta) + Bg_{ij}(\theta) + Cr^{\frac{1}{2}} h_{ij}(\theta) + \text{- - -} \tag{8.1}$$

where r = distance from the crack tip and the functions $f_{ij}(\theta)$, $g_{ij}(\theta)$, and $h_{ij}(\theta)$ are functions of the angle of orientation θ of a point with respect to the polar axis set located at the crack tip and the parameters A, B, C are related to the remotely applied load or stress. For example:

$$A = \frac{K}{\sqrt{2\pi}}$$

The terms B and C are dependent on the geometry of the cracked body while the first singular term which dominates the crack tip stress as $r \to 0$ depends only on K. Larsson and Carlsson [8.2] analyzed several crack configurations by the finite element method and showed that the discrepancies between the stress fields of different geometries can be largely reconciled by considering the contribution of the first non-singular term on the right-hand side of Equation (8.1) and neglecting the others. They also noted that this term contributes only and uniformly to the direct stress $\sigma_{11}$ which acts parallel to the plane of the crack as follows:

$$\sigma_{11}(r,\theta) = \frac{K}{\sqrt{2\pi r}} f_{11}(\theta) + T \quad \text{as } r \to 0 \tag{8.2}$$

To describe the observation that the T-stress contributes only to the $\sigma_{11}$ stress near the crack tip, Equation (8.2) can be written as:

$$\sigma_{ij}(r,\theta) = \frac{K}{\sqrt{2\pi r}} f_{ij}(\theta) + T \delta_{1i} \delta_{ij} \tag{8.2a}$$

where $\delta_{1i}$ and $\delta_{ij}$ are the Kroncecker delta. T is known as the elastic T-stress; it is proportional to the applied stress and it is dependent on the geometry and size of the body. For example, in a semi-infinite body with a center crack of length 2a subjected to a remote uniform stress $\sigma$ (Griffith's problem), $T = -\sigma$. For other geometries, it may be written as [8.3]:

$$T = \sigma_{11}(r,\theta) - \frac{K}{\sqrt{2\pi r}} f_{11}(\theta) \tag{8.3}$$

where

$$f_{11}(\theta) = \cos\frac{\theta}{2}\left(1 - \sin\frac{\theta}{2} \sin\frac{3\theta}{2}\right) \tag{8.4}$$

as derived earlier in Chapter 3.

An expression for T-stress was suggested by Leevers and Radon [8.4] of the following form:

$$T = \frac{BK}{\sqrt{\pi a}} \tag{8.5}$$

since

$$T = \sigma BY(a/W) \qquad\qquad K = \sigma\sqrt{\pi a}\, Y(a/W)$$

or

$$T/\sigma = BY(a/W) = Y_1(a/W) \qquad (8.6)$$

Figures 8.1 to 8.4 show the $Y_1$ (a/W) for the four most commonly used specimen geometries. A compendium of T-stress solutions for 2 and 3 dimensional cracked geometries is given by A.H. Sherry et al. [8.3].

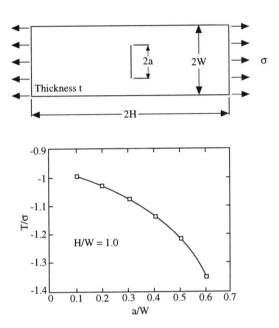

*Figure 8.1* *The function $Y_1$ (a/W) for estimating elastic T-stress in the CCT specimen (adapted from Ref. 8.3). $T/\sigma = Y_1(a/W) = -0.997 + 0.283(a/W) - 3.268(a/W)^2 + 6.622(a/W)^3 - 5.995(a/W)^4$.*

Figure 8.5 shows a plot of numerically estimated crack tip opening stress ($\sigma_y$) inside the plastic zone for several positive and negative values of $T/\sigma_0$ from the work of Kirk et al. [8.5]. The different conditions of $T/\sigma_0$ were obtained by an analysis technique referred to as the modified boundary layer (MBL) analysis. In this technique, a circular model is used containing an edge crack and stresses according to Equation 8.2a are applied on the circular boundary, Figure 8.5. A plastic zone is allowed to develop at the crack tip but its size is kept small relative to the size of the model to ensure the validity of the boundary conditions. This configuration is able to simulate the near crack tip conditions for several values of T-stress which would otherwise require analyzing a variety of specimen geometries and crack sizes. The Ramberg-Osgood exponent in the Kirk et al. [8.5] analysis was assumed to be 10. The distance ahead of the crack tip, r, is normalized by $J/\sigma_0$ (estimated value of CTOD). The reference condition is designated by T = 0 which corresponds to the small-scale-yielding (SSY) limit in which the plastic zone is negligible in comparison to the pertinent dimensions of the cracked body. The positive T values shift the stresses to values above those predicted by the SSY limit while the negative T values shift the stesses below those predicted by the SSY limit. Also, the influence of negative T values is much

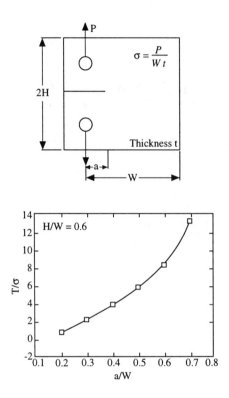

*Figure 8.2* *The function $Y_1$ (a/W) for estimating elastic T-stress in the CT specimen (adapted from Ref. 8.3). $T/\sigma = Y_1(a/W) = 6.063 - 78.987(a/W) + 380.46(a/W)^2 - 661.7(a/W)^3 + 428.45(a/W)^4$.*

more pronounced than the positive values. This effect can also be shown clearly by plotting the x and y extents of the plastic zone boundary for different values of T as shown in Figures 8.7a and 8.7b from the work of Shih et al. [8.6]. If we consider the forward sector (x, y ≥ 0 or $0 \le \theta \le \pi/2$) of the crack tip region because this is where damage leading to fracture occurs, it is clear that T < 0 has a much more dramatic influence on the plastic zone size than T > 0. Since T ≥ 0 refer to high constraint situations in which there is a high degree of triaxiality in the crack tip stress, it is expected that a single parameter characterization of fracture is more appropriate. Further, as the constraint decreases, such as for negative T values, the single-parameter characterization of toughness will become less and less likely. This theme will be developed further in the later sections of this chapter.

The elastic T-stress approach is based on a linear-elastic analysis and may be extended to situations, as described above, for conditions where plastic deformation is limited to a small region ahead of the crack tip. When the plastic deformation is wide-spread, alternate approaches must be considered. Since ductile fracture is often accompanied by large scale plasticity, it becomes essential to explore an alternate framework to the one described above. An ideal alternative framework will be one which will extend the elastic T-stress concept into the high plasticity regime but will map one-to-one to the T-stress concept in the elastic and SSY regimes. The J-Q approach is one such approach.

*8.1.2 The J-Q Approach*
O'Dowd and Shih [8.6 - 8.9] have proposed a J-Q framework which extends the T-stress concept into the elastic-plastic and fully plastic situations. Based on this new framework, fracture in low constraint

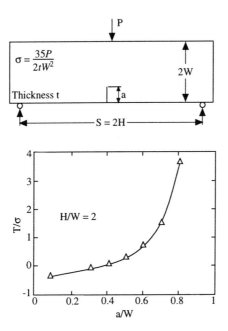

**Figure 8.3** *The function $Y_1(a/W)$ for estimating elastic T-stress in SENB specimen (adapted from Ref. 8.3). $T/\sigma = Y_1(a/W) = 0.111 - 8.982(a/W) + 53.610(a/W)^2 - 109.32(a/W)^3 + 78.977(a/W)^4$.*

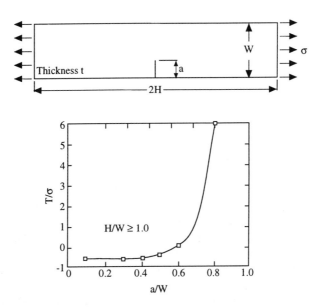

**Figure 8.4** *The function $Y_1(a/W)$ for estimating elastic T-stress in SENT specimen (adapted from Ref. 8.3). $T/\sigma = Y_1(a/W) = 0.639 - 27.332(a/W) + 233.160(a/W)^2 - 981.98(a/W)^3 + 2162(a/W)^4$.*

226  Nonlinear Fracture Mechanics for Engineers

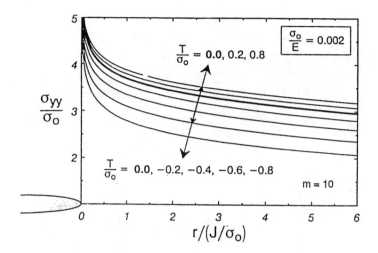

**Figure 8.5** Crack tip opening stresses as function of distance from the crack tip for different values of $T/\sigma_0$ obtained from a modified boundary layer analysis from Kirk et al. (Ref. 8.5).

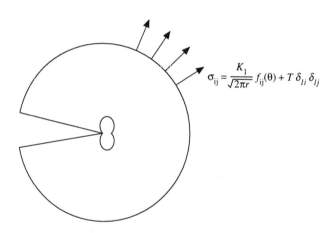

**Figure 8.6** Model for the modified boundary layer analysis.

crack geometries can be rationalized. While the application of the J-Q theory for describing fracture in low-constraint geometries is discussed later in this chapter, this section deals with the fundamental basis of this approach and shows how this theory provides a unified quantitative measure of crack-tip constraint under a wide range of deformation conditions.

From a full-field analysis, O'Dowd and Shih [8.7, 8.8] have proposed the following representation of the plane strain crack tip stresses valid only with the small displacement assumptions. In other words, finite deformations are said to occur only in the small zone ahead of the crack tip:

$$\frac{\sigma_{ij}}{\sigma_0} = \left(\frac{J}{\alpha\varepsilon_0\sigma_0 I_m r}\right)^{\frac{1}{m+1}} \hat{\sigma}_{ij}(\theta,m) + Q\left(\frac{r}{J/\sigma_0}\right)^q \tilde{\sigma}_{ij}(\theta,m) \qquad (8.7)$$

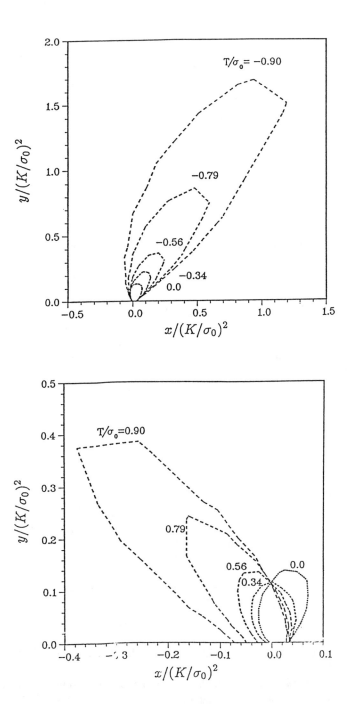

**Figure 8.7** *Plastic zone calculation from modified boundary layer analysis (a) for negative T-stress values and (b) for positive T-stress values from Shih et al. (Ref. 8.6).*

Here, $r$, $\theta$ are polar coordinates with the origin located at the crack tip, $\tilde{\sigma}_{ij}(\theta, m)$ and $\hat{\sigma}_{ij}(\theta, m)$ are both angular functions, and $q$ can be envisioned as a fitting parameter. Note that the first term on the right-hand side of equation (8.7) is the HRR term introduced in Chapter 4 and corresponds to the condition of $Q = 0$. O'Dowd and Shih also noted that $q \ll 1$ which is essentially equivalent to stating that the second order stress term is independent of the radial distance. If we normalize the angular term $\tilde{\sigma}_{ij}$ in such a way that $\hat{\sigma}_{\theta\theta}(\theta = 0)$ is unity, $Q$ is equal to the amplitude of the second order hoop stress term in equation (8.7). They further noted that $\hat{\sigma}_{rr} \approx \hat{\sigma}_{\theta\theta} \approx$ constant and $|\hat{\sigma}_{r\theta}| \ll |\hat{\sigma}_{\theta\theta}|$ in the forward sector. Thus, the second order term essentially corresponds to a uniform hydrostatic stress in that sector [8.6]. $Q$ can then be interpreted as a stress triaxiablity parameter. If $Q$ is negative, the stress triaxiality is reduced compared to the $Q = 0$ state and if $Q$ is positive, the stress triaxiality is increased compared to the $Q = 0$ state. Equation (8.7) can be rewritten as:

$$\sigma_{ij} = (\sigma_{ij})_{HRR} + Q\sigma_0 \delta_{ij} \tag{8.8}$$

Equation (8.8) is an approximation of the more rigorous equation (8.7) to simplify the calculation of $Q$. Since both these equations have been defined for the small-displacement condition, they are not valid near the crack-tip in the region $r \sim J/\sigma_0$ or $r/(J/\sigma_0) \sim 1$. The validity of this assumption was verified numerically by O'Dowd and Shih [8.9] as shown in Figure 8.8 in which the $\sigma_{\theta\theta}/\sigma_0$ values as a function of $r/(J/\sigma_0)$ are plotted from computations in which the small-displacement assumption was used with ones based on finite strain theory. For $r/(J/\sigma_0) > 1.5$, the results from the two approaches are very similar which is in agreement with the previous observation.

Based on the observation from Figure 8.8 that in the annulus $J/\sigma_0 < r < 5J/\sigma_0$, the second term in equation (8.8) corresponds to an effectively, spatially uniform hydrostatic stress-state, $Q$ can be defined by:

$$Q = \frac{\sigma_{\theta\theta} - (\sigma_{\theta\theta})_{HRR}}{\sigma_0} \quad \text{for } \theta = 0, r = 2J/\sigma_0 \tag{8.9}$$

The chosen value of $r = 2J/\sigma_0$ for the definition of $Q$ is somewhat arbitrary but it is necessary to specify a single $r$ value for consistency [8.7 - 8.9]. It has been shown that the mean variation in $Q$ with $r$ in the range $J/\sigma_0 < r < 2J/\sigma_0$ is less than 0.1 from its value at $r = 2J/\sigma_0$ [8.9]. O'Dowd and Shih have also proposed an alternate definition of $Q$ based on using the small-scale-yielding field for $T = 0$ as the reference state instead of $(\sigma_{\theta\theta})_{HRR}$. According to this definition:

$$Q = \frac{\sigma_{\theta\theta} - (\sigma_{\theta\theta})_{SSY, T=0}}{\sigma_0} \tag{8.10}$$

The above definition is more attractive because $Q$ can then be directly related to $T/\sigma_0$ as is evident by comparing equations (8.3) and (8.1).

The expressions for estimating $Q$ for different geometries of finite width are given in various publications [8.9]. A common approach used is to plot the value of $Q$ as function of $J/(b\sigma_0)$ where $b = W - a$, the uncracked ligament or as a function of $J/(a\sigma_0)$. Figures 8.9 to 8.11 provide the values of $Q$ for various $a/W$ and $m$ values for center crack tension (CCT) and single edge notch bend (SENB) specimens [8.9].

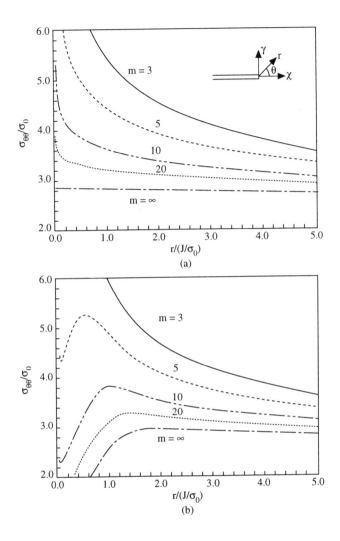

**Figure 8.8** *Plane strain hoop stress reference fields for n = 3, 5, 10, 20, and ∞ (E/$\sigma_0$ = 500, ν = 0.3) (a) from small-strain assumption and (b) accounting for finite strains. From O'Dowd and Shih (Ref. 8.9).*

### 8.1.3 T - Q Relationship

A strict one-to-one correspondence exists between Q and T for small-scale-yielding conditions. Thus, if we take the Q values for low values of $J/b\sigma_0$ and plot them with $T/\sigma_0$, a relationship given in Figure 8.12 is obtained for different m values. This relationship can be given for any geometry under small-scale yielding, by [8.9]:

$$Q = a_1 \left(\frac{T}{\sigma_0}\right) + a_2 \left(\frac{T}{\sigma_0}\right)^2 + a_3 \left(\frac{T}{\sigma_0}\right)^3 \tag{8.11}$$

where the values of $a_i$ for different values of m are given in Table 8.1.

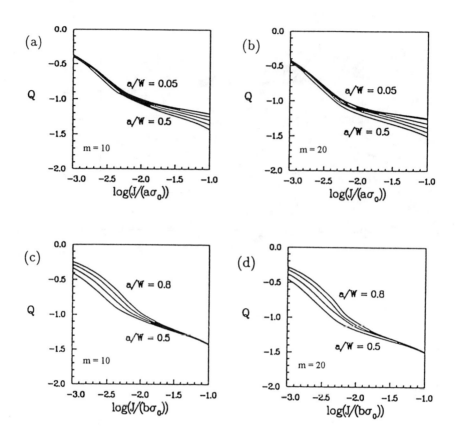

**Figure 8.9** Evolution of Q fields with increasing J for CCT specimens for $a/W = 0.05, 0.1, 0.2, 0.3, 0.4, 0.5$ (a) $m = 10$ and (b) $m = 20$, and for $a/W = 0.5, 0.6, 0.7,$ and $0.8$ for (c) $m = 10$ and (d) $m = 20$. Note that in (a) and (b) the J value is normalized with crack size, a, and in (c) and (d) the J value is normalized by the remaining ligament $W - a = b$. From O'Dowd and Shih (Ref. 8.9).

Two observations are made from Figure 8.12. The condition of $T/\sigma_0 = 0$ corresponds to $Q = 0$ as one would expect from the definition of Q, equation (8.10). It is also observed that Q increases monotonically with $T/\sigma_0$ and the crack tip stress triaxiality can be significanty lowered with respect to the reference state of T or $Q = 0$, but it cannot be increased substantially above the reference state. In view of this result, it is not surprising that under the dominantly linear-elastic conditions, the influence of specimen geometry on fracture toughness is negligible, but the differences become more prominent as we deviate from the small-scale-yielding condition.

We conclude from the above discussion on the T-stress and the triaxiality parameter Q that the latter is more general because it rigorously accounts for the influence of plasticity on the triaxial nature of the crack-tip fields and therefore has an advantage over the former. On the other hand, the T-stress approach is easier to apply because it is based on elastic analysis. In the SSY regime, the two approaches are equivalent. This whole area of research on constraint effects on fracture is still very fertile and new ideas are continuing to evolve. In the subsequent sections, we will use these concepts to explain some important aspects of cleavage and ductile fracture.

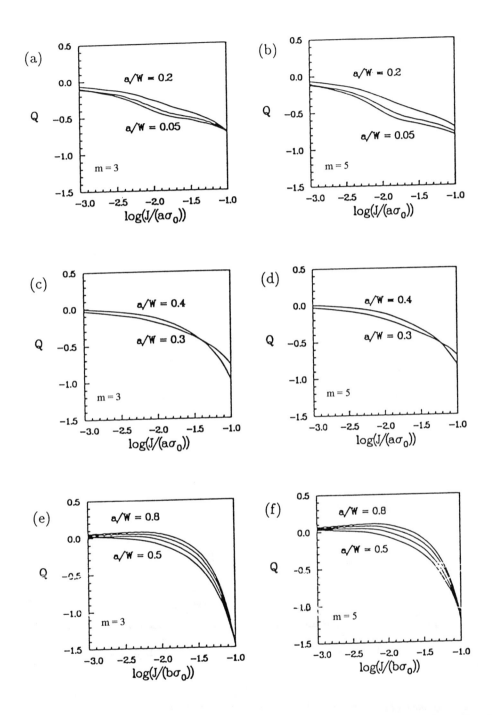

*Figure 8.10* Same as Figure 8.8 except from SENB specimens (3-point loading) and m = 3 and 5.

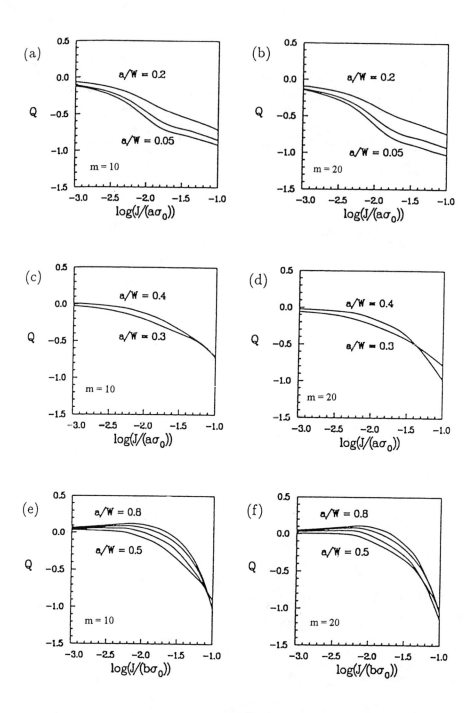

**Figure 8.11** Same as Figure 8.9 except for m values of 10 and 20.

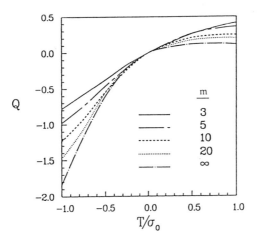

**Figure 8.12** Relationship between Q and $T/\sigma_0$ for m = 3, 5, 10, 20 and ∞. From O'Dowd and Shih (Ref. 8.9).

**Table 8.1** Polynomial Expression for Q in Terms of T-Stress (8/9)

| m | $a_1$ | $a_2$ | $a_3$ |
|---|---|---|---|
| 3 | 0.6438 | -0.1864 | -0.0448 |
| 5 | 0.7639 | -0.3219 | -0.0906 |
| 10 | 0.7594 | -0.5221 | 0 |
| 20 | 0.7438 | -0.6673 | 0.1078 |
| ∞ | 0.6567 | -0.8820 | 0.3275 |

**Example Problem 8.1**

Compare the T-stress values for SENB and CCT specimens for a/W = 0.1 and 0.6 using the CT specimens with a/W = 0.6 as reference. Assume that the applied K levels are identical and small-scale-yielding conditions are satisfied in all specimens.

*Solution:*

From Figures 8.1, 8.2, and 8.3 the T-stress can be calculated for CCT, CT, and SENB geometries, respectively, as follows:

| Specimen Geometry | a/W = | $T/\sigma_0$ 0.1 | 0.6 |
|---|---|---|---|
| CT |  | - | 8.236 |
| CCT |  | -0.995 | -1.350 |
| SENB |  | -0.353 | 0.643 |

Compared to the deeply cracked CT specimen, the other two geometries are low constraint geometries. However, T for deeply cracked SENB is > 0 and, therefore, it can be considered as a high constraint geometry. The CCT specimen has uniformly low constraint.

## 8.2 Cleavage Fracture

### 8.2.1 Microscopic Aspects of Cleavage Fracture

Cleavage fracture in crystalline materials occurs by separation along definite crystallographic planes. Although cleavage fracture has a brittle appearance, it can be preceded by substantial amounts of plastic deformation. For example, it is common during fracture testing of steels for specimens to sustain large-scale plastic deformation prior to the test being suddenly interrupted by cleavage fracture. Recall in Chapter 6 we had defined several measures of cleavage fracture toughness. Among those, the critical J at fracture, $J_c$, and the critical CTOD, $\delta_c$, were two measures described as representing fracture toughness values which were independent of specimen geometry and size. We find that $J_c$ or $\delta_c$ values exhibit quite a bit of scatter, the reasons for which cannot be explained by the preceding discussion. In this section, we will explore mechanisms of cleavage fracture to improve our understanding of this phenomenon and subsequently describe phenomenological models to quantify the scatter in its behavior for engineering purposes.

Since the orientation of crystallographic planes is different in neighboring grains, when cleavage fracture travels from one grain to the next, locally it must change angles with respect to the nominal fracture direction as shown schematically in Figure 8.13. The fracture proceeds along planes from within a family of planes such as {100} for BCC materials, that are most suitably oriented within the grain. Examples of cleavage fracture in a ferritic steel are shown in Figure 8.14 at a high magnification [8.10]. The fractures are flat with occasional ledges (also called cleavage steps) as shown in Figure 8.14a. To an unaided eye, cleavage fracture appears shiny and flat. Other features characterisitc of cleavage are shown in Figure 8.14b and include river patterns, tongues, and herringbone structure.

River patterns are caused by merging of several cleavage steps in the vicinity of grain boundaries. When a cleavage crack approaches a twist boundary and is forced to reinitiate in the new crystal, it frequently does so at several locations. This gives rise to several cleavage steps giving the appearance of a river pattern, especially when the cleavage steps appear to converge downstream. Thus, the local crack propagation direction can also be inferred from these pictures. Tongues are formed in local regions where deformation twinning occurs. When a growing cleavage crack encounters a deformation twin, the cleavage fracture continues along the twin/matrix interface. The fracture in the twinned region occurs latently along a different plane and sticks out like a 'tongue'.

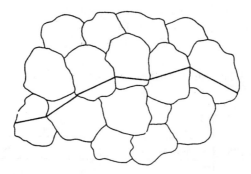

*Figure 8.13* Cleavage fracture spreading through several grains.

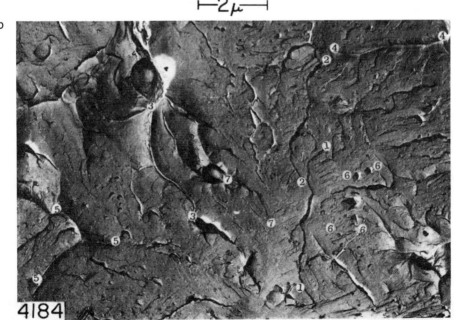

*Figure 8.14* Photomicrograph of a cleavage fracture showing (a) cleavage steps and (b) cleavage tongues and river patterns (Ref. 8.10).

Cleavage fracture occurs along planes with low atomic density because fewest atomic bonds are required to be broken along these planes for the fracture to proceed. In BCC metals these are {100} planes. Stresses must exceed the theoretical cohesive strength of these planes for cleavage fracture to occur. Figure 8.15 shows the normal stress, $\sigma_{yy}$, as a function of distance from the crack tip in small-scale yielding under plane strain conditions estimated by finite element analysis for m = 10 [8.11]. The peak stress is between 3 to 4 times the yield strength, $\sigma_0$, and it occurs one CTOD distance away from the crack tip. If we assume a high strength steel with $\sigma_0$ = 1000 MPa, the peak stress is expected to be between 3000 to 4000 MPa. The theoretical cohesive strength, estimated to be about E/10 where E = Young's modulus, is about 210,000 MPa which is much higher than the calculated stress. Therefore, for cleavage to proceed, local stress concentrations must occur. Sources of local stress concentrations are (i) the inclusions and the second phase particles, (ii) lack of appropriate number of slip systems in polycrystalline materials, and (iii) presence of microcracks ahead of the crack tip. While the first source for stress concentration is obvious, the other two need more explanation.

As we already know, polycrystals are an aggregate of several single crystals bonded together at the grain boundaries. Therefore, when a polycrystalline sample is deformed to a specified strain, the deformation in the individual grains is constrained by the deformation in the surrounding grains to maintain continuity. For example, consider the schematic in Figure 8.16. If all grains in the sample Figure 8.16a are free to deform as unrestrained single crystals, the deformation pattern will look as schmatically shown in Figure 8.16b. Since slip planes are oriented differently in each grain, the amount of shear strain on each of the active slip planes will be different to accomodate the applied deformation. There will be regions in which the neighboring grains will overlap with each other and others in which there will be gaps. Since material continuity cannot be maintained in this scheme, we must conclude that

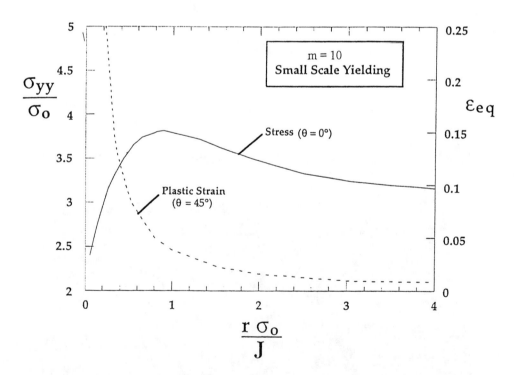

*Figure 8.15* Crack tip stress and strain ahead of the crack tip for small-scale yielding from finite element analysis by McMeeking and Parks (Ref. 8.11).

the deformation in each grain is constrained by the neighboring grains and, therefore, each grain must be capable of sustaining a general state of strain allowing it to take on any required shape. Also, the deformation in the neighboring grains will not be homogeneous. A minimum of five independent slip systems are required for each grain to be able to accomodate a general state of strain. Recall there are six independent components of strain, but during plastic deformation, the volume is conserved giving rise to the condition that the sum of all normal strains must vanish. Therefore, only five of the six strain components are independent. If five independent slip systems are not available, stress concentration will occur raising the local stress levels, especially along the grain boundaries, giving rise to conditions for cleavage fracture to occur; micro-cracks can also form to accomodate the differential strain from one grain to the next.

From the above arguments it is clear that the lack of available slip systems will promote cleavage fracture. This is consistent with the observation that face-centered cubic (FCC) materials which have 12 independent slip systems are not known to exhibit cleavage fracture. Body-centered cubic (BCC) materials exhibit cleavage fracture at low temperatures (in the lower shelf and the transition region) because sufficient slip systems are not available at low temperatures. At higher temperatures, more slip systems (up to 48 independent primary slip systems) are available in BCC materials, therefore, cleavage fracture does not occur. Hexagonal closed packed (HCP) metals also frequently exhibit cleavage fracture because there are only three independent slip systems operative in these materials at low temperatures.

Figure 8.17 shows a schematic of how cleavage fracture conditions can occur due to the presence of microcracks ahead of the main crack. Here, the crack tip stress is further amplified by the presence of the microcracks raising it to levels which exceed the theoretical cohesive strength. As explained in the previous paragraph, microcracks can be formed to accomodate strain inhomogenieties in neighboring grains. This is frequently observed in brittle materials such as in ceramics and in intermetallics, both of which have limited slip systems.

### 8.2.2 RKR Model for Cleavage Fracture

Ritchie, Knott, and Rice [8.12] developed a simple model for predicting cleavage fracture toughness which captures some of the salient microscopic observations described in the previous discussion. They postulated that cleavage fracture is stress-controlled and it occurs when the stress at the crack tip exceeds a critical stress, $\sigma_c$, over a critical distance $x_c$. The idea of the critical distance $x_c$ was explained on the basis that in order for the cleavage crack to first initiate at an inclusion and propagate into a ferrite grain, the stress must be sufficient to continue propagation into the next grain. Thus, they argued that $x_c$ must be equal to at least two grain diameters. However, no consistent relationships between grain size and the critical distance, $x_c$, have been observed. Curry and Knott [8.13] argued that a finite volume of material must be sampled to find a particle that is sufficient in size to trigger cleavage fracture. This critical volume can be related to the average distance between such particles, and the critical distance can then be linked to this spacing. This proposal also provides an explanation for why fracture toughness data in the cleavage regime are so scattered. One can argue that since the distance between the crack front and the nearest cleavage trigger can vary considerably from sample to sample, the fracture toughness is also expected to vary similarly. Also, a thicker sample is more likely to have a large fracture trigger along the crack front than a thinner sample, the fracture toughness of the larger sample is much more likely to be lower [8.14].

### 8.2.3 Mathematical Description of Scatter in Cleavage Fracture Toughness

It is now well known that fracture toughness data in the cleavage regime can have large amounts of scatter. Part of the reason is purely statistical as explained in the preceding discussion. However, it is also true that considerable scatter in the data is caused by the enhanced effect of constraint. This is particularly the case when the data are based on specimens from different size and geometry. Also, constraint effects are more important in cleavage fracture as compared to ductile fracture because, as ex-

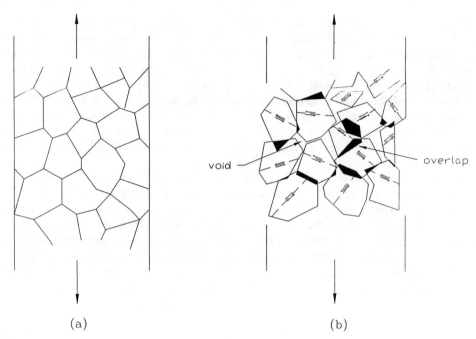

**Figure 8.16** *Deformation in polycrystals (a) underformed polycrystalline sample and (b) deformation in individual grains in response to a remotely applied strain if each grain could act as an unrestrained single crystal.*

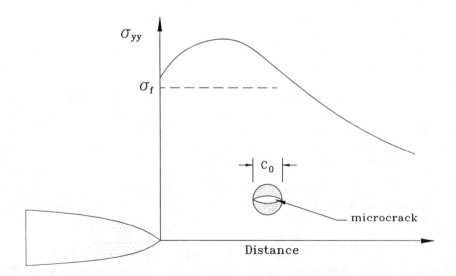

**Figure 8.17** *Initiation of cleavage microcrack on a second phase particle ahead of a crack tip. The microcrack raises the local stress to sufficiently high levels to initiate cleavage fracture.*

plained before, cleavage fracture is stress-dominated while ductile fracture is more strain controlled, and constraint has more influence on the peak stress. In this section, we will consider two models which have been developed by Anderson, Dodds, and co-workers [8.15-8.18] which provide a frame-work for treating data scatter in the cleavage regime. The first model treats effects due to constraint while the second describes the statistical aspects of cleavage fracture.

**Constraint Effects in Explaining Data Scatter** Anderson and Dodds have made use of the weakest link statistics to model cleavage failure. The weakest link model assumes that cleavage fracture is controlled by the largest or the most favorable oriented fracture triggering particle. Suppose a principal stress, $\sigma_1$, is required to trigger cleavage fracture associated with a particle. Then the probability of cleavage fracture in a cracked specimen can be expressed in the following form:

$$F = F[V(\sigma_1)] \qquad (8.12)$$

where F = failure probabilty, $V(\sigma_1)$ = cummulative volume sampled in which the principal stress exceeds $\sigma_1$. Dimensional analysis shows that for small-scale yielding, the principal stress ahead of the crack tip at a point specified by $(r, \theta)$ is given by:

$$\frac{\sigma_1}{\sigma_0} = f\left(\frac{J}{\sigma_0 r}, \theta\right) \qquad (8.13)$$

where $\sigma_0$ = yield strength and J = J-integral. The above equation can be inverted and written as:
The area, A, inside a specific principal stress contour is given by:

$$r(\sigma_1/\sigma_0, \theta) = \frac{J}{\sigma_0} g(\sigma_1/\sigma_0, \theta) \qquad (8.14)$$

where

$$A(\sigma_1/\sigma_0) = \frac{J^2}{\sigma_0^2} h(\sigma_1/\sigma_0) \qquad (8.15)$$

$$h(\sigma_1/\sigma_0) = \frac{1}{2} \int_{-\pi}^{\pi} g^2(\sigma_1/\sigma_0, \theta) \, d\theta \qquad (8.16)$$

For a given value of $\sigma_1$, the area scales with $J^2$ for the case of small-scale yielding. Under large-scale yielding, the specimen experiences a loss of constraint and therefore the area inside a given principal stress contour is less than predicted from small-scale yielding.

$$A(\sigma_1/\sigma_0) = \phi \frac{J^2}{\sigma^2} h(\sigma_1/\sigma_0) \qquad (8.17)$$

240   *Nonlinear Fracture Mechanics for Engineers*

where $\phi$ can be regarded as the constraint-loss factor with a value less than 1. By comparing equations (8.15) and (8.17) and designating the J value corresponding to small-scale-yielding as $J_{SSY}$, we can write:

$$\frac{J}{J_{SSY}} = \sqrt{\frac{1}{\Phi}} \tag{8.18}$$

$J_{SSY}$ can be viewed as the effective driving force for cleavage for the condition of Q = 0 and J can be regarded as the apparent driving force. As an example, if a finite size specimen fails at $J_c$ = 200 kJ/m², and has a $J/J_{SSY}$ = 2.0, a very large specimen from the same material may fail at a J value as low as 100 kJ/m². Using this approach, the fracture toughness values can be adjusted for constraint. Anderson et al. [8.18] have performed finite element analysis of single edge notch bend specimens of different a/W ratios as well as different m values to estimate the value of $J/J_{SSY}$. In these studies, $\alpha$ = 1.0, $\varepsilon_0$ = .002, and $\sigma_0$ = 414 MPa were selected as other material parameters. These results are shown in Figures 8.18a and 8.18b. The values of $J/J_{SSY}$ were estimated to be as high as 4 for shallow crack specimens at high values of J. These results clearly show the significant loss of constraint at small a/W values.

Anderson et al. [8.18] used the above calculations to adjust the $J_c$ values measured from specimens of different size and a/W values. Figure 8.19a shows the uncorrected data on A515 Grade 70 steel at room temperature generated by Kirk et al. [8.19] showing the overall large variation in $J_c$, including a systematic variation with a/W. Correcting these data for constraint by using them to calculate the $J_{SSY}$ value shows in Figure 8.19b that the scatter is considerably reduced and there are no systematic variations with a/W or with thickness.

From Figures 8.18a and b it appears that to measure the geometry and size independent value of fracture toughness, the following size criterion must be satisfied:

$$B, W - a, a \geq \frac{200 J_c}{\sigma_y}$$

This criterion is consistent with the most recent ASTM recommended criterion, equation (6.8). Specimens not following the above requirement may yield cleavage fracture toughness values that are not very useful in practical applications. Also note that the a/W value should be $\geq$ 0.5 in addition to meeting the minimum size requirement. The above criterion is not sufficient for specimens with shallow cracks.

**Example Problem 8.2**

The average value of $J_c$ measured for a ferritic steel is 50 kJ/m² using three-point bend specimens with a/W = 0.15. The specimen thickness was 25mm, width was 50mm, and the span was 200mm. What is the most likely average $J_c$ value from a specimen with an a/W = 0.5 if the exponent, m, in this material = 10 and $\sigma_0$ = 400 Mpa?

*Solution:*

For the specimen with a/W = 0.15, $a\sigma_0/J$ = 60 at fracture from Figure 8.18a, the value of $J/J_{SSY}$ ≈ 2.95, thus, $J_{SSY}$ = 17.1 kJ/m². For specimen with a/W = 0.5, the $J/J_{SSY}$ vs. $a\sigma_0/J$ relationship is also provided in Figure 8.18a. To determine the J at fracture, we need to calculate the value of $J/J_{SSY}$ and $a\sigma_0/J$ for several values of J and make a plot of this relationship. The J value we are looking for will be determined by the intersection between the plot in Figure 8.18a and the above plot. For a/W = 0.5:

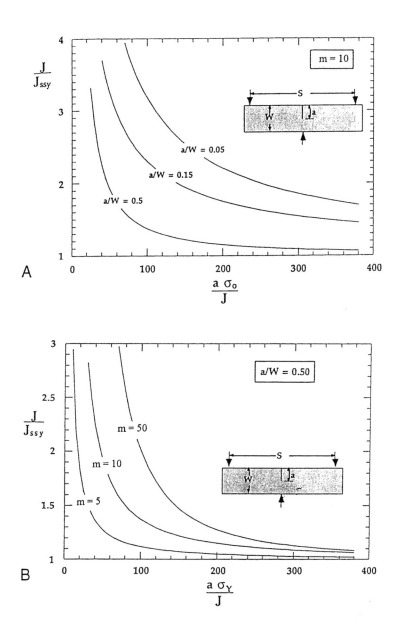

**Figure 8.18** $J/J_{SSY}$ ratio as function of specimen size and (a) crack size for m = 10 and (b) m for a crack size given by a/W = 0.5 from Anderson et al. (Ref. 8.18).

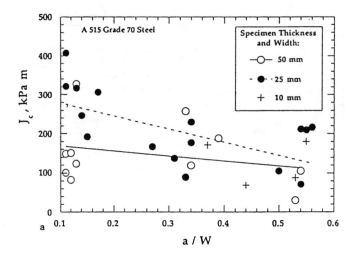

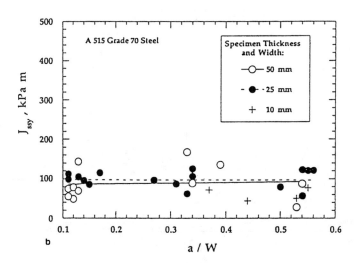

*Figure 8.19* Fracture toughness data for an A515 Grade 70 steel at room temperature from the work of Kirk et al. (Ref. 8.19). (a) Experimental values of $J_c$; (b) $J_c$ values corrected for constraint loss from Anderson et al. (Ref. 8.18).

| J (KJ/m²) | $a\sigma_0/J$ | $J/J_{SSY}$ |
|---|---|---|
| 17.1 | 584 | 1.0 |
| 18 | 555 | 1.052 |
| 19 | 526 | 1.111 |
| 20 | 500 | 1.169 |
| 25 | 400 | 1.462 |
| 30 | 333 | 1.75 |

The intersection occurs at $J/J_{SSY} = 1.0$ as shown in the plot below. Hence, the estimated $J_c \approx J_{SSY} = 17.1$ kJ/m² for a specimen with a/W = 0.5.

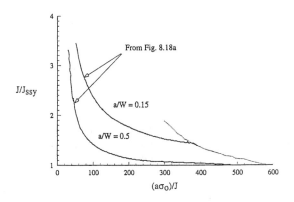

## Example Problem 8.3

Starting from equation (8.8) and using the RKR criterion, derive an expression for cleavage toughness, $J_c$, as a function of Q, cleavage stress, $\sigma_c$ and $J_{SSY}$ = cleavage toughness for the condition of Q = 0 (or T = 0).

*Solution:*

RKR criterion states that fracture occurs by cleavage when the stress, $\sigma_{\theta\theta}$, exceeds the critical stress, $\sigma_c$, at the critical distance, $r_c$. From equation (8.8), we can write at cleavage:

$$\frac{\sigma_c}{\sigma_0} = \left( \frac{J_c}{\alpha\varepsilon_0\sigma_0 I_m r_c} \right)^{\frac{1}{m+1}} \tilde{\sigma}_{\theta\theta}(\theta=0) + Q$$

For Q = 0, $J_c = J_{SSY}$, hence:

$$\frac{\sigma_c}{\sigma_0} = \left( \frac{J_{SSY}}{\alpha \varepsilon_0 \sigma_0 I_m r_c} \right)^{\frac{1}{m+1}} \tilde{\sigma}_{\theta\theta} (\theta = 0)$$

From the above equation, we solve for $r_c$ and substitute into the first equation to obtain:

$$J_c = J_{SSY} (1 - Q\sigma_0 / \sigma_c)^{m+1}$$

The above equation yields the J-Q locus for cleavage fracture toughness and is an alternative way to the one used in Figure 8.19 to account for constraint. To apply this approach, a value of $\sigma_c/\sigma_0$ is needed which for m = 10 is 3.8 (see Figure 8.15). For other m values, this value may have to be back calculated from the data. In such a case, this equation only predicts the J-Q trend. In Exercise Problem 8.18, you are required to plot this trend for an A515 steel.

**Models for Statistical Scatter** The models presented in this section known as the weakest link model is due to concepts developed by several researchers, but its present form is due to efforts by Anderson and Dodds [8.18]. If we consider a volume of material V, with ρ critical size particles per unit volume. In the weakest link model, the probabiltiy of failure, F, can be inferred from the Poisson distribution:

$$F = 1 - \exp(-\rho V) \tag{8.19}$$

The exponential term in the above equation is the probability of finding zero critical particles in V, therefore, F is the probability of finding at least one critical particle in V capable of triggering cleavage fracture. The Poisson distribution can be derived from the binomial distribution because ρ is small and V is large. Since cleavage is stress controlled, and ρ increases with stress, $\sigma_1$, because more particles are capable of causing cleavage at higher stresses, we can write:

$$\rho = \rho(\sigma_1) \tag{8.20}$$

We know that the front of crack, $\sigma_1$, varies with distance from the crack tip. Therefore, failure probability must be integrated over individual volume elements ahead of the crack tip:

$$F = 1 - \exp\left[ -\int_V \rho \, dV \right] \tag{8.21}$$

dV can be obtained by differentiating equation (8.15) after multiplying the area A by the thickness B. Thus:

$$dV(\sigma_1) = \frac{BJ^2}{\sigma_0^2} \frac{\partial h}{\partial \sigma_1} d\sigma_1 \tag{8.22}$$

or $\quad F = 1 - \exp\left( - \frac{BJ^2}{\sigma_0^2} \int_{\sigma_u}^{\sigma_{max}} \rho(\sigma_1) \frac{\partial h}{\partial \sigma_1} d\sigma_1 \right) \tag{8.23}$

where $\sigma_{max}$ is the maximum stress which is likely to occur ahead of the crack tip and $\sigma_u$ is the threshold fracture stress corresponding to the largest fracture-triggering particle the material is likely to contain. $\sigma_u$ can also be interpreted as the lowest stress at which cleavage fracture is likely to occur in this material. By setting $J = J_c$ in the above equation, we can obtain an expression for the statistical distribution of critical J values which can be rewritten in the following form:

$$F = 1 - \exp\left[-\frac{B}{B_0}\left(\frac{J_c}{\theta_J}\right)^2\right] \qquad (8.24)$$

In the above equation, $B_0$ is a reference thickness. When $B = B_0$, $\theta_J$ is the 63$^{rd}$ percentile $J_c$ value. Equation (8.24) can be written in terms of K as follows:

$$F = 1 - \exp\left[-\frac{B}{B_0}\left(\frac{K_{Ic}}{\theta_K}\right)^4\right] \qquad (8.25)$$

The above equations are in the form of a two-parameter Weibull distribution having a fixed slope of 4, if plotted in terms of $K_{IC}$. In Figure 8.20, the fracture toughness data, $K_{IC}$, is plotted for A508 steel at -75° in a Weibull plot. Recall, $K_{Jc}$ is the cleavage fracture toughness derived from $J_c$ and converted by the formula $K_{Jc} = (EJ_c/(1-v^2))^{1/2}$. The as-measured data as well as the data after applying the constraint correction described in the previous section are plotted in this figure. As expected, the constraint corrected fracture toughness values have less scatter. The Weibull slope of 4 is shown in the figure. From this plot, it is clear that the slope of 4 is far less than the actual slope of the data. Since a higher Weibull slope represents less scatter, it can be concluded that this approach overpredicts the amount of scatter in this data. Thus, the weak link model is not a satisfactory description of the scatter in the cleavage toughness data.

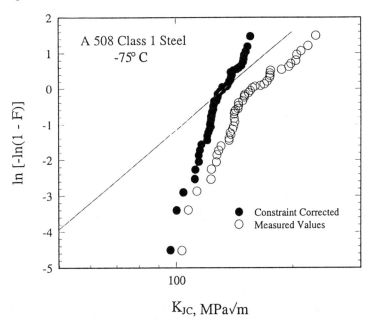

**Figure 8.20** *Weibull plot of the A508 Class 1 steel at -75°C from Anderson et al. (Ref. 8.18).*

There is also an additional weakness in the weakest-link model. This equation predicts the minimum toughness to be zero which is incorrect. A modified statistical model has been proposed by Anderson et al. [8.18] to overcome some of these problems. However, this model lacks the detailed physical basis and should be more correctly classified as a model for representing the data. The characteristic equation in this model is:

$$F = 1 - \exp \left[ \frac{B}{B_0} \left( \frac{K_{Jc} - K_{min}}{\theta_K - K_{min}} \right)^4 \right] \quad (8.26)$$

In the above equation, a third Weibull parameter, $K_{min}$, is introduced which is the minimum value of critical K at which fracture is expected to occur. This equation describes the experimental data quite accurately as shown in Figure 8.21. It appears that the 3 parameter Weibull representation has sufficient flexibility to represent cleavage fracture toughness data. However, as mentioned before this equation lacks a rigorous physical basis and it merely provides a good fit to the experimental data. It is nevertheless quite useful for engineering purposes. As can be seen in equation (8.25) as well as in equation (8.26), the failure probability is dependent on the thickness of the specimen. Thus, both equations correctly predict the size effect in the data scatter. In order to fit the equation, large amounts of data are needed under each test condition. This is a significant drawback in applications.

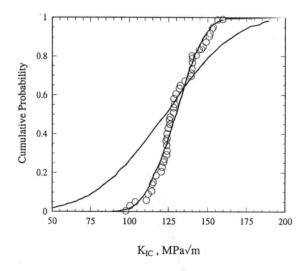

*Figure 8.21* *Comparison of experimental fracture toughness data in Figure 8.20, except using the three-parameter Weibull fit. The predictions from the two-parameter Weibull fit (of the weakest link model) is also shown for comparison. The experimental data are adjusted for constraint loss. From Anderson et al. (Ref. 8.18).*

## 8.3 Ductile Fracture

### 8.3.1 Microscopic Aspects of Ductile Fracture

As mentioned in previous chapters, ductile fracture is accompanied by extensive amounts of plastic deformation. The classic example of ductile fracture is seen in FCC pure metals in which polycrystalline tensile specimens neck down to essentially a point prior to separation as schematically seen in Figure

8.22. FCC systems exhibit such phenomenon because they have a 12 slip system and thus can avoid cleavage fracture. However, engineering alloys contain particles in the form of inclusions, dispersoids, and precipitates which considerably alter the ductile fracture behavior when compared to that of pure metals. Several of these particles are added deliberately to metals to strengthen them while others are present as impurities. Regardless of their origin, these particles participate in the fracture process. During cleavage fracture, as mentioned in the previous section, they provide sites for nucleating microcracks that subsequently propagate through the matrix by raising the local stress levels. By contrast, during ductile fracture, microvoids form at the particles due to matrix/particle interface decohesion which is followed by void growth when the deformation is continued. Fracture occurs when these voids become large enough such that the ligaments between them rupture and large-scale void coalescence occurs. Consider a cylindrical tensile specimen at the onset of necking. Necking produces a localized triaxial state of stress in the interior of the specimen which promotes void initiation, growth, and coalescence in that region. The near surface region of the specimen does not experience the same stress triaxiality and as a result does not undergo the same amount of void nucleation and growth as the interior. Thus, a penny-shaped crack can be envisioned to form in the center of the specimen from which 45° shear deformation bands extend to the specimen surface. The subsequent fracture propagates from the interior of the specimen to the surface along the shear bands giving rise to the classic cup and cone fracture as shown in Figure 8.23. Besides explaining why cup and cone fractures are produced during tensile testing, this example also points to the role of stress triaxiality in the initiation, growth and coalescence of voids.

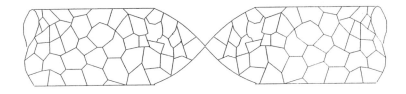

*Figure 8.22* Fracture in tensile specimens of pure FCC metals showing necking down to virtually a point prior to fracture.

Second phase particles in structural metals are present at three size levels. Some particles, which are largely impurities, can be as large as 100 microns in size and others can be as small as 1 micron. Such particles typically start out smaller and are spherical in shape in the as-cast billet but become elongated in the direction of plastic deformation during the subsequent steps in the fabrication process. For example, rolling or extrusion can cause the particles to become needle-like and much larger in the axial direction compared to the other dimensions. Figure 8.24 shows an example of how these particles participate in the stable crack growth process in a 303 stainless steel at different levels of J. The crack is seen to grow along these particles. The next size is that of intermediate size particles which can range from 50 to 500 nanometers (1 nm = $10^{-9}$ m). These particles are complex compounds which are quite often oxides and are deliberately added to provide dispersion strengthening and creep deformation resistance. The smallest particles are coherent precipitates and are typically 5 to 50 nanometers in size. These precipitates have cyrstal structure which can be coherent with the crystal structure of the matrix and provide strengthening of the matrix, typically at the expense of fracture toughness.

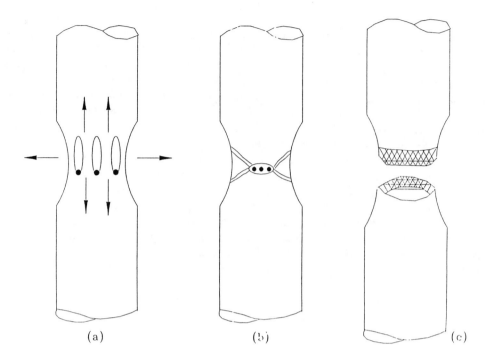

***Figure 8.23*** *Evolution of cup and cone fracture in tensile specimens of engineering alloys, (a) the triaxial stress state in the interior during necking helps to nucleate voids on inclusions resulting in (b) a penny-shaped crack from which 45° shear deformation bands emanate leading to (c) a cup and cone fracture.*

When voids grow on particle sites, dimples are formed on the fracture surface which can be observed under the electron microscope. Comparing the size of the dimple at failure with the size of particle nucleating it can give some clues about the fracture toughness. When voids are able to grow substantially compared to their initiation size, considerable plastic deformation accompanies fracture and the toughness can be expected to be high, and vice-versa.

At the microscopic level, dimples are visible on the fracture surface as shown in Figures 8.25a and b. The dimples in Figure 8.25a are equiaxed in nature and are formed by predominantly tensile loading while the elongated dimples, Figure 8.25b, are indicative of a shear process [8.12]. For example, in the central region of the cylindrical tensile specimen one would expect equiaxed dimples while as on the slant surface, the dimples are expected to be elongated.

### 8.3.2 Models for Predicting the $J_R$-Curve

The $J_R$-curve, which is also known as the stable ductile tearing curve, is a classic example of ductile fracture in action. In this section, we will consider a simple model proposed by Saxena et al. [8.22, 8.23] based on void initiation, growth, and coalescence to derive the shape of the $J_R$-curve. We assume the following physical process occuring at the crack tip. Ahead of the fatigue precrack, inclusions capable of initiating voids are present, Figure 8.26. These inclusions have an average diameter of $2R_0$ and an average spacing of $\lambda$ between them. As we increase the load on the specimen, voids first initiate and subsequently grow under the stresses in the vicinity of the crack tip. When the void nearest to the crack tip becomes large enough to coalesce with the crack tip, the crack can be considered to have advanced. Since this is the first physical evidence of tearing, the J value corresponding to this point can be identified with $J_{Ic}$. Stable crack growth continues as the second, third, and subsequent voids coalesce with the crack tip giving rise to the R-curve phenomenon. This picture seems consistent with the stable crack growth process shown in Figure 8.24 [8.22] in a 303 stainless steel. Thus, in order to mathematically describe

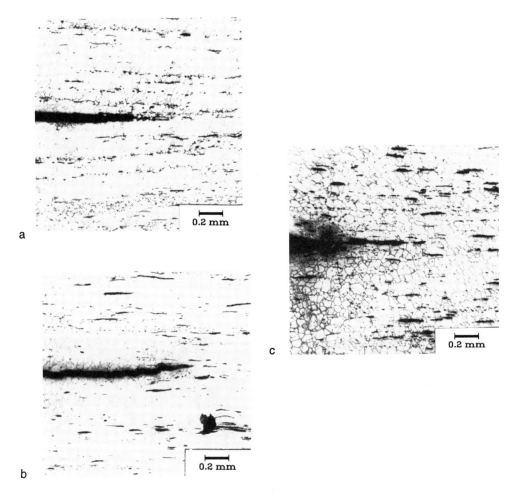

*Figure 8.24* Interaction betwwen particles and ductile crack growth in 303 stainless steel (a) $J = 43$ kJ/m², (b) $J = 55$ kJ/m², and (c) $J = 70$ kJ/m². From Saxena et al. (Ref. 8.22).

this phenomenon, we need to develop models for describing void nucleation, void growth, and a criterion for void coalescence. We will first describe models suitable for representing these steps in the stable crack growth process and subsequently use them to estimate the $J_R$-curve.

The most widely used model for void nucleation is one proposed by Argon et al. [8.24]. They proposed that the interfacial stress at the particle/matrix interface responsible for decohesion is equal to the sum of the applied hydrostatic stress and the effective von Mises stress. When the sum of these stresses exceeds a critical value, decohesion occurs at the particle/matrix interface and the void can be considered to have nucleated. Their model further states that nucleation strain decreases as the hydrostatic stress increases. Since the crack tip is under a severe hydrostatic state of stress and the interface strength between the particle and matrix is low, at least for the large particles, we can assume that void nucleation occurs quite early in the fracture process; in fact, to simply assume that voids nucleate in the crack tip region prior to the value of J approaching $J_{Ic}$. Therefore, this step can be conveniently neglected as being inconsequential in determining the $J_{Ic}$ fracture toughness and the development of the $J_R$-curve.

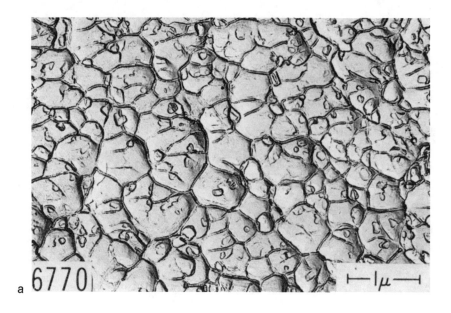

*Figure 8.25* Formation of dimples on the fracture surface during ductile fracture (a) equiaxed dimples and (b) elongated dimples. From Beachem and Pelloux (Ref. 8.10).

Constraint Effects and Microscopic Aspects of Fracture 251

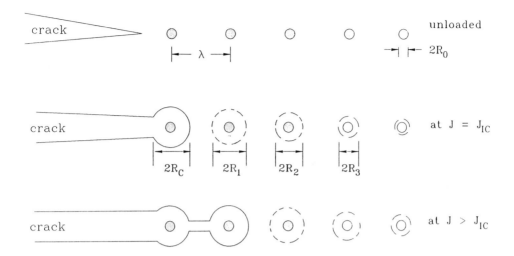

*Figure 8.26 Schematic one-dimensional portrayal of the void growth and coalescence process at the crack tip (Ref. 8.22).*

The void growth model used is the simple Rice and Tracey [8.25] model which considers an isolated void of radius $R_0$ in an infinite media consisting of nonhardening material being subjected to a 3-d state of stress. In their derivation they varied the stress states considerably and provided the following simple relationship which relates the increment in the void radius to the increment in the applied strain:

$$\frac{dR}{R_0} = .322 \exp\left(\frac{1.5\sigma_m}{\sigma_e}\right) d\varepsilon_e \qquad (8.27)$$

where dR is the average increase in void radius, $\sigma_e$ and $\varepsilon_e$ are effective von Mises stress and strain, respectively, and $\sigma_m$ is the hydrostatic stress. Applying the above equation to the crack tip region where plane strain conditions dominate, $\sigma_e = 1.15\ \sigma_0$, so equation (8.27) can be written as:

$$\frac{dR}{R_0} = .280 \exp\left(\frac{1.5\sigma_m}{\sigma_0}\right) d\varepsilon_e \qquad (8.27a)$$

If we assume that the coalescence of the void nearest to the crack tip occurs when its radius reaches a critical diameter, $R_c$, we can estimate the corresponding critical strain, $\varepsilon_e^c$, by the following equation:

$$\varepsilon_e^c = \frac{\ln(R_c/R_0)}{.280\exp(1.5\sigma_m/\sigma_0)} \qquad (8.28)$$

Recalling from Chapter 4 and considering only the proportional term in equation (4.28), we get the following equation:

$$\frac{d\varepsilon_e}{\varepsilon_e} = \frac{m}{m+1} \frac{dJ}{J} \tag{8.29}$$

Using the HRR equation (4.18), we can write:

$$\varepsilon_e = \alpha\varepsilon_0 \left[ \frac{J}{I_m \sigma_0 \varepsilon_0 \alpha r} \right]^{\frac{m}{(1+m)}} \hat{\varepsilon}_e(\theta) \tag{8.30}$$

Substituting equation (8.30) into equation (8.29) and rearranging the terms and integrating the left-hand side between limits of 0 and $\varepsilon_e^c$ and the right-hand side between the limits of 0 and $J_{ic}$, we get:

$$J_{Ic} = \left( \frac{\varepsilon_e^c}{\alpha\varepsilon_0 \hat{\varepsilon}_e(\theta)} \right)^{\frac{(m+1)}{m}} (\alpha\varepsilon_0\sigma_0 I_m \lambda) \tag{8.31}$$

In the above equation, a further assumption is made that $r = \lambda$ which is the distance between the first void and the crack tip. This assumption is valid for $R_0 \ll \lambda$. Further substituting equation (8.28) into equation (8.31), we get the following result:

$$J_{Ic} = \left[ \frac{\ln(R_c/R_0)}{.280\, \alpha\varepsilon_0 \hat{\varepsilon}_e(\theta)\exp(1.5\sigma_m/\sigma_0)} \right]^{\frac{(m+1)}{m}} (\alpha\sigma_0\varepsilon_0 I_m \lambda) \tag{8.32}$$

In the above equation, all quantities with the exception of $R_c$ can be explicitly obtained. The HRR field quantities were described in Chapter 4 and are tabulated in Appendix 1 (8.26). The ratio $\sigma_m/\sigma_0$ can be considered as the stress elevation factor. The value of $\sigma_m/\sigma_0$ as a function of distance from the crack tip for different values of m was reported by O'Dowd and Shih [8.9]. They carried out the calculations using finite element analysis using the MBL approach using both small strain and finite strain approaches.

As expected, beyond the value of $r/(J/\sigma_0)$ of 1, the results from the two analyses were essentially equal. A value of $\sigma_m/\sigma_0$ is selected for $r/(J/\sigma_0)$ of 2 to use in the model. At this distance from the crack tip, $\sigma_m/\sigma_0$ becomes relatively constant. Also, since $\sigma_m/\sigma_0$ value is highly related to the triaxiality at the crack tip, it is best to be consistent with the definition of Q, the triaxiality parameter which is also defined at the same location. For $r/(J/\sigma_0) = 2$, the value of $\sigma_m/\sigma_0$ can be accurately approximated by the following equation:

$$\sigma_m/\sigma_0 = 2.266 + 5.754\left(\frac{1}{m}\right) - 3.24\left(\frac{1}{m}\right)^2 \tag{8.33}$$

$R_0$ and $\lambda$ are microstructural parameters representing the average inclusion radius and spacing, respectively. These can be obtained by metallography. The estimation of $R_c$ is somewhat indirect. If the ductility of the material is available from a tensile test, the following procedure may be used to estimate $R_c$. In a tensile test, void coalescence in the interior of the specimen begins to occur at the point of necking. In a large gage length specimen, this strain can be approximated by the percent elongation. Alternatively, if the full stress-strain curve is available, the strain, $\varepsilon_u$, corresponding to the ultimate tensile point, $\sigma_u$, should be used. If this strain is larger than 10 percent, it should be converted to true strain and engineering stress must be converted to true stress before using their values. In a tensile test, $\varepsilon_e = \varepsilon_1$ and $\sigma_m = \sigma_1/3$. Thus, integrating equation (8.27) and applying the result to the ultimate tensile point, we get:

$$\varepsilon_u = \frac{\ln(R_c/R_0)}{.322 \exp(1.5 \sigma_m/\sigma_e)}$$

Since $\sigma_e = \sigma_u$ and $\sigma_m = (1/3) \sigma_u$ during a tensile test at the ultimate tensile point, we can derive the following relationship for $R_c$:

$$R_c = R_0 \exp(.531 \varepsilon_u) \qquad (8.34)$$

Substituting for $R_c$ in equation (8.32), we can write:

$$J_{Ic} \approx \left[ \frac{.532 \varepsilon_u}{.280 \alpha \varepsilon_0 \hat{\varepsilon}_e(\theta) \exp(1.5 \sigma_m/\sigma_0)} \right]^{\frac{m+1}{m}} (\alpha \sigma_0 \varepsilon_0 I_m \lambda) \qquad (8.35)$$

By recognizing that while the first void grew and coalesced with the crack tip, the 2$^{nd}$, 3$^{rd}$, 4$^{th}$, and 5$^{th}$ --- voids also grew. The magnitude of their growth can be estimated by the following equation:

$$\ln(R_i/R_0) = \left[ .280 \alpha \varepsilon_0 \hat{\varepsilon}(\theta) \exp(1.5 \sigma_m/\sigma_0) \right] \left[ \frac{J_{Ic}}{\alpha \sigma_0 \varepsilon_0 I_m \lambda i} \right]^{\frac{m}{m+1}} \qquad (8.36)$$

where $i$ refers to number of the void. While the second void coalesced with the crack tip, the third, fourth, and fifth voids grew further and this process continued on. Since at the time the second void becomes the nearest void to the crack tip, its size is greater than the size of the first void when it was at that position relative to the crack tip. Hence, the increment in J needed to grow that void to critical size will be less than $J_{Ic}$. As this process continues, the amount of J increment required to grow the crack by another void spacing continuously decreases leading to a $J_R$-curve with a continuously decreasing slope. An example of such a calculation is shown in Figure 8.27 from the recent work of Saxena and Cretegny [8.23]. The predicted $J_R$-curve is shown in this figure along with experimental data for 303 steel [8.22]. The agreement between the predicted and observed trends is excellent. The details of this calculation will be illustrated as part of Example Problem 8.4.

The $J_{Ic}$ estimated from equation (8.35) is somewhat different from the operational definition given by ASTM [8.26] and also described in Chapter 4. In the ASTM definition, the $J_{Ic}$ is referenced to the

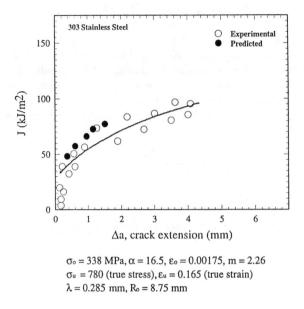

*Figure 8.27  Comparison between predicted and measured $J_R$-curves for 303 stainless steel (Ref. 8.23).*

point of intersection between the .2 mm crack extension offset line and the $J_R$-curve. In the definition used in this section, $J_{Ic}$ is referred to as the point of first crack extension equal in size to the interparticle spacing. If the interparticle spacing is 0.2 mm, the definitions are essentially the same but not otherwise. However, since the model predicts the entire $J_R$-curve, the predicted curve can be analytically fitted and the J corresponding to 0.2 mm crack extension can be calculated. When this was done for the 303 stainless steel used in reference [8.22], a $J_{Ic}$ of 20.9 kJ/m² was calculated which compares very favorably with the measured value of 21.8 kJ/m². This provides considerable confidence in the model.

Recent studies by Dodds and Ruggieri [8.27] have attempted to apply the more advanced Gurson void growth model [8.28] in an elaborate finite element scheme. They have also accounted for the crack tip constraint effects more accurately than can be represented in a simple model presented above. However, the implementation of their model is far more complex and it contains parameters such as area fraction occupied by voids ($f_o$) and the process zone size (D). A predicted $J_R$-curve from their results of A 533B steel is shown in Figure 8.28 along with experimental data. The influence of crack size on fracture toughness due to differences in constraint is very nicely predicted in their calculations which the simple model is not capable of predicting.

### Example Problem 8.4

A 303 stainless steel has the following material properties:

| $\sigma_0 = 338$ MPa | $\alpha = 16.5$ | $\varepsilon_0 = .00175$ | $m = 2.26$ |
|---|---|---|---|
| $\sigma_u = 780$ MPa | $\varepsilon_u = .165$ | $\lambda = .285$ mm | $R_0 = 8.75$ um |

Estimate the $J_{Ic}$ and the $J_R$-curve for this material.

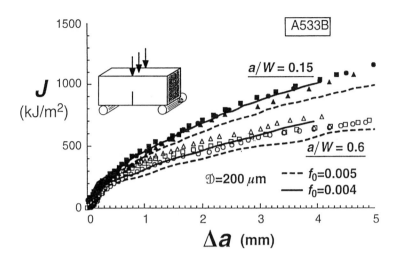

**Figure 8.28** *Predicted and measured $J_R$-curves for A533B steel at 23°C for shallow and deep crack SENB specimen. From Dodds and Ruggieri (Ref. 8.27).*

*Solution:*

$J_{Ic}$ can be estimated by substituting the various parameters in equation (8.35). The only parameters in the equation not included in the above list are the $\hat{\varepsilon}_e(\theta)$ for $\theta = 0°$, $I_m$, and $\sigma_m/\sigma_0$. The value of $\sigma_m/\sigma_0$ can be estimated from equation (8.33). For $m = 2.26$, $q_m/q_0 = 4.177$. $I_m$ and $\hat{\varepsilon}_e(\theta)$ are quantities in the HRR field equations. The former is given in equation (4.19a) and has a value of 5.69 for $m = 2.26$ under plane strain. In the Appendix, the values of the angular functions $\hat{\varepsilon}_{rr}(m,0)$ and $\hat{\varepsilon}_{\theta\theta}(m,\theta)$ are listed for different $\theta$ and $m$ values. For $m = 2.26$ and $\theta = 0°$, values of $\hat{\varepsilon}_{\theta\theta}$ and $\hat{\varepsilon}_{rr}$ for plane strain are extrapolated as $\hat{\varepsilon}_{\theta\theta} = -\hat{\varepsilon}_{rr} = .0084$ and $\hat{\varepsilon}_z = 0$. For proportional loading (see Chapter 2 for definition):

$$\varepsilon_e = [\ \frac{2}{3}(\varepsilon_1^2 + \varepsilon_2^2 + \varepsilon_3^2)\ ]^{\frac{1}{2}}$$

Hence:

$$\varepsilon_e = [\ \frac{2}{3}(\varepsilon_{rr}^2 + \varepsilon_{\theta\theta}^2)\ ]^{\frac{1}{2}}$$

Since:

$$\varepsilon_e = \alpha\varepsilon_0 [\ \frac{J}{\alpha\sigma_0\varepsilon_0 I_m r}\ ]^{\frac{m}{m+1}} \hat{\varepsilon}_e(\theta)$$

$$= \alpha\varepsilon_0 [\ \frac{J}{\alpha\sigma_0\varepsilon_0 I_m r}\ ]^{\frac{m}{m+1}} \{\frac{2}{3}(\hat{\varepsilon}_{rr}^2 + \hat{\varepsilon}_{\theta\theta}^2)\}^{\frac{1}{2}}$$

Comparing terms on the left- and right-hand sides of the above equation, we get:

$$\hat{\varepsilon}_e = \frac{2}{\sqrt{3}} \hat{\varepsilon}_{\theta\theta} = 1.15 \hat{\varepsilon}_{\theta\theta} = .00966$$

Thus, all terms are now complete for substitution in equation (8.35) to estimate $J_{Ic}$. The value calculated = 47,273 J/m². We now proceed to estimate the $J_R$-curve.

From equation (8.35), the critical radius, $R_c$ is calculated as 9.55 µm for $R_0$ = 8.75 µm. Using equation (8.36), Table 8.2 is constructed providing the void sizes at the various instants during stable crack growth. The void numbering is referenced to the original crack tip. The numbers in the two right most columns in the above table give the $J_R$-curve. This predicted curve is repre-sented by the following equation:

$$J_R = 70,445 \, (\Delta a)^{.3166}$$

Substituting $\Delta a$ = 0.2mm in the above equation, the $J_{Ic}$ can be estimated by the ASTM procedure. This value is 42,321 J/m². The $J_R$-curve is plotted in Figure 8.27 along with the measured values for this material.

*Table 8.2 Void Sizes During Stable Crack Growth*

| Instant | Void Radii (µm) | | | | | dJ (J/m²) | Δa (mm) | J (J/m²) |
|---|---|---|---|---|---|---|---|---|
| | $R_1$ | $R_2$ | $R_3$ | $R_4$ | $R_5$ | | | |
| Initial | 8.75 | 8.75 | 8.75 | 8.75 | 8.75 | 0 | 0 | 0 |
| 1st Void becomes critical | 9.55 | 9.238 | 9.116 | 9.048 | 9.005 | 47,273 | .285 | 47,273 |
| 2nd Void becomes critical | - | 9.55 | 9.3056 | 9.1904 | 9.12 | 11,709 | .57 | 58,982 |
| 3rd Void becomes critical | - | - | 9.55 | 9.339 | 9.23 | 8,183 | .855 | 67,164 |
| 4th Void becomes critical | - | - | - | 9.55 | 9.36 | 6,586 | 1.14 | 73,750 |
| 5th Void becomes critical | - | - | - | - | 9.55 | 4,561 | 1.425 | 78,311 |

*8.3.3 Effects of Specimen Geometry on the $J_R$-Curve*

Before we leave the topic of ductile fracture and stable crack growth, it is important to revisit the issue of the effect of geometry on the $J_R$-curve or the $\delta_R$-curve in light of the discussion on the effects of constraint. From the very first experiments reported by Landes and Begley [8.29] proposing J-integral as a criterion for ductile crack initiation toughness, the effects of geometry on the $J_R$-curve have been recognized. In Chapter 4, these effects were explained by an arguement based on loss of J-dominance. Additional clarifications can be provided by the two parameter J-Q theory discussed in Section 8.1. Figure 8.29 schematically shows the influence of specimen geometry and crack size on the $J_{Ic}$ and the $J_R$-curve. Fortunately, the influence of constraint on $J_{Ic}$ is minimal, therefore, if the toughness must be represented by a single parameter, $J_{Ic}$ is the correct parameter. However, the slope of the $J_R$-curve is dependent on the constraint which is geometry and crack size dependent. As the constraint or the value of Q decreases, the slope of $J_R$-curve rises. Thus, the $J_R$-curve is now considered to be a family of curves instead of a single unique curve. In applications, it is not sufficient to specify a value of J. Instead, to accurately predict stable crack growth, it is also necessary to specify the value of Q.

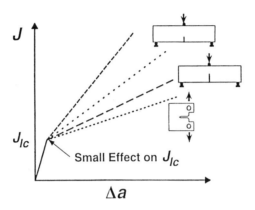

**Figure 8.29** *Schematic representation of the influence of specimen geometry on the $J_R$-curve. From Dodds and Gruggeiri (Ref. 8.27).*

## 8.4 Ductile-Brittle Transition

In the ductile to brittle transition region, fracture is controlled by the competition between ductile tearing as described in Section 8.3 and cleavage fracture as described in Section 8.2. Recall that ductile crack growth occurs by void growth and coalescence process which is driven by the increasing strain (as per the Rice and Tracey model) and cleavage fracture occurs by a stress controlled process. Since a high crack tip constraint can promote cleavage fracture conditions and low constraint can promote ductile void growth mechanism at temperatures in the mid ductile to brittle transition regime, either type of fracture can occur depending on the chosen geometry and size of the specimen. This is illustrated in Figure 8.30 [8.5] in which the critical J is schematically plotted against decreasing constraint. The fracture toughness locus is plotted in the shape of a typical three region behavior where the upper-shelf behavior corresponds to ductile fracture and the lower-shelf to cleavage fracture and the transition region where both types of fracture are observed. The x-axis in this plot must not be confused with temperature. The purpose of this plot is to merely demonstrate that even at a fixed temperature, the fracture behavior can range from pure cleavage for high constraint geometries such as a deeply cracked SENB specimen to purely ductile for the very low constraint CCT geometry. This is illustrated by the J-Q driving force curves which are distinct for each geometry.

Another very important point made in Figure 8.30 is about the possible variation in toughness values measured from specimen of the same geometry, which can be particularly high in the transition region. This is due to loss of constraint due to the increased level of plasticity. To understand this point, consider the following situation. In the transition region, the onset of cleavage fracture is typically triggered by a weak point ahead of the crack tip following some amount of stable crack growth, $\Delta a_0$. This process is schematically illustrated in Figure 8.31. The amount of stable crack growth which precedes cleavage fracture depends on the distance between the trigger point and the original fatigue precrack [8.30]. The further the trigger point is from the precrack front, the higher will be the amount of stable crack growth preceding cleavage. Therefore, the value of J at which cleavage occurs will accordingly be higher. Since the constraint also decreases with the value of J, the cleavage fracture event gets pushed to a higher J. For example, if we follow the intersection of the fracture locus and the driving force curve for the SENT specimen in Figure 8.30, it is clear that the $J_c$ values measured will exhibit a large amount of scatter even within specimens of the same size and geometry depending on the distance between the cleavage trigger point and the precrack.

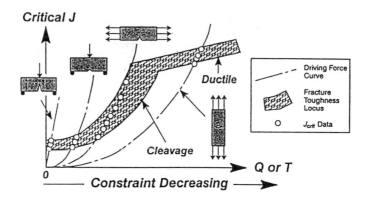

*Figure 8.30* Schematic illustration of fracture toughness locus constructed in the mid-transition temperature region for specimens of various geometries over a range of constraint conditions. From Kirk et al. (Ref. 8.5).

Iwadate and Yokobori [8.30] have tested multiple CT specimens of different sizes of Ni-Cr-Mo-V steel at room temperature to quantify the data scatter in the transition region. The specimen sizes ranged from 150 mm thick and 300 mm wide (6T) to 12.5 mm thick and 25 mm wide (.5T). They have also followed their fracture tests with detailed fractography and measured the distance between the precrack front and the trigger point. The 2T, 4T, and 6T data were valid $K_{Ic}$ tests while the .5T specimens were elastic-plastic in nature in which the $K_{Jc}$ was estimated from the measured $J_c$ values. The values of $K_{Jc}$ for the various tests are plotted in Figure 8.32 as a function of distance between the precrack and the cleavage trigger point, $\Delta a_0 + x$, where $\Delta a_0$ is the amount of stable crack growth and x is the distance between the trigger point and the stable crack front. Two observations can be made from these data, the first being that the scatter among the data from the large specimens (6T to 2T) is much lower than the scatter among the smaller specimens (0.5T) data. Also, a definite correlation is obtained between $K_{Jc}$ and $\Delta a_0 + x$. Based on these observations, Iwadate and Yokobori [8.30] have proposed a method to obtain the lower bound cleavage toughness value in the transition region which is the $K_{Jc}$ value corresponding to the intersection between the ordinate and the best-fit line that can be drawn through the $K_{Jc}$ vs. $\Delta a_0 + x$ data. Several other authors have also proposed methods for determining the lower bound fracture toughness curves in the transition region [8.30-8.33]. These methods are still evolving and are a topic of considerable debate.

In closing this section, it is important to state that fracture toughness measurements in the transition region especially using elastic-plastic methods are not very meaningful unless a large number of tests are

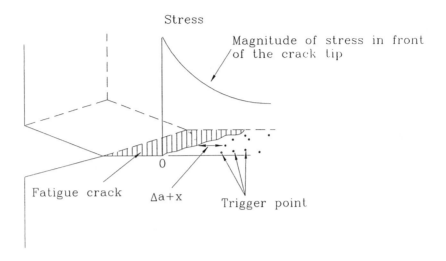

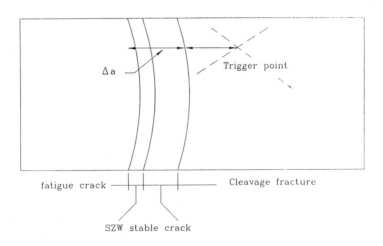

*Figure 8.31* Schematic showing the trigger points of cleavage fracture following some stable crack growth in the transition region. From Iwadate and Yokobori (Ref. 8.30).

conducted. Uncertainties can result from the loss of constraint and from varying amounts of stable crack growth which precedes fracture. A small sample bears little resemblance to the distribution of the whole population. The picture is somewhat better for larger specimens in the SSY regime because the results of these tests tend to lie toward the lower end of the scatter band. It must also be emphasized that transition region data must be adjusted for constraint loss and ductile crack extension prior to applying any type of statistical analysis to determine the lower bound toughness. Methods of adjusting for con-

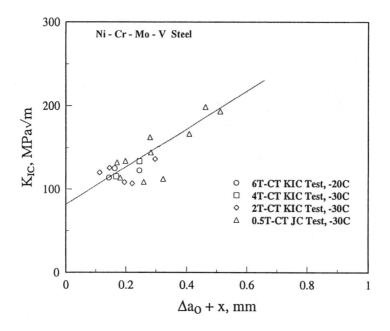

***Figure 8.32*** $K_{Jc}$ *vs. distance between the precrack front and the trigger point for Ni-Cr-Mo-V specimens of different sizes. From Iwadate and Yokobori (Ref. 8.30).*

straint loss were discussed in Section 8.2. In addition to the method presented above, a method for adjusting fracture toughness for stable crack growth is also proposed by Wallin [8.34] which can be followed. This method does not require fractography to measure the trigger point which can sometimes be difficult.

## 8.5 Summary

In this chapter, a framework for including constraint effects in elastic, elastic-plastic, and fully plastic cracked bodies is discussed. In the elastic region, constraint effects are adequately described by the T-stress concept. In the elastic-plastic and fully plastic regions, it is necessary to describe constraint by another parameter Q (due to limitations of the T-stress approach). It was shown that in the elastic region, T and Q are equivalent parameters. The effects of constraint were considered in explaining the influence of specimen geometry and crack size on the cleavage fracture toughness using the RKR model. A statistical model was developed to represent scatter in the cleavage fracture toughness data after correction for constraint effects. In the subsequent section, the effects of constraint were considered in explaining the influence of specimen geometry and crack size on the $J_R$-curve. A model for predicting the $J_{Ic}$ and $J_R$ curve from first principles was presented and shown to predict the behavior of 303 stainless steel very accurately. The last section of the chapter dealt with explanations for why extensive scatter exists in the fracture toughness data in the ductile to brittle transition range. It was shown that this scatter resulted from several factors including (i) due to constraint which varies more significantly in this region because of varying amounts of plasticity, (ii) due to varying amounts of stable crack growth which precedes cleavage fracture in the mid to upper transition region, and (iii) due to statistical variability in the distance of the trigger points relative to the fatigue precrack front between samples. Thus, the effects of constraint which were not considered in earlier chapters are introduced and thoroughly discussed in this chapter.

## 8.6 References

8.1  M.L. Williams, "On the Stress Distribution at the Base of a Stationary Crack", Journal of Applied Mechanics, Vol. 24, 1957, pp. 109-114.

8.2  S.G. Larsson and A.J. Carlsson, "Influence of Non-Singular Stress Terms and Specimen Geometry on Small-Scale Yielding at Crack Tips in Elastic-Plastic Materials", Journal of Mechanics and Physics of Solids, Vol. 21, 1973, pp. 263-277.

8.3  A.H. Sherry, C.C. France, and L. Edwards, "Compendium of T-Stress Solutions for Two and Three Dimensional Cracked Geometries", Fatigue and Fracture of Engineering Materials and Structures, The International Journal, Vol. 18, 1995, pp. 141-155.

8.4  P.S. Leevers and J.C. Radon, "Inherent Biaxiality in Various Fracture Specimen Geometries", International Journal of Fracture, Vol. 19, 1982, pp. 311-325.

8.5  M.T. Kirk, R.H. Dodds, Jr., and T.L. Anderson, "An Approximate Technique for Predicting Size Effects on Cleavage Fracture Toughness ($J_c$) Using Elastic T Stess", in Fracture Mechanics: Twenty-Fourth Volume, ASTM STP 1207, American Society for Testing and Materials, Philadelphia, 1994, pp. 62-86.

8.6  C.F. Shih, N.P. O'Dowd, and M.T. Kirk, "A Framework for Quantifying Crack Tip Constraint", in Constraint Effects in Fracture, ASTM STP 1171, American Society for Testing and Materials, Philadelphia, 1993, pp. 2-20.

8.7  N.P. O'Dowd and C.F. Shih, "Family of Crack-Tip Fields Characterized by a Triaxiality Parameter - I Structure of Fields", Journal of Mechanics and Physics of Solids, Vol. 39, 1991, pp. 989-1015.

8.8  N.P. O'Dowd and C.F. Shih, "Family of Crack-Tip Fields Characterized by a Triaxiality Parameter - II Fracture Applications", Journal of Mechanics and Physics of Solids, Vol. 40, 1992, pp. 939-963.

8.9  N.P. O'Dowd and C.F. Shih, "Two-Parameter Fracture Mechanics: Theory and Applications", in Fracture Mechanics: Twenty-Fourth Volume, ASTM STP 1207, American Society for Testing and Materials", Philadelphia, 1994, pp. 21-47.

8.10 C.D. Beachem and R.M.N. Pelloux, "Electron Fractography - A Tool for Study of Micromechanisms of Fracturing Processes", in Fracture Toughness Testing and Applications, ASTM STP 381, American Society for Testing and Materials, 1965, pp. 210-245.

8.11 R.M. McMeeking and D.M. Parks, "On Criteria for J-Dominance of Crack Tip Fields in Large Scale Yielding", in Elastic-Plastic Fracture, ASTM STP 668, American Society for Testing and Materials, Philadelphia, 1979, pp. 175-194.

8.12 R.O. Ritchie, J.F. Knott, and J.R. Rice, "On the Relationship Between Critical Tensile Stress and Fracture Toughness in Mild Steel", Journal of Mechanics and Physics of Solids, Vol. 21, 1973, pp. 395-410.

8.13 D.A. Curry and J.F. Knott, "Effects of Microstructure on Cleavage Fracture Stress in Steel", Metal Science, 1978, pp. 511-514.

8.14 J.D. Landes and D.H. Schaffer, "Statistical Characterization of Fracture in the Transition Region", Fracture Mechanics: Twelfth Conference, ASTM STP 700, American Society of Testing and Materials, Philadelphia, 1980, pp. 368-382.

8.15 T.L. Anderson and R.H. Dodds, "Specimen Size Requirements for Fracture Toughness Testing in the Ductile-Brittle Transition Region", Journal of Testing and Evaluation, Vol. 19, 1991, pp. 123-134.

8.16 R.H. Dodds, T.L. Anderson, and M.T. Kirk, "A Framework to Correlate a/W effects on Elastic-Plastic Fracture Toughness ($J_c$)", International Journal of Fracture, Vol. 88, 1991, pp. 1-22.

8.17 T.L. Anderson and D. Stienstra, " A Model to Predict the Sources and Magnitude of Scatter in Toughness Data in the Transition Region", Journal of Testing and Evaluation, Vol. 17, 1989, pp. 46-53.

8.18 T.L. Anderson, David Stienstra, and R.H. Dodds, "A Theoretical Framework for Addressing Fracture in the Ductile-Brittle Transition Region", in Fracture Mechanics: Twnety Fourth Volume, ASTM STP 1207, American Society for Testing and Materials, Philadelphia, 1994, pp. 186-214.

8.19 M.T. Kirk, K.C. Koppenhoeffer, and C.F. Shih, "Effect of Constraint on Specimen Dimensions Needed to Obtain Structurally Relevant Toughness Measures", in Constraint Effects in Fracture, ASTM STP 1171, American Society for Testing and Materials, Philadelphia, 1993, pp. 79-103.

8.20 K. Wallin, T. Saario, and K Törrönen, "Statistical Model for Carbide Induced Brittle Fracture in Steel", Metal Science, Vol. 18. 1984, pp. 13-16.

8.21 T. Lin, A.G. Evans, and R.O. Ritchie, "Statistical Model of Brittle Fracture by Transgrannular Cleavage", Journal of Mechanics and Physics of Solids, Vol. 34, 1986, pp. 477-496.

8.22 A. Saxena, D.C. Daly, H.A. Ernst, and K. Bauerji, "Microscopic Aspects of Ductile Tearing Resistance in AISI 303 Stainless Steel", Fracture Mechanics: Twenty First Symposium, ASTM STP 1074, American Society for Testing and Materials, Philadelphia, 1990, pp. 378-395.

8.23 A. Saxena and L. Cretegny, "A Model for Estimating $J_{Ic}$ and the $J_R$-Curve from Microstructural Parameters", manuscript submitted to Materials and Metallurgical Transactions, 1997.

8.24 A.S. Argon, J. Im, and R. Sofoglu, "Cavity Formation from Inclusions in Ductile Fracture", Metallurgical Transactions, Vol. 6A, 1975, pp. 825-837.

8.25 J.R. Rice and D.M. Tracey, "On Ductile Enlargement of Voids in Triaxial Stress Fields", Journal of Mechanics and Physics of Solids, Vol. 17, 1969, pp. 201-217.

8.26 "Standard Method for Measurement of Fracture Toughness", Draft of Proposed ASTM Standard, September 1994, Committee E-08 on Fatigue and Fracture, ASTM, Philadelphia.

8.27 R.H. Dodds and C. Ruggieri, "Modeling of Constraint Effects of Ductile Crack Growth", Paper Presented at the 27th National Symposium on Fatigue and Fracture, Williamsburg, VA, June 1995.

8.28 A.L. Gurson, "Continuum Theory of Ductile Rupture by Void Nucleation and Growth: Part I - Yield Criteria and Flow Rules for Porous Ductile Media", Journal of Engineering Materials and Technology, Vol. 99, 1977, pp. 2-15.

8.29 J.D. Landes and J.A. Begley, "The Effect of Specimen Geometry on $J_{Ic}$", Fracture Toughness, ASTM 514, American Society for Testing and Materials, 1972, pp. 24-39.

8.30 T. Iwadate and T. Yokobori, "Evaluation of Elastic-Plastic Fracture Toughness Testing in the Transition Region Through Japanese Interlaboratory Tests", Fracture Mechanics: Twenty Fourth Volume, ASTM STP 1207, American Society for Testing and Materials, Philadelphia, 1994, pp. 233-263.

8.31 J.D. Landes, U. Zerbst, J. Heerens, B. Petrovski, and K.H. Schwalbe, "Single-Specimen Test Analysis to Determine Lower-Bound Toughness in Transition", Fracture Mechanics: Twenty Fourth Volume, ASTM STP 1207, American Society for Testing and Materials, Philadelphia, 1994, pp. 171-185.

8.32 D.E. McCabe, J.G. Merkle, and R.K. Nanstad, "A Perspective on Transition Temperature and $K_{Jc}$ Data Characterization", Fracture Mechanics: Twenty Fourth Volume, ASTM STP 1207, American Society for Testing and Materials, Philadelphia, 1994, pp. 215-232.

8.33 M.T. Miglin, L.A. Oberjohn, and W.A. Van Der Sluys, "Analysis of Results from the MPC/JSPS Round Robin Testing Program in the Ductile-to-Brittle Transition Region", Fracture Mechanics: Twenty Fourth Volume, American Society for Testing and Materials, Philadelphia, 1994, pp. 342-354.

8.34 K. Wallin, "The Effects of Ductile Tearing on Cleavage Fracture Probability in Fracture Toughness Testing", Engineering Fracture Mechanics, Vol. 32, 1989, pp. 523-531.

## 8.7 Exercise Problems

8.1 Provide arguments for why the following form is the correct form for expressing the elastic T-stress:

$$T = \frac{BK}{(\pi a)^{1/2}}$$

8.2 Repeat the calculations in the Example Problem 8.1.1 for SENT specimen.

8.3 What are the limits of applicability of the elastic T-stress in explaining the effects of geometry on fracture toughness? Provide complete explanations for your answer.

8.4 Explain why the value of Q referenced to the small-scale-yielding crack tip stress field for T = 0 is a better definition for Q.

8.5 Using the relationship between Q and T, estimate the value of Q for SSY for a CCT specimen with a/W = 0.5 and the Ramberg-Osgood exponent, m = 5. Compare this value of Q with that for m = 10.

8.6 Explain in your own words the advantages of the J-Q appraoch vs. the elastic T-stress approach for quantifying the effects of constraint on fracture.

8.7 Explain why FCC metals do not undergo a ductile-to-brittle transition while the BCC and HCP metals do.

8.8 What is the mechanism of river pattern formation on fracture surfaces and what type of fracture causes these patterns to develop. If these features are formed on the fracture surface of a specimen mixed with dimples, what can you conclude about the measurements from the test?

8.9 Describe how cleavage cracks form and propagate in cracked specimens. How can this mechanism be used to explain unusual amounts of data scatter in the lower-shelf and the lower transition region dominated by cleavage fracture.

8.10 Using the data from the average trend lines in Figure 8.19a for a/W values of 0.2, 0.4, and 0.5 for 50 mm and 25 mm thicknesses, estimate the average values of $J_{SSY}$ for this material separately for both thicknesses. Discuss the implications for your answer regarding thickness effects. Assume $m = 10$ and $\sigma_0 = 400$ MPa for this material.

8.11 Explain why it is necessary to correct for constraint prior to using the Weibull 3-parameter model for representing data scatter in the measured $J_c$ value in the lower-shelf region. What other corrections are necessary if the $J_c$ measurements are made in the upper transition region where the cleavage is preceded by ductile crack extension?

8.12 What are the limitations of the Rice and Tracey model for void growth in the context of its applicaton to ductile crack growth in precracked specimens?

8.13 If you are given the following material properties, estimate the $J_{Ic}$ and the $J_R$-curve for the material:

$$\sigma_0 = 400 \text{MPa}, \; \varepsilon_0 = 1.9 \times 10^{-3}, \; \alpha = 4.6, \; m = 5, \; \varepsilon_u = .15$$

Metallography of the material to characterize the microstructure has revealed that there are MnS inclusions with average diameter of 10 μm and an average spacing between them is 100 μm.

8.14 Equation (8.31) provides an expression for calculating $J_{ic}$. Write down all the steps in the derivation of this equation starting from equation (8.29).

8.15 Why is void nucleation not considered in models for estimating $J_{Ic}$ and the $J_R$-curve?

8.16 What drawbacks do you see in estimating the critical void radius, $R_c$, from purely tensile data as in equation (8.34)?

8.17 Explain the additional source of scatter in cleavage fracture toughness data in the upper transition region compared to the factors affecting scatter in the data from the lower-shelf region.

8.18 In Example Problem 8.3 an equation is provided for relating $J_c$ and $J_{ssy}$ for cleavage fracture. For a A 515 steel, you are given the following parameters:

$$\sigma_c = 3.5\sigma_0, \ J_{ssy} = 40 \text{ kJ/m}^2 \text{ and } m = 5$$

Plot the J-Q fracture locus for cleavage fracture toughness for this material.

# CHAPTER 9

# FATIGUE CRACK GROWTH UNDER LARGE-SCALE PLASTICITY

In Chapter 3, we briefly described the linear-elastic fracture mechanics approach for characterizing the rate of fatigue crack growth. In this approach, the rate of crack growth per cycle, da/dN, was correlated with the cyclic stress intensity parameter, $\Delta K$. There are several excellent treatments of the subject of fatigue crack growth under linear-elastic conditions in the literature [9.1 - 9.5]. These treatments also address test methods, fatigue crack growth models, and crack growth under variable amplitude loading in considerable detail. It is assumed that the reader is familiar with the topic at this basic level.

In this chapter, the focus will be on fatigue crack growth behavior under conditions where fundamental limitations exist in correlating da/dN with $\Delta K$. Examples of such behavior include fatigue crack growth under gross-plasticity conditions and behavior of small cracks.

In engineering applications, fatigue crack growth under large-scale plasticity conditions can be a consideration under the following circumstances: (i) during testing of low-strength, high-toughness materials, (ii) in high temperature applications where the yield strength of the material is substantially reduced, (iii) during growth of small fatigue cracks from notches in which the local stresses exceed the yield strength of the material, and (iv) during thermal-mechanical fatigue and thermal-fatigue crack growth. Transient thermal stresses can quite often be considerably higher than the yield strength of the material. In addressing these engineering problems, it is necessary to develop an alternate approach which (i) accounts for large-scale plasticity under cyclic loading and (ii) is consistent with the well developed and widely accepted linear-elastic fracture mechanics approach when the extent of plasticity is negligible, similar to the approach adopted for characterizing ductile fracture in the previous five chapters.

## 9.1 Crack Tip Cyclic Plasticity, Damage, and Crack Closure

Prior to launching into a detailed discussion on crack tip parameters for characterizing fatigue crack growth under large-scale plasticity conditions, it is important to understand the nature of deformation and damage in the crack tip region. Additionally, the phenomenon of crack closure which occurs in the crack tip region is also relevant to the discussion of crack tip parameters for uniquely characterizing fatigue crack growth rates. These topics are discussed in this section.

### 9.1.1 Crack Tip Cyclic Plasticity

When metals are subjected to cyclic deformation in the plastic regime, they initially tend to either cyclically harden or soften but soon attain a steady-state called the saturation stage which lasts through the majority of their cyclic life. The phenomena of cyclic hardening, softening, and saturation are schematically illustrated in Figure 9.1. Figure 9.1a shows that variation in the control variable, strain, as a function of time. The amplitude of the strain cycle is constant and is large enough to induce plastic deformation during the positive as well as the negative halves of the fatigue cycles. Figures 9.1b and 9.1c show how the cyclic stress response, represented by its amplitude $\Delta\sigma/2$, changes as a function of cycles for cyclic hardening and softening materials, respectively. Both types of material show a transient region during which the stress amplitude changes from cycle to cycle and the saturation region during which the stress amplitude remains constant from cycle to cycle. There are a few metal alloys that do not exhibit a cyclic hardening or softening behavior and exhibit the saturation type behavior from the beginning. Such alloys are called cyclically stabilized materials. Typically, metals reach the saturation stage relatively early, during approximately the first 10% of their fatigue life. It is therefore convenient in a majority of fatigue analyses in the inelastic regime to simply neglect the transient fatigue behavior. In this chapter, for the most part, we will be treating the cyclic behavior of metals by idealizing them as elastic, satura-

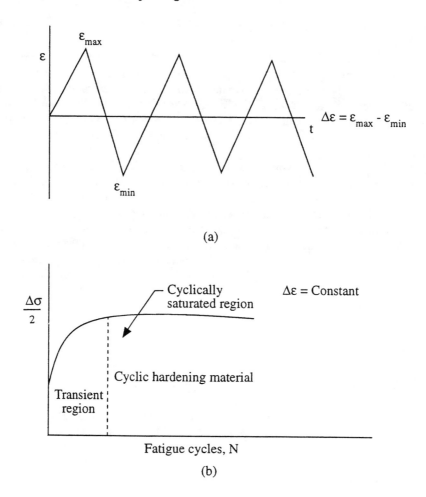
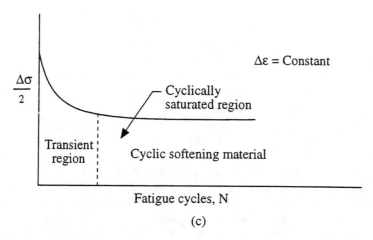

**Figure 9.1** (a) The controlled variable strain, as a function of time. (b) Response stress range ($\Delta\sigma$) a function of cycles for cyclic hardening material. (c) Same for a cyclic softening material.

tion-hardened materials. Figure 9.2 shows the stress-strain behavior of metals after they have reached the saturated stage. This cyclic stress-strain relationship is known as the hysteresis loop. The total strain range and stress range are represented by $\Delta\varepsilon$ and $\Delta\sigma$, respectively. The plastic and elastic parts of the strain range are represented by $\Delta\varepsilon_p$ and $\Delta\varepsilon_e$, respectively. The following relationship relates the various strain ranges:

$$\Delta\varepsilon = \Delta\varepsilon_p + \Delta\varepsilon_e \qquad (9.1)$$

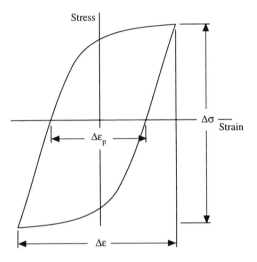

**Figure 9.2** *Cyclic stress-strain behavior in the cyclically saturated condition.*

Figure 9.3 shows the relationship between the cyclic stress amplitude and cyclic strain amplitude for elastic, saturation-hardened materials. For comparison, the monotonic stress-strain response is also shown. Figure 9.3a shows the behavior of cyclic hardening materials and Figure 9.3b the behavior of a cyclic softening material. The stress amplitude corresponding to a plastic strain amplitude of .002 (.2%) is termed as the cyclic yield strength, $\sigma_0^c$, Figure 9.3. The relationship between the cyclic-stress amplitude and the cyclic-strain amplitude is called the cyclic stress-strain curve and is mathematically represented by the following equation:

$$\frac{\Delta\varepsilon}{2} = \frac{\Delta\sigma}{2E} + \alpha'\left(\frac{\Delta\sigma}{2\sigma_0^c}\right)^{m'} \qquad (9.2)$$

$$\text{or} \quad \Delta\varepsilon = \frac{\Delta\sigma}{E} + 2\alpha'\left(\frac{\Delta\sigma}{2\sigma_0^c}\right)^{m'} \qquad (9.2a)$$

The above equation is analogous to the Ramberg-Osgood relationship with $\Delta\varepsilon$ and $\Delta\sigma$ replacing $\varepsilon$, $\sigma$; $2\sigma_0^c$ replacing $\sigma_0$; $2\alpha'$ replacing the product of $\alpha\varepsilon_0$; and m′ replacing m. We will be making considerable use of this analogy in defining crack tip parameters later in this chapter.

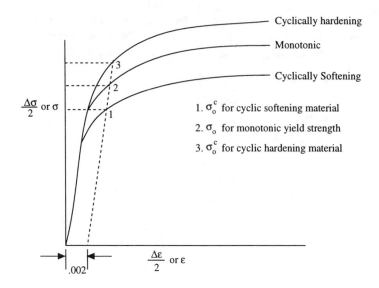

***Figure 9.3*** *Monotonic and cyclic stress-strain curves and the definition of cyclic yield strength: (a) cyclic hardening material; (b) cyclic softening material.*

Figure 9.4 shows the stress-strain behavior of an element of material as it is being approached by a growing crack tip under small scale yielding conditions. The element starts in region 1 where the deformation is elastic. In region 2 it enters the monotonic plastic zone where yielding occurs during the first cycle and subsequently the material is subjected to elastic loading and unloading. In region 3, which is located at the boundary of the monotonic and cyclic plastic zones, yielding takes place during each loading and unloading cycle and the material can be assumed to deform according to the cyclic stress-strain curve. Inside the cyclic plastic zone, region 4, increasing amounts of cyclic plasticity occur as the element approaches the crack tip. A first order approximation of the size of the cyclic plastic zone for a non-hardening material will result in the following expression:

$$r_p^c = \frac{1}{\pi}\left(\frac{\Delta K}{2\sigma_0^c}\right)^2 \tag{9.3}$$

where $r_p^c$ = cyclic plastic zone size. When $r_p^c$ becomes comparable to the crack size and the remaining ligament size, small-scale yielding conditions no longer hold. Thus, $\Delta K$ is no longer expected to correlate the fatigue crack growth behavior. The size of the cyclic plastic zone for nonhardening materials is approximately one-fourth of the monotonic plastic zone as one can see from equation (9.3). Also, the $\Delta K$ levels at which fatigue crack growth is characterized are always less than $K_{IC}$ and in most instances substantially less than $K_{IC}$. Therefore, compared to the situation during fracture toughness testing, large-scale plasticity considerations are less frequently encountered during fatigue crack growth testing and applications.

Nevertheless, when the cyclic plastic zone size is no longer negligible in comparison to the uncracked ligament size, the load-displacement (load-line) behavior exhibits substantial hysteresis as shown schematically in Figure 9.5. In this figure, the load-displacement behavior of a specimen that is

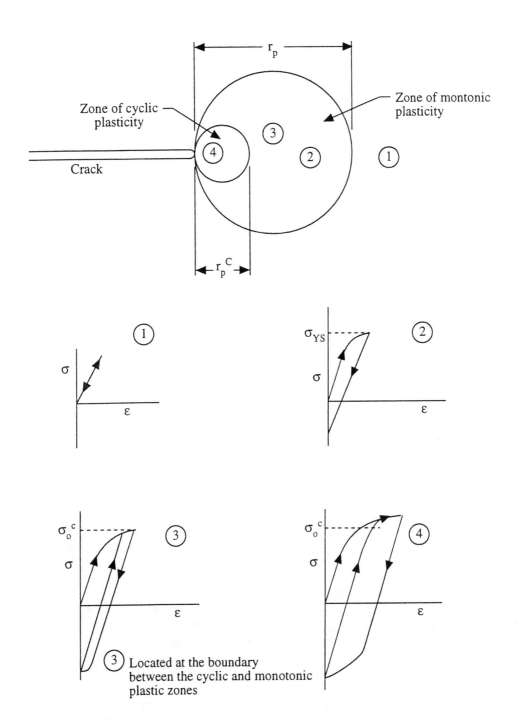

*Figure 9.4 Stress-strain behavior of an element of material approaching the crack tip.*

cycled between a maximum deflection, $V_{max}$, and a minimum deflection, $V_{min}$, is shown. The non-linearity in the load-deflection behavior necessitates the use of nonlinear fracture mechanics parameters for characterizing crack growth and the crack tip conditions.

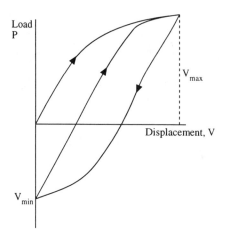

*Figure 9.5* *The load-deflection behavior of a specimen cycled under deflection controlled conditions showing large-scale cyclic plasticity.*

### 9.1.2 Crack Tip Damage

Fatigue damage is caused by accumulation of reversed plastic deformation. At low applied stress levels, reversed plastic deformation takes place in localized slip bands called persistent slip bands (PSBs). Therefore, fatigue damage accumulates on crystallographic planes on which slip preferentially takes place. Thus the initiation and growth of fatigue crack also occurs along the crystallographic planes. As the strain range is increased, the deformation becomes more homogeneous and the heterogeneous slip bands give way to dislocation cell structures due to cross-slip of screw dislocations [9.6 - 9.8]. If we apply these observations to the crack tip region, we can explain why the crack path at low $\Delta K$ levels is zig-zag and crystallographic in nature, while at higher $\Delta K$ levels it is smoother and normal to the direction of loading. The zig-zag and crystallographic crack growth at low $\Delta K$, also known as stage I crack growth, occurs due to damage concentration in the slip bands that form ahead of the crack tip as shown in Figure 9.6. In this case, the plastic deformation extends only in the grain nearest to the crack tip. On the other hand, at higher $\Delta K$ levels, the plastic deformation extends over several grains and the deformation is spread much more homogeneously. As a result, the crack path follows the direction normal to the loading direction. This is also known as stage II crack growth. Since microstructure influences slip, the influence of microstructure can be expected to be particularly strong during stage I crack growth but not as strong during stage II. Such trends have, in fact, been observed.

During stage II fatigue, a number of materials show fatigue striations on the fracture surface, Figure 9.7. Depending on the $\Delta K$ and the crack growth rate regime, there is also a one-to-one correlation between da/dN and the striation spacing. A mechanism for formation of fatigue striations was proposed by Laird and Smith [9.9] which is schematically depicted in Figure 9.8. In this model, the increment of crack extension per cycle is envisioned due to plastic blunting of the crack tip. The crack tip blunts due to plastic deformation during the loading portion of the cycle, while the slip concentrates along two slip bands as shown in Figures 9.8b and c. When the loading direction is reversed, the slip direction reverses

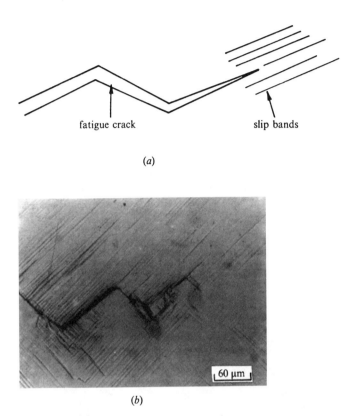

***Figure 9.6*** *(a) A schematic of stage I fatigue crack growth. (b) An example of state I fatigue crack growth in a single crystal of Mar M-200 nickel base superalloy (courtesy of Aswath and Suresh, Ref. 9.1).*

with it, the crack tip resharpens, and the crack simultaneously extends in the blunted region by virtue of the material being squeezed in that region. A striation is formed in the process as shown in Figures 9.8d and e because resharpening cannot recover all the deformation caused during blunting. The entire process begins again with the application of the next load cycle as shown in Figure 9.8f.

It should be noted that the one-to-one correspondence between da/dN and striation spacing is not always observed. Also, not all materials exhibit striation formation during stage II crack growth. Striations are more commonly observed in ductile materials. Aluminum alloys typically exhibit striations but they are less frequently observed in steels. Striation formation is influenced by the crack growth rate, environment, and the alloy content.

It is clear that the driving force for the crack tip deformation and damage mechanisms leading to fatigue crack growth is the strain range experienced by the near crack tip material. The strain range in the crack tip region is both a function of the distance from the crack tip and the magnitude of the applied $\Delta K$ value. We may write the following relationship between applied $\Delta K$ and the crack tip strain range for an elastic material:

$$\Delta \varepsilon_{ij} = \frac{\Delta K}{E\sqrt{2\pi r}} f_{ij}(\theta) \tag{9.4}$$

Thus, other crack tip phenomena which can change the relationship between the crack tip strain range and the applied $\Delta K$ will also be expected to influence the fatigue crack growth behavior. One such phenomenon is crack closure described in the following section.

*9.1.3 Crack Closure*

Crack closure is a phenomenon by which the crack surfaces remain closed at the crack tip during a portion of the fatigue cycle. This portion of the loading cycle is ineffective in growing the crack and thus the corresponding load must be subtracted from the applied cyclic load, $\Delta P$, to determine the value of the loading parameter such as $\Delta K$. This phenomenon was first discovered by Elber [9.10 - 9.11] who argued that a zone of residual tensile deformation is left in the wake of a growing fatigue crack. The residual deformation causes the opposing fracture surfaces in the crack vicinity to come in contact with one another prior to the applied load being completely removed during unloading. The load at which the opposing crack surfaces first come in contact is called the crack closure load. Similarly, upon reloading, the crack tip remains at least partially closed until a load equal to the crack opening load is applied. Thus, the region ahead of the crack tip remains less stressed until the load exceeds the crack opening load. The crack opening and closing loads are nearly identical. This type of crack closure is very often referred to as the plasticity-induced crack closure in the literature. There are other forms of crack closure which will be discussed later in this section.

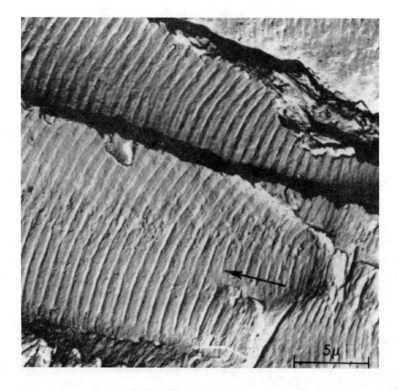

*Figure 9.7* Example of fatigue crack growth striation in a high strength Al alloy. From C.D. Beachem and R.M. Pelloux, in Fracture Toughness Testing and its Applications, ASTM STP 381. Reproduced with permission of American Society for Testing and Materials.

In Figure 9.9 the phenomenon of crack closure is shown schematically. We assume a nonhardening material for simplicity and plot the stress and strain in the y-direction vs. distance from the crack tip at various points during the fatigue cycle. The point of maximum load is designated by A and the point of minimum load is designated B. The stress profiles at points A and B are plotted in Figure 9.9a and the strain profile is plotted in Figure 9.9b. At point A, all the stress is borne by the uncracked ligament ahead of the crack tip. At point B, a small region in the wake of the crack is subjected to compressive stresses due to the residual plastic deformation. The region ahead of the crack tip within the cyclic plastic zone is subjected to a compressive stress equal to $\sigma_0$. This stress decays gradually outside of the cyclic plastic zone and to maintain equilibrium, a residual tensile stress field is established in the remainder of the uncracked ligament. It should be mentioned that, in this representation, the size of the cyclic plastic zone, even in an ideal case, will be different from the one predicted by equation (9.3) because crack closure was not considered in the derivation of equation (9.3). Figure 9.9b shows schematics of the deformation profiles of the crack tip region at loading points A and B. The profile corresponding to point B provides the residual deformation and the profile corresponding to point A shows the deformations imposed by the applied load on top of the residual deformations at point B. From this schematic, one can clearly see that the residual plastic deformation in the crack wake will promote contact between the opposing fracture surfaces prior to reaching the minimum load. If the relationship between applied K and the crack mouth deflection is plotted as shown in Figure 9.10a, the stiffness (or compliance) of the specimen begins to change as the crack closure process begins. This point is labeled C and the K corresponding to this point is labeled $K_{cl}$. At some load level below point C, the point, labeled B in Figure 9.9a, the crack is completely closed. During the loading portion of the cycle, the crack begins to open at point B but is not completely open until point C. The K level corresponding to point C is called the opening K-level or $K_{op}$. Typically, $K_{op}$ and $K_{cl}$ are the same.

The same phenomenon is more clearly illustrated in Figure 9.10b in which the elastic deflection, $V_e$, is subtracted electronically from the actual deflection, V, and then the offset deflection, $V_{off}$, is plotted as a function of K. The elastic deflection is calculated as follows. We assume that the crack is fully open at $K_{max}$. Then, from the measured deflection at $K_{max}$, we can find the proportionality factor between displacement and K as follows:

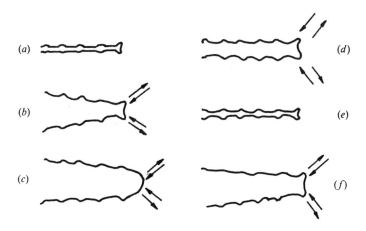

*Figure 9.8* Mechanism of striation formation proposed by Laird and Smith, Ref. 9.9.

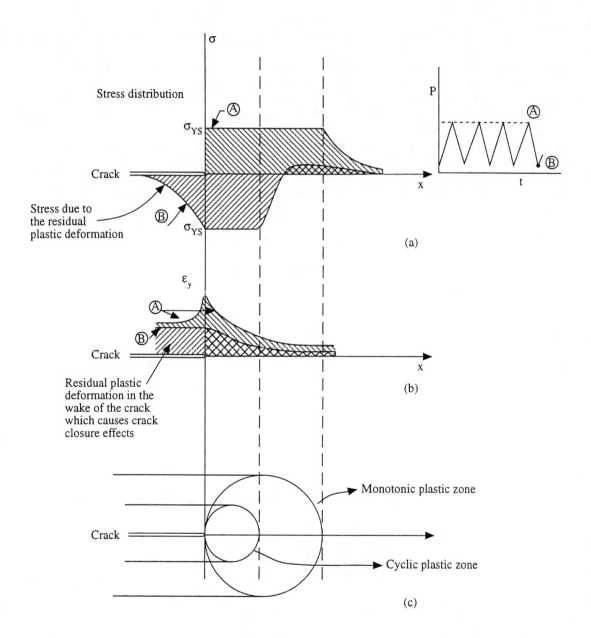

*Figure 9.9 Schematic illustration of fatigue crack closure.*

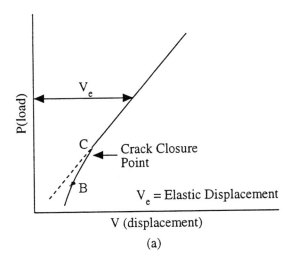

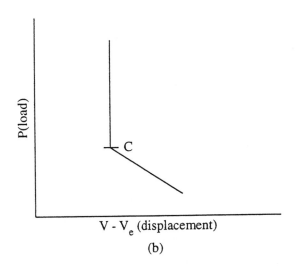

*Figure 9.10 Illustration of crack closure point in (a) a load-displacement diagram and (b) in a plot of load-displacement in which the theoretical elastic displacement has been subtracted from the measured value and the difference is amplified to highlight crack closure.*

$$V_{max} = \alpha K_{max}$$

$$V_e = \alpha K$$

Therefore, $V_e$ corresponds to the displacement of a fully open crack at any K. The value $V-V_e$ is the change in displacement caused by crack closure. This signal can be easily amplified 10 to 100 times and then recorded. It is considerably easier to detect the crack opening and closing levels from this plot as seen in Figure 9.10b.

When dealing with large-scale plasticity, the effects of plasticity-induced closure can be quite important. Therefore, a detailed discussion of this topic is necessary. As several researchers have noted, crack closure can also occur due to a variety of additional factors. Among them are oxide-induced crack closure, roughness-induced crack closure, and viscous-fluid-induced crack closure. These mechanisms of crack closure, though very important to the fatigue crack growth process, are somewhat peripheral to this discussion which is focused on the effects of large-scale plasticity during fatigue crack growth. The interested readers are referred to the publications and books on these topics [9.12 - 9.18].

The most obvious way to account for crack closure in estimating the crack driving force[1] is to define an effective value of $\Delta K$, $\Delta K_{eff}$, as follows [9.11]:

$$\Delta K_{eff} = K_{max} - K_{cl} = U \Delta K \tag{9.5}$$

where $\Delta K$ = applied value of $\Delta K$ and U is a value less than 1. The value of U is very often determined empirically by considering fatigue crack growth rate data at several load ratios, R. At high load ratios, U = 1 and at low load ratios, it is less than 1.

Budiansky and Hutchinson [9.19] were the first to propose a detailed crack closure model with the ability to predict $K_{cl}$ and since this is truly an analytical model, it is worthwhile to discuss it in detail.

**Budiansky and Hutchinson Model for Predicting Crack Closure** The assumptions in this model include (i) steady-state fatigue crack growth occurs under small-scale yielding conditions; (ii) the opening profile of the crack is described by the Dugdale analysis at the maximum load and a residual deformation of previously yielded material is attached to the crack faces; (iii) upon unloading, this residual deformation leads to contact between the crack faces at a K level higher than 0; (iv) in the immediate vicinity of the crack tip, the residual displacements are determined by the reversed plastic deformation zone; and (v) in the region ahead of the crack tip beyond the cyclic plastic zone, the deformation created at $K = K_{max}$ is unaltered by unloading. From the Dugdale analysis, the plastic zone size, $r_p$, and the crack tip opening displacement, $\delta_t$, are given by:

$$r_p = \frac{\pi}{8} \left( \frac{K}{\sigma_0} \right)^2 \tag{9.6a}$$

$$\delta_t = \frac{K^2}{\sigma_0 E} \tag{9.6b}$$

If the coordinate origin is located at the crack tip, the variation in the plastic deformation, $\delta$, at $K = K_{max}$, is given by :

---

[1] The term crack driving force is quite frequently used in the fracture mechanics literature to refer to the magnitude of the correlating crack tip parameters such as K or J. This term must not be confused with the thermodynamic driving forces which are related to the free energies that drive reactions.

$$\frac{\delta}{\delta_t} = g\left(\frac{x}{r_p}\right) \tag{9.7}$$

where

$$g = \sqrt{1-\phi} - \frac{\phi}{2}\log\left(\frac{1+\sqrt{1-\phi}}{1-\sqrt{1-\phi}}\right) \tag{9.8}$$

$$\phi = x/r_p$$

If we assume that reversed plastic yielding occurs over the cyclic plastic zone, $r^c_p = (1/4)\, r_p$, the residual plastic deformation at $K = 0$ (the minimum load) over the interval of x between 0 and $r_p/4$ is:

$$\frac{\delta}{\delta_t} = g\left(\frac{x}{r_p}\right) - \frac{1}{2} g\left(\frac{4x}{r_p}\right) \tag{9.9}$$

The term ½ in the second term of the right-hand side of equation (9.9) comes from the fact that the effective yield strength doubles for reversed cyclic plasticity. Also note that cyclic plasticity occurs only over a distance $r_p/4$ and the crack tip displacement at $K_{min} = 0$ is half of its value at $K_{max}$.

Budiansky and Hutchinson, in their derivation, assumed that a plastic deformation of magnitude $\delta_R/2$ must be appended to the upper and lower crack surfaces in order to achieve crack closure all along the crack length. Using complex potentials, they suggested the following forms for $\delta_R$:

$$\frac{\delta_R}{\delta_t} = \frac{\pi^2\alpha}{4} + \int_0^1 \sqrt{\frac{\phi-\alpha}{\phi}}\, h(\phi)d\phi \tag{9.10a}$$

and

$$\frac{\delta_R}{\delta_t} = -\frac{\pi^2\alpha}{4} + \int_0^1 \sqrt{\frac{\phi}{\phi-\alpha}}\, h(\phi)d\phi \tag{9.10b}$$

where

$$h(\phi) = \frac{1}{2}\log\left(\frac{1+\sqrt{1-\phi}}{1-\sqrt{1-\phi}}\right) \tag{9.10c}$$

$\alpha$ = ratio of the cyclic plastic zone size for the growing crack to the monotonic plastic zone size, $r_p$. Equating (9.10a) and (9.10b), we can show that:

$$\int_0^1 \frac{h(\phi)d\phi}{\sqrt{\phi(\phi-\alpha)}} = \frac{\pi^2}{2} \qquad (9.11)$$

The above equation can be solved numerically to yield the value of $\alpha = 0.09286$ and $\delta_R / \delta_t = 0.8562$. Thus, the cyclic plastic zone for a growing crack is less than 10% of the monotonic plastic zone ($\approx$ 40% of the cyclic plastic zone for stationary crack with no plastic deformation in the wake region). Further, the residual deformation in the crack tip region is as high as 86% of the deformation at $K_{max}$. When the applied K is reduced during unloading from its peak value $K_{max}$, the crack opening displacement for $x < 0$ in the vicinity of the crack tip at any value K during unloading is given by:

$$\frac{\delta}{\delta_t} = g\left(\frac{x}{r_p}\right) - \left(\frac{2\bar{r}_c}{r_p}\right) g\left(\frac{x}{\bar{r}_c}\right) \qquad (9.12a)$$

where

$$\bar{r}_c = \frac{r_p}{4}\left(1 - K/K_{max}\right)^2 \qquad (9.12b)$$

is the instantaneous size of the plastic zone at any point during unloading.

The K level at which the first contact occurs, $K_{cl}$, can be obtained by recognizing that at $K_{cl}$, $\delta = \delta_R$ at the point where $\delta/\delta_t$ is minimum. These conditions yield the following equations:

$$g\left(\frac{x}{r_p}\right) - 2\left(\frac{\bar{r}_c}{r_p}\right) g\left(\frac{x}{\bar{r}_c}\right) = \frac{\delta_R}{\delta_t} \qquad (9.13a)$$

and

$$g'\left(\frac{x}{r_p}\right) - 2g'\left(\frac{x}{r_c}\right) = 0 \qquad (9.13b)$$

Substituting $\delta_R/\delta_t = 0.8562$ and $\alpha = 0.09286$ leads to:

$$\frac{K_{cl}}{K_{max}} = 1 - \sqrt{1 - \left(\frac{\delta_R}{\delta_t}\right)^2} = 0.483 \qquad (9.14)$$

If this analysis is extended to estimate the $K_{op}$, the ratio $K_{op}/K_{max}$ is calculated as 0.557.

The analysis provides a theoretical justification for use of $\Delta K_{eff}$ as a correlating parameter for fatigue crack growth rate. The model captures several important physical features of the crack closure phenomenon such as predicting that da/dN vs. $(\Delta K)_{eff}$ trends are independent of load-ratio, R. However, the actual closure trends measured in the laboratory are considerably more complex. For example it, is well known that the value of U which is predicted to be constant by this model, does depend on the $\Delta K$ level. Further, the Dugdale model applies to plane stress conditions during fatigue crack growth; plane strain conditions are more likely to be present. Finally, crack closure levels are also influenced by oxidation, fracture debris, and fracture surface roughness in the crack tip region, thereby reducing the influence of plasticity-induced crack closure. Various modifications of this model to account for some of these deficiencies have been proposed by Newman [9.20 - 9.21], by Sehitoglu and co-workers [9.22 - 9.23], and by McClung [9.24] and have been implemented in finite element analysis and in life prediction codes.

## 9.2 ΔJ-Integral

Making use of the analogy between the cyclic stress-strain curve, equation (9.2a), and the Ramberg-Osgood relationship representing the monotonic stress-strain curve, the following path-independent integral was defined by Lamba [9.25]:

$$\Delta J = \int_\Gamma \Delta W dy - \Delta T_i \frac{\partial \Delta u_i}{\partial x} ds \qquad (9.15)$$

where

$$\Delta W = \int_0^{\Delta \varepsilon_{ij}} \Delta \sigma_{ij} \, d(\Delta \varepsilon_{ij}) \qquad (9.16)$$

$\Gamma$ is a contour beginning on the lower crack surface and ending on the upper crack surface traveling counterclockwise. Since $\Delta \varepsilon_{ij}$ is a single valued function of $\Delta \sigma_{ij}$ all through the loading portion of the fatigue cycle, it can be shown by analogy with Rice's J-integral that $\Delta J$ is path-independent provided its value is computed during the loading portion of the fatigue cycle only. Thus, $\Delta W$ represents the stress work-per-unit volume performed during loading, rather than the stress work during the complete cycle. Consider a point in front of the crack tip which during a complete fatigue cycle goes through the stress and strain relationship given by Figure 9.11a. The corresponding load vs. load-line displacement relationship is shown in Figure 9.11b. The $\Delta \sigma_{ij}$, $\Delta \varepsilon_{ij}$, $\Delta u_i$, and $\Delta T_i$ in equations (9.15) and (9.16) are given by the following relationships:

$$\Delta \sigma_{ij} = \sigma_{ij}^{max} - \sigma_{ij}^{min}$$

$$\Delta \varepsilon_{ij} = \varepsilon_{ij}^{max} - \varepsilon_{ij}^{min}$$

$$u_i = u_i^{max} - u_i^{min}$$

$$\Delta T_i = T_i^{max} - T_i^{min}$$

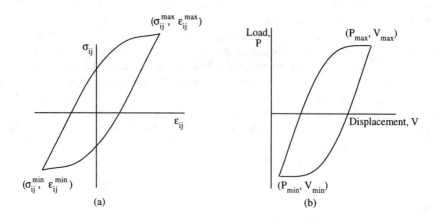

**Figure 9.11** *(a) Schematic cyclic stress-strain behavior at point ahead of a growing fatigue crack within the cyclic plastic zone. (b) The load-displacement behavior of a specimen undergoing gross cyclic plasticity during fatigue crack growth.*

Alternatively, $\Delta J$ can be defined using the load-displacement diagram as per the independent work of Dowling and Begley [9.26 - 9.27]:

$$\Delta J = \frac{\eta}{Bb} \int_0^{\Delta V} \Delta P \, d(\Delta V) \qquad (9.17)$$

The above equation is equivalent to the following:

$$\Delta J = \frac{\eta}{Bb} \int_{V_{min}}^{V_{max}} (P - P_{min}) \, dV \qquad (9.17a)$$

where, as usual, B and b are the thickness and uncracked ligament of the specimen, and $\eta$ is a geometry and crack size-dependent function and has the same value as for the monotonic case (see Chapter 5). $\Delta P$ and $\Delta V$ are the load and displacement ranges and $P_{max}$, $P_{min}$, $V_{max}$, $V_{min}$ are as shown in Figure 9.11b. In their original experimental work, Dowling and Begley [9.26] found high levels of crack closure accompanying fatigue crack growth under large-scale plasticity conditions. Therefore, they defined a $\Delta J$ which is equivalent to the effective value of $\Delta J$, $(\Delta J)_{eff}$, given in Figure 9.12. In this definition $\Delta P$ and $\Delta V$ are defined by $P_{max} - P_{cl}$ and $V_{max} - V_{cl}$, respectively, as shown in the above figure. Thus, $P_{min}$ and $V_{min}$ are replaced by $P_{cl}$ and $V_{cl}$ in equations (9.8) and (9.8a).

In both definitions of $\Delta J$, equations (9.6) and (9.8), if we assume the $\Delta\sigma$ and $\Delta\varepsilon$, and $\Delta P$ and $\Delta V$, are linearly related as would be the case for linear-elastic materials, $\Delta J$ reduces to its elastic value, $\Delta J_e$, given by the following equation:

$$\Delta J_e = \frac{\Delta K^2}{E} \qquad (9.18)$$

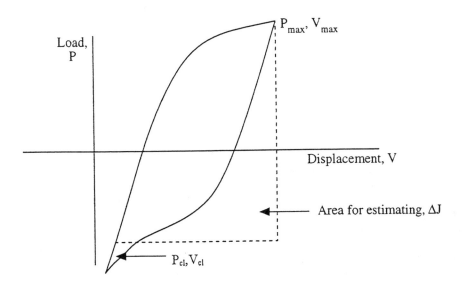

**Figure 9.12** *Estimation of $\Delta J$ in the presence of significant crack closure.*

It should be pointed out that:

$$\Delta J_e \neq \mathcal{G}_{max} - \mathcal{G}_{min}$$

where $\mathcal{G}_{max}$ and $\mathcal{G}_{min}$ are the Griffith's crack extension force corresponding to the maximum and minimum load points. In elastic-plastic bodies, the total value of J-integral is obtained by combining the elastic and plastic parts, much like the J-integral in the following way:

$$\Delta J = \Delta J_e' + \Delta J_p \tag{9.19}$$

where $\Delta J_e'$ is the portion of $\Delta J$ based upon the linear elastic stress intensity factor $\Delta K$, and including a first-order plasticity correction based on an effective crack length; the actual crack length is increased by a fraction of the estimated crack tip plastic zone size. Recall that this correction and the procedure for estimating J for elastic-plastic loading was discussed in detail in Chapter 5. Thus, if $\Delta K_p$ is the plastic zone size corrected value of $\Delta K$, $\Delta J_e' = \Delta K_p^2/E$.

*9.2.1 Relationship Between $\Delta J$ and Crack Tip Stress Fields*

Following the same steps used in deriving the relationship between the J-integral and crack tip stress field, equation (4.18), one can derive the relationship between $\Delta J$ and the crack tip stress and strain ranges as follows:

$$\Delta \sigma_{ij} = 2\sigma_0^c \left( \frac{\Delta J}{4\alpha' \sigma_0^c I_{m'}} \right)^{\frac{1}{1+m'}} \hat{\sigma}_{ij}(\theta, m') \qquad (9.20a)$$

$$\Delta \varepsilon_{ij} = 2\alpha' \left( \frac{\Delta J}{4\alpha' \sigma_0^c I_{m'} r} \right)^{\frac{m'}{1+m'}} \hat{\varepsilon}_{ij}(\theta, m') \qquad (9.20b)$$

From equation (9.20), one can make the argument that since $\Delta J$ characterizes the stress and strain ranges in the crack tip region, it must uniquely characterize the fatigue crack growth rates over a wide range of conditions ranging from dominantly elastic to elastic-plastic and fully-plastic conditions. In the linear-elastic regime, $\Delta J$ is by definition equivalent to $\Delta K$, but unlike $\Delta K$, its application is not limited just to the linear-elastic conditions.

Experimental evidence to demonstrate the use of $\Delta J$ for characterizing fatigue crack growth under a wide range of conditions was provided by Dowling and Begley [9.27] and by Brose and Dowling [9.28] using A533 steel and 304 stainless steel, respectively. These results are reproduced in Figures 9.13a and b, respectively. In their studies, they tested large compact type specimens which were 204 to 408 mm wide in order to maintain elastic conditions for high $\Delta K$ values and they also tested small specimens (25.4 mm wide) which exhibited essentially fully plastic behavior at high $\Delta K$ values. Included in the tests were other intermediate size specimens which yielded a range of elastic-plastic conditions. All fatigue crack growth data correlated nicely with $\Delta J$ or with its linear-elastic equivalent, $(E\Delta J)^{1/2}$, demonstrating the utility of $\Delta J$. In the correlation shown in Figure 9.13a using A533 steel, the data obtained on center-crack-tension specimen geometry are also included providing further proof of the versatility of the $\Delta J$ parameter.

### 9.2.2 Methods of Determining $\Delta J$

Using the analogy between J and $\Delta J$, we can easily develop ways of determining $\Delta J$. All the methods described in Chapter 5 for determining J-integral can be modified to determine $\Delta J$. Therefore, instead of describing the methods in detail which will be a repeat of the discussion in Chapter 5, we will consider three examples which show us how to derive expressions for $\Delta J$, if we are given an expression for determining J.

**Example Problem 9.1**

A standard compact type (CT) specimen is fatigue-cycled under deflection-control conditions in which the deflection is varied from 0 to a maximum value. The load-displacement diagram is given by Figure 9.14a. Derive an expression for estimating $\Delta J$ for this configuration.

*Solution:*

Since the load-displacement diagram is recorded during a fatigue crack growth test, we will develop an expression for determining $\Delta J$ which makes use of these measurements. The equations for determining J for this configuration are given by equations (5.33) and (5.34). Of these, equation (5.34) is much simpler to implement. Using the analogy, we simply write:

$$\Delta J_p = \frac{\Delta A_p}{Bb}\left(2 + 0.522 \frac{b}{W}\right) \qquad (1)$$

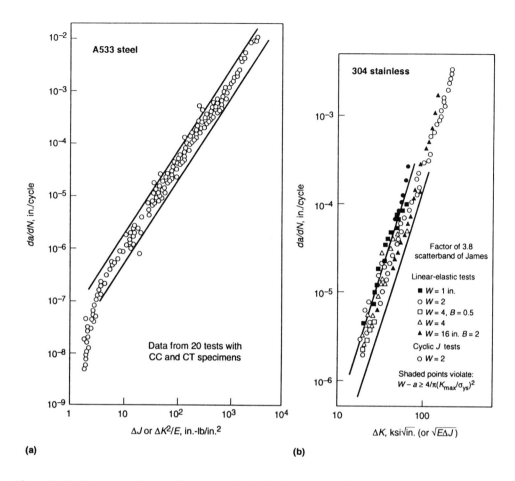

**Figure 9.13** *Fatigue crack growth rates under high plasticity conditions compared with fatigue crack growth rates obtained under linear-elastic conditions employing large specimens (a) for A533 steel (Ref. 9.26) and (b) for 304 stainless steel (Ref. 9.28).*

where $\Delta A_p$ is the area under the load vs. the plastic part of the load-line displacement diagram as shown in Figure 9.14b. The total value of $\Delta J$ is thus given by:

$$\Delta J = \frac{\Delta K^2}{E} + \frac{\Delta A_p}{Bb}\left(2 + 0.522\frac{b}{W}\right) \qquad (2)$$

The method for determining $V_p$ for plotting in Figure 9.14b is graphically shown in Figure 9.14c.

For deeply cracked compact specimens ($a/W \geq 0.5$), a further simplification can be made. The first term on the left-hand side of equation (2) can be written as:

$$\frac{\Delta K^2}{E} = \frac{\Delta A_e}{Bb}\alpha(a/W)$$

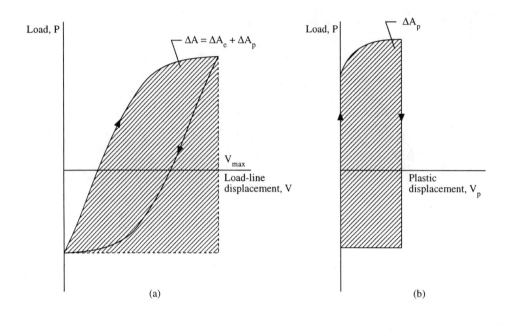
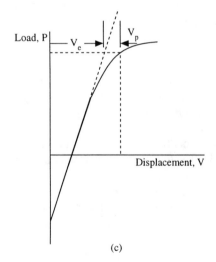

*Figure 9.14* (a) The load-displacement diagram for a full cycle of a compact specimen loaded under displacement controlled conditions. (b) The plot of load vs. plastic deflection showing the plastic portion of the area under the load-displacement diagram. (c) Schematic illustration of obtaining plastic displacement from the load-displacement diagram.

where $\Delta A_e$ is the area under the load vs. elastic displacement curve. We should also note that any inelastic displacement is counted as plastic displacement. Therefore, there is no need to correct the elastic term using the plastic zone adjusted crack size when dealing with experimentally measured displacements. That correction is much more important when analytically calculating the values of $\Delta J$ as will be discussed later. In CT specimens for $a/W \geq 0.5$, the value of $\alpha$ is approximately equal to $(2 + 0.522\, b/W)$ as was previously explained in Chapter 5. Therefore, the expression for total $\Delta J$ can be written as:

$$\Delta J = \frac{\Delta A_e + \Delta A_p}{Bb}\left(2 + 0.522\,\frac{b}{W}\right)$$

or

$$\Delta J = \frac{\Delta A}{Bb}\left(2 + 0.522\,\frac{b}{W}\right) \qquad (3)$$

where $\Delta A$ is the area under the load vs. total deflection curve as shown in Figure 9.14a. It is left as an exercise problem to show that $\alpha \approx 2 + 0.522\, b/W$ for $a/W \geq 0.5$ in CT specimens. In the early studies of Dowling and Begley on crack growth under large-scale plasticity, the above simplification was used for determining $\Delta J$.

**Example Problem 9.2**
Derive an expression for $\Delta J_p$ in a DEN specimen (Figure 5.12) under plane strain conditions in terms of the applied stress range, the width, W, of the specimen, crack size, a, and the cyclic stress-strain properties of the material.

*Solution:*
From Example Problem 5.3, the expression for $J_p$ is given by:

$$J_p = \alpha\, \sigma_0\, \varepsilon_0\, a\, (1 - a/W)\, h_1 \left[\frac{\sigma\,(2W)}{0.72 + 1.82\,(1-a/W)\,\sigma_0\, W}\right]^{m'+1}$$

making the substitutions that $\sigma_0 \to 2\sigma_0^c$, $\sigma \to \Delta\sigma$, $\alpha\varepsilon_0 \to 2\alpha'$, and $m \to m'$, we get:

$$\Delta J_p = 4\alpha'\sigma_0^c\, a(1-a/W)\, h_1 \left[\frac{\Delta\sigma(2W)}{.72 + 1.82\,(1-a/W)\,(2\sigma_0^c\, W)}\right]^{m'+1}$$

where $h_1$ is a function of $a/W$ and $m'$ given in Table 5.6.

**Example Problem 9.3**
Derive an expression for $\Delta J$ for an edge crack in an infinite body subjected to a remote cyclic stress $\Delta\sigma$.

*Solution:*

The expression for $J_p$ for this configuration is given by equation (5.62). Making the appropriate substitutions in this equation, we can write:

$$\Delta J_p = 1.21 \ \pi \ \sqrt{m'} \ \frac{(2\alpha')a}{(2\sigma_0^c)^{m'}} \left( \frac{\sqrt{3}}{2} (\Delta \sigma) \right)^{m'+1}$$

$$\Delta J_e = 1.21 \ \Delta \sigma^2 \ (\pi a)/E$$

The total $\Delta J$ can be obtained by adding $\Delta J_e$ and $\Delta J_p$ as follows:

$$\Delta J = 1.21 \ \pi \ a \left[ \sqrt{m'} \frac{2\alpha'}{(2\sigma_0^c)^{m'}} \left( \frac{\sqrt{3}}{2} (\Delta \sigma)^{m'+1} \right) + \frac{\Delta \sigma^2}{E} \right]$$

The reference stress method described in Section 5.4.5 for determining J-integral can be adapted for a number of geometries, particularly for 3-d cracks, to approximate the values of $\Delta J$. This extends the utility of this approach to a large range of cracked configurations relevant to applications [9.29].

The use of $\Delta J$ for characterizing fatigue crack growth under the large-scale plasticity conditions has been criticized, at times even severely, by researchers. One of the strongest criticisms has been that any crack tip parameter based on deformation theory of plasticity is fundamentally unsuited for application to fatigue loading due to presence of unloading. Recall that unloading is not permitted during the use of deformation theory of plasticity because the stress-strain relationships are different during loading and unloading. Thus, for situations involving both loading and unloading such as fatigue, the stress-strain relation is no longer unique. This was the reason why the use of J-integral was restricted to conditions of monotomically increasing loads and also to only small amounts of crack extension. The latter condition was imposed to ensure that unloading occurs only in a small region near the crack tip. However, the cyclic stress-strain relationships, equation (9.2), is unique provided the cyclic stress and cyclic strain range increase continuously, much like the Ramberg-Osgood relationships. This condition is met during the definition of $\Delta J$ by defining it for only the increasing load portion of the load-deflection diagram. Recall that $\Delta J \neq J_{max} - J_{min}$; it is rather the increment in J in going from the condition corresponding to the minimum load to one corresponding to the maximum load. Since the crack increment during one loading cycle is small ($\sim$ 0.1 mm in the extreme), the restriction in the use of $\Delta J$ for small amounts of crack extension is automatically satisfied. Thus, suitable arguments are made to counter the strongest criticisms against the use of $\Delta J$. Nevertheless, other limitations on the use of $\Delta J$ still exist and are described in the subsequent section.

### 9.2.3 Limitations of $\Delta J$

There are three limitations on the use of $\Delta J$ for characterizing the fatigue crack growth behavior under large-scale plasticity in metals. The first of these arises from the idealization of the material as an elastic, cyclically-saturated material. There are several metals in which the saturation conditions are reached rapidly, or they are cyclically stable. $\Delta J$ is the correct crack tip parameter for characterizing fatigue crack growth in these materials. However, as argued by Yoon and Saxena [9.30], there are large numbers of structural materials, such as nickel base alloys, cold-worked steels, and copper alloys that do not achieve true saturation stage even at failure. Since there is no single equation which can describe the stress-strain behavior in the transient regime of cyclic-hardening or softening, the uniqueness of the cyclic stress-cyclic strain relationship cannot be assured. Figure 9.15 schematically illustrates this particular problem [9.30]. Therefore, $\Delta J$-integral is neither path-independent nor is it uniquely related to the crack tip stress and strain range. Thus, the parameter measured at the load-points is not uniquely related to the crack tip conditions and, therefore, cannot be expected to uniquely correlate fatigue crack growth rate. The limits of the application of $\Delta J$ for materials that do not saturation-harden have not been fully explored. The readers are referred to reference [9.30] for some preliminary ideas on how this problem can be approached. The same limitation is also a concern in the use of $\Delta J$ for thermal-mechanical fatigue in the presence of temperature gradients. The temperature gradients will result in property gradients across the cracked body because material flow properties change with temperature. This leads to concerns similar to those for materials that do not quickly attain the saturated state.

The second limitation in the use of $\Delta J$ is the uncertainty about the varying load ratio in tests in which the maximum and minimum deflections are specified as compared to maximum and minimum loads which are more conventional. A systematic study comparing the fatigue crack growth rates at a variety of load (or deflection) ratios under load-controlled and deflection-controlled conditions has not been made. Most of the data generated thus far has all been at low load (or deflection) ratios. Since load ratio

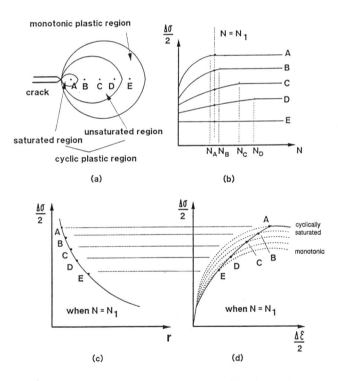

**Figure 9.15** *Schematic representation of stress-strain behavior in the crack tip plastic region for a cyclic hardening material: (a) Schematic of the crack tip deformation zones; (b) Relationship between stress amplitudes and the number of fatigue cycles for strain amplitudes corresponding to material elements A to E; (c) The stress amplitude as a function of distance from crack tip to cycle, $N_1$; and (d) Cyclic stress-strain behavior characterizing the unsaturated material behavior at cycle, $N_1$.*

is only a secondary variable in Region II fatigue crack growth behavior, this is not a significant consideration.

The third limitation is due to the increased importance of crack closure under large-scale plasticity conditions. Since there still is considerable uncertainty in accurately predicting crack closure, it is a limitation in the use of $\Delta J$. One engineering approach to circumvent this problem is to choose the load at which the crack begins to close as the closure point for determining $\Delta J$ in tests. This will provide a realistic estimate of the fatigue crack growth resistance. However, if we choose the entire load or stress range without adjusting for closure in estimating $\Delta J$, we can be assured that the resulting crack growth rate predictions will be conservative.

## 9.3 Test Methods for Characterizing FCGR Under Large Plasticity Conditions

In this section, we will discuss special considerations in generating fatigue crack growth data at very high stress intensities. Under these conditions, as discussed earlier, small-scale-yielding conditions cannot be assured without using large specimen sizes which are inconvenient and expensive to test. We begin by demonstrating the inadequacy of the conventional load-controlled and displacement-controlled test methods used successfully for obtaining fatigue crack growth rate data at the lower $\Delta K$ levels.

Consider a cracked specimen subjected to cyclic loads sufficiently high to produce significant levels of plasticity, Figure 9.16a. The load is cycled between a maximum value $P_{max}$ and a minimum value $P_{min}$. The nonlinearity in the load and deflection behavior due to plasticity is shown in Figure 9.16b. If $P_{min} > 0$, there will be a residual deflection at the end of each cycle as shown in the plot of the maximum and minimum deflections as a function of fatigue cycles, Figure 9.16c. The maximum and minimum deflections will increase rapidly with each cycle, resulting ultimately in instability. Thus, little or no fatigue crack growth data can be obtained. On the other hand, if the test is conducted under conditions of fixed deflection limits, the resultant load-deflection behavior will be as shown in Figure 9.17. There is considerable nonlinearity and therefore, hysteresis between the loading and unloading segments of the load-deflection curve in the initial portion of the test. However, as the crack size increases, the extent of plasticity decreases as seen in Figure 9.17b and the magnitude of $\Delta J$ values are encountered in the initial portions of the test requiring high load capacities for the test machine and the clevises.

To address the above shortcomings, Dowling [9.31] developed a fatigue crack growth test in which he maintained displacement-controlled conditions but increased the displacement amplitude gradually as the crack extended. The load-deflection behavior in such a test is shown in Figure 9.18. On the load-deflection diagram, the line OA represents an envelope for the load and displacement limits. A device was set up to detect the peak load in each cycle. If the peak load falls below the point of intersection between line OA and the hysteresis loop, the deflection range is automatically incremented to bring the peak load back on the envelope line. By selecting the slope of the line OA, the rate of increase in $\Delta J$ with crack size can be selected. Note, that the $d(\Delta J)/da$ will by no means be a constant. However, an envelope line closer to the horizontal will result in more rapid increases in $\Delta J$ if all other factors such as the material and the specimen geometry and size are the same. Dowling designed special circuits to conduct these tests. However, with computer control becoming standard in servo-hydraulic equipment these days, the control part of these tests is quite straightforward. The clevis design for such tests is more complex because even though the two deflection limits are positive, the resulting minimum load is compressive. Therefore, the standard pin assemblies with universal joints to accommodate small misalignments are no longer adequate. All joints in these tests must be rigid and considerably more care is then necessary to avoid misalignment.

The crack length, a, vs. fatigue cycles, N, data can be obtained from such a test to obtain da/dN values at various crack sizes. The load-deflection hystersis can be used to calculate the value of $\Delta J$ during different portions of the test.

# Fatigue Crack Growth Under Large-Scale Plasticity 291

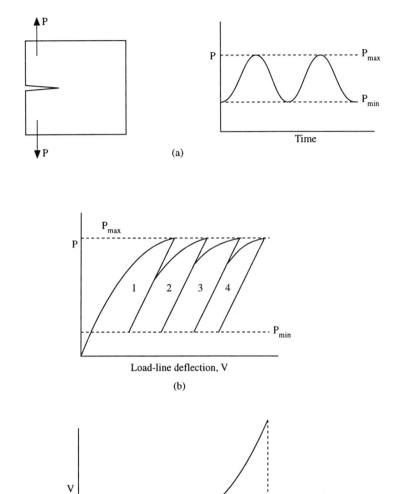

**Figure 9.16** (a) Cracked body shown as being subjected to fatigue under load-controlled condtions. (b) Load vs. load-line deflection behavior. (c) Maximum and minimum deflections as a function of fatigue cycles.

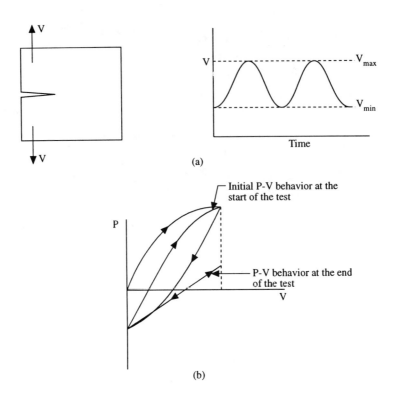

*Figure 9.17* (a) Cracked body shown as being subjected to fatigue under displacement-controlled conditions. (b) Load vs. deflection behavior during inital and last stage of the test.

### 9.4 Behavior of Small Fatigue Cracks

For safe design or remaining life calculations, the initial flaw size in components is assumed to be the minimum size that can be reliably detected by the selected nondestructive inspection (NDI) technique. The size of these critical cracks are often small for several components especially in aerospace applications where the operating stresses are high. With advances in NDI techniques, the limits of crack detection will only get smaller. Small cracks have been shown to grow at rates faster than those predicted by the LEFM theory and are therefore an important concern.

A schematic of small vs. large crack growth rate behavior at low load ratios (R ≤ 0.5) is illustrated in Figure 9.19. Initially, the small cracks propagate at rates faster than the large cracks at nominally the same $\Delta K$. As crack size increases, the growth rates of these cracks decelerate to a minimum, accelerate, and eventually equal the large crack propagation rates, as shown by the curve ABC in Figure 9.19. The above small crack effect is more pronounced in the near-threshold region than in Region II of the fatigue crack growth behavior. However, small cracks emanating from notches are known to also exhibit this trend in Region II. Further, if the applied stress amplitudes are very low, small cracks emanating from

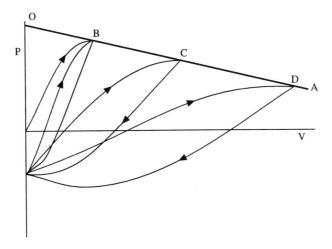

***Figure 9.18*** *Schematic of a load-deflection behavior during a defection-controlled test in which the defelction range is incremented with increase in crack size (Ref. 9.31).*

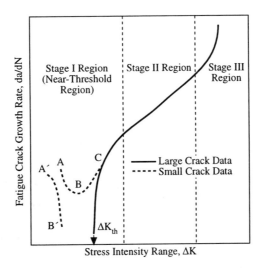

***Figure 9.19*** *Schematic of a large and small fatigue crack propagation behavior illustrating the small crack effects.*

notches may propagate from the notch root at a decreasing rate and eventually stop growing, curve A', B' in Figure 9.19. Small cracks have also been shown to propagate at $\Delta K$ levels below the threshold stress intensity range, $\Delta K_{th}$, for large cracks as also shown in Figure 9.19.

If we relate the endurance limit for fatigue, $\Delta \sigma_e$, to crack size, a, the plot will schematically appear like one shown in Figure 9.20 known as the Kitagawa diagram [9.32]. When the crack size is large, one can expect the following relationship between $\Delta K_{th}$ and $\Delta \sigma_e$:

$$\Delta K_{th} \approx 1.12 \, \Delta \sigma_e \sqrt{\pi a} \tag{9.21}$$

Since we are dealing with small cracks, we can assume that the crack size correction factor is the surface correction factor 1.12. If $\Delta K_{th}$ is a material constant, the relationship between log $\Delta \sigma_e$ and log a is linear with a slope of -1/2. However, as the crack size decreases, $\Delta \sigma_e$ must asymptotically approach the smooth specimen endurance limit $\Delta \sigma_e^0$ corresponding to zero crack size. Thus, a crack size $a_0$ can be defined below which the cracks are no longer expected to be characterized by LEFM:

$$a_0 = \frac{1}{\pi} \left( \frac{\Delta K_{th}}{1.12 \, \Delta \sigma_e^0} \right)^2 \tag{9.22}$$

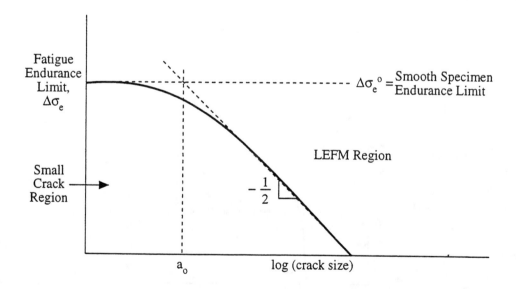

**Figure 9.20** *Plot of fatigue endurance limit as a function of crack size (Kitagawa diagram).*

Equation (9.22) provides a definition of small crack which is also known as the physically short crack. The term "short" is used in place of "small" because, in this case, only the crack depth needs to be small. The length dimension can be large. The term "small" is reserved for the case when all pertinent dimensions of the crack are small in relation to some characteristic length scale. In the case of microstructurally small cracks, this length scale is on the order of the dimensions of microstructural periodicity such as grain size. Such small cracks are often referred to as microstructurally small cracks. Physically small cracks are ones in which the limitations of LEFM are important. In the subsequent discussion, the regime of physically small or short crack is explored further. Several factors have been suggested to account for the non-unique da/dN vs. $\Delta K$ behavior of small cracks. Collectively, these factors can be considered as limitations of LEFM.

### 9.4.1 Limitations of LEFM for Characterizing Small Fatigue Crack Growth Behavior

There are several conditions listed below which must be met for da/dN to uniquely correlate with $\Delta K$:

- The crack tip stress field must be uniquely characterized by K in a region which is large compaared to size of the crack tip plastic zone.

- The material at the length scale over which fatigue damage develops is isotropic and homogeneous.

- The cracks must grow under dominantly mode I conditions.

- Scales of plasticity associated with the growth of fatigue cracks must be small not only in relation to the dimensions of the cracked body but the crack size itself.

- Crack closure levels must be independent of crack size.

Collectively, the five conditions listed above are necessary for ensuring that similitude exists in the conditions at the tips of two cracks that are loaded to the same nominal $\Delta K$ but may differ in size, geometry, loading, configuration, etc. Only when all of these conditions are met, a unique da/dN vs. $\Delta K$ relationship can be expected. Each of these limitations are discussed further in this section. Models and approaches to account for these shortcomings in LEFM are discussed in the next section.

Talug and Reifsnider [9.33] have determined the elastic stress distribution ahead of cracks that are 0.25mm and 6.25mm deep in 25.0mm wide single-edge-notch panels loaded to uniform tension. Their results were presented earlier in Figure 3.4. To solve this problem, they determined several coefficients of the Williams stress function, equation (8.1) [9.34], by boundary collocation. Thus, full crack tip stress field solutions including several terms beyond the singular term were obtained for two cracks. Readers are reminded that the coefficient of the first singular term in the Williams stress function is related to K, and yields results which are identical to the results given by Irwin which were based on Westergaard's method [9.35]. Figure 3.4 plots the stress $\sigma_{yy}$ as a function of distance from the crack tip for both cracks at a constant K value of 22.2 Mpa$\sqrt{m}$. The solid line shows the full solution for the small crack and the dashed line shows the same for the long crack. The Irwin solution for either crack based only on the first singular term is indistinguishable from the dashed line corresponding to the full-field solution of the larger crack. Thus, the long crack stress distribution is completely dominated by K, but considerable difference exists between the single term solution and full-field solution for the short crack. When the same stress distributions are plotted as a function of distance which is normalized by its crack size, as shown in Figure 3.4, the deviation between the full-field solution and that of Irwin's single term solution occurs at r/a ≈ 0.15 for both cracks. This establishes that the region of K-dominance is on the order of 15% for both cracks in this geometry. It is well established that the magnitude of the coefficients of the second and higher order terms in Williams's series are very dependent on geometry and loading configuration. Thus, if the K-dominated region for small cracks is not large in comparison to the process zone in which fatigue damage develops, ΔK cannot be expected to uniquely characterize the crack growth rate. Further, it is necessary that the K-dominated region be much larger in size than the plastic zone or else the crack tip plastic zone size and shape will no longer be uniquely determined by K. The correlation between da/dN and ΔK is expected to break down under these circumstances.

Linear elastic fracture mechanics (LEFM) is based on the principles of elasticity which also assume that the material is homogeneous and isotropic. When crack sizes are smaller than the average grain diameter, it cannot be assumed that the material is isotropic. Also, the crack tip experiences considerable non-uniformity in the microstructural features that it encounters. For example, the crack tip stress fields can be influenced significantly by the presence of grain boundaries in the proximity of the crack tip or second phase particles such as inclusions. The back stresses from these microstructural entities can cause considerable variability in the crack growth rates which cannot be addressed within the framework of linear elasticity theory.

The third condition that the crack growth occur in mode I is also violated frequently when considering the growth of small cracks. For example, small cracks at low ΔK levels often grow along slip bands which form preferentially along certain crystallographic directions. Thus, the crack growth is no longer in pure mode I and mixed-mode growth becomes a consideration leading to deviation from long crack behavior which is predominantly in mode I.

In order for LEFM conditions to dominate, one can write the following condition from ASTM standard E-399 [9.36] for fracture toughness testing:

$$a \geq 2.5 \left( \frac{K}{\sigma_0} \right)^2 \tag{9.23}$$

It can be argued whether the factor 2.5 on the right-hand side of the inequality in (9.23) is applicable for fatigue. Unfortunately, no equivalent crack size requirements valid for fatigue crack growth have been established, thus as a first approximation, we will use the one in equation (9.23). We can clearly see that as the crack size decreases, this condition may not be met and elastic-plastic analysis may be necessary.

The crack closure level can be different for large and small cracks. Since plasticity induced crack closure occurs due to interference in the plastically deformed wake of the crack, differences can occur in the crack closure levels between very small and long cracks. Therefore, for the same applied nominal ΔK, the $\Delta K_{eff}$ for the small crack can be higher and thus lead to a higher crack growth rate. Since crack closure is more important in the near-threshold fatigue crack growth region, it follows from this hypothesis that the small crack effect will also be more important in this region. Since this prediction is in agreement with the observed trends, it lends credence to the hypothesis that small crack effects are caused by differences in crack closure levels between small and large cracks.

**Example Problem 9.4**

0.5mm deep surface cracks, oriented normal to the axis, are found in a cylindrical rod subjected to a fatigue stress of 200 Mpa. If the yield strength of the rod is 350MPa, the smooth specimen endurance limit is 250 Mpa and the $\Delta K_{th}$ is 8 Mpa√m, can the growth of this crack be treated using the conventional LEFM analysis?

*Solution:*

The critical crack size, $a_0$, which defines the size below which cracks can be considered small, equation (9.22), is given by:

$$a_0 = \frac{1}{\pi}\left(\frac{8}{1.12 \times 350}\right)^2 = .00041\,m = 0.41\,mm$$

Since $a_0$ is smaller than the crack size, the crack will not be considered small by this condition. The smallest crack size for which the condition in equation (9.23), based on plasticity considerations, is satisfied is given by:

$$2.5\left(\frac{1.12 \times 200\sqrt{\pi \times 0.5 \times 10^{10^{-3}}}}{350}\right)^2 = 1.6 \times 10^{-3}\,m = 1.6\,mm$$

Since the above crack size is larger than the crack size in the component, this crack should be considered small and LEFM should not be used for predicting its growth rate. Note that to use LEFM, the crack size must be larger than the crack sizes predicted by both equations (9.22) and (9.23).

### 9.4.2 Models for Predicting the Growth of Small Fatigue Cracks

Models are available to account for some aspects of small crack behavior. These models address not only LEFM limitations in the presence of plasticity which exceeds the permissible limits of theories based on small-scale yielding, but also ones which account for the influence of crack size on plasticity induced crack closure. Before describing these models, it is useful to first scope the overall problem of modeling small crack behavior so the reader can clearly understand the limitation of the models and also appreciate that much research is needed in areas outside the conventional boundaries of linear and nonlinear fracture mechanics to develop a fundamentally sound approach for accurately predicting small crack behavior.

Figure 9.21 is a schematic of how crack growth rates vary with crack size for long and small cracks for low and high applied stress amplitudes as proposed by McDowell [9.37]. The rates of propagation of small cracks are shown to possess an oscillatory character which is particularly pronounced at the low stress amplitudes. To fully understand this behavior, several factors must be considered. We begin with the assumption that small cracks of the size of characteristic microstructural scale are present in the material as a result of thermo-mechanical processing. For illustration we assume that this size is equal to the grain size, Figure 9.22a. At low stress amplitudes, even though the scale of plasticity may be small compared to both the crack size and the grain size, the characteristics of the slip bands, etc., are very dependent on the orientation of the grain located next to the crack tip. Thus, the growth rate of the cracks will also vary considerably with the characteristics of the grains as shown in Figure 9.21. At high stress amplitudes, plasticity is more extensive and can engulf several grains, Figure 9.22b, thus homogenizing slip and reducing the influence of microstructural barriers on the crack growth rate. This results in relatively lower levels of variability in the crack growth rate curves shown in Figure 9.21. Nevertheless, because crack size is small relative to the microstructural length scale, considerable scatter still exists in the data.

Note that distinction is made in Figure 9.21 between a regime in which the microcrack length is on the scale of the grain size and that in which the crack size sufficiently exceeds the grain size. In the former, the microstructural characteristics dominate the crack growth behavior, more so at the low stress amplitudes than at high stress amplitudes. Since the damage process zone in the form of the slip bands in the regime is on the same order of magnitude as the crack size itself, it is inappropriate to use elastic-plastic fracture mechanics (EPFM) concepts based on crack tip stress singularity arguments. This regime is termed as the Microstructural Fracture Mechanics (MFM) [9.38] and is not well developed. However,

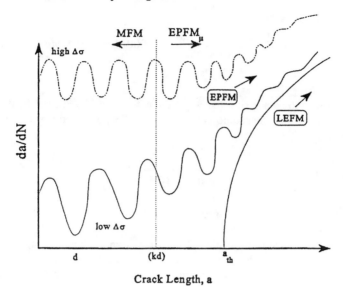

*Figure 9.21* Schematic of typical progression of crack growth rate behavior as a function of crack length and stress amplitude for microstructurally small, physically small, and long cracks (courtesy of D.L. McDowell).

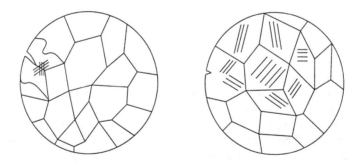

*Figure 9.22* Schematic of a microstructurally small crack under (a) low stress amplitude fatigue showing slip bands restricted to single grains and (b) high stress amplitude fatigue showing slip extendig into several grains.

there are promising models available for this regime which are based upon extending the models for predicting crack nucleation. For further background on these models, the reader is referred to reference [9.37] to [9.39]. In the second case, where the crack size is on the order of several grain diameters, the process zone is reasonably well confined to the crack tip region and spreads over several grain diameters. This is the regime which has been previously called as the regime of physically small cracks. As we will

*Fatigue Crack Growth Under Large-Scale Plasticity* 299

show later in this section, EPFM is very relevant to characterizing the behavior of cracks in this regime. The transition from the MFM to the EPFM regimes is made at a crack size several times the grain diameter designated by kd in Figure 9.21.

**Models for Considering Large-Scale Plasticity** The growth rate of physically small cracks can be described by the following equation proposed by Dowling [9.27] and others [9.40 - 9.42].

$$\frac{da}{dN} = c_1 (\Delta J_{eff})^{n_0} \tag{9.24}$$

Where $\Delta J_{eff}$ is the effective value of $\Delta J$ corrected for crack closure, and $c_1$ and $n_0$ are regression constants directly related to the constants in the Paris equation. This relationship can be easily derived by replacing $\Delta J_{eff}$ by $(\Delta K_{eff})^2/E$. Examples of plots of da/dN vs. $\Delta J_{eff}$ are shown in Figure 9.23 for AISI 1026 steel [9.42] obtained by testing axial fatigue specimens subjected to various strain ranges and using plastic replica techniques for measuring the growth rates of small fatigue cracks. By recording the stress-strain hysteresis loops, crack closure can be conveniently measured. However, the displacement cannot be directly measured in components; therefore, we must rely on accurate methods for estimating $\Delta J$ in components which account for plasticity and crack closure. In the subsequent discussion, such methods are described.

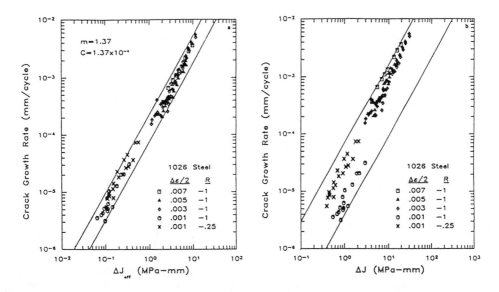

*Figure 9.23 Correlations of constant amplitude crack growth rate data with estimate of $\Delta J$ based on (a) $\Delta J_{eff}$ and (b) based on $\Delta J$ calculated by using the full stress range. From Ref. 9.42.*

Methods for determining $\Delta J$ were described in Section 9.2.2. In this section, this discussion will be augmented to include special considerations for small cracks. When the displacement or strain is directly measured, $\Delta J$ can be obtained as [9.27, 9.42]:

$$\Delta J = \Delta J_e + \Delta J_p = 2\pi F^2 a [\Delta W_e + f(n) \Delta W_p] \tag{9.25}$$

Where $\Delta W_e$ and $\Delta W_p$ are elastic and plastic components of remote strain energy densities, respectively, and F is the boundary correction factor (1.12 for surface cracks and 1 for interior cracks). It is left as a problem to derive the function f(n) for single edge and center cracks. To calculate the values of $\Delta W_e$ and $\Delta W_p$, it is recommended that $\Delta \sigma_{eff}$ be used. When measured values of displacements are not available, it is strongly recommended that in calculating the elastic contribution to $\Delta J$, the plasticity corrected value of crack size be used. Newman [9.43] recommends that the crack length be increased by 0.25 of the plastic zone size based on the Dugdale estimate. Thus:

$$\Delta K_p = \Delta \sigma_{eff} \sqrt{\pi d}\, F_1 \tag{9.26}$$

where

$$d = a + \gamma\, r_p \tag{9.27}$$

where $\gamma \sim 0.25$ and $r_p$ is plastic zone size for plane stress as estimated by Dugdale. For further details on the importance of his plasticity correction, the readers are referred to reference [9.43]

Newman [9.43] has studied the influence of crack closure on growth behavior of small fatigue cracks by conducting a series of numerical simulations using his crack closure model. He calculated the crack opening stress as function of crack size emanating from a small defect of height 2h. The initial defect or void has a size, $a_i$ of 3 μm, $c_i = 12$ μm and the value of h, the void height, was varied. The results of this analysis are presented in Figure 9.24a. The crack opening stress for h = 0 and h ≥ 0.4 μm are plotted in this figure for the test material which was 2024-T3 Al alloy. These simulations were performed for R = -1 and $\sigma_{max}/\sigma_0 = 0.15$ in a center crack panel as shown in the figure. Results shown in this figure clearly show that the defect height, 2h, had a large influence on the closure behavior of small cracks. For h ≥ 0.4 μm, the initial defect surfaces did not close, even under compressive loading. However, as the crack grew, it did close according to the trend shown in Figure 9.24a and the opening stress reached a steady-state value. The crack growth rate is predicted to reach a minimum value and subsequently increase to the level predicted by the long crack data. On the other hand, for h = 0, the defect surfaces made contact during compressive loading, and the contacting surfaces greatly influenced the amount of residual plastic deformation left behind as the crack grew. The calculated crack-opening stresses stabilized very quickly at the steady-state value, as shown in Figure 9.24a. These results suggest that part of the small crack effect may be due to an initial defect height that is sufficient to prevent closure over the initial defect surfaces.

Initially, the low crack-opening stresses give rise to high effective stress ranges and, consequently, high growth rates. However, as the crack grows the crack opening stresses rise much more rapidly than the stress intensity factor causing a reduction in the effective value of $\Delta K$, causing a minimum in the crack growth rate at a crack size of 20 μm for h ≥ 0.4 μm. Thus, crack closure is another mechanism by which the behavior of the type shown in Figure 9.19 can be explained as discussed earlier. Newman [9.43] made further calculations of opening stresses as a function of crack size for different load ratio values ranging from -2 to +0.5 as shown in Figure 9.24b. The corresponding values of $\sigma_{max}/\sigma_0$ are also noted on the figure. These results clearly show that as the load ratio increases, the crack opening stress reaches its saturation value much more rapidly. Thus, we expect the small crack behavior to be much more "suppressed" at high R values which agrees with the trends of the data in the literature. Similarly, higher stress amplitude will decrease the extent of the crack closure transient and will contribute toward suppressing the small crack behavior.

Figure 9.25a shows the predicted trends for small crack behavior from Newman's analysis of 2024-T3 Al panels for R = -1 for various stress amplitudes. The large crack data trend as a function of $\Delta K$ and $\Delta K_{eff}$ are shown for comparison. The predicted trends resemble the shape of the actual data as can be seen in Figure 9.25b which includes the experimental data. The stress amplitude used in the predicted trend was identical to one used for conducting the tests. The crack growth rates in the tests were measured using the replication technique. In the above figure it is also seen that the da/dN vs. $\Delta K_{eff}$ curve for long cracks appears to provide the upper bound for all crack growth behavior from small or large crack behavior. This also is a very good argument for the importance of the role played by crack closure in determining the small crack behavior.

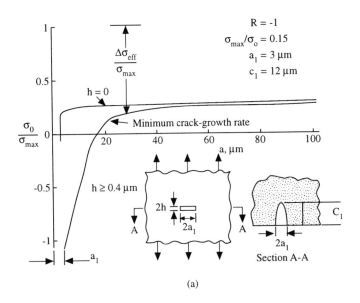

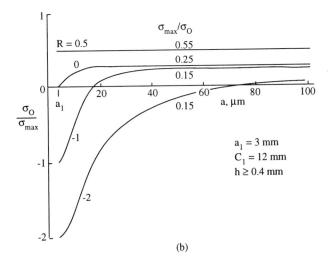

**Figure 9.24** (a) Influence of defect height on the predicted crack opening stress. (b) Influence of load ratio on the crack opening stress. From Ref. 9.43.

## 9.5 Summary

In this chapter, the plasticity aspects of the fatigue crack growth process were discussed in detail. The chapter began with a description of crack tip plasticity which accompanies a growing fatigue crack and its role in determining the stress (or load) levels at which cracks open or close during cyclic loading. An analytical model was presented for predicting the crack opening load and the concept of the effective stress intensity factor was introduced and shown to have a fundamental meaning in determining the fatigue crack growth rate. The inadequacy of the LEFM concepts was demonstrated when crack tip plasticity was no longer contained. The cyclic J-integral, $\Delta J$, was defined and shown to be the correct crack tip parameter for characterizing fatigue crack growth rates in the presence of extensive plasticity. Methods

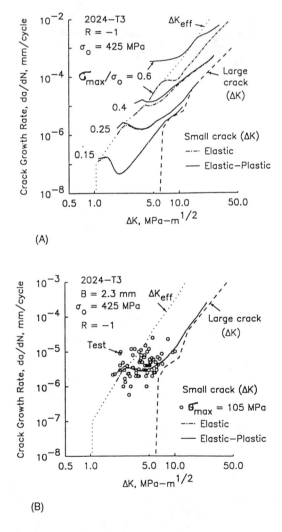

**Figure 9.25** *(a) Predicted influence of crack closure on the growth of small crack of 2024-T3 aloy. (b) Comparison of predicted and actual small crack data in 2024-T3 Al alloy (courtesy of Newman, Ref. 9.43).*

for determining ΔJ were described. These methods are based on analogy with the J-integral under monotonic loading conditions. This analogy makes it possible to apply the ΔJ approach to a wide variety of cracked component geometries and loading configurations. Experimental techniques for measuring fatigue crack growth rates under high stress intensities were discussed. The final section in this chapter deals with the behavior of small cracks and the role of elastic-plastic fracture mechanics (EPFM) in describing the behavior of physically small cracks. It is argued that while EPFM concepts are inadequate for characterizing the behavior of cracks which are smaller than or equal to the microstructural length scales of the material, they seem to provide an engineering basis for the treatment of physically short cracks. Crack closure and the scale of plasticity appear to play significant roles in the behavior of physically small cracks. The most promising engineering approaches for dealing with the small crack behavior are discussed.

## 9.6 References

9.1 S. Suresh, "Fatigue of Materials", Cambridge University Press, 1991.

9.2 D. Broek, "Elementary Engineering Fracture Mechanics", Kluwer Academic Publishers, Fourth Edition, 1991.

9.3 T.L. Anderson, "Fracture Mechanics - Fundamentals and Application", Second Edition, CRC Press, 1995.

9.4 A. Saxena and C.L. Muhlstein, "Fatigue Crack Growth Testing and Data Analysis", ASM Handbook, Vol.19, ASM International, Metals Park, 1996.

9.5 N.E. Dowling, "Mechanical Behavior of Materials", Prentice Hall, 1993.

9.6 C.E. Feltner and C. Laird, "Cyclic Stress-Strain Response of FCC Metals and Alloys - I. Phenomenological Experiments", Acta Met., Vol. 15, 1967, pp. 1621-1632.

9.7 C.E. Feltner and C. Laird, "Cyclic Stress-Strain Response of FCC Metals and Alloys - II. Dislocation Structures and Mechanisms", Acta Met., Vol. 15, 1967, pp. 1633-1653.

9.8 A. Saxena and S.D. Antolovich, "Low Cycle Fatigue and Fatigue Crack Propagation in Polycrystaline Cu-Al Alloys", Metallurgical Transactions, 1975, pp. 1809-1823.

9.9 C. Laird and G.C. Smith, "Crack Propagation in High Stress Fatigue", The Philosophical Maganize, Vol. 7, 1962, pp. 847-853.

9.10 W. Elber, "Fatigue Crack Closure Under Cyclic Tension", Eng. Fracture Mechanics, 1970, pp. 573-584.

9.11 W. Elber, "The Significance of Fatigue Crack Closure", in Damage Tolerance in Aircraft Structures", ASTM STP 486, American Society of Testing and Materials, Philadelphia, 1971, pp. 230-242.

9.12 P.C. Paris, R.J. Bucci, E.T. Wessel, W.G. Clark, and T.R. Mager, "Extensive Study of Low Cycle Fatigue Crack Growth Rates in A533 and A508 Steels" in Stress Analysis and Growth of Cracks, ASTM STP 513, American Society of Testing and Materials, Philadelphia, 1971, pp. 141-176.

9.13 R.P. Skelton and J.R. Haigh, "Fatigue Crack Growth Rates and Thresholds in Steels Under Oxidizing Conditions", Materials Science and Engineering, Vol. 36, pp. 17-25.

9.14 S. Surerh, G.F. Zaminski, and R.O. Ritchie, "Oxide-Induced Crack Closure: An Explanation for Near Threshold Corrosion-Fatigue Crack Growth Behavior", Metallurgical Transactions, Vol. 12A, 1981, pp. 1435-1443.

9.15 S. Purushothaman and J.K. Tien, "A Fatigue Crack Growth Mechanism for Ductile Materials", Scripta Metallurgica, Vol. 9, 1975, pp. 923-926.

9.16 M.D. Halliday and C.J. Beevers, "Some Aspects of Crack Closure in Two Contrasting Ti-Alloys", Journal of Testing and Evaluation, Vol. 9, pp. 195-201.

9.17 R.O. Ritchie and S. Suresh, "Some Considerations on Fatigue Crack Closure at Near Threshold Stress Intensities Due to Fracture Surface Morphologies", Metallurgical Transactions, Vol. 13A, 1981, pp. 101-110.

9.18 K. Endo, T. Okada, and T. Hariya, "Fatigue Crack Growth in Bearing Metals Lining of Steel Plates in Lubricating Oil", Bulletin of Japan Society of Mechanical Engineers, Vol. 15, 1972, pp. 439-445.

9.19 B. Budiansky and J.W. Hutchinson, "Analysis of Crack Closure in Fatigue Crack Growth", Journal of Applied Mechanics, Vol. 45, 1978, pp. 267-276.

9.20 J.C. Newman and P.R. Edwards, "Short Crack Behavior in an Aluminum Alloy - An AGARD Cooperative Test Program", AGARD R-73, 1988.

9.21 J.C. Newman, "A Review of Modelling Small-Crack Behavior and Fatigue-Life Predictions for Aluminum Alloys", Engineering Fracture Mechanics, Vol. 17, 1994, pp. 429-439.

9.22 R.C. McClung and H. Schitoglu, "On the Finite Element Analysis of Fatigue Crack Closure - 1. Basic Modeling Issues", Engineering Fracture Mechanics, Vol. 23, 1989, pp. 237-252.

9.23 R.C. McClung and H. Sehitoglu, "Closure Behavior of Small Cracks Under High Strain Fatigue Histories", Mechanics of Fatigue Crack Growth, ASTM STP 982, American Society for Testing and Materials, 1988, pp. 279-299.

9.24 R.C. McClung, "Applications of a Finite Element Analysis of Fatigue Crack Closure", Advances in Fracture Research, ICF-7, Pergamon Press, 1989, pp. 1257-1264.

9.25 H.S. Lamba, "The J-Integral Applied to Cyclic Loading", Engineering Fracture Mechanics, Vol. 7, 1975, pp. 693-703.

9.26 N.E. Dowling and J.A. Begley, "Fatigue Crack Growth During Gross Plasticity and the J-Integral", in Mechanics of Crack Growth, ASTM STP 590, American Society for Testing and Materials, 1976, pp. 82-103.

9.27 N.E. Dowling, "Crack Growth During Low-Cycle Fatigue of Smooth Axial Specimens", in Cyclic Stress-Strain and Plastic Deformation Aspects of Fatigue Crack Growth", ASTM STP 637, American Society for Testing and Materials, 1977, pp. 97-121.

9.28 W.R. Brose and N.E. Dowling, "Fatigue Crack Growth Under High Stress Intensities in 304 Stainless Steel", in Elastic-Plastic Fracture, ASTM STP 668, American Society for Testing and Materials, 1979, pp. 720-735.

9.29 R.C. McClung, G.C. Chell, D.A. Russell and G.E. Orient, "A Practical Methodology for Elastic-Plastic Fatigue Crack Growth", Paper Presented at the ASTM 27th National Symposium on Fatigue and Fracture Mechanics, June 27-29, 1995, Williamsburg, VA.

9.30 K.B. Yoon and A. Saxena, "An Interpretation of $\Delta J$ for Cyclically Unsaturated Materials", International Journal of Fracture, July 1991.

9.31 N.E. Dowling, "Fatigue Crack Growth Rate at High Stress Intensities", in Cracks and Fracture, ASTM STP 601, American Society for Testing and Materials, 1976, pp. 19-32.

9.32 H. Kitagawa and S. Takahashi, "Applicability of Very Small Cracks or Cracks in Early Stage", Proceedings of the Second International Conference on Mechanical Behavior of Materials", ASM International, Metals Park, 1976, pp. 627-631.

9.33 A. Talug and K. Reifsnider, "Analysis and Investigation of Small Flaws", in Cyclic Stress-Strain and Plastic Deformation Aspects of Fatigue Crack Growth, ASTM STP 637, American Society for Testig and Materials, 1977, pp. 81-96.

9.34 M.L. Williams, "On the Stress Distribution at the Base of a Stationary Crack", Journal of Applied Mechanics, Vol. 24, 1957, pp. 109-114.

9.35 H.M. Westergaard, "Bearing Pressures and Cracks", Journal of Applied Mechanics, Vol. 61, 1939, A49-53.

9.36 ASTM Standard E399-90, "Standard Test Method for Plane-Strain Fracutre Toughness of Metallic Materials", Book of ASTM Standards, Vol. 03.01, American Society for Testing and Materials, 1992, pp. 506-536.

9.37 D.L. McDowell, "An Engineering Model for Propagation of Small Cracks in Fatigue", Paper submitted for publication to Engineering Fracture Mechanics, 1996.

9.38 K.J. Miller, "Materials Science Perspective of Metal Fatigue Resistance, Materials Science and Technology, Vol. 9, 1987, pp. 169-189.

9.39 K.J. Miller, "The Two Thresholds of Fatigue Behavior", Fatigue and Fracture of Engineering Materials and Structures, Vol. 16, 1993, pp. 931-939.

9.40 D.L. McDowell and J.Y. Berard, "A $\Delta J$-based Approach to Biaxial Low-Cycle Fatigue of Shear Damaged Materials", in Fatigue Under Biaxial and Multiaxial Loading, ESIS 10, Mechanical Engineering Publications, London, 1990, pp. 413-431.

9.41 J.C. Newman, "A Review of Modelling Small-Crack Behavior and Fatigue Life Predictions for Aluminum Alloys", Fatigue and Fracture of Engineering Materials and Structures, Vol. 17, 1994, pp. 429-439.

9.42 R.C. McClung and H. Sehitoglu, "Closure Behavior of Small Cracks Under High Strain Fatigue Histories", in Mechanics of Fatigue Crack Closure, ASTM STP 982, American Society for Testing and Materials, 1988, pp. 279-299.

9.43 J.C. Newman, "Fracture Mechanics Parameters for Small Fatigue Cracks", in Small Crack Test Methods, ASTM STP 1149, American Society for Testing and Materials, 1992, pp. 6-33.

## 9.7 Exercise Problems

9.1 List all the simplifying assumptions in the Budiansky-Hutchinson Model for estimating the stress intensity parameter for crack closure. What are the implications of these assumptions on the ability to predict realistic crack closure levels?

9.2 The cyclic stress-strain relationship for several engineering materials is cataloged in handbooks in the following form:

$$\frac{\Delta \sigma}{2} = K' \left( \frac{\Delta \varepsilon_p}{2} \right)^{n'}$$

where $\Delta \varepsilon_p$ = plastic strain range and $\Delta \sigma$ = stress range. $K'$ and $n'$ are material constants. Derive a relationship between $K'$, $n'$ and the constants $\alpha'$ and $m'$ in equation (9.2).

9.3 Provide a mathematical basis for Elber's proposal that fatigue crack growth rate is uniquely determined by the effective value of $\Delta K$, $\Delta K_{eff}$.

9.4 Show by analogy to J-integral that for a material which obeys equation (9.2), $\Delta J$ is rigorously path-independent.

9.5 List reasons why $\Delta J$ can be used as a unifying crack tip parameter for charcterizing fatigue crack growth behavior under conditions ranging from small-scale yielding to gross plasticity.

9.6 Derive the equations for crack tip stress and strain ranges for a cyclically loaded cracked body in terms of $\Delta J$.

9.7 Prove that under linear-elastic conditions, the following equation is true for plane stress conditions:

$$\Delta J = \frac{\Delta K^2}{E}$$

9.8 For an elastic cyclically stable material, derive an expression for estimating $\Delta J$ for a center crack panel subjected to a cyclic stress range of $\Delta \sigma$. The corresponding deflection range is not available but you are given the material constants, $\sigma_0^c$, E, $\alpha'$, and $m'$. Do the same for an edge crack specimen subjected to bending.

9.9 The expression for estimating $J_p$ for a single edge crack in a semi-infinte body loaded in tension is:

$$J_p = 1.12 \, \pi\sqrt{m} \, \sigma \, \varepsilon_p a$$

Based on the above expression, derive an approximate expression for estimating $\Delta J$ in an axial fatigue specimen containing a very shallow surface crack with its plane oriented normal to the loading axis.

9.10 The deformation of stainless steel at 600°C follows a relationship of the form given by equation (9.2) with the following constants: E = 235 GPa, $\sigma_0° = 200$ Mpa, $m' = 5$, and $\alpha' = 2 \times 10^3$. A compact type specimen, 25mm thick and 50mm wide of this material is to be used to obatin crack growth rates for a $\Delta K$ up to 100 Mpa √m. Assuming that the maximum value of $\Delta K$ is reached when a/W = 0.6, estimate the load and deflection ranges (load-line) necessary to obtain this data.

9.11 Suppose that in Problem 9.10 you are given an additional condition that the machine available for testing has a maximum load capacity of ± 40 kilo-newtons. What size CT specimen and starting crack size would you recommend to obtain data in the range of $\Delta K$ = 50 to 100 MPa √m?

9.12 Thermal stresses in heavy section components are often transient and also high enough to be in the inelastic range. A potential application of $\Delta J$ is to characterize crack growth behavior due to thermal-mechanical fatigue (TMF). Discuss limitations of $\Delta J$ for use in TMF.

9.13 Clearly state all reasons why linear elastic and elastic-plastic fracture mechanics concepts are inadequate for characterizing the behavior of small cracks.

9.14 The threshold value of $\Delta K$, $\Delta K_{th}$, for mild steel is 7 Mpa √m at R = -1 and the endurance limit is 175 Mpa. Draw a Kitagawa diagram and estimate the size of the smallest crack for which LEFM concepts can be applied. Assume that the average grain diameter for the material is 10 μm.

CHAPTER 10

# ANALYSIS OF CRACKS IN CREEPING MATERIALS

Several components of power-plants, chemical reactors, and land-, air-, and sea-based gas turbines operate at temperatures where creep deformation and fracture is a design concern. Creep deformation is said to occur as a function of time due to sustained stress at elevated temperature. Creep becomes a design concern when service temperatures exceed 35% of the melting point of the alloy in degrees Kelvin from which the component is made.

Failures due to creep can be classified either as resulting from widespread creep damage, or resulting from localized creep damage. The structural components which can be damaged by widespread creep are those that are typically subjected to uniform temperatures and stress during service, such as thin-wall pipes. The life of these components can be estimated from creep rupture data, an approach that has been used in engineering analyses for several decades. However, frequently high temperature components, especially those containing thick sections, are subjected to stress and temperature gradients and do not fail by creep rupture. It is more likely that at the end of the predicted creep rupture life, a crack develops at a high stress location which propagates and ultimately causes failure. The discrepancy between the actual life and that predicted from the rupture data can be the crack propagation period which was not considered. Failures can also result from pre-existing defects, in which case the entire life is consumed by crack propagation. Therefore, it is important to develop the capability to predict crack propagation life at elevated temperatures in the presence of creep deformation.

The linear-elastic and elastic-plastic fracture mechanics concepts discussed in previous chapters are unable to predict crack growth in the presence of significant creep strains. Thus, this chapter will focus on developing the concepts of time-dependent fracture mechanics (TDFM) which possesses such capabilities. In looking to extend the fracture mechanics concepts to conditions where time-dependent deformation is no longer a limitation, we will be taking advantage of considerable analogies that exist between TDFM and the elastic-plastic fracture mechanics (EPFM) with which the reader should be quite familiar from Chapters 4 to 9.

Figure 10.1 shows the creep deformation as a function of time in uniaxial specimens of 1 Cr-1Mo-0.25V steel subjected to a constant stress load. The deformation rates increase as the applied stress increases, and the creep behavior can be divided into three regions (see Figure 2.9). The first region is called the primary creep region in which the strain rate continuously decreases with time until it reaches a steady-state value, from here it enters into the second region. This latter region is called the steady-state creep region. In the tertiary creep region which follows the steady-state region, the creep strain rates begin to rise with time as necking develops and is followed by rupture. Since the object of engineering design is to avoid rupture, tertiary creep regime is not significant in practical applications. Primary creep in several materials is short-lived, thus, making the steady-state creep region the most important in design considerations. As mentioned earlier, the strain rates in this region remain constant with time; however, they change significantly with stress and are described by a power-law, (also know as the Norton's Law) equation (10.1):

$$\dot{\varepsilon}_{ss} = A\, \sigma^n \qquad (10.1)$$

where $\dot{\varepsilon}_{ss}$ is the steady-state strain rate and $\sigma$ = applied stress (see Figure 2.10). In Sections 10.1 and 10.2, we will be using equation (10.1) to describe the creep deformation kinetics. In Section 10.3, we will generalize the creep deformation equation to also include primary creep in our consideration of crack tip parameters.

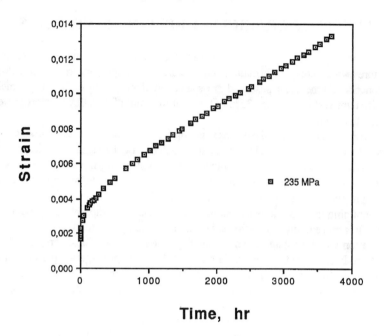

*Figure 10.1* Creep deformation behavior of 1Cr-1Mo-0.25V steel at a stress level of 235 MPa.

## 10.1 Stress Analysis of Cracks Under Steady-State Creep

To identify relevant field parameters for characterizing crack growth at elevated temperature, the following process should be considered [10.1]. Figure 10.2 shows a schematic of the deformation zones ahead of a stationary crack tip subjected to load in the creep regime. A load, P, is assumed to be applied instantaneously and is sustained indefinitely. Upon loading, a plastic zone is formed at the crack tip. The size of the plastic zone depends on the applied K level and the yield strength of the material. Also shown in the figure is a K-zone in which the K-controlled elastic stress and deformation fields hold, if small-scale yielding conditions are maintained during the initial loading. If extensive plastic deformation occurs during initial loading, K loses its significance as the dominating crack tip parameter. Under such conditions, the crack tip stresses and strains are characterized by the J-integral. With time, in either case, the stresses in the vicinity of the crack begin to relax due to creep deformation and the size of the relaxation (or creep) zone increases with time if the crack is assumed to remain stationary. Neither K nor J is expected to uniquely characterize the crack tip stress relaxation (or redistribution) behavior within the creep zone because creep deformation is not admitted in their formulation. However, if the creep zone is small, K or J will continue to characterize the crack tip stresses outside the creep zone. When the creep zone becomes comparable to the dimensions of the cracked body, K and J completely lose their significance as crack tip parameters. Therefore, when creep is present we have to look for new crack tip parameters.

Figure 10.3 schematically shows the regimes of small-scale creep (SSC), the transition creep (TC), and the extensive creep (EC). Under SSC, the creep zone size is small in comparison to the crack length and pertinent dimensions of the body. On the other hand, under EC conditions the creep zone completely engulfs the uncracked ligament. The TC condition represents the intermediate regime. The SSC and TC conditions are transient conditions characterized by the crack tip stresses which vary with time as the stress redistributes within the uncracked ligament. On the other hand, the EC condition is a steady-state condition because the crack tip stresses remain constant with time.

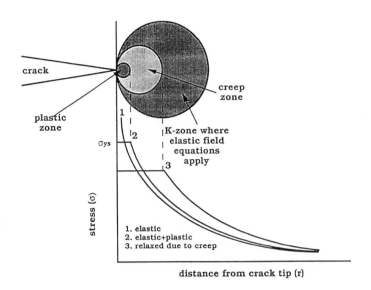

*Figure 10.2 Schematic of the deformation zones ahead of the crack tip and the associated crack tip stress fields, from Ref. 10.1.*

The steady-state, extensive creep region is the simplest for analysis purposes, hence, we will begin the crack tip stress analysis under those conditions. The small-scale creep and transition creep conditions will be considered in Section 10.2. In developing these analyses, it must be pointed out that the crack tip is assumed to remain stationary. This assumption places fundamental limitations in the use of crack tip parameters resulting from these analyses. These limitations will be discussed in detail when the crack tip parameters are formulated.

*10.1.1 The $C^*$-Integral*

We assume that a cracked body is subjected to a static load at elevated temperature and the load has been applied for sufficiently long time for steady-state creep to have engulfed the entire remaining ligament. The stress and strain rates everywhere in the body are related by equation (10.1) which is analogous to the relationship between plastic strain and stress, equation (4.17) in the subcreep temperature regime. In the analogy, strain is replaced by strain rate, $\dot{\varepsilon}$, the term $\alpha\varepsilon_0/\sigma_0^m$ is replaced by the constant, A, and m is replaced by n.

Recognizing the above analogy between the power-law descriptions of plasticity and creep, Landes and Begley [10.2] and Nikbin, Webster, and Turner [10.3] independently defined an integral analogous to Rice's J-integral, equation (4.1). Landes and Begley called this new integral $C^*$ and defined it in the following manner:

$$C^* = \int_\Gamma W^* \, dy - T_i \left( \frac{\partial \dot{u}_i}{\partial x} \right) ds \qquad (10.2)$$

312   *Nonlinear Fracture Mechanics for Engineers*

where

$$W^* = \int_0^{\dot{\varepsilon}_{ij}} \sigma_{ij}\, d\dot{\varepsilon}_{ij} \qquad (10.3)$$

$\Gamma$ is a line contour shown in Figure 10.4 taken counterclockwise from the lower crack surface to the upper crack surface. W* is the strain energy rate density associated with the point stress, $\sigma_{ij}$, and strain rate $\dot{\varepsilon}_{ij}$. $T_i$ is the traction vector defined by the outward normal, $n_j$, along $\Gamma$, Figure 10.4. Thus, $T_i = \sigma_{ij}n_j$. The displacement vector is noted by ui and s is the arc length along the contour.

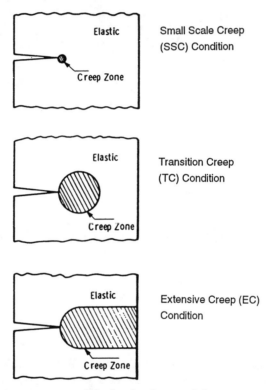

**Figure 10.3** *Schematic representation of the levels of creep deformation under which creep crack growth can occur.*

The analogy between J and C* is very clear by comparing terms between equations (4.1) and (10.2) and between equations (4.2) and (10.3). The strain and displacement quantities are replaced with their respective time rates and stress remains as stress.

Like J, the C*-integral is also path-independent. The proof of path-independence is identical to the proof in Section 4.2 and is left as an exercise problem. Two other very important consequences of the analogy between J and C* also follow. These are (1) the energy-rate interpretation of $\mathcal{C}$ and (2) the relationship between the crack tip stress fields and C*. These are discussed below.

**Energy Rate Interpretation of C*** The energy rate definition of C* was given by Landes and Begley [10.2] as follows. Consider two cracked bodies which are identical in all respects with one exception.

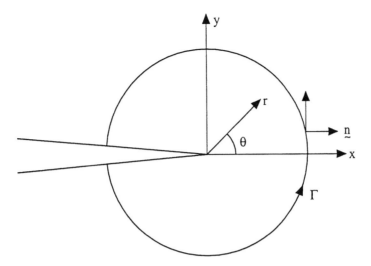

**Figure 10.4** *Crack tip coordinate system and arbitrary contour along which C*-integral is evaluated.*

The crack length of one body is $a$ and the crack length of the other is incrementally longer and is $a + \Delta a$. Both bodies are subjected to identical static loads, $P_1$; their load deflection rates $\dot{V}_c$, are monitored after sufficient time has been allowed for steady-state conditions to develop in the specimens. The subscript c is added to V to designate that this deflection is entirely due to creep. Recall that the crack sizes are assumed to be constant so the entire deflection beyond the instant of load application ($t = 0^+$) is due to creep. Next, we consider several other identical pairs of cracked bodies and load them to different levels of loads, $P_2$, $P_3$, -- etc. We can then plot the relationship between the load (P) and the displacement rate ($\dot{V}_c$) separately for all specimens of crack size a and those with crack size $a + \Delta a$. The area under the load-displacement rate diagrams will be U*(a) and U*($a + \Delta a$) for a fixed displacement rate as shown in Figure 10.5. The difference between the two areas is $\Delta U^*$. U* is the energy rate or the stress-power input into the cracked bodies and $\Delta U^*$ is the stress-power difference between the two bodies. As shown previously in Section 4.2.2 for J, equation (4.13), the C*- integral can be shown to be equal to:

$$C^* = -\frac{1}{B}\frac{dU^*}{da} \qquad (10.4)$$

where B = thickness of the body. Again, the proof of the above relationship is left as an exercise. Equation (10.4) provides the basis for measuring the value of C* at the loading pins.

**Relationship Between C*-Integral and the Crack Tip Stress Fields** Goldman and Hutchinson [10.4] used the analogy with the HRR crack tip stress fields for elastic-plastic and fully-plastic conditions, equation (4.18), and derived the following relationships between crack tip stress and strain rate and the C*-integral for extensive steady-state creep:

$$\sigma_{ij} = \left(\frac{C^*}{I_n A r}\right)^{\frac{1}{1+n}} \tilde{\sigma}_{ij}(\theta, n) \qquad (10.5a)$$

$$\dot{\varepsilon}_{ij} = A \left( \frac{C^*}{I_n A r} \right)^{\frac{n}{n+1}} \hat{\varepsilon}_{ij}(\theta, n) \qquad (10.5b)$$

where r = distance from the crack tip, $I_n$ can be obtained from equations (4.19a) and (4.19b) by substituting $n$ for $m$ and $\hat{\sigma}_{ij}$ and $\hat{\varepsilon}_{ij}$ are angular functions listed in the Appendix. The above equations establish the unique relationship between crack tip stress and $C^*$. $C^*$-integral has been shown to have the following properties. (1) It is a path-independent integral which can be computed along contours remote from the crack tip, (2) it can be measured at the loading points, and (3) it can be uniquely related to the magnitude of crack tip stress and strain rate. Thus, it is an attractive parameter for characterizing the creep crack growth rate under conditions of extensive steady-state creep. There is, therefore, considerable interest in methods for determining its value.

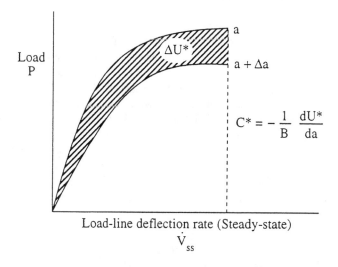

**Figure 10.5** *Schematic illustration of the energy rate interpretation of the $C^*$-integral.*

### 10.1.2 Methods of Determining $C^*$

The methods for determining $C^*$ follow the methods for determining J-integral. We will see that by making use of appropriate analogies, we can derive a $C^*$ expression for any geometry and loading configuration for which we have an expression available for J-integral. However, as a caution, the following must always be kept in mind while using analogies to develop expressions for $C^*$. The analogy between J and $C^*$ is limited to the plastic part of the J-integral, $J_p$. Thus, we will need to separate the plastic part of J from the expression for total J before proceeding with the analogy.

$C^*$ can be obtained analytically for some very limited geometries such as ones shown in Section 5.1. The methods that can be applied to complex geometries and loading conditions include (1) experimental method, (2) semi-empirical methods, (3) from numerical solutions. These methods are discussed further.

**Experimental Methods for Determining C*** This method is based on the energy rate (or stress-power) interprelation of C* given equation 10.4 and was developed by Landes and Begley [10.2] in their original work for evaluating C* for characterizing creep crack growth rate behavior. The method requires several sets of identical specimens. The crack size is varied between different sets of specimens. Let us arbitrarily choose 5 specimens in a set, totaling 5 sets. The crack sizes of specimens in the various sets are $a_1$, $a_2$, $a_3$, $a_4$, $a_5$. Thus, a total of 25 specimens are machined, 5 specimens each of the 5 crack sizes. We select 5 load levels, $P_1$, $P_2$, -- $P_5$, and subject a specimen from each set to one of the five load levels, measuring the steady-state deflection rates. This is repeated for the other four load levels. We will use the designation $\dot{V}_{ss}$ to represent steady-state deflection rate which is determined as shown in Figure 10.6. Note that the steady-state conditions are realized in the specimen only after some time. Next, the $P$ - $\dot{V}_{ss}$ relationship for each crack size can be plotted and the area under the curve for different values of $\dot{V}_{ss}$ can be obtained to determine U*, as shown in Figure 10.7. We then plot U* as a function of crack size for fixed values of $\dot{V}_{ss}$ as shown in Figure 10.8. The slope of the U* vs. a curve can be related to C*. This method of determining C* is not a practical method due to the extensive amount of testing that is required. The C* values obtained are not only for a specific geometry and loading configuration, they are also specific to the material being tested. Although this method was used for determining C* in the early studies of creep crack growth [10.1, 10.2], it was quickly abandoned in favor of more direct methods. These are the semi-empirical methods described next.

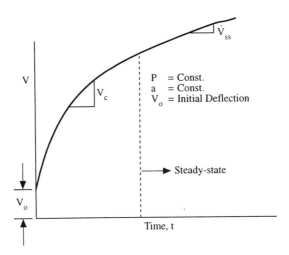

*Figure 10.6 The load-line deflection as a function of time for a cracked body for a fixed applied load. The determination of $\dot{V}_{ss}$ is also shown in the figure.*

**Semi-empirical Methods of Determining C*** Expressions analogous to equations (5.8b) and (5.10b) can be written for estimating C*. In these equations, we must substitute $\dot{V}_{ss}$ in place of the displacement, V, to obtain the following relationships:

$$C^* = -\frac{1}{B}\int_0^{\dot{V}_{ss}}\left(\frac{\partial P}{\partial a}\right)_{\dot{V}_{ss}} d\dot{V}_{ss} \tag{10.6a}$$

$$C^* \frac{1}{B}\int_0^{P}\left(\frac{\partial \dot{V}_{ss}}{\partial a}\right)_{P} dP \tag{10.6b}$$

316  *Nonlinear Fracture Mechanics for Engineers*

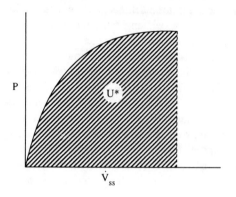

**Figure 10.7.** $P - \dot{V}_{ss}$ *relationship for a fixed crack size and the definition of stress-power, U\*.*

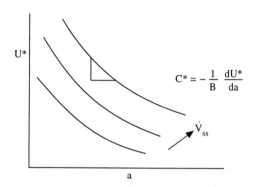

**Figure 10.8** $U^*$ *as a function of crack size for various values of $\dot{V}_{ss}$*

During the extensive steady-state creep conditions, the following equation holds between load and $\dot{V}_{ss}$:

$$\dot{V}_{ss} = \phi\,(a/W, n)\,P^n \tag{10.7}$$

Combining equation (10.7) with equation (10.6a) leads to an equation for $C^*$ which is analogous to equation (5.34), for compact type (CT) specimens for a/W > 0.40:

$$C^* = \frac{P\dot{V}_{ss}}{B(W-a)} \left(2 + .522\,(1-a/W)\right) \frac{n}{n+1} \tag{10.8}$$

where W = width of the specimen. The equivalent expression for center crack tension (CCT) specimens is:

$$C^* = \frac{P\dot{V}_{ss}}{2BW(1-a/W)} \frac{n-1}{n+1} \tag{10.9}$$

where W = half width of the specimen. The detailed derivations of equations (10.8) and (10.9) are left as exercises. Similar expressions can also be derived for other specimen geometries such as the SENB and the DEN specimens.

**Example Problem 10.1**

A 50mm-wide and 25mm-thick CT specimen of 304 stainless steel is subjected to a constant load of 18kN at 594°C. If the A and n values of this material at 594°C are $2.13 \times 10^{-18}$ and 6, respectively, for stress in MPa and strain rate in $hr^{-1}$, calculate the value of $C^*$ when a/W = 0.5 and the measured load-line deflection rate is $2.5 \times 10^{-6}$ m/hr.

*Solution:*

The value of $C^*$ can be obtained by direct substitution of values into equation (10.9). Thus:

$$C^* = \frac{18\ (kN) \times 2.5 \times 10^{-6}(m/hr)}{(0.25m)\,(.025m)} \left(2 + .522 \times .5\right) \frac{6}{7}$$

$$= 0.139\ \frac{kNm}{m^2 hr} = .139\ kJ/m^2\ hr$$

Note that the units of $C^*$ are $kJ/m^2$ hr. An alternate set of units can also be k Watts/$m^2$.

**$C^*$ Based on Numerical Solutions** In Section 5.4.2, numerical solutions were presented for estimating the fully-plastic J-integral values. These equations can be modified to estimate $C^*$ [10.1, 10.5]. Such equations are necessary for estimating $C^*$ in components where no measurements of the load-line deflection rate, $V_{ss}$, are available. As with the J-integral, these solutions are restricted to either plane-stress or plane-strain conditions.

In this section, we will fully develop the equations for estimating $C^*$ in the CT specimen. The derivation of equations for other geometries will be left as an exercise. The equations for estimating fully-plastic J in CT specimens are given in equation (5.50a). Substituting the value of $P_0$ from equation (5.51) into equation (5.50a) and replacing $\alpha\varepsilon_0/\sigma^m{}_0$ by A and m by n, we get the following equation:

$$C^* = A\,(W-a)\,h_1\,(a/W,n) \left(\frac{P}{1.455\,\eta_1\,B(W-a)}\right)^{n+1} \tag{10.10}$$

where $\eta_1$ is defined in equation (5.53). The above equation is for plane-strain condition. By substituting the $P_0$ value for plane stress from equation (5.52), we can derive an expression for $C^*$ for plane stress. The corresponding expressions for load-line displacement rate and the CTOD rate due to creep, $\dot{\delta}_{ss}$, given by the following equations:

$$\dot{V}_{ss} = Aah_3\,(a/W,n)\left[\frac{P}{1.455\,\eta_1\,B(W-a)}\right]^n \tag{10.11}$$

$$\dot{\delta}_{ss} = Aah_2\,(a/W,n)\left[\frac{P}{1.455\,\eta_1\,B(W-a)}\right]^n \tag{10.12}$$

The functions of $h_1$, $h_2$, and $h_3$ in the above equations are the same as ones listed in Tables 5.2. If we substitute equation (10.11) into equation (10.10) and compare the terms with equation (10.8), we can show that:

$$\frac{h_1}{h_3}\frac{(1-a/W)}{1.455\,\eta_1(a/W)} \approx (2+.522\,(1-a/W))\frac{n}{n+1} \tag{10.13}$$

**Example Problem 10.2**
Show that for a CT specimen of standard proportions, equation (10.13) is approximately correct for a/W values of 0.5 and 0.625 for n = 7.

*Solution:*
From Table 5.2, $h_1$ (0.5, 7) and $h_1$ (0.625, 7) are 0.685 and 0.752, respectively. Also from Table 5.2, $h_3$ (0.5, 7) and $h_3$ (0.625, 7) are 1.41 and 1.37, respectively. $\eta_1$ (0.5) = 0.162 and $\eta_1$ (0.625) = 0.114. Substituting the values into equation (10.13).
For a/W = 0.5, the left side of equation (10.13) yields:

$$\frac{0.685}{1.41}\frac{.5}{1.455\times.162\times.5} = 2.061$$

The right side of equation (10.13) yields:

$$(2+.261)^{7/8} = 1.978$$

which is within 5% of the calculated value for the left-hand side of equation (10.13). For a/W = 0.625, the left side of equation (10.13) yields:

$$\frac{0.752}{1.37}\frac{.375}{1.455\times.114\times.625} = 1.985$$

The right-hand side of equation (10.13) yields:

$$(2 + .522\,(.375))^{7/8} = 1.92$$

which is within 3% of the value of the left-hand side of equation (10.13).

### 10.1.3 Correlation Between Creep Crack Growth Rates and $C^*$

To demonstrate the validity of $C^*$ for characterizing creep crack growth rates, it is necessary to show that the crack growth rates measured from specimens of the same geometry and size but loaded to different load levels as well as those from specimens of different geometry and size uniquely correlate with $C^*$. The early attempts by Landes and Begley [10.2] to correlate da/dt with $C^*$ under creep conditions met only with limited success. These experiments were conducted on A286 creep resistant alloy in which it was difficult to obtain the conditions of extensive steady-state creep within the time period of the experiment. Similar experiments were then conducted by Saxena [10.1] on 304 stainless steel at 594°C to promote extensive steady-state creep in the specimens. These experiments were conducted using CT and CCT specimen geometries. Figure 10.9 shows the results of this study which demonstrated the validity of $C^*$ for correlating creep crack growth rates. Similar correlations have been obtained by several other investigators since these early experiments and the validity of $C^*$ for characterizing creep crack growth rates is now widely accepted.

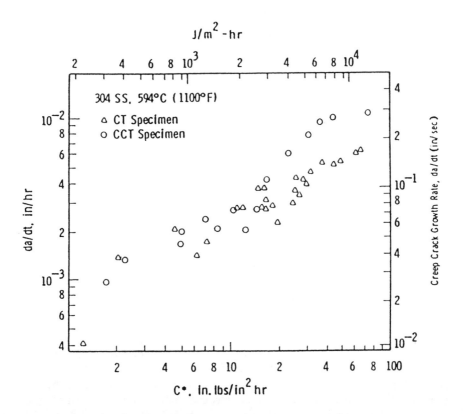

*Figure 10.9* Creep crack growth rate as a function of the $C^*$-integral for 304 stainless steel, Ref. 10.1.

## 10.2 Analysis of Cracks Under Small-Scale and Transition Creep

The validity of the C*-integral presented in the previous section is limited to extensive steady-state creep conditions. In practice, this condition may not always be realized because components contain stress and temperature gradients and are designed to resist widespread creep deformation. Therefore, in this section, we will derive the crack tip stress fields for the conditions of small-scale creep (SSC) and the transition creep (TC). In these analyses, we will continue to assume that creep deformation occurs by power-law creep and that the crack is stationary. However, the elastic stress redistributions due to creep deformation will be considered. Thus, the material is modeled as elastic and power-law creep material for which the uniaxial version of the constitutive equation is given by:

$$\dot{\varepsilon} = \frac{\dot{\sigma}}{E} + A\sigma^n \tag{10.14}$$

It is further assumed that the elastic and creep strains are additive, thus:

$$\varepsilon = \varepsilon_e + \varepsilon_c \tag{10.15}$$

where $\varepsilon_e$ = elastic strain and $\varepsilon_c$ = accumulated time-dependent strain. Integrating equation (10.15) gives:

$$\varepsilon = \frac{\sigma(t)}{E} + \int_0^t A\left[\sigma(\tau)\right]^n d\tau \tag{10.16}$$

where $\tau$ is a dummy integration variable and $\sigma(t)$ is the time-dependent stress at any time, t.

### 10.2.1 Crack Tip Stress Fields in Small-Scale Creep

Riedel and Rice [10.5] and Ohji, Ogura, and Kubo [10.6] independently derived the nature of the crack tip stress fields under small-scale creep condtions and reached identical results. In the following discussion, we will follow the Ohji et al. derivation because of its simplicity.

Equation (10.16) can be written as:

$$\varepsilon = \frac{\sigma(t)}{E} + Af(t)(\sigma(t))^n \tag{10.17}$$

where

$$f(t) = \int_0^t (\sigma(\tau)/\sigma(t))^n d\tau \tag{10.18}$$

Since f(t) is a function of time only, in other words, independent of stress, it is perfectly valid to include it inside the integral and to also apply it to all the area in the crack tip region. For constant time, the form of equation (10.17) is identical to the Ramberg-Osgood equation used in elastic-plastic fracture. We note also that the relationship between stress and strain is univalued for a constant time. This would imply that the J-integral is path-independent. Hence, we can write the HRR stress fields, equations (4.18a) and (4.18b), as:

$$\sigma_{ij} = \left[\frac{J}{I_n A f(t) r}\right]^{\frac{1}{1+n}} \hat{\sigma}_{ij}(\theta, n) \tag{10.19a}$$

$$\varepsilon_{ij} = A f(t) \left[\frac{J}{I_n A f(t) r}\right]^{\frac{n}{n+1}} \hat{\varepsilon}_{ij}(\theta, n) \tag{10.19b}$$

Note that to use equation (4.18), we had to substitute n for m and Af(t) for $\alpha \varepsilon_0/\sigma_0^n$.

In equation (10.19a), $\sigma_{ij}(t)$ is propsortional to $(1/f(t))^{1/(n+1)}$. Thus, equation (10.18) can be written as:

$$f(t) = \int_0^t [f(t)/f(\tau)]^{\frac{n}{n+1}} d\tau \tag{10.20}$$

By trial and error, we find that the following solution fits the integral equation (10.20):

$$f(t) = (n+1) t \tag{10.21}$$

If we substitute equation (10.21) into equation (10.19a) and also recognize that for small-scale-yielding and elastic conditions $J = (K^2/E)(1 - v^2)$, the result can be written as follows:

$$\sigma_{ij} = \left[\frac{K^2 (1-v^2)}{E I_n A (n+1) tr}\right]^{\frac{1}{1+n}} \hat{\sigma}_{ij}(\theta, n) \tag{10.22a}$$

$$\varepsilon_{ij} = \left[\frac{K^2 (1-v^2)}{E I_n A (n+1) tr}\right]^{\frac{n}{n+1}} A (n+1) t \hat{\varepsilon}_{ij}(\theta, n) \tag{10.22b}$$

$$\dot{\varepsilon}_{ij} = A \left[\frac{K^2 (1-v^2)}{E I_n A (n+1) tr}\right]^{\frac{n}{n+1}} \hat{\varepsilon}_{ij}(\theta, n) \tag{10.22c}$$

In the above equations, the time-dependence of the crack tip stress, strain, and strain rate fields is quite evident. The above analysis lends itself to the estimation of the creep zone size and transition time, $t_T$, which is the time needed for extensive creep conditions to develop from SSC conditions. These are discussed in subsequent sections.

*10.2.2 Estimation of the Creep Zone Size*

Riedel and Rice [10.5] arbitrarily defined the creep zone boundary as the locus of points where time-dependent effective creep strains equal the instantaneous effective elastic strains in the cracked body. This leads to:

$$r_c(\theta,t) = \frac{1}{2\pi}\left(\frac{K}{E}\right)^2 \left[\frac{(n+1) I_n E^n A t}{2\pi(1-v^2)}\right]^{\frac{2}{n-1}} F_{cr}(\theta,n) \quad (10.23)$$

This equation can also written as:

$$r_c(\theta,t) = \frac{1}{2\pi}\left(\frac{(n+1)^2}{2n\alpha_n^{n+1}}\right)^{\frac{2}{n-1}} K^2 (EAt)^{\frac{2}{n-1}} F_{cr}(\theta,n) \quad (10.23a)$$

where $\alpha_n^{n+1}$ has an approximate value of 0.69 for $3 \le n \le 13$ and $F_{cr}$ is a function dependent on $\theta$, n and the state of stress as shown in Figure 10.10. The creep zone size is dependent on time as $t^{2/(n-1)}$ and on K as the second power.

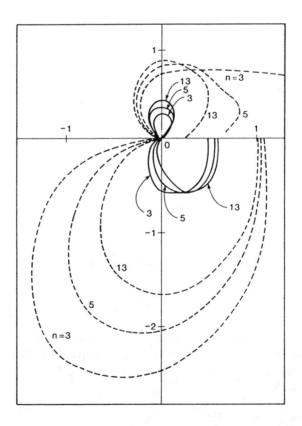

**Figure 10.10** *The function $F_{cr}(\theta, n)$ for plane stress and plane strain for various values of n. From Ref. 10.5.*

Adefris [10.7] gave an alternate definition for the creep zone boundary by choosing the creep zone boundary to be a locus of points where the time-dependent strain has a value of 0.2%. This leads to:

$$r_c(\theta,t) = \left[\frac{(n+1)A}{.002}\right]^{\frac{n+1}{n}} \frac{(1-v^2)K^2}{(n+1)AEI_n} t^{\frac{1}{n}} \tilde{r}_c(\theta,n) \tag{10.24}$$

Equation (10.23) becomes unstable for n = 1, and for n < 3 it predicts very rapidly growing creep zone size. Both these results are an artifact of the Riedel-Rice definition of the creep zone. When the alternate creep zone definition is considered, such as in equation (10.24), this problem no longer exists. For n = 1, a linear-viscous material, the creep zone size increases linearly with time which is a reasonable prediction. Thus, the Riedel-Rice definition of creep zone size is not recommended for n values approaching 3, or less than 3. For most metals n is significantly greater than 3 and the difference between the two definitions of creep zone size is not of major significance.

### 10.2.3 Transition Time ($t_T$)

Riedel and Rice as well as Ohji et al. in their original analyses presented a concept of transition time, $t_T$. They defined the transition time as the time when the small-scale-creep stress fields equal the extensive steady-state creep fields characterized by $C^*$. Therefore, equating equations (10.5a) and (10.22a) at time, $t_T$, we get:

$$t_T = \frac{K^2(1-v^2)}{E(n+1)C^*} \tag{10.25}$$

We can then state that if $t << t_T$, equation (10.22) describes the crack tip conditions and for $t >> t_T$, equation (10.5) determines the crack tip conditions. An interpolation scheme for determining the crack tip stress fields for $t/t_T \sim 1$ will be described later in this section.

### 10.2.4 C(t) - Integral and the Stress Fields in the Transition Creep Region

Bassani and McClintock [10.8] recognized that the crack tip stress fields under SSC can also be characterized by a time-dependent C-integral, whose value is determined along a contour taken very close to the crack tip:

$$C(t) = \int_{\Gamma-0} W^* \, dy - T_i \frac{\partial \dot{u}_i}{\partial x} \, ds \tag{10.26}$$

Note that C(t) is the same as $C^*$ but its value is determined close to the crack tip within a region where the creep strains dominate over the elastic strains. Recall that the value of $C^*$ could be determined along any contour which originated at the lower crack surface, ended on the upper crack surface, and enclosed the crack tip. Thus, determining the C-integral requires accurate solutions of stress and strain near the crack tip. Bassani and McClintock [10.8] further related the value of C(t) with the HRR type stress fields as follows:

$$\sigma_{ij} = \left(\frac{C(t)}{AI_n r}\right)^{\frac{1}{n+1}} \hat{\sigma}_{ij}(\theta,n) \tag{10.27a}$$

$$\dot{\varepsilon}_{ij} = A \left( \frac{C(t)}{AI_n r} \right)^{\frac{n}{n+1}} \hat{\varepsilon}_{ij}(\theta, n) \qquad (10.27b)$$

Comparing these equations with the crack tip stress equations given earlier, equation (10.22 a-c), we get for the conditions of small-scale creep:

$$C(t) \approx \frac{K^2(1-v^2)}{E(n+1)\,t} \qquad (10.28)$$

The reason for the approximate equality in equation (10.28) is that both sets of equations, (10.22) and (10.26), are valid only as r → 0. For finite r values, they are only approximately valid. Equation (10.28) has been verified by numerical anlaysis and has been shown [10.8, 10.9] to hold true.

The validity of the C(t)-integral is not simply limited to the small-scale creep conditions. As a matter of fact, C(t) becomes equal to $C^*$ for extensive steady-state creep with the additional property that its value becomes path-independent. Hence, C(t) can be said to be the amplitude of the HRR field for all conditions ranging from small-scale to extensive secondary-state creep and also including the transition creep conditions in between. Ehlers and Riedel [10.9] and Bassani and McClintock [10.8] independently proposed the same interpolation formula for approximating the HRR stress fields in the transition creep region:

$$C(t) \approx \frac{K^2(1-v^2)}{E(n+1)\,t} + C^* \qquad (10.29)$$

Equation (10.29) can be combined with equation (10.25) to develop an alternate expression for C(t):

$$C(t) = (1 + t_T/t)\, C^* \qquad (10.30)$$

The accuracy of equation (10.29) has been proven by several finite element studies as seen in Figure 10.11 from the work of Bassani, Hawk, and Saxena [10.10] and that of Leung, McDowell, and Saxena [10.11] and also in the original work of Ehlers and Riedel [10.9].

The C(t)-integral, by virtue of its ability to characterize the HRR fields from small-scale creep to extensive steady-state creep, is an attractive parameter for correlating creep crack growth rate data. However, it has a major drawback in that it cannot be measured at the loading pins under small-scale and transition creep conditions. Its value must always be calculated from equation (10.29). The calculated value is very sensitive to the accuracy of the creep constants, A and n and the assumptions regarding the state of stress. Figure 10.12 shows the creep crack growth rate plotted with C(t) for several CT specimens of 1 Cr-1Mo-V steel showing considerable variablity in the data from different specimens. It cannot be determined with certainty if the poor correlation is due to inaccuracy of equation (10.29) in estimating the value of C(t) or if the parameter itself is not suitable for characterizing creep crack growth. Nonetheless, it is a shortcoming of the approach which makes it less desirable in practical applications.

Analysis of Cracks in Creeping Materials    325

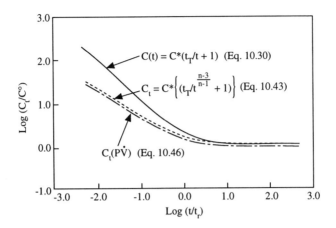

**Figure 10.11** *Comparison between the numerically calculated values of the C-integral and those predicted by the interpolation formula, equation (10.29).*

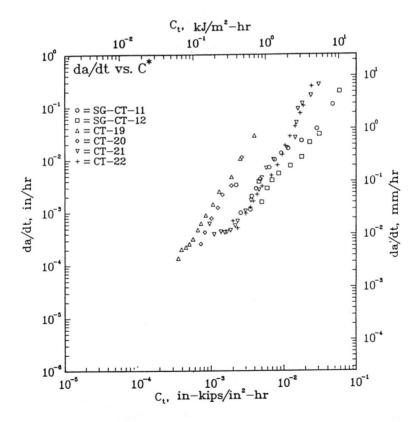

**Figure 10.12** *Creep crack growth rate as a function of the $C_t$ parameter for 1Cr-1Mo-0.25V steel at 593°C.*

**Example Problem 10.3**
Estimate the value of the C(t)-integral for the compact specimen described in Example Problem 10.1 as a function of time. Discuss the implications of the results.

*Solution:*
The expressions for calculating C(t) are given in equations (10.29) and (10.30). Since it is required that we express the value of C(t) as a function of $t/t_T$, we can rearrange equation (10.30) as:

$$\frac{C(t)}{C^*} = 1 + \frac{1}{t/t_T}$$

$C^*$ was calculated as 0.139 kJ/m² hr in Example Problem 10.1. The $t_T$ can be caculated from equation (10.25):

$$K = \frac{P}{BW^{\frac{1}{2}}} F(a/W)$$

where

$$F(a/W) = \frac{2 + a/W}{(1-a/W)^{-3/2}} \left[ 0.886 + 4.64\,(a/W) - 13.32\,(a/W)^2 + 14.72\,(a/W)^3 - 5.6\,(a/W)^4 \right]$$

For a/W = 0.5, F (a/W) = 9.65

Hence:

$$K = \frac{18\ (kN)}{(0.25m)\,(.05m)^{1/2}} \times 9.65 = 31{,}102\ kPa\sqrt{m}$$

$$\approx 31.1\ Mpa\ \sqrt{m}$$

$$t_T = \frac{(31.1)^2\,(1-.09) \times 10^6}{172 \times 10^6\,(7) \times (.139)} = 5.26\ hrs$$

Note that Poison's ratio has been chosen as 0.3 and the elastic modulus is chosen as $172 \times 10^6$ kPa. The values of C(t) as a function $t/t_T$ are shown in the Table 10.1.

We note from the above calculations that during the small-scale and transition creep, the crack tip stresses are significantly higher than under extensive steady-state creep conditions. The crack growth rates during that period are expected to be higher as well. Therefore, ignoring the small-scale creep regime during analysis of components can be quite nonconservative.

**Table 10.1** The Values of C(t) as a Function $t/t_T$

| $t/t_T$ | t (hrs) | C(t)/C* | C(t) |
|---|---|---|---|
| 0.1 | .526 | 11 | 1.529 |
| 0.2 | 1.052 | 6 | 0.834 |
| 0.5 | 2.63 | 3 | 0.417 |
| 1 | 5.26 | 2 | 0.278 |
| 2 | 10.52 | 1.5 | 0.208 |
| 5 | 26.3 | 1.2 | 0.166 |
| 10 | 52.6 | 1.1 | 0.152 |
| ∞ | ∞ | 1.0 | 0.139 |

### 10.2.5 $C_t$ Parameter

From the above example problem, it is evident that small-scale creep and transition creep regions cannot be ignored in component analysis. Also, recall that the results from the applications of the C(t) integral to characterize crack growth rates were not very successful. Therfore, alternate parameters are needed to characterize creep crack growth rates under a wide range of conditions from small-scale to extensive steady-state creep. The $C_t$ parameter proposed by Saxena [10.12] fulfills this need and is discussed in this section.

In searching for a new parameter, it is important to keep in mind that the candidate parameter must be identical to C* in the extensive creep regime because of its proven value. If we further impose the condition that the parameter be measurable at the load-line, it seems reasonable to generalize the stress-power definition of C* into the small-scale and transition creep region. Once such a parameter is defined, its relationship to crack tip conditions can be explored.

Consider several identical pairs of cracked specimens. Within each pair, one specimen has a crack length, a, and the other has an incrementally differing crack length a + Δa. The specimens of each pair are loaded to various load levels - $P_1, P_2, P_3, \text{---} P_i$, etc. - at elevated temperatures, and the load-line deflection due to creep, $V_c$, is recorded with time Figure 10.13a. It is assumed that no crack extension occurs in any of the specimens and the instantaneous response is linear elastic. In this discussion, let us first limit our discussion to the SSC condition characterized by $t/t_T \ll 1$. At a fixed time, t, the load vs. creep deflection rate, $\dot{V}_c$, behavior is plotted for all these specimens. A schematic of the expected behavior is shown in Figure 10.13b; several such plots for varying time can be generated from the above tests by varying time. When steady-state conditions are reached, the plots will become identical and will be similar to ones shown earlier in Figure 10.7.

The area between the P - $\dot{V}_c$ curves for specimems with crack length a and a + Δa is called $\Delta U_t^*$. Physically, $\Delta U_t^*$ (the subscript denotes that this value is at a fixed time, t) represents the instantaneous difference in the stress-power supplied to two cracked bodies with identical creep deformation histories as they are loaded to different deflection-rate levels. The $C_t$ parameter is given by:

$$C_t = -\frac{1}{B} \frac{\partial U_t^*(a,t,\dot{V}_c)}{\partial a} \quad (10.30)$$

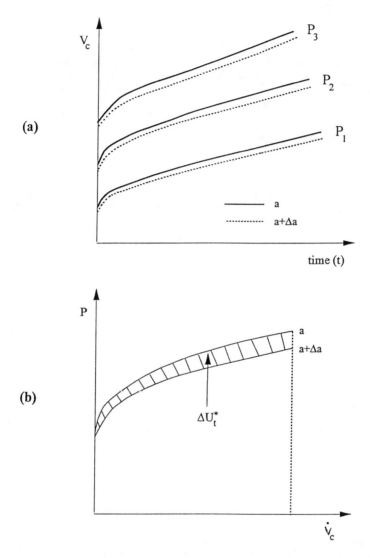

**Figure 10.13** (a) Load-line deflection due to creep as a function of time for bodies with crack lengths $a$ and $a + da$ at constant loads. (b) Definition of the $C_t$ parameter.

As $t/t_T \rightarrow \infty$, $C_t = C^*$ by definition. Thus, the $C_t$ parameter satisfies one of the conditions that we specified earlier. However, it still remains to be seen what relationship exists between $C_t$ and the crack tip stress and deformation fields. Further, we also need to develop expressions for calculating $C_t$ in small-scale creep and transition creep condtions. Since $C_t$ is equal to $C^*$ in the extensive creep regime, methods of estimating $C_t$ in this regime have already been covered.

**Estimation of $C_t$ for Small-Scale Creep** In order to derive an expression for estimating $C_t$, it is necessary to derive a relationship between the applied load, P, load-line deflection rate, $\dot{V}_c$, crack size, $a$, and the time t. This will allow us to determine $U_t^*$ as a function of $a$ and t for given values of P and $\dot{V}_c$. Subsequently, $U_t^*$ can be differentiated to determine the value of $C_t$ as shown below [10.12].

We invoke the Irwin concept [10.13] of effective crack length, $a_{eff}$, for calculating load-line deflection as a function of P, a, and t. We write:

$$a_{eff} = a_0 + \beta r_c \qquad (10.31)$$

where $r_c$ is the creep zone size as defined in equation (10.23) and $\beta$ is a scaling factor. Since the creep zone can sustain stresses, the entire creep zone does not act as a crack. Therfore, the value of $\beta$ is less than one. If we choose the value of $r_c$ corresponding to $\theta = 90°$, it has been shown by finite element analysis [10.10] that $\beta = 1/3$. The additional load-line displacement due to creep, $V_c$, at any time, t, is given by:

$$V_c = V - V_o = P \frac{dC}{da} \beta r_c \qquad (10.32)$$

where P = applied load, V = total displacement, $V_o$ = instantaneous deflection, and C is the elastic compliance of the cracked body. From equation (3.16) the following relationship exists between compliance and K for plane strain:

$$(1-v^2) \frac{K^2}{E} = \frac{P^2}{2B} \frac{dC}{da}$$

or

$$\frac{dC}{da} = \frac{2BK^2}{EP^2} (1-v^2) \qquad (10.33)$$

Substituting equation (10.23a) and (10.33) into equation (10.32) and differentiating with time, we get:

$$\dot{V}_c = \frac{4\alpha(1-v^2)}{E(n-1)} \beta F_{cr} \left(\frac{P}{B}\right)^3 \frac{F^4}{W^2} (EA)^{2/(n-1)} t^{-(n-3)/(n-1)} \qquad (10.34)$$

where

$$\alpha = \frac{1}{2\pi} \left( \frac{(n+1)^2}{2n\alpha_n^{n+1}} \right)^{\frac{2}{n-1}} \qquad (10.35)$$

and

$$F(a/W) = (K/P) BW^{\frac{1}{2}} \qquad (10.36)$$

Equation (10.35) in the functional form can be written as:

$$\dot{V}_c = \phi\,(a/W, t)\left(\frac{P}{B}\right)^3 \qquad (10.37)$$

where

$$\phi = \frac{4\alpha(1-\nu^2)}{E(n-1)}\frac{F^4}{W}\,(EA)^{\frac{2}{n-1}}\, t^{-\frac{n-3}{n-1}} \qquad (10.38)$$

The $C_t$ parameter can be determined by:

$$(C_t)_{ssc} = -\frac{1}{B}\frac{\partial U_t^*}{\partial a} = \frac{1}{Bda}\left[\int_0^{\dot{V}_c}\left(\frac{\dot{V}_c}{\phi}\right)^{1/3} d\dot{V}_c - \int_0^{\dot{V}_c}\left(\frac{\dot{V}_c}{\phi + \frac{\partial \phi}{\partial a}da}\right)^{1/3} d\dot{V}_c\right] \qquad (10.38a)$$

$$= \frac{1}{Bda}\int_0^{\dot{V}_c}\left(1 - \left(1 + \frac{1}{\phi}\frac{\partial \phi}{\partial a}da\right)\right)^{-1/3}\left(\frac{\dot{V}_c}{\phi}\right)^{1/3} d\dot{V}_c$$

$$= \frac{1}{Bda}\int_0^{\dot{V}_c}\left(\frac{\dot{V}_c}{\phi}\right)^{1/3}\frac{1}{3\phi}\frac{\partial \phi}{\partial a}da\, d\dot{V}_c$$

$$= \frac{1}{B}\frac{1}{3\phi}\frac{\partial \phi}{\partial a}\frac{3}{4}\left(\dot{V}_c\right)^{4/3}\left(\frac{1}{\phi}\right)^{1/3}$$

$$= \frac{1}{B}\frac{1}{\phi}\frac{\partial \phi}{\partial a}\frac{1}{4}P\dot{V}_c$$

Substituting for $1/\phi\,(\partial \phi / \partial a)$ from equation (10.38), we get:

$$(C_t)_{ssc} = \frac{P\dot{V}_c}{BW}\frac{F'}{F} \qquad (10.39)$$

where $F' = dF/d(a/W)$. Substituting for $\dot{V}_c$ in equation (10.39) from equations (10.32) and (10.34), we get equations (10.40) and (10.41), respectively, which are alternate expressions of determining $C_t$:

$$(C_t)_{ssc} = \frac{P^2}{BW} \frac{dC}{da} \beta \dot{r}_c F'/F$$

$$= \frac{2 K^2 (1-v^2)}{EW} \beta F'/F \dot{r}_c \qquad (10.40)$$

$$(C_t)_{ssc} = \frac{4 \alpha (1-v^2)}{E(n-1)} \beta F_{cr}(\theta,n) \frac{K^4}{W} (EA)^{\frac{2}{n-1}} (F'/F) t^{-\frac{n-3}{n-1}} \qquad (10.41)$$

Equation (10.40) relates $(C_t)_{ssc}$ to the creep zone expansion rate which is a crack tip quantity. This relationship qualifies $(C_t)_{ssc}$ as a crack tip parameter, so the next step in the evaluation of $(C_t)_{ssc}$ will be to conduct an experimental study. Equation (10.41) is a method by which $(C_t)_{ssc}$ can be computed for any geometry and loading configuration provided the K-calibration expression is available and also the creep constants A and n are known. If we choose, $\theta = 90°$, as mentioned before, $\beta = 1/3$ [10.10] and $F_{cr}(\theta,n)$ ~ 0.4 for n values of approximately 10. In general, $F_{cr}(\theta)$ will depend on n, but, the dependence is weak as seen in Figure 10.10.

**Example Problem 10.4**
A plate of width W loaded with uniform stress at elevated temperature has an edge crack of length a. Derive an expression for determining $(C_t)_{ssc}$ if the load-line displacment measurement is available. Assuming that $E = 175 \times 10^3$ MPa, A is $2.13 \times 10^{-18}$ (MPa)$^{-n}$ hr$^{-1}$ and n = 7, derive an expression for determining $(C_t)_{ssc}$ as a function of time for a stress of 35 MPa and an a/W = 0.5.

*Solution:*
The K-calibration for this specimen is given by:

$$K = \sigma \sqrt{\pi a} [ 1.122 - 0.231 (a/W) + 10.55 (a/W)^2 -$$

$$21.71 (a/W)^3 + 30.382 (a/W)^4 ] \qquad (1)$$

Equation (10.39) for this configuration can be written as:

$$(C_t)_{ssc} = \sigma \dot{V}_c F'/F \qquad (2)$$

$$F(a/W) = (K/P) BW^{1/2} = (K/\sigma)(1/\sqrt{W})$$

From equation (1), we can write:

$$F = \sqrt{\frac{\pi a}{W}}\, g\,(a/W)$$

where $g\,(a/W) = [1.122 - 0.231\,(a/W) + 10.55\,(a/W)^2 - 21.71\,(a/W)^3 + 30.382\,(a/W)^4]$:

$$F' = \sqrt{\frac{\pi a}{W}}\, g'\,(a/W) + \frac{1}{2}\sqrt{\frac{\pi}{a/W}}\, g\,(a/W)$$

$$F'/F = g'/g + \frac{1}{2(a/W)}$$

$$g' = -.231 + 21.1\,(a/W) - 65.13\,(a/W)^2 + 121.528\,(a/W)^3$$

Thus:

$$(C_t)_{ssc} = \sigma\, \dot{V}_c \left[ g'/g + \frac{1}{2(a/W)} \right]$$

By substituting $\sigma = 35$ MPa, $A = 2.13 \times 10^{-18}$, $n = 7$, $\beta \approx 1/3$, $F_{cr}(\theta, n) \approx 0.4$, $\alpha = .299$ from equation (10.35), $v = 0.3$, $E = 172 \times 10^3$ MPa into equation (10.41), we get:

$$(C_t)_{ssc} = 8.6 \times 10^{-10}\,(F'/F)\,\frac{t^{-.667}}{W}\, K^4 \; MJ/m^2\, hr$$

$$g\,(0.5) = 2.829 \qquad g'\,(0.5) = 9.227$$

$$F'/F\,(0.5) = 4.262$$

$$K = \sigma \sqrt{\frac{\pi a}{W}}\, \sqrt{W}\, g\,(a/W) = 124.1\, \sqrt{W}\; MPa\,\sqrt{m}$$

$$(C_t)_{ssc} = .0102\, W t^{-.667}\; MJ/m^2 hr$$

or $\quad (C_t)_{ssc} = 10.2\, W t^{-.669}\; kJ/m^2 hr$

Equation (10.39) is suitable for determining $(C_t)_{ssc}$ in test specimens where the load is applied and $\dot{V}_c$ is measured. Conversely, equation (10.41) is more suitable for estimating $(C_t)_{ssc}$ in components where load-line deflection cannot be measured.

**Determining $C_t$ for a Wide Range of Conditions** It is convenient to have a single expression for estimating $C_t$ for conditions ranging from small-scale to extensive creep. An approach similar to the one used for estimating the elastic-plastic part of J-integral has also been used successfully for $C_t$. Thus $C_t$ can be expressed as [10.10, 10.13]:

$$C_t = (C_t)_{ssc} + C^* \qquad (10.42)$$

This is illustrated in the following example.

**Example Problem 10.5**
For the specimen in Example Problem (10.4), calculate $C_t$ for SSC to extensive creep conditions and the transition time. Assume that $W = 10$ cm.

*Solution:*
The expression for estimating $C^*$ can be assembled from the expression for estimating $J_p$ given in equations (5.54) and (5.59) for plane strain. Substituting equation (5.59) into (5.54) gives:

$$J_P = \frac{\alpha \varepsilon_0}{(1-a/W)^n} \left( \frac{\sigma}{1.455 \eta_2} \right)^{n+1}$$

where $\eta_2$ is given by equation (5.61) for $a/W = 0.5, \eta_2 = 0.414$. The corresponding expression for $C^*$ will be:

$$C^* = \frac{A a h_1}{(1-a/W)^2} \left( \frac{\sigma}{1.455 \times .414} \right)^{n+1}$$

$h_1 (0.5, 7) = 0.514$ for plane strain from Table 5.5a. Thus:

$$C^* = \frac{2.13 \times 10^{-18} \, (.05) \, (0.514)}{(0.5)^7} \left( \frac{35}{1.455 \times 0.414} \right)^8$$

$$= 9.10 \times 10^{-4} \, MJ/m^2 hr = 0.910 \, kJ/m^2 hr$$

Therefore:

$$C_t = 1.02 t^{-.667} + 0.91 \qquad kJ/m^2 hr$$

$$t_T = K^2(1-v^2)/(E(n+1)C^*)$$

$$= \frac{(39.24)^2(0.91)}{172 \times 10^3 \times 8 \times 9.1 \times 10^{-4}} = 1.12 \text{ hrs.}$$

Another approximate expression for estimating $C_t$ over a wide range was proposed by Bassani et al. [10.10] as follows:

$$C_t \approx \left(1 + \left(\frac{t_T}{t}\right)^{\frac{n-3}{n-1}}\right) C^* \tag{10.43}$$

As an exercise, you can substitute the values of $C^*$ and $t_T$ from Example Problem 10.3 and show that equation (10.43) provides a reasonably accurate estimate of $C_t$. Equation (10.42) can also be implemented for test specimens where the load-line displacement is available by recognizing that:

$$\dot{V}_c \approx \dot{V}_{ssc} + \dot{V}_{ss} \tag{10.44}$$

where $\dot{V}_{ss}$ is the small-scale-creep deflection rate, $\dot{V}_c$ is the measured deflection rate, and $\dot{V}_{ss}$ is the steady-state deflection rate. Let us express $C^*$ for any crack geometry in the following form:

$$C^* = \frac{P\dot{V}_{ss}}{BW}\eta(a/W,n) \tag{10.45}$$

The term $\eta$ can be derived for CT and CCT specimens easily by comparing equations (10.8) and (10.9), respectively, with equation (10.45). Then $C_t$ can be written as [10.14]:

$$C_t = \frac{P\dot{V}_c}{BW}\frac{F'}{F} - C^*\left(\frac{F'}{\eta F} - 1\right) \tag{10.46}$$

The values of $F'/F$ and $\eta$ for CT and CCT geometries are calculated and compared in Table 10.2. For CT geometries, the difference between $F'/F$ and $\eta$ is no more than 20% for $0.4 \leq a/W \leq 1$ and for $3 \leq n \leq 20$. Thus, for this geometry, $C_t$ can be estimated by either equation (10.45) or (10.39) within an accuracy of approximately 20%. However, for the CCT geometry, equation (10.46) must be used.

In SSC, $C_t$ is direclty related to the rate of expansion of the creep zone size as it evolves with time. We have also shown in equation (10.42) that:

$$(C_t)_{ssc} \propto \frac{K^4}{(t)^{(n-3)/(n-1)}} \tag{10.47}$$

**Table 10.2** Comparison of $F'/F$ and $\eta\ (a/W,n)$ for CT and CCT Specimens

| a/W | CT Specimen | | | | | | CCT Specimen | | | | | |
|-----|---|---|---|---|---|---|---|---|---|---|---|---|
| | | $\eta(a/W, n)$ | | | | | | $\eta(a/W, n)$ | | | | |
| | $F'/F$ | n = 3 | 5 | 7 | 10 | 20 | $F'/F$ | n = 3 | 5 | 7 | 10 | 20 |
| 0.2 | 2.976 | - | - | - | - | - | 2.755 | 0.312 | 0.417 | 0.468 | 0.511 | 0.565 |
| 0.3 | 2.590 | - | - | - | - | - | 2.067 | 0.357 | 0.476 | 0.537 | 0.584 | 0.646 |
| 0.4 | 2.644 | 2.725 | 3.211 | 3.373 | 3.504 | 3.67 | 1.821 | 0.417 | 0.555 | 0.625 | 0.682 | 0.753 |
| 0.5 | 3.078 | 3.274 | 3.767 | 3.957 | 4.111 | 4.307 | 1.785 | 0.500 | 0.667 | 0.750 | 0.818 | 0.905 |
| 0.6 | 3.920 | 4.063 | 4.60 | 4.83 | 5.020 | 5.259 | 1.914 | 0.625 | 0.833 | 0.937 | 1.022 | 1.131 |
| 0.7 | 5.330 | 5.340 | 5.99 | 6.29 | 6.536 | 6.847 | 2.255 | 0.833 | 1.111 | 1.250 | 1.363 | 1.508 |
| 0.8 | 7.949 | 7.856 | 8.763 | 9.205 | 9.563 | 10.019 | 3.042 | 1.25 | 1.667 | 1.875 | 2.045 | 2.262 |

On the other hand:

$$(C(t))_{ssc} \propto \frac{K^2}{t} \qquad (10.48)$$

Thus, $(C_t)_{ssc}$ and $(C(t))_{ssc}$ are not uniquely related to each other. This is a negative for $(C_t)_{ssc}$ as a candidate parameter for characterizing creep crack growth. Figure 10.14 shows a plot of da/dt vs. $C_t$ using the same data as shown in Figure 10.12. The consolidation of all data into a single trend is evident, validating the use of $C_t$ for creep crack growth.

Figure 10.15 shows creep crack growth data obtained on CT specimens of 1Cr-1Mo-0.25V which were 254mm wide and 63.5mm thick [10.15]. The purpose of testing large specimens is to ensure significant crack extension occurs under both small-scale and extensive creep conditions. The crack growth rates in these specimens begin at a higher rate, decrease progressively until they reach a minimum, and then increase. During both the increasing and decreasing portions of the test, the growth rates correlate with $C_t$. The decreasing portion of the test is dominated by small-scale and transition creep. Some decrease in $C_t$ may also have been caused by primary creep deformation in the early part of the test. The increasing crack growth rate portion is dominated by extensive creep. These results nicely support the validity of $C_t$ for uniquely characterizing the creep crack growth behavior under a wide range of creep conditions. Figure 10.16 shows similar data from CT and CCT geometries obtained from tests conducted under constant load conditions and also from tests conducted under constant deflection rate conditions [10.12]. Again these results strongly support the viability of $C_t$ as a crack tip parameter.

## 10.3 Consideration of Primary Creep

Several high temperature materials exhibit significant primary creep at stress levels which are typical of service operation. As can be seen in Figure 10.1, the extent of primary creep increases with decreasing stress level. Therefore, consideration of primary creep in estimating $C_t$ can be significant. In the subsequent discussion, we will first consider primary creep in isolation and then consider the combination of primary and secondary creep and elastic stress redistribution all together.

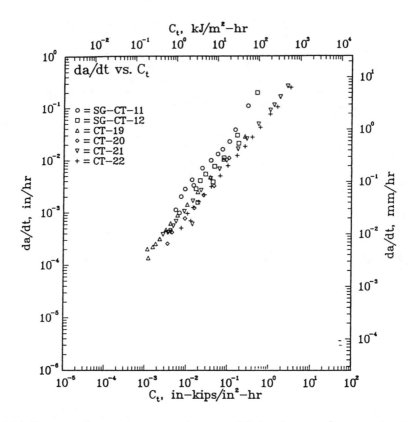

**Figure 10.14** *Creep crack growth rate as a function of $C_t$ for the same data correlated with $C(t)$ in Figure 10.12.*

### 10.3.1 Creep Constitutive Law Including Primary Creep

A creep constitutive law which combines elastic deformation, primary creep, and steady-state creep can be written as follows:

$$\dot{\varepsilon} = \dot{\sigma}/E + A_1 \, \varepsilon^{-p} \, \sigma^{n_1(1+p)} + A \, \sigma^n \tag{10.49}$$

In the above equation, the first and third terms on the right-hand side are the elastic deformation and the secondary-creep deformation terms, respectively, described in the previous section. The middle term accounts for creep strains and contains 3 constants, p, $A_1$, and $n_1$, which are derived from regression of the creep data. Figure 10.17 shows a plot of the fitted curves from equation (10.49) and the creep data for a Cr-1Mo-0.25V showing the ability of equation (10.49) to accurately represent the creep data at different stress levels. We will be exclusively using this form of equation in our discussion regarding primary creep.

### 10.3.2 Crack Tip Parameters for Extensive Primary Creep

If we consider extensive primary creep condition to dominate or, in other words, primary creep in isolation, we can simplify equation (10.49) to:

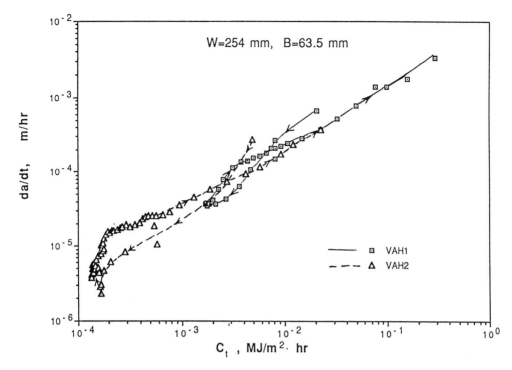

**Figure 10.15** *Creep crack growth rate as a function of $C_t$ for 63.5 mm thick and 254 mm wide CT specimens. The arrows indicate the direction of change in $C_t$ and da/dt from the beginning of the test to its end (Ref. 10.15).*

$$\dot{\varepsilon} = A_1 \varepsilon^{-p} \sigma^{n_1 (1+p)} \tag{10.50}$$

which can also be expressed as:

$$\varepsilon = \left[ (1+p) A_1 t \right]^{1/(1+p)} \sigma^{n_1} \tag{10.50a}$$

and as:

$$\dot{\varepsilon} = A_1 \left[ (1+p) A_1 t \right]^{\frac{-p}{(1+p)}} \sigma^{n_1} \tag{10.50b}$$

If we hold time as constant in equation (10.50b), Riedel [10.16] was first to recognize that the form of this equation is identical to equation (10.1) for pure steady-state creep. That is, the strain rate is only a function of stress at a fixed time. Therefore, $C^*$-integral is path-independent for a fixed time and its value can be calculated at a time, t, by simply replacing A with $A_1 [(1 + p) A_1 t]^{-p/(1+p)}$ and n with $n_1$ in the ex-

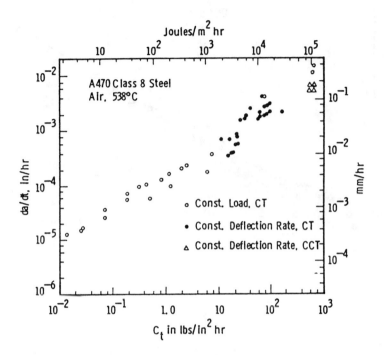

**Figure 10.16** *Creep crack growth rate as a function of $C_t$ from several tests conducted on CT and CCT specimens under conditions of constant load and constant deflection rate (Ref. 10.12).*

pressions for calculating the $C^*$-integral. By examining equation (10.50a), we can see that for a fixed time, the stress-strain relationship is also unique. That is, strain is only a function of stress. Therefore, the J-integral is also path-independent and can be obtained by replacing $(\alpha\varepsilon_0)/\sigma_o^m$ in the equation for estimating J (see, for example, Chapter 5) with $[(1+p) A_1 t]^{1/(1+p)}$ and m with $n_1$. Riedel [10.16] defined yet another path-independent integral, $C_h^*$, which can be estimated by replacing $(\alpha\varepsilon_0/\sigma_o^m)$ in the equations for estimating J by $[(1+p) A_1]^{1/(1+p)}$ and m by $n_1$. The relationship between $C_h^*$ and $J$ is then given by:

$$J = C_h^* \, t^{1/(1+p)} \qquad (10.51)$$

The value of $C_h^*$ is independent of time. Therfore, the HRR crack tip field equation for stress is given by:

$$\sigma_{ij} = \left[ \frac{C_h^* \, t^{1/(1+p)}}{I_{n_1} \, r \, ((1+p) A_1 t)^{1/(1+p)}} \right]^{1/(n_1+1)} \hat{\sigma}_{ij}(\theta, n_1) \qquad (10.52) \qquad (10.52)$$

or

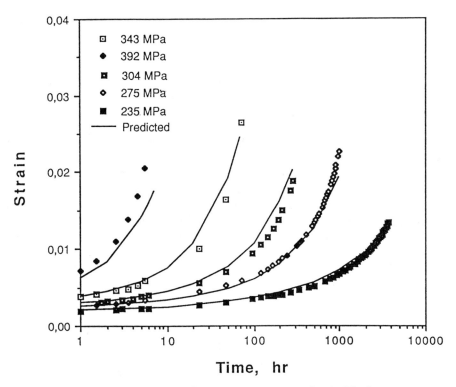

**Figure 10.17** *Creep strain as a function of time at various stress levels. The line represents predicted values from equation (10.49) with $A_1 = 1.49 \times 10^{-31}$ (for $\sigma$ MPa and $\dot{\varepsilon}$ in $hr^{-1}$), $n_1 = 3.54$, $p = 1.55$, $A = 9.51 \times 10^{-41}$ (for $\sigma$ in MPa and $\dot{\varepsilon}$ in $hr^{-1}$), and $n = 14.35$ (Ref. 10.15).*

$$\sigma_{ij} = \left[ \frac{C_h^*}{I_{n_1} r ((1+p) A_1)^{1/(1+p)}} \right]^{1/(n_1+1)} \hat{\sigma}_{ij}(\theta, n_1) \qquad (10.52a)$$

or by

$$\sigma_{ij} = \left[ \frac{C^*(t)}{I_{n_1} r A_1 [(1+p) A_1 t]^{-p/(1+p)}} \right]^{1/(n_1+1)} \hat{\sigma}_{ij}(\theta, n_1) \qquad (10.53)$$

where $C^*(t)$ is the time-dependent $C^*$-integral for primary creep. By comparing equations (10.52a) and (10.53), one can show that:

$$C^*(t) = \frac{C_h^*}{(1+p)\, t^{p/(1+p)}} \qquad (10.54)$$

The above analogies have worked because of the separability of stress and time dependence in strain and strain rate in the formulation of primary creep. Therfore, by a judicious choice of the form of constitutive equations, the solutions available in the literature for estimating J-integral and $C^*$ can be directly used for estimating $C^*(t)$ for primary creep. Also note in equation (10.52a), that the crack tip stress field for primary creep is independent of time and is only a function of spatial coordinates; its amplitude is characterized by $C_h^*$.

Since the units of $C_h^*$ are different from $C^*(t)$, it is preferable to characterize creep crack growth rates with $C^*(t)$, which has the same units and stress-power dissipation rate iterpretation as does $C^*$ for steady-state creep [10.13]. During extensive primary creep conditions, even though the crack tip stress field is constant, the rate of creep strain accumulation varies with time and it is found that the creep crack growth rates also vary with time. Thus, no unique correlations between $C_h^*$ and da/dt can be expected. On the other hand, good correlations between $C^*(t)$ and da/dt have been observed in the presence of primary creep [10.13]. The relevance of $C_h^*$ in including primary creep is mostly convenience in the computation of $C^*(t)$. Another important observation is that $C^*(t)$, because of its relationship to the instantaneous stress-power dissipation rate, is identical to $C_t$. Thus, for extensive primary creep:

$$C_t \equiv C^*(t) \qquad (10.55)$$

Following the analogy further, the load-line deflection rate due to primary creep, $\dot{V}_{pc}$, can be given by making the same substitutions for creep constants as for estimating $C^*(t)$, as illustrated in the following example.

**Example Problem 10.6**
Assume that a compact specimen of width, W, and thickness, B, is undergoing extensive primary creep. Derive expressions for estimating the load-line deflection rate and for estimating $C^*(t)$ if the deflection rate is measured in a test.

*Solution:*
Substituting $A_1 [A_1(1+p)t]^{-p/(1+p)}$ for A and $n_1$ for $n$ in equations (10.10) and (10.11), we get:

$$C^*(t) = \frac{A_1^{1/(1+p)} (W-a)}{((1+p)t)^{p/(p+1)}} h_1(a/W, n_1) \left( \frac{P}{1.455 \eta_1 B(W-a)} \right)^{n_1+1}$$

and

$$\dot{V}_{pc} = \frac{A_1^{1/(1+p)} a}{((1+p)t)^{p/(1+p)}} h_3(a/W, n_1) \left( \frac{P}{1.455 \eta_1 (W-a)} \right)^{n_1}$$

Combining the two equations above, we get:

$$C^*(t) = \frac{P\dot{V}_{pc}}{BW} \frac{h_1}{h_3} \frac{1}{1.455(a/W)\, \eta_1}$$

or

$$C^*(t) = \frac{P\dot{V}_{pc}}{B(W-a)} (2 + .522(1-a/W)) \frac{n_1}{1+n_1}$$

from equation (10.13). This equation is identical to the one used for calculating $C^*$ with the exception that $n$ is relaced by $n_1$.

### 10.3.2 Small-Scale Primary Creep

In this discussion, we will consider the conditions of small-scale primary creep. Therefore, in equation (10.49), we neglect the third term on the right-hand side and retain the other two while also assuming the creep zone size is small in comparison to the other charcteristic dimensions of the cracked body. We go back to equation (10.39) and equation (10.40) which were derived for small-scale creep conditions. In deriving either of those equations, no material constitutive laws were assumed. In other words, the validity of these equations hold for any general creep-law and is not restricted to power-law creep, even though the equations were derived in the context considering only power-law creep. In the case of equation (10.39), the correct creep constitutive law is automatically included in the measured value of $\dot{V}$. Therfore, equation (10.39) can be used to estimate $(C_t)_{ssc}$ under primary creep also. Similarly, equation (10.40) is valid for any creep-law. However, an analytical expression for estimating $\dot{r}_c$ for elastic-primary creep must be derived. This equation is given by [10.17]:

$$\dot{r}_c(\theta,t) = \frac{K^2}{2\pi} \left[ \frac{I_{n_1} E}{2\pi(1-v^2)} \right]^{2/(n_1-1)} \left[(1+n_1)(1+p)A_1\right]^{2/(1+p)(n_1-1)}$$

$$x \ \frac{2F_{cr}(\theta,n_1)}{(1+p)(n_1-1)} \ t^{[(2/(1+p)(n_1-1))-1]} \quad (10.56)$$

Substituting equation (10.56) into equation (10.40) will give us an analytical expression for estimating $(C_t)_{ssc}$ under elastic and primary creep deformation.

### 10.3.3 Primary and Secondary Creep

If we assume that extensive creep conditions exist in the cracked body but it consists of a mixture of primary plus secondary creep, the constitutive equation will consist of the second and third terms on the right-hand side of equation (10.49). In this equation the stress and time-dependence of strain rate are not separable, and a unique relationship between strain rate and stress cannot be defined. Therefore, no path-independent integral exists which uniquely characterizes the crack tip stress and strain fields. However, approximate path-independence (within 2%) of C*(t) has been shown for one case [10.14, 10.17]. This result, although derived for one material and geometry, is expected to be quite general. This extends the validity of C*(t) and, therefore $C_t$, as crack tip parameters for the case of any extensive creep condition regardless of whether it is steady-state creep, primary creep, or a combination of both.

A transition time may be defined for extensive secondary creep conditions to develop from extensive primary creep conditions. This transition time, $t_2$, can be estimated by equating the value of C*(t) for primary creep, equation (10.54) to $C_s^*$ [10.16]. The latter is the C*-integral corresponding to steady-state condition. Since this value is fixed, it is prudent to use a special designation, $C_s^*$, for its value which is distinct from the time-dependent value of C*. Thus:

$$t_2 = \left[ \frac{C_h^*}{(1+p)C_s^*} \right]^{1+1/p} \tag{10.57}$$

In this regime the values of $C_t$ and C*(t) can be estimated by the following equations:

$$C^*(t) = C_t = [1 + (t_2/t)^{p/(p+1)}] C_s^* \tag{10.58}$$

In test specimens in which the load-line deflection rate can be measured, expressions can be derived for estimating C*(t) and $C_t$. For example, for a CT specimen equation (10.8) can be modified in the following way [10.13]:

$$C^*(t) = \frac{P\dot{V}_c}{B(W-a)} (2 + .522 (1-a/W)) f(n,n_1) \tag{10.59}$$

$$\text{or} \quad C^*(t) = \frac{P\dot{V}_c}{BW} \eta (a/W, n_1, n) \tag{10.59a}$$

$$f(n,n_1) = \frac{n}{n+1} (t/t_2) + \frac{n_1}{1+n_1} (1 - t/t_2) \tag{10.60}$$

### 10.3.4 Transition From Small-Scale To Extensive Primary Creep

The extensive primary creep and small-scale primary creep conditions were considered separately in sections 10.3.1 and 10.3.2, respectively. The transition time, $t_{Tp}$, from small-scale to extensive primary creep can be obtained by following the analysis of Riedel [10.16] which equates the amplitudes of the HRR fields predicted by the two limiting conditions:

$$t_{Tp} = \left[ \frac{K^2 (1-v^2)}{E C_h^*} \right]^{1+p} \frac{1}{1+n_1} \tag{10.61}$$

## 10.3.5 Elastic, Primary, and Secondary Creep Combined

The value of $C_t$ can be estimated by adding the $C_t$ for small-scale creep, $(C_t)_{ssc}$ with the $C_t$ for extensive creep, $C^*(t)$. Obtaining the value of $C^*(t)$ for combined primary and secondary creep is quite straightforward from equation (10.58). Analytical expressions for estimating $(C_t)_{ssc}$ for combined primary and secondary creep have not been derived. In the case of experiments, equation (10.39) remains valid and combined contributions from secondary and primary creep will be represented in the measured value of $V_c$. One approach to analytically estimate $(C_t)_{ssc}$ for combined primary and secondary creep is to add the values of the creep zone expansion rates, $\dot{r}_c$, for primary and secondary creep. For secondary creep, $\dot{r}_c$, is given by differentiating equation (10.23a) and for primary creep, it is given by equation (10.56). These combined rates can be substituted into equation (10.40) to obtain $(C_t)_{ssc}$.

## Example Problem 10.7

A 50mm wide and 25mm compact specimen has a 25mm long crack. The specimen is made from a Cr-Mo-V steel which has the following creep constants at 538°C:

E = 160 x 10³ MPa
$A_1$ = 1.49 x 10⁻³¹ (MPa)⁻ⁿ, hr⁻¹
$n_1$ = 3.54
p = 1.55
A = 9.51 x 10⁻⁴¹ (MPa)⁻ⁿ x hr⁻¹
n = 14.35

The specimen is loaded with a force of 20kN. Estimate the three transition times $t_T$, $t_{Tp}$, and $t_2$. Also estimate the value of $C_t$ at t = 100 hrs. Assume plane strain conditions.

## Solution:

We begin by calculating the values of K, $C^*$, and $C_h^*$ for this specimen. The transition times and $C_t$ can be estimated from these values. From Example Problem 10.3, we can obtain K as follows:

$$K = (20 / (0.025 \times .05^{0.5})) (9.65) = 34,525 \text{ KPa}\sqrt{m}$$

For estimating $C^*$, we will use an $h_1$ for a/W = 0.5 and n = 16. Similarly, for calculating $C_h^*$, we will use an $h_1$ for $n_1$ of 3. From Table 5.2, we obtain $h_1 (0.5, 16) = 0.216$ and $h_1 (0.5, 3) = 1.24$, $\eta_1 (0.5) = 0.162$ from equation (5.52):

$$C_s^* \text{ (from equation (10.10))} = 9.51 \times 10^{-41} (0.025) (0.216) \left( \frac{20 \times 10^{-3} (MN)}{1.455 \times .162 \times .025 (.025)} \right)^{15.35}$$

$$= 0.28 \times 10^{-9} \text{ MJ/m}^2 \text{ hr}$$

$$\left[ A_1 (1+p) \right]^{\frac{1}{(1+p)}} = \left[ 1.49 \times 10^{-31} (2.55) \right]^{\frac{1}{2.55}} = 1.176 \times 10^{-12}$$

$$C_h^* = 1.176 \times 10^{-12} (.025) (1.24) \left( \frac{20 \times 10^{-3}}{1.455 \times .162 \times .025 \times .025} \right)^{4.54}$$

$$= 1.756 \times 10^{-4}$$

$$t_T \text{ (from equation 10.25)} = \frac{(34.525)^2 \times .91}{160 \times 10^3 \times (15.35) \times .28 \times 10^{-9}}$$

$$= 1.577 \times 10^6 \text{ hrs}$$

$$t_{T_p} \text{ (equation (10.61))} = \left(\frac{(34.525)^2 \times .91}{160 \times 10^3 \times 1.756 \times 10^{-4}}\right)^{2.55} \frac{1}{4.54}$$

$$= 2{,}449 \text{ hrs}$$

$$t_2 \text{ (from equation (10.57))} = \left(\frac{1.756 \times 10^{-4}}{2.55 \times .28 \times 10^{-9}}\right)^{1.645}$$

$$= 7.373 \times 10^8 \text{ hrs}$$

$C^*(t)$ for $t = 100$ hrs. from equation (10.58):

$$= [\,1 + (7.337 \times 10^6)^{0.607}\,] \, .28 \times 10^{-9} = 4.12 \times 10^{-6} \text{ MJ/m}^2 \text{ hr}$$

By comparing $t_T$ and $t_{T_p}$ it is clear that the deformation in this specimen is dominated by primary creep. From equation (10.56), we get the following:

$$\dot{r}_c(\theta,t) = \frac{(34.525)^2}{2\pi} \left[\frac{I_{n_1}(160 \times 10^3)}{2\pi(.91)}\right]^{.787} [\,4.54 \times 2.55 \times 1.49 \times 10^{-31}\,]^{.3087}$$

$$\times \frac{2 \times 0.3}{(2.55)(2.54)} \, t^{-.691}$$

$I_{n_1}$ from equation (4.19a) = 5.33. We have chosen $F_{cr}(\theta, n_1) \approx 0.3$ for $\theta = 90°$ and $n_1 = 3.54$ from Figure 10.10. Therefore:

$$\dot{r}_c(\theta,t) = 189.7 \ [\ 11793.5\ ]\ [\ 6.487 \times 10^{-10}\ ]\ [\ .00384\ ] = 5.57 \times 10^{-6} \text{ m/hr}$$

The above creep rate is for primary creep alone. For secondary creep, the creep zone size is given by equation (10.23a) which can be differentiated with time with the following result:

$$\dot{r}_{cr}(\theta,t) \text{ for steady-state creep} =$$

$$\frac{2\alpha}{n-1} K^2 (EA)^{\frac{2}{n-1}} t^{-(n-3)/(n-1)} F_{cr}(\theta,n)$$

where $\alpha$ is given by equation (10.35). $F_{cr}(90, 14.35) \approx 0.4$ from Figure 10.10, $\alpha = 0.230$:

$$\dot{r}_{cr} = (0.0344)(1191.97)(160 \times 10^3 \times 9.51 \times 10^{-41})^{0.15}(100)^{-.85}$$

$$= 1.19 \times 10^{-7} \text{ m/hr}$$

Comparing the creep zone expansion rates, it is also clear that primary creep dominates the specimen behavior at $t = 100$ hrs. By adding the two creep rates and then substituting into equation (10.40), we get:

$$(C_t)_{ssc} = \frac{2 \times .91 \ (34.525)^2}{160 \times 10^3 \times .050} (0.33)(F'/F)(5.68 \times 10^{-6})$$

$F'/F$ for CT specimens for a/W = 0.5 can be obtained from Table 10.1 as 3.078. Thus:

$$(C_t)_{ssc} = 1.56 \times 10^{-6} \text{ MJ/m}^2 \text{ hr}$$

$$C_t = (C_t)_{ssc} + C^*(t) = 5.68 \times 10^{-6} \text{ MJ/m}^2 \text{ hr} = 5.68 \text{ J/m}^2 \text{ hr}$$

## 10.4 Effects of Crack Growth on the Crack Tip Stress Fields

Crack growth is expected to perturb the stress fields for stationary cracks. Since the analysis in the previous sections was limited to stationary cracks, an obvious question which must be addressed is what restrictions must be placed on stationary crack parameters such as $C_t$, $C(t)$, and $C^*(t)$ for characterizing crack growth rates. The stress fields in the vicinity of a crack growing at a velocity $\dot{a}$ in an elastic, power-law creeping material was addressed by Hui and Riedel [10.18]. They argued that stresses and elastic strains near the moving tip will be different from the HRR field due to the instantaneous elastic response of the material. They derived what is termed as the HR (Hui-Riedel) field, and found that two distinctly different singularities develop depending on steady-state creep exponent, n. For n < 3, the asymptotic stress field is dominated by elastic strain rates so that $r^{-1/2}$ singularity develops. For $n \geq 3$, which is typical of metals, the competing effects of stress elevation due to crack growth and stress relaxation due to creep give rise to a field with an amplitude related to the crack growth rate:

$$\sigma_{ij} = \alpha_n \left( \frac{\dot{a}}{AEr} \right)^{\frac{1}{n-1}} \tilde{\sigma}_{ij}(\theta, n) \tag{10.62}$$

where $\tilde{\sigma}_{ij}(\theta,n)$ is an angular function and $\alpha_n$ depends on and is approximately unity for plane strain. For $n = 4$ and 6, the value of $\alpha_n = 0.815$ and 1.064, respectively. The amplitude of the HR field is completely determined by the crack growth rate and is independent of the loading and crack growth history. The domain of dominance of the HR field is estimated by equating the effective stress predicted by the HRR field with that corresponding to equation (10.62). Such nesting of the singular crack tip solutions (HR, HRR, and K for small-scale creep conditions) was found to be accurate by Hawk and Bassani [10.19] who conducted numerical analysis to verify the concept. We will first derive the equations for extensive secondary creep conditions and then for small-scale creep.

### 10.4.1 Crack Growth Under Extensive Steady-State Creep

The HRR field for extensive steady-state creep conditions are given by equation (10.5a). Equating equation (10.5a) and (10.62) and solving for $r$ gives the size of the domain in which the HR field dominates [10.20]:

$$r_{HR} = \frac{1}{A} \left( \frac{\dot{a}}{E} \right)^{\frac{n+1}{2}} \left( \frac{I_n}{C^*} \right)^{\frac{n-1}{2}} \beta(\theta, n) \tag{10.63}$$

where $\beta$ is an angular function.

In order for $C^*$ to be a valid parameter for characterizing creep crack growth, $r_{HR} << r_{HRR}$. In extensive creep, the extent of $r_{HRR}$ is on the order of the crack size and the value of $r_{HR}$ can be calculated. For example, for 304 stainless-steel the $\dot{a}$ vs. $C^*$ relationship is shown in Figure 10.9 and the A and $n$ values are given in Example Problems 10.1 and 10.2, E = 172 x $10^3$ MPa. If one estimates a crack growth rate of 2.5 x $10^{-4}$ m/hr at a $C^*$ value of 15 kJ/m$^2$ hr and A = 2.13 x $10^{-18}$ and n = 6, we get:

$$r_{HR} \approx \frac{1}{2.13 \times 10^{-18}} \left( \frac{2.5 \times 10^{-4}}{172 \times 10^3} \right)^{3.5} \left( \frac{I_n}{.015} \right)^{2.5} \beta(\theta, n)$$

$I_n$ for n = 6 from equation (4.19) is 4.9. If we assume that the angular term is on the order of unity, we get $r_{HR} \sim 10^{-6}$ m. The crack sizes in the specimen is on the order $10^{-9}$ m. Therefore, we can make the argument that the HR field is not significant or the effects of the growing crack are not important in the case of these tests. In general, if crack growth rate occurs under extensive creep conditions, the effects of crack growth are not likely to be significant. Therfore, $C^*$ is a valid parameter for characterizing creep crack growth under the conditions of extensive secondary creep.

## 10.4.2 Crack Growth Under Small-Scale Creep

In this section, we will examine the limits of $(C_t)_{ssc}$ derived for stationary crack conditions for characterizing creep crack growth if small-scale creep conditions exist in the specimen (or component). We postulate that for $(C_t)_{ssc}$ to charcterize $\dot{a}$, we must meet the following condition:

$$\dot{r}_c >> \dot{a} \tag{10.64}$$

From differentiating equation (10.32), we can write:

$$\dot{V}_c = P \frac{dC}{da} \beta \dot{r}_c$$

By combining the above equation with equation (10.33), we get:

$$\dot{V}_c = \frac{2BK^2}{PE}(1-v^2)\beta \dot{r}_c \tag{10.65}$$

The above equation provides a relationship between the deflection rate due to creep deformation and the creep zone expansion rate. In the case of a growing crack in the absence of creep, the relationship between elastic compliance and deflection, $V_e$, can be expressed by the following relationship:

$$V_e = PC$$

For constant load, P:

$$\dot{V}_e = P\frac{dC}{dt} = P\dot{a}\frac{dC}{da}$$

Combining the above equation with equation (10.33), we get:

$$\dot{V}_e = P\dot{a}\left(\frac{2BK^2}{EP^2}(1-v^2)\right)$$

or

$$\dot{V}_e = \frac{2B\dot{a}}{P}\frac{K^2}{E}(1-v^2) \tag{10.66}$$

Comparing equation (10.65) to (10.66) and noting that $\beta$ for $\theta = 0°$ is approximately unity, we can meet the condition in equation (10.64) by requiring that $\dot{V}_c \gg \dot{V}_e$. The total measured deflection rate in a test is:

$$\dot{V} = \dot{V}_c + \dot{V}_e$$

Thus, the condition of $\dot{V}_c \gg \dot{V}_e$, can also be written as:

$$\dot{V} \gg \dot{V}_e = \frac{2B\dot{a}}{P} \frac{K^2}{E} (1 - v^2) \tag{10.67}$$

The above equation, known as deflection-rate partitioning, was first derived by Saxena et al. [10.21] and proposed as a condition for validating da/dt vs. $C_t$ correlations of the type shown in Figures 10.15 and 10.16.

If the condition of equation (10.67) is not met, the stationary crack tip parameters such as $C_t$ should not be used for correlating creep crack growth rate. The materials in which creep crack growth occurs while satisfying equation (10.67) are know as creep-ductile materials and the ones in which this condition is not satisfied are known as the creep-brittle materials. In the latter, the crack grwoth effects are quite significant in determining the crack tip stress fields. The next section discusses approaches for characterizing creep crack growth in such materials.

## 10.5 Crack Growth in Creep-Brittle Materials

Creep-brittle behavior is said to occur when $\dot{a}$ becomes comparable in magnitude to $\dot{r}_c$ such that the growing crack effects on the crack tip stress field are no longer negligible. As mentioned in the previous section, the stationary crack tip parameters, $C^*$ and $C_t$ are no longer valid for characterizing the creep crack growth behavior under these circumstances. In this section, we will focus our discussion on crack tip parameters in the presence of crack growth effects. There are several high temperature materials such as high temperature Al alloys, titanium alloys, nickel base superalloys, intermetallics, and ceramics whose behavior at their intended service temperatures and environments leads to creep-brittle crack growth conditions. Therefore, modeling of such behavior is of considerable technological interest.

Figure 10.18 shows the development of creep deformation of the crack tip in specimens of two hypothetical materials. The specimens are identical in all respects and are subjected to the same loads. The creep deformation behavior of the two materials is also the same. In one material, the $\dot{a}$ is ten times larger than the other. Hall [10.22] analyzed the specimens by conducting finite element analysis which simulated the stress fields at various times in the presence of growing cracks, Figure 10.18. In the specimen with slower crack growth rate, the creep-ductile conditions prevailed and extensive creep conditions developed progressively. In the specimen with higher crack growth rate, creep-brittle conditions developed and the specimen always remained in the small-scale creep condition. This simulation brings out the role of crack extension in determining the crack tip conditions. One can imagine a steady-state condition for which $\dot{a}$ and $(\partial r_c / \partial t)$ are equal along $\theta = 0$. In this case, to an observer stationed at the moving crack tip, the creep zone size is constant along the x-axis. Under these circumstances, da/dt is likely to correlate with K. However, prior to reaching the steady-state conditions, considerable crack growth can occur under transient conditions. Also, even if $\dot{r}_c = \dot{a}$ at $\theta = 0$, the self-similarity in the growth of the creep zone is not necessarily assured. In the subsequent sections, both the steady-state and transient conditions are explored further.

Analysis of Cracks in Creeping Materials 349

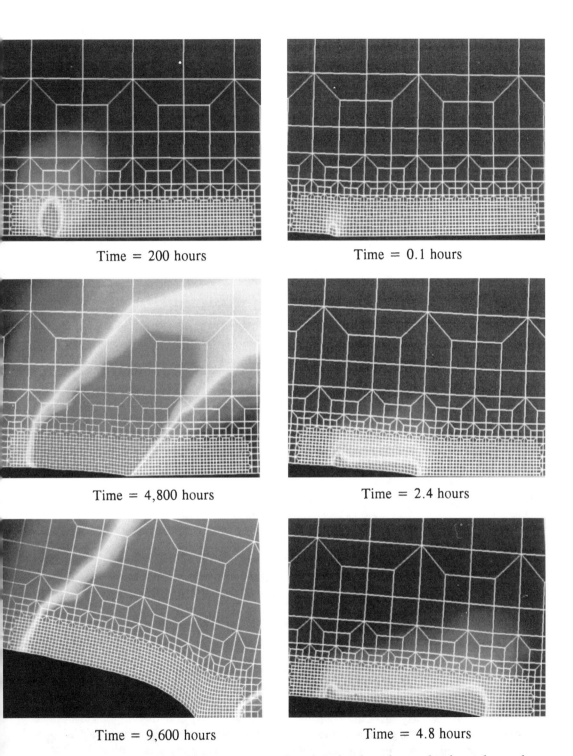

**Figure 10.18** Contour plots of effective creep strain for creep-ductile and creep-brittle crack growth at three times during crack growth. From Ref. 10.22.

## 10.5.1 Steady-State Creep Crack Growth Under Small-Scale Creep

Steady-state creep crack growth under small-scale creep in elastic, power-law creeping material has been analytically treated by Hui [10.23, 10.24]. He assumed that away from the immediate vicinity of the crack tip, the stress field is characterized uniquely by K. By dimensional analysis, Hui [10.24] showed that the crack tip stress field in the immediate vicinity of the crack tip is described by:

$$\sigma_{ij} = \left[ \frac{\dot{a}}{EAK^2} \right]^{(1/n-3)} \Sigma_{ij}(R,\theta) \tag{10.68}$$

where the dimensionless radical coordinate R is given by:

$$R = r \left( \frac{EAK^{n-1}}{\dot{a}} \right)^{-2/(n-3)} \tag{10.69}$$

A measure of the characteristic length over which the asymptotic field, equation (10.68), dominates is obtained by choosing R = 1. For K to be a valid parameter, this distance must be small in comparison to the crack length and the uncracked remaining ligament. Figure 10.19 shows da/dt vs. K for an ASTM SA-106C carbon steel at 360°C [10.25]. Thus, in principle, steady-state crack growth in small-scale creep is likely when these special conditions are met. However, steady-state is likely to be preceded by transient crack growth which is discussed next.

## 10.5.2 Transient Crack Growth Under Small-Scale Creep

Due to the difficulty of obtaining analytical solutions for the transient crack growth under small-scale creep accounting for crack growth effects, this problem has been simulated numerically [10.26]. Experimentally observed crack growth histories of four compact type specimens (BCH-1,4,6,7) of aluminum alloy 2519-T87 tests at 135°C were numerically enforced to study the evolution of crack tip fields and far-field fracture parameters.

The time-dependent deformation characteristics for the aluminum alloy 2519-T87 at 135°C were modeled using primary and steady-state creep material constants. This model was incorporated in the finite element analyses conducted. A constant external load of 10.5 kN was applied to one of the specimens, BCH-6, whose width and thickness were 5.08 cm and 2.21 cm, respectively. Crack growth began at a crack length of 2.25 cm and ended 105 hours later at a crack length of 2.52 cm when the specimen became unstable. A plot of the crack growth history, Figure 10.20, reveals a relatively long incubation period during which the crack grew either very slowly or not at all. The incubation period was followed by a period of crack growth. Figure 10.21 compares the experimentally observed deflection as a function of time with that predicted by the finite element analysis for this specimen. The good agreement between the experimental results and the finite element analysis results are clear indications of the accuracy of the analysis.

Interpretation of the finite element results can be enhanced by utilization of a visualization program to simulate the evolution of the various crack tip parameters at different stages of crack growth. Figure 10.22 shows the contour plots of the effective creep strain and effective stress at three stages of crack growth for this specimen. Deflections in all of these plots are magnified by 60. Color is used to illustrate the creep zone boundaries. The yellow areas mark the boundary of where the creep strains exceeded .005. In the stress plots, the yellow areas mark the region where the stress exceeded 310 MPa and blue areas where the stress exceeded 103 MPa.

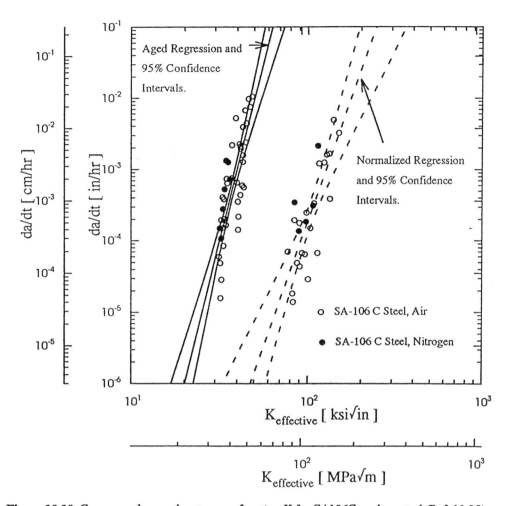

*Figure 10.19* Creep crack growth rates as a function K for SA106C carbon steel (Ref. 10.25).

Significant levels of creep deformation are confined to a thin region adjacent to the crack growth path. The characteristic "knob" observed near the initial crack tip is due to the initially elastic crack tip fields which produce high creep strain rates while the crack growth is slow due to incubation. As the stresses near the crack tip relax, the driving force for creep deformation decreases resulting in a corresponding decrease in the creep zone size. Both residual stress and residual creep strain are left in the wake of the growing crack. As the crack grows, the crack tip fields become more intense resulting in larger creep zone expansion rates. However, the extent of the creep zone is limited by the accelerating crack growth, since accumulating creep deformation is quickly left in the wake of the growing crack.

Profiles of effective stress along a radial line from the crack tip at $\theta = 90°$ reveal that stresses in a region just outside the creep zone scale with $r^{-1/2}$ throughout the crack growth history, Figure 10.23. Consequently, small-scale creep conditions exist and K describes the crack tip fields in an annular region around the crack tip. Both the amplitude and the strength of the stress singularity within the creep zone ($r < \approx 0.25$ mm) increase with time. It is not clear whether the amplitude scales with K or with the crack growth rate (à) since both quantities increase significantly during the test. The strength of the singularity is given by $r^{-1/16}$ just after incubation (after one crack growth increment) and approximately by $r^{-1/12}$

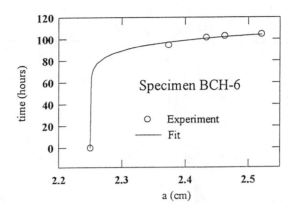

*Figure 10.20* Experimentally determined crack growth history of an aluminum alloy 2519-T87 CT specimen tested at 135°C and the curve fit used to describe the behavior in a finite element study.

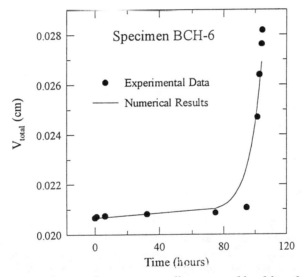

*Figure 10.21* Comparison between the experimentally measured load-line displacement as a function of time and those predicted by the finite element analysis (Ref. 10.26).

throughout most of the remainder of the test. Similar trends are predicted analytically when a stationary crack tip field (HRR type field) gives way to a growing crack tip field (HR type field). Thus, the results presented here seem to support the notion of "nested" crack tip fields, although the particular form of the fields or the character of the nesting cannot be direclty compared with theory due to the complex constitutive law employed in this analysis.

**Correlation of Crack Growth Rate with K** From an analytical viewpoint, successful correlation of crack growth with a fracture parameter requires that the deformation and damage fields surrounding the

Analysis of Cracks in Creeping Materials 353

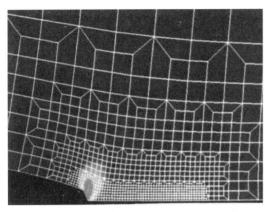 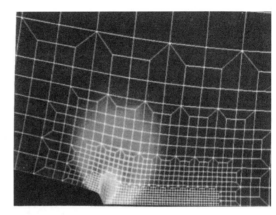

Time = 74.5 hours;  Crack Length = 2.26 cm

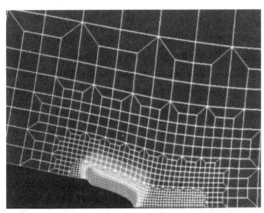 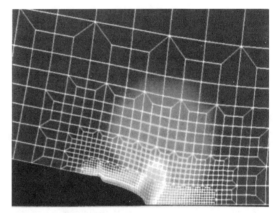

Time = 97.8 hours;  Crack Length = 2.39 cm

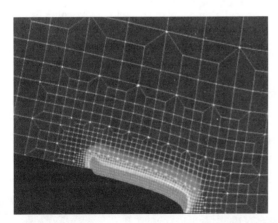 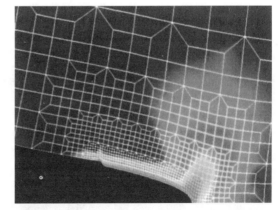

Time = 104.2 hours;  Crack Length = 2.52 cm

**Effective Creep Strain**  **Effective Stress**

*Figure 10.22* Contour plots of effective stress and creep strain at various stages of crack growth in a finite element simulation of a real test on 2519-T87 aluminum alloy (Ref. 10.26).

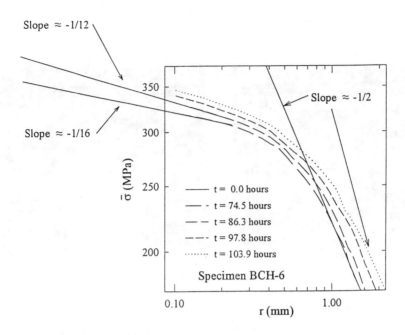

*Figure 10.23* Profiles of effective stress along θ = 90° showing a slope equal to (-1/2) outside the creep zone in the simulation on 2519-T87 aluminum specimen (Ref. 10.26).

crack tip must scale with the parameter in a roughly "self-similar" fashion. For example, the creep zone for a stationary crack under small-scale yielding expands in a self-similar fashion such that its shape does not change with time (i.e., $r_c(\theta = 90°) / r_c(\theta = 0°)$ is constant. Figure 10.24 indicates that the shape of the creep zone for specimen BCH-6 changes significantly during the early stages of crack growth and later assumes a somewhat constant shape. Thus, crack growth for this specimen appears to approach self-similar conditions following a transient period where the character of the creep zone changes significantly (the other three specimens follow similar trends).

The applicability of small-scale creep conditions, coupled with the existence of similitude, suggest that the later stages of crack growth can be correlated with K, Figure 10.25. This figure essentially represents smoothed out experimental data since K is uniquely determined by the current crack length for a given specimen geometry and loading and is unaffected by the results of the finite element analyses. Specimens BCH-4 and BCH-7 have a thickness of 2.21 cm, while specimen BCH-1 has a thickness of 0.635 cm. Figure 10.25 reveals good correlation of K with crack growth rates after the initial period of transience for the thicker specimens. Crack growth rates also appear to have a power law dependence on K for the thinner specimen, although the dependence is shifted relative to the thicker specimens. The higher resistance to creep crack growth demonstrated by the thinner specimen is attributed to reduced constraint, since it is believed that crack tip damage processes accelerate under increasing triaxiatly levels, particularly for creep-brittle materials.

**Transition Times for Growing Cracks** The initial stages for crack growth in aluminum alloy 2519-T87 are characterized by highly transient crack tip fields, Figure 10.24, and crack growth rates that do not correlate well with K, Figure 10.25. These initially transient conditions are expected, since the stationary crack tip fields (HRR type fields) which develop during the incubation period are replaced by growing crack tip fields (possibly HR type fields) as the crack extends through the initial creep zone. The numerical results indicate that the period of time after which crack growth becomes self-similar and crack

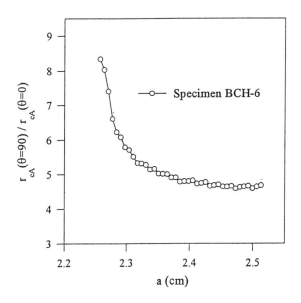

**Figure 10.24** *The ratio of the creep zone size at $\theta = 90°$ with its size at $\theta = 0°$ as check for similitude in the crack tip deformation processes (Ref. 10.26).*

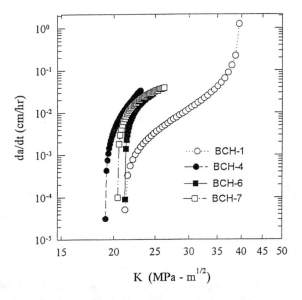

**Figure 10.25** *Creep crack growth rate as a function of K for 2519-T87 aluminum alloy (Ref. 10.26).*

growth rates correlate with K roughly corresponds to the time requried for the crack to grow through the inital creep zone. Transition times which relate to the passage of the initially transient crack tip fields are presented in this section.

Due to nesting of the HR, HRR, and K fields [10.19], the expansion of the creep zone during the initial stage of crack growth is given by equation (10.23) for a material deforming according to an elastic-power law creep constitutive relation. Differentiating equation (10.23) with respect to time shows that $\dot{r}_c$ decreases with $(1/t)^{(n-3)/(n-1)}$. Regardless of the crack growth rate $\dot{a}$, a unique reference time, $t'_g$, will eventually be reached where $\dot{a} = \dot{r}_c$ even when $\dot{a}$ varies with time. The creep zone size at this reference time, denoted by $r_g$, approximates the distance that the crack must extend to leave the initial creep zone in the wake of the growing crack. The time required for the crack to extend a distance $\dot{r}_g$ may be considered as a transition time for growing cracks, designated here as $t_g$.

Examples of the evolution of $\dot{a}$ and $\dot{r}_c$ are shown in Figure 10.26 for constant and varying crack growth rates. Notice that $\dot{r}_c$ for the growing crack is expected to approximately follow the stationary crack solution for $t < t'_g$. However, as the crack tip approaches the reference distance $r_g$, $\dot{r}_c$ becomes dominated by crack growth effects and approaches $\dot{a}$ at time $t_g$. The creep-brittle character of the material ensures that $\dot{a} \approx \dot{r}_c$ during the quasi-steady state crack growth regime which occurs after time $t_g$, assuming the rate of expansion of the creep zone is measured relative to a fixed position along $\theta = 0°$. Here, the quasi-steady state crack growth regime is characterized by crack tip fields which change slowly with increasing crack length.

Closed form expressions for $t'_g$ and $t_g$ can be developed when the crack growth rate is constant. The reference time $t'_g$ is defined as the time when $\dot{r}_c$ is first equal to $\dot{a}$. Equating $\dot{a}$ with $\dot{r}_c$ and solving for time results in:

$$t'_g = \left(\frac{1}{\pi(n-1)}\right)^{\frac{n-1}{n-3}} \left(\frac{EA(n+1)I_n}{2\pi(1-v^2)}\right)^{\frac{2}{n-3}} \left(\frac{1-2v}{\bar{\sigma}_e(\theta=0°)}\right)^{\frac{2(n+1)}{n-3}} \left(\frac{K^2}{\dot{a}}\right)^{\frac{n-1}{n-3}} \quad (10.70)$$

Since the creep zone radius along $\theta = 0°$ is larger than the local amount of crack extension at $t'_g$, additional crack growth is necessary for the crack to grow out of the influence of the initial creep zone. It follows that $\dot{r}_c = (2/(n-1)) r_c / t$. Since $\dot{r}_c = \dot{a}$ and $r_c = r_g$ at time $t_g$ the initial creep zone size can be written in terms of $t_g$ as $r_g = ((n-1)/2) \dot{a} t'_g$. The transition time for growing cracks ($t_g$) corresponds to the time required for the crack to extend a distance $r_g$ through the initial creep zone. For a constant crack growth rate, $r_g = \dot{a} t_g$ resulting in:

$$t_g = \left(\frac{n-1}{2}\right) t'_g \quad (10.71)$$

While equations (10.70) and (10.71) apply to elastic, power-law creeping materials, similar expressions may be derived for materials which deform by primary creep.

The duration of the transient regime of crack growth can also be estimated by qualitatively examining the finite element results. Figure 10.24 indicates that the crack has grown through the inital creep zone at a crack length of 2.3 cm or after 84.5 hours, since the creep zone radius along $\theta = 90°$ changes slowly with crack length after this time. Thus, the finite element results indicate that $t_g = 87.4$ hours is a good estimate of the time required for the onset of quasi-steady state crack growth. Repeating the anlaysis for the other specimens and comparing the finite element results with computed values of $t_g$ yields Figure 10.27. This comparison shows good agreement between $t_g$ and the time required for the crack to grow

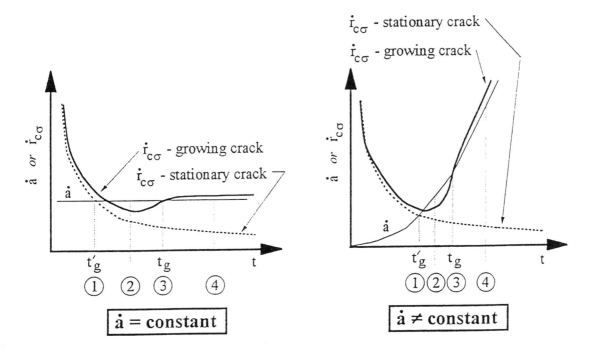

*Figure 10.26* Relationship between $\dot{a}$ and $\dot{r}_c$ for creep-brittle materials: (a) constant crack growth rate and (b) varying crack growth rate. $\dot{r}_c$ is measured relative to a fixed reference. Steady-state conditions are achieved when $\dot{a} \sim \dot{r}_c$ (Ref. 10.26).

through the initial creep zone. Examining $r_g$ for specimens BCH-1, BCH-4, BCH-6, and BCH-7 indicates that quasi-steady state crack growth occurs after 0.89, 0.58, 0.39, and 0.43 mm of crack growth, respectively.

The reference time $t_R$ of Hawk and Bassani [10.19] can also be used to mark the onset of quasi-steady state crack growth conditions. The formulation of this reference time is based on equating the radius of the creep zone for growing cracks (the HR zone which depends on K and ($\dot{a}$)) to the radius of the creep zone for stationary cracks (the HRR zone which depends on K and t) and solving for the time. When the crack growth rate is constant, the expression for $t_R$ is similar to the expresssion for $t_g$ above, since both quantities scale with $(K^2/\dot{a})^{(n-1)(n-3)}$. However, interpretation of $t_R$ is not clear for varying crack growth rates. Moreover, $t_T$ may not exist when crack growth is preceded by an incubation period, since the creep zone predicted from stationary crack mechanics may always be greater than the creep zone predicted from growing crack mechanics.

The aim of examining these transition times is to define the region where quasi-steady state crack growth occurs and $\dot{a}$ correlates well with K. Computing the transition times for all of the aluminum 2519 specimens and plotting only the (K, $\dot{a}$) pairs which occur after the transition time for each specimen results in Figure 10.25. Notice that the tails which occur during the initial stages of crack growth are re-

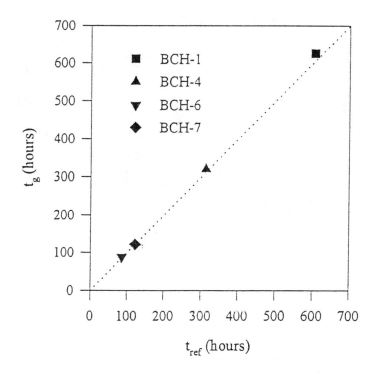

**Figure 10.27** *Comparison of $t_{ref}$ and $t_g$, where $t_{ref}$ is the time for the crack to grow through the initial creep zone obtained from the finite element results (Ref. 10.26).*

moved when compared to Figure 10.28. Consequently, the transition time $t_g$ presented here provides a useful estimate of the time after which crack growth can be better correleated with K, at least for aluminum 2519. The utility of this approach is that the duration of the highly transient regime of creep-brittle crack growth can be estimated based on material properties and the crack growth history, without detailed information from growing crack finite element analyses.

## 10.6 Summary

In this chapter, the crack tip stress fields have been derived in the presence of creep deformation. Initially, the deformation has been modeled with a power-law relationship between strain-rate and stress, ignoring elastic stress redistribution near the crack tip, primary creep, and also effects due to crack growth. The C*-integral has been identified for such conditions as the relevent crack tip parameter for characterizing creep crack growth rates. Methods for estimating C* in test specimens and in components were described. When the additional effects of stress redistribution and primary creep are considered, the C-integral has been shown to characterize the amplitude of the crack tip stress field. However, C-integral cannot be measured at the loading pins under small-scale creep and transition creep conditions. Therefore, an alternative crack tip parameter, $C_t$, was defined. It was shown that under extensive creep conditions, C*-integral, C-integral, and $C_t$ are identical. Under small-scale and transition creep, C* is not defined and $C_t$ and C-integral become distinct from each other. $C_t$ has been shown to uniquely characterize creep crack growth rates for a variety of materials in experimental studies. Methods to incorporate primary creep in the estimation of $C_t$ for test specimens and components were also described.

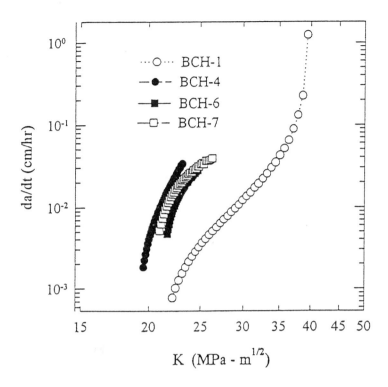

*Figure 10.28* Creep crack growth rate as a function of K for 2519-T87 aluminum after excluding the data up to $t_g$ (Ref. 10.26).

In Sections 10.4 and 10.5, the effects of crack growth on the crack tip stress fields were considered. In Section 10.4, the behavior of creep-ductile materials was considered and it was shown that even though $C_t$ and $C^*$ are derived for stationary cracks, they are quite suitable for characterizing creep crack growth, provided the crack growth is slow, such as in creep-ductile materials. In Section 10.5, the behavior of creep-brittle materials was considered. It was shown that after an initial transient period, the creep crack growth rates may be correlated with K. The growth of cracks in creep-brittle materials still remains a subject in which considerable further reserach is needed.

## 10.6 References

10.1 A. Saxena, "Evaluation of $C^*$ for Characterization of Creep Crack Growth Behavior of 304 Stainless-Steel", in Fracture Mechanics: Twelfth Conference, ASTM STP 700, American Society for Testing and Materials, Philadelphia, 1980, pp. 131-151.

10.2 J.D. Landes and J.A. Begley, "A Fracture Mechanics Approach to Creep Crack Growth", in Mechanics of Crack Growth, ASTM STP 590, American Society for Testing and Materials, 1976, pp. 128-148.

10.3 K.M. Nikbin, G.A. Webster, and C.E. Turner, "Relevance of Nonlinear Fracture Mechanics to Creep Crack Growth", Crack and Fracture, ASTM STP 601, American Society for Testing and Materials, 1976, pp. 47-62.

10.4 N.L. Goldman and J.W. Hutchinson, "Fully Plastic Crack Problems: the Center Cracked Strip Under Plane Strain", International Journal Solids and Structures, Vol. 11, 1975, pp. 575-591.

10.5 H. Riedel and J.R. Rice, "Tensile Cracks in Creeping Solids", Fracture Mechanics: Twelfth Conference, ASTM STP 700, American Society for Testing and Materials, Philadlphis, 1980, pp. 112-130.

10.6 K. Ohji, K. Ogura, and S. Kubo, "Stress-Strain Fields and Modified J-Integral in the Vicinity of the Crack Tip Under Transient Creep Conditions", Japan Society of Mechanical Engineering, No. 790-13, 1979, pp. 18-20 (in Japanese).

10.7 N. Adefris, "Creep-Fatigue Crack Growth Behavior of 1Cr-1Mo-0.25V Rotor Steel", Ph.D. dissertation, Georgia Institute of Technology, 1993.

10.8 J.L. Bassani and F.L. McClintock, "Creep Relaxation of Stress Around a Crack Tip", International Journal of Solids and Structures, Vol. 17, 1981, pp. 79-89.

10.9 R. Ehlers and H. Riedel, "A Finite Element analysis of Creep Deformation in a Specimen Containing a Macrosopic Crack", Advances in Fracture Research, Vol. 2 (edited by D. Francois), ECF-5, Pergamon Press, 1991, pp. 691-698.

10.10 J.L. Bassani, D.E. Hawk, and A. Saxena, "Evaluation of $C_t$ Parameter for Characterizing Creep Crack Growth Rate in the Transient Regime", Nonlinear Fracture Mechanics: Time-Dependent Fracture Mechanics, Vol. I, ASTM STP 995, American Society for Testing and Materials, Philadelphia, 1989, pp. 7-29.

10.11 C. Leung, D.L. McDowell, and A. Saxena, "A Numerical Study of Nonsteady-State Creep at Stationary Crack Tips", Nonlinear Fracture Mechanics: Time-Dependent Fracture Mechanics, Vol. I, ASTM STP 995, American Society for Testing and Materials, Philadelphia, 1989, pp. 141-158.

10.12 A. Saxena, "Creep Crack Growth Under Nonsteady-State Conditions", in Fracture Mechanics: Seventeenth Volume, ASTM STP 905, American Society for Testing and Materials, 1986, pp. 185-201.

10.13 A. Saxena, "Creep Crack Growth in Ductile Materials", Engineering Fracture Mechanics, Vol. 40, 1991, pp. 721-736.

10.14 A. Saxena, "Mechanics and Mechanisms of Creep Crack Growth", in Fracture Mechanics: Microstructure and Micromechanisms S.V. Nair et al., editors, ASM International, Metals Park, Ohio, 1989, pp. 283-334.

10.15 A. Saxena, K. Yagi, and M. Tabuchi, "Crack Growth Under Small Scale and Transition Creep Conditions in Creep-Ductile Materials", Fracture Mechanics: Twenty-Fourth volume, ASTM STP 1207, American Society for Testing and Materials, Philadelphia, 1994, pp. 481-497.

10.16 H. Riedel, "Creep Deformation of Crack Tips in Elastic-Viscoplastic Solids", Journal of Mechanics and Physics of Solids, Vol. 29, 1981, pp. 35-49.

10.17   C.P. Leung and D.L. McDowell, "Inclusion of Primary Creep in the Estimation of the $C_t$ Parameter", International Journal of Fracture, Vol. 46, 1990, pp. 81-106.

10.18   C.Y. Hui and H. Riedel, "The Asymptotic Stress and Strain Field Near the Tip of a Growing Crack Under Creep Conditions", International Journal of Fracture, Vol. 17, 1981, pp. 409-425.

10.19   D.E. Hawk and J.L. Bassani, "Transients Crack Growth Under Creep Conditions", Journal of Mechanics and Physics of Solids, Vol. 34, 1986, pp. 191-212.

10.20   H. Riedel and W. Wagner, "The Growth of Macroscopic Cracks in Creeping Materials", Advances in Fracture Research, Pergamon Press, 1981, pp. 683-690.

10.21   A. Saxena, H.A. Ernst, and J.D. Landes, "Creep Crack Growth Behavior of 316 Stainless Steel", International Journal of Fracture, Vol. 23, 1983, pp. 245-257.

10.22   D.E. Hall, "Analysis of Crack Growth in Creep-Brittle Materials", Ph.D. Dissertation, Georgia Institute of Technology, 1995.

10.23   C.Y. Hui, "The Mechanics of Self-Similar Crack Growth in an Elastic Power-Law Creeping Material", International Journal of Solids and Structures, Vol. 22, 1986, pp. 357-372.

10.24   C.Y. Hui, "Steady-State Crack Growth in Elastic Power Law Creeping Materials", Elastic Plastic Fracure, Vol. I, ASTM STP 803, American Society for Testing and Materials, Philadelphia, 1983, pp. 573-593.

10.25   Y. Gill, A. Saxena, and J.M. Bloom, " Creep Crack Growth Behavior of a SA-106C Carbon Steel", unpublished research, Georgia Institute of Technology, Altanta, Georgia, 1994.

10.26   D.E. Hall, B.C. Hamilton, D.L. McDowell, and A. Saxena, "Creep Crack Growth Behavior of Aluminum Alloy 2519: Part II - Numerical Analysis", in Effects of Temperature on Fatigue and Fracture, ASTM STP 1297, 1997, pp. 19-36.

## 10.7 Exercise Problems

10.1   Descirbe the development of crack tip stress fields for a stationary crack subjected to a constant load at elevated temperature. Assume that the initial response is in the regime of small-scale yielding.

10.2   Show by analogy to J, a path-independent integral $C^*$ can be defined for creeping cracked bodies. What are the assumptions for the validity of $C^*$ and why is it suitable for characterizing creep crack growth? What is the significance of path-independence?

10.3   Prove that:

$$C^* = -\frac{1}{B}\frac{dU^*}{da}$$

where B = thickness and $U^*$ is the stress-power.

**10.4** Derive equations (10.8) and (10.9) for estimating $C^*$ for CT and CCT specimens, respectively.

**10.5** Why is $C^*$ path-dependent in the small-scale and transition creep region? What is the implication of this path-dependency?

**10.6** Show that the transition time, $t_T$, is given by:

$$t_T = \frac{K^2 (1-\upsilon^2)}{E(n+1)C^*}$$

State all conditions for which the above equation is appropriate.

**10.7** Derive, on your own, the equations for crack tip stress and strain rate fields as a function of time for small-scale creep conditions assuming that the crack is stationary and the material undergoes elastic and power-law creep behavior.

**10.8** Derive an expression for estimating the transition time in an elastic, power-law creep material for a compact specimen of width, W, thickness, B, and crack size, a. Assume:

Applied load is P
Creep deformation constants are A and n
Elastic modulus is E

**10.9** Compare the transition times of CT and CCT specimens loaded to the same value of K assuming that the a, W, and B are also the same for the specimens of the two geometries. Assume elastic, power-law creep material.

**10.10** What is the $C_t$ parameter and in what way is it different from $C^*$? Why is $C_t$ suitable for characterizing creep crack growth rate under a wide range of deformation conditions?

**10.11** Show on your own that $C_t$ under small-scale-creep, $(C_t)_{ssc}$, is given by the following equation:

$$(C_t)_{ssc} = \frac{P\dot{V}_c}{BW} F'/F$$

where the symbols have their usual meaning.

**10.12** Derive an expression for estimating $C_t$ for double-edge-notch specimen subjected to uniform tension for an elastic plus power-law creep material.

**10.13** What modifications are necessary to the result in Problem 10.12 to account for primary creep?

**10.14** Repeat Example Problem 10.7 for a SENT specimen assuming the same width and thickness. The remote stress is 150 MPa and the a/W = 0.5.

# CHAPTER 11

# CREEP CRACK GROWTH

Creep crack growth in metals is an important design consideration for several high temperature components. In the previous chapter, we had identified the $C_t$ parameter, $C^*$-integral, and the stress intensity parameter, K, as crack tip parameters suitable for characterizing creep crack growth in metals. The $C_t$ parameter and the $C^*$-integral are suitable for characterizing crack growth in creep-ductile metals in which the crack growth is accompanied by substantial amounts of creep deformation and the effects due to crack growth can be neglected. Conversely, K is more suitable for characterizing steady-state creep crack growth in creep-brittle materials in which the crack tip and the creep zone boundary move at the same rate. In this chapter, we will be concerned with techniques for measurement of creep crack growth rates in creep-ductile and creep-brittle materials and the micro mechanics of steady-state and early crack growth in metals. We will also examine some special considerations for creep crack growth in weldments. In our discussion of creep-ductile materials, we will be correlating the creep crack growth rate behavior with $C_t$ because $C^*$ is a subset of $C_t$, valid only for extensive creep conditions. Since $C_t$ was shown to unify the behavior under small-scale, transition and extensive creep conditions, it will be the parameter of choice for the discussion in this chapter that focuses on material behavior.

Figure 11.1 shows the creep crack growth rates as a function of $C_t$ for a variety of Cr-Mo and Co-Mo-V steels at several temperatures [11.1]. These data were collected from several sources [11.2-11.7] in the literature and re-analyzed so they could be plotted on a common basis. Several interesting trends are observed from the data. The most important of these trends is influence of temperature. Usually, one associates higher creep rates with higher temperatures. Then, why is the da/dt vs. $C_t$ relationship, at least to a first degree of approximation, independent of temperature? Figure 11.2 (a and b) shows data on 1 Cr-1Mo-0.25V and 304 stainless steel, respectively, at various temperatures. These figures show little influence of temperature on the creep crack growth behavior [11.8,11.9]. Thus, this trend is quite general. The second question, perhaps equally important, does chemical composition and microstructure influence the da/dt vs. $C_t$ relationship? Judging from the somewhat limited data in Figure 11.1, one may conclude that the influence of chemical composition and microstructure within a given alloy system is also minimal. The engineering significance of these correlations is phenomenal because if a single da/dt vs. $C_t$ trend applies to all Cr-Mo-V and Cr-Mo steels and at different temperatures, the engineering analysis of creep crack growth is considerably simplified. However, the limitations of such trends must be clearly understood. These will be explored in Section 11.2.

The other observation that is evident from Figures 11.1 and 11.2 is that da/dt and $C_t$ are related by the following relationship:

$$da/dt = b C_t^q \qquad (11.1)$$

where $b$ and $q$ are material constants obtained from regression of the data and are related to the intercept and slope, respectively, of the da/dt vs. $C_t$ relationship on a log-log plot. Such a relationship between da/dt and $C_t$ is now widely accepted. The values of $b$ and $q$ can change from material to material.

Figure 11.1 also demonstrates that there is remarkable consistency between the measurements of creep crack growth from a variety of laboratories. Thus, the correlation between da/dt and $C_t$ must be robust enough to withstand minor variations in experimental procedures that may have been present in the test techniques used by the various laboratories. All the data included in Figures 11.1 and 11.2 date prior to the development of the ASTM standard [11.10] on creep crack growth testing. Thus, the availability of the ASTM standard should promote an even higher level of comparability within the data developed by different laboratories.

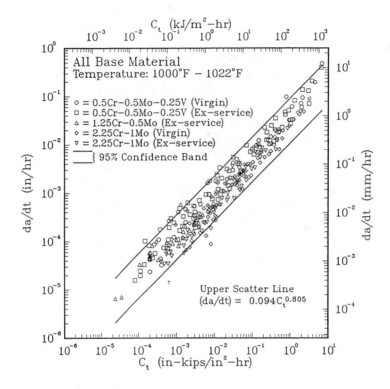

**Figure 11.1** *Creep crack growth behavior of Cr-Mo and Cr-Mo-V materials. The plot includes ex-service and new materials data (Ref. 11.1).*

## 11.1 Test Methods for Characterizing Creep Crack Growth

The detailed test procedure for characterizing creep crack growth for creep-ductile materials is described in ASTM standard E-1457-92 [11.11]. In this section, some of the important steps in testing are described. Data reduction steps for creep-ductile and creep-brittle materials are also described.

### 11.1.1 Overall Test Method

To characterize creep crack growth behavior, precracked specimens of compact type (CT) geometry or other geometries are heated to the desired test temperature and then subjected to a constant load. The crack length and the load-line deflection are both recorded as a function of time for the duration of the test. The crack length is determined by either dc potential drop method or by compliance change. The latter is regarded as an alternate method when testing high electrical conductivity materials (such as Al alloys) which can give rise to fluctuations in the output signal during the potential drop measurement. Also, the elastic compliance method is valid only for creep-brittle materials because, for creep-ductile materials, the specimens must be partially unloaded to measure elastic compliance. Such unloading is not possible when using dead-weight creep machines which are preferred over servo-hydraulic machines for creep crack growth testing. The test is continued until either the specimen fails by rapid fracture or until sufficient crack growth has occurred. Interrupting the test prior to failure allows the opportunity for the final visually measured crack length to be compared with the nonvisual techniques. If the specimen fails, the terminal point of creep crack growth may not be discernable from the fast fracture surface and this comparison may not be possible. The details of different steps are presented in the following sections.

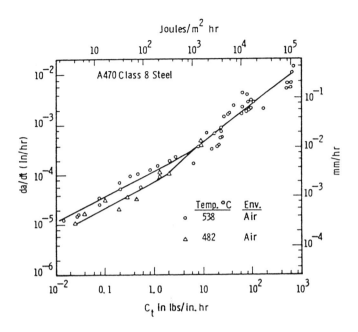

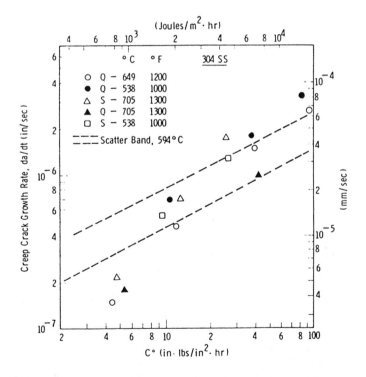

*Figure 11.2* Creep crack growth behavior of: (a) 1Cr-1Mo-0.25V steel at various test temperatures (Ref. 11.8) and (b) 304 stainless steel also at various temperatures (Ref. 11.9).

## 11.1.2 Fatigue Precracking

Following the ASTM standard E1457 [11.11], the specimens are precracked at room temperature utilizing a fatigue load. The precracking equipment should provide a symmetrical load distribution with respect to the machine notch and control the maximum K-level to within ±5%. Any load frequency which meets these requirements may be used for precracking. The precrack length should be at least 2.54mm (0.1 in), and the selected K-levels for precracking should be less than the intended test K-levels. The precrack length is measured from the tip of the machine notch on the front and back surfaces of the specimens to within 0.1mm (0.004 in). Precrack front requirements are outlined in detail in the ASTM standard. Following precracking, the specimens are side grooved 10% of the thickness on each side of the crack plane.

## 11.1.3 Test Equipment

Any machine which can maintain a constant, axially applied load for extended time periods is suitable for creep crack growth testing. Direct lever, dead weight type creep machines are most suitable for this application. If these creep machines are unavailable, servohydraulic machines may be used; however, these machines have a greater risk of overloading the specimens during the test. The temperature must not vary by more than ±1°C from the desired test temperature for the duration of the test. A setup for creep crack growth testing is schematically shown in Figure 11.3.

## 11.1.4 Loading and Unloading Procedure

In accordance with E1457, a small portion of the load, no more than 10%, may be applied while heating the specimens to maintain proper axial alignment. The load should be added carefully and incrementally to avoid shock loading the specimens or overloading them due to inertia. The entire load should be added within a few minutes to minimize the amount of time-dependent deformation accumulating ahead of the crack tip.

## 11.1.5 Measurements During Testing

**Load-Line Deflection** The change in load-line deflection with time must be accurately monitored during the test. This can be accomplished using a LVDT (linear voltage differential transducer) or a DVDT (direct voltage displacement transducer) extensometer attached to knife blades which are inserted into the machined knife edges of the notch. The signal from the extensometer can be continuously recorded utilizing a strip chart recorder or can be digitally recorded using a suitable data acquisition device (e.g., voltmeter). The frequency of data collection must be such that deflection measurements can be made in increments of 0.01mm (0.0004 in) or less.

**Measurement of Crack Length** Both the dc potential drop method and compliance change have been successfully used to measure the crack length during creep crack growth testing [11.11-11.12]. Any non-visual technique which can resolve crack extensions of 0.1mm (0.0004 in) or less is suitable for creep crack growth testing. If the dc potential drop method is selected, then the suggested electrical lead locations are shown in Figure 11.4. The leads are attached to the specimens with screws inserted into machined holes. Welding the leads to the specimens is a possibility for steels but is not easy for Al and Ti alloys. Output leads should be nearly equivalent in length to account for the resistance of the wire and should have a diameter near 2mm (0.08 in). Nickel and copper wires have been successfully used as lead material. Oxidation behavior, however, is an important consideration when choosing the lead material. A complete circuit diagram of the potential drop method is provided in Figure 11.5. The circuit can be modified to accommodate several specimens receiving current from a single source. In this situation, when a specimen fails, a trip switch should replace the specimen with a short circuit so that the other tests can continue. A constant dc current source with a stability of ±0.1% is required. A dc current level of 2A

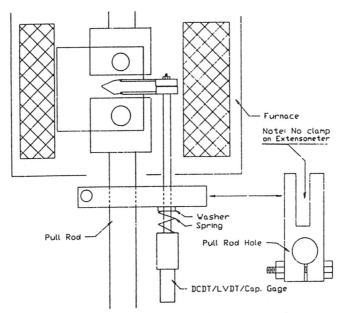

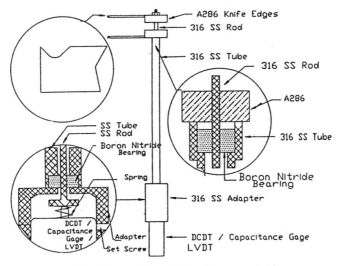

**Figure 11.3** *(a) A schematic of creep crack growth test set-up. (b) Details of the extensometer for measuring load-line deflection.*

has been used for testing, but higher current levels may be pursued; however, these higher levels can heat the lead wires and accelerate the oxidation process. The current level should be routinely monitored during the test and adjustments should be made if any variation occurs. For low electrical resistance materials, the reversed dc potential technique is highly recommended. In the reversed potential technique, the sign of the input current is reversed using a solid-state switch. The range of the potential change is measured as output. This technique not only removes intervening effects of thermal voltages and amplifier drifts that occur over long periods, but also multiplies the output voltage by two for the same magnitude of current.

In Figure 11.4, the leads are configured to relate the output voltage to the crack size using Johnson's formula [11.11]:

$$a_i = \frac{2W}{\pi} \arccos\left\{ \frac{\cosh(\pi y/2W)}{\cosh\left[\left(\frac{U_i}{U_0}\right) \text{arccosh}\left[\frac{\cosh(\pi y/2W)}{\cos(\pi a_0/2W)}\right]\right]} \right\} \qquad (11.2)$$

where $a_i$ is the instantaneous crack length, W is the specimen width, y is the half distance between output leads, $a_0$ is the reference crack size, $U_i$ is the instantaneous voltage output corresponding to the crack length $a_i$, minus $V_{th}$, and $U_0$ is the voltage output corresponding to the crack length $a_0$ minus $V_{th}$. A digital voltmeter or strip chart recorder can be used to measure the output voltage. $V_{th}$ is the thermal voltage, which is the voltage output in the absence of an input current. The thermal voltage is simply measured by turning off the power supply for a short period and recording the output voltage. The thermal voltage should be measured at the beginning of a test and at frequent intervals during the test, particularly during the latter stages. Fluctuation in the thermal voltage is observed for high conductivity materials. These fluctuations can be of the same magnitude as the voltage changes which accompany crack extension in such materials and could mask this information. Use of the reversing potential method, as mentioned earlier, eliminates this problem.

The change in elastic compliance may also be used to calculate the crack length [11.13]:

$$a_i/W = 1.000196 - 4.06319 u_{LL} + 11.242 u_{LL}^2 - 106.043 u_{LL}^3 + 464.335 u_{LL}^4 - 650.677 u_{LL}^5 \qquad (11.3)$$

where

$$U_{LL} = \frac{1}{[B_e E C_i]^{1/2} + 1} \qquad (11.4)$$

$B_e$ = effective specimen thickness:

$$(B_e = B - (B - B_N)^2 / B) \qquad (11.5)$$

$C_i$ is the elastic compliance of the specimen at crack length $a_i$, B is the specimen thickness, and $B_N$ is the net specimen thickness. The compliance may be measured by partially unloading the specimen or by assuming that all deflection changes are elastic for creep resistant, creep-brittle materials. In the latter case, compliance change may be more effective for determining crack extension rather than the crack length itself. For more detail on using elastic compliance, the reader is referred to the Annex of ASTM standard E647 [11.14].

Creep Crack Growth 369

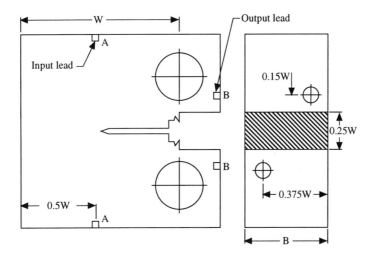

*Figure 11.4* Suggested lead locations for the input current and measurement of the output potential for measuring crack length by electric potential technique.

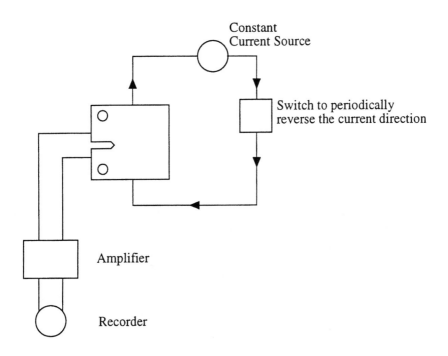

*Figure 11.5* Schematic of a circuit diagram for implementing the electric potential technique.

## 11.1.5 Post Test Measurements

Whichever nonvisual technique is used, visual measurements of the final crack length should be made as frequently as possible to ensure the accuracy of the nonvisual mehthod used. Specimens which do not fracture can be opened by fatigue loading at room temperature. The fatigue fracture will clearly delineate the boundary between the creep crack and the fatigue crack such that the final crack length can be measured. Any loading frequency and load amplitude which is convenient may be used in this role. Similarly, the precrack length should be measured. The length of the precrack and the final crack should be measured at nine points along the specimen thickness and an average should be computed.

## 11.1.6 Analysis of Data for Creep-Ductile Materials

The crack length from the electric potential reading is determined using equation (11.2) or an equivalent calibration function derived for the specimen geometry being tested and the specific input and output lead locations. Thus, a table of values consisting of time, load-line deflection, and crack size are available as the raw data for further processing. For detailed validation requirements, the reader is referred to ASTM standard [11.10]. Since elastic compliance is not recommended for use with creep-ductile materials, we are restricting our discussion in this section to the electric potential method.

**Calculation of Crack Growth Rates and Deflection Rates** Crack length and deflection vs. time data were numerically processed to obtain rates. The two techniques used for calculating rates were the secant method and the seven-point incremental polynomial method recommended in ASTM standard E-647 [11.14] for calculating rates from fatigue crack growth data. In the secant method, the crack growth rate, da/dt (or $\dot{a}$), and the deflection rate, dV/dt (or $\dot{V}$), are calculated as follows:

$$\left(\frac{da}{dt}\right)_i = \frac{a_{i+1} - a_{i-1}}{t_{i+1} - t_{i-1}} \tag{11.6}$$

$$\left(\frac{dV}{dt}\right)_i = \frac{V_{i+1} - V_{i-1}}{t_{i+1} + t_{i-1}} \tag{11.7}$$

The seven-point incremental polynomial method involves fitting a second order polynomial through a group of seven successive data points. The rates are then obtained by differentiating the fitted equation. The characteristic equations for this procedure are given below:

$$\hat{a}_i = b_0 + b_1 \left(\frac{t_i - C_1}{C_2}\right) + b_2 \left(\frac{t_i - C_1}{C_2}\right)^2 \tag{11.8}$$

where

$$-1 < \left(\frac{t_i - C_1}{C_2}\right) < +1$$

where $\hat{a}_i$ is the fitted value of $a_i$ corresponding to time $t_i$. $b_0$, $b_1$, and $b_2$ are regression parameters which are determined by the least-squares method (that is the minimization of the square of the deviations between the observed and the fitted values of crack length) over the range, $a_{i-3} < a < a_{i+3}$. The parameters $C_1 = \frac{1}{2}(t_{i-3} + t_{i+3})$ and $C_2 = \frac{1}{2}(t_{i-3} - t_{i-3})$, are used to scale the input data, thus avoiding numerical difficulties in determining the regression parameters. The crack growth rate at time, $t_1$, is obtained from the derivative of the above parabola which is given by the following expression:

$$\frac{da_i}{dt} = \dot{a}_i = \frac{b_1}{C_2} + \frac{2b_2(t_i - C_1)}{C_2^2} \tag{11.9}$$

An identical set of equations can also be written for obtaining the deflection rates, $\dot{V}$. In using the seven-point incremental polynomial method above, the first three and last three data points are excluded in the estimation of the rates. These data are frequently too valuable to discard. Hence, it is common to calculate the rates from these data by the secant method and include them in the plots of da/dt vs. $C_t$.

The dV/dt data obtained from the above step is further processed to calculate $dV_c/dt$, the deflection rate due to creep deformation, using the following equation [11.15-11.16]:

$$\dot{V}_c = \dot{V} - \frac{\dot{a}B}{P}\left[\frac{2K^2}{E} + (m+1)J_p\right] \tag{11.10}$$

Equation (11.10) reduces to equation (10.66) when the plastic part of J, $J_p$, is negligible. $m$ is the plasticity exponent as defined in Chapter 3. If the instantaneous loading results in considerable plasticity, the full equation (11.10) must be used. The methods for obtaining $J_p$ were discussed in Chapter 5 for various geometries.

**Determination of $C_t$** For $t \ll t_T$ (transition time), equation (10.39) is used to calculate the value of $C_t$. Recall that this equation is valid for any type of creep constitutive law. In other words, the contribution from both primary and secondary creep are included in the measured value of $\dot{V}_c$. For $t \geq t_T$, equation (10.59) can be used to calculate $C_t$ which is identical to the value of $C^*(t)$ for CT specimen. Similar expressions for other geometries can be derived using methods discussed in Chapter 10. However, for conditions ranging from small-scale to extensive creep, the derivation of an appropriate equation is presented. This derivation is strictly valid only for elastic, power-law creep but arguments for extending it to materials with significant primary creep will also be made.

Equations (10.43) and (10.46) are two forms of equations that can be used to estimate $C_t$ for a wide range of creep conditions. Equating the two yields:

$$\frac{P\dot{V}_c}{BW}(F'/F) - \left(\frac{F'/F}{\eta} - 1\right)C^* = \left[\left(\frac{t_T}{t}\right)^{(n-3)/(n-1)} + 1\right]C^*$$

or

$$\frac{P\dot{V}_c}{BW}(F'/F) = \left[\frac{F'/F}{\eta} + \left(\frac{t_T}{t}\right)^{\frac{n-3}{n-1}}\right]P\frac{\dot{V}_{ss}}{BW}$$

$$\dot{V}_{ss} = \frac{(F'/F)\dot{V}_c}{(F'/F)/\eta + (t_T/t)^{(n-3)/(n-1)}}$$

Thus:

$$C_t = \frac{P\dot{V}_c}{BW}(F'/F)\left[1 - \left(\frac{F'/F}{\eta} - 1\right)\frac{1}{F'/F/\eta + (t_T/t)^{(n-3)/(n-1)}}\right] \qquad (11.11)$$

Although equation (11.11) has been derived for elastic, power-law creep, it is also approximately valid for materials in which primary creep deformation is significant. This is because $\eta$ is a much stronger function of a/W than the value of $n$ and $\dot{V}_c$ includes any contributions from primary creep deformation to the value of $C_t$. Thus, equation (11.11) can be used over the entire range of deformation levels to estimate $C_t$. The value of $t_T$ can be calculated using equation (10.25).

Only data which satisfy the condition:

$$\dot{V}_c / \dot{V} \geq 0.8 \qquad (11.12)$$

are considered valid for correlating with $C_t$. Recall, the rationale for this condition from Section 10.4.2. The da/dt vs. $C_t$ data can be plotted on a log-log plot and the values of constants b and q can be obtained from a linear regression of the data. This procedure is illustrated with an example.

**Example Problem 11.1**

A 25.4mm thick and 50.8mm wide CT specimen of 1Cr-1Mo-0.25V steel is tested under constant load conditions. The initial crack size after precracking is 26.25mm and the applied load is 26.7kN. The measurements in Table 11.1 were made during the test. The output voltage from electric potential is represented by U and is expressed in micro-volts ($\mu$V). The thermal voltage during the test was constant and was found to be 3$\mu$V. The deformation constants for this material at 538°C are given as follows:

E = 183.4 GPa       $\sigma_0$ = 457.8 MPa    m = 21.6          $\alpha$ = 0.686

$\varepsilon_0$ = 2.49 x $10^{-3}$      A = 1.16 x $10^{-24}$ $MPa^{-8}$ $hr^{-1}$        n = 8

Reduce the data in the form of da/dt vs. $C_t$. You may assume that primary creep is negligible.

*Solution:*

Implementing equations (11.6) and (11.7) for the first three and last three data points and equations (11.8) and (11.9) for the rest of the data set, we get the results shown in Table 11.2. Knowing the value of $\dot{a}$, we next calculate $J_p$ and then combine it with the calculated value of K from 11.2 to es-

*Table 11.1 Measurements Made During the Test in Example Problem 11.1*

| Time (hr) | U (μV) | a(mm) | V(mm) |
|---|---|---|---|
| 0 | 384 | 26.25 | 0 |
| 120 | 386 | 26.36 | 0.254 |
| 380 | 388 | 26.47 | 0.305 |
| 450 | 390 | 26.59 | 0.330 |
| 550 | 392 | 26.71 | 0.356 |
| 620 | 395 | 26.88 | 0.381 |
| 830 | 399 | 27.11 | 0.432 |
| 880 | 402 | 27.28 | 0.457 |
| 950 | 405 | 27.45 | 0.483 |
| 1000 | 407 | 27.57 | 0.508 |
| 1120 | 413 | 27.91 | 0.584 |
| 1170 | 415 | 28.02 | 0.609 |
| 1290 | 425 | 28.56 | 0.737 |
| 1340 | 428 | 28.73 | 0.775 |
| 1390 | 430 | 28.84 | 0.838 |
| 1460 | 433 | 29.00 | 0.927 |
| 1530 | 440 | 29.36 | 1.054 |
| 1630 | 450 | 29.88 | 1.372 |
| 1680 | 456 | 30.19 | 1.626 |
| 1720 | 475 | 31.13 | 2.66 |

timate $\dot{V}_c$ from equation (10.66). These calculations are presented in Table 11.3, where:

$$K_{\it eff} = \left(\left(\frac{K^2}{E}(1-v^2) + J_p\right)\frac{E}{(1-v^2)}\right)^{1/2}, \quad v = 0.3$$

From Table 11.3, several observations are made. It appears that data sets 1 to 15 meet the requirement that $\dot{V}_c / \dot{V} \geq 0.8$. Therefore, $C_t$ can be estimated for those points using equation (11.11). For points 16 to 18, the above condition is not satisfied. Thus, these data are not considered further. The negative $\dot{V}_c / \dot{V}$ value for data point #18 needs further explanation. Note that $\dot{V}_c$ is an experimentally measured value, and in order to estimate $\dot{V}_c$ we subtract from $\dot{V}_c$ the analytically calculated values of $\dot{V}_e + \dot{V}_p$. There can often be discrepancies between calculated and experimentally measured deflection rates and if the calculated deflection rates are over predicted, negative values of $\dot{V}_c$ can result when the instantaneous deflection rates due to elastic and plastic de-

**Table 11.2** First Part of the Solution for Example Problem 11.1

| Time (hr) | $\dot{V}_c$ (m/hr) | a (mm)[1] | da/dt (m/hr) | K (MPa√m) |
|---|---|---|---|---|
| 60 | $2.11 \times 10^{-6}$ | 26.305 | $4.49 \times 10^{-7}$ | 47.64 |
| 250 | $1.92 \times 10^{-7}$ | 26.401 | $1.69 \times 10^{-6}$ | 47.93 |
| 550 | $2.74 \times 10^{-7}$ | 26.71 | $1.31 \times 10^{-6}$ | 48.88 |
| 620 | $2.95 \times 10^{-7}$ | 26.82 | $1.56 \times 10^{-6}$ | 49.22 |
| 830 | $3.53 \times 10^{-7}$ | 27.18 | $2.00 \times 10^{-6}$ | 50.39 |
| 880 | $4.26 \times 10^{-7}$ | 27.27 | $2.2 \times 10^{-6}$ | 50.68 |
| 950 | $4.74 \times 10^{-7}$ | 27.43 | $2.3 \times 10^{-6}$ | 51.22 |
| 1000 | $5.38 \times 10^{-7}$ | 27.55 | $2.72 \times 10^{-6}$ | 51.63 |
| 1120 | $7.19 \times 10^{-7}$ | 27.90 | $3.2 \times 10^{-6}$ | 52.86 |
| 1170 | $8.05 \times 10^{-7}$ | 28.07 | $3.3 \times 10^{-6}$ | 53.47 |
| 1290 | $1.02 \times 10^{-6}$ | 28.50 | $3.25 \times 10^{-6}$ | 55.08 |
| 1340 | $1.17 \times 10^{-6}$ | 28.69 | $3.45 \times 10^{-6}$ | 55.82 |
| 1390 | $1.50 \times 10^{-6}$ | 28.84 | $3.78 \times 10^{-6}$ | 56.41 |
| 1460 | $1.89 \times 10^{-6}$ | 29.05 | $3.78 \times 10^{-6}$ | 57.26 |
| 1530 | $3.83 \times 10^{-6}$ | 29.25 | $5.53 \times 10^{-6}$ | 58.09 |
| 1580 | $5.2 \times 10^{-6}$ | 29.88 | $6.274 \times 10^{-6}$ | 60.84 |
| 1655 | $6.2 \times 10^{-6}$ | 30.19 | $6.15 \times 10^{-6}$ | 62.28 |
| 1700 | $2.60 \times 10^{-5}$ | 31.13 | $2.35 \times 10^{-5}$ | 67.02 |

Note: The times and crack sizes for the first two and last three sets of data are averages of the intervals over which the measurements were made and rates were calculated.

[1] The values of $a$ in this column are predicted values from the fit at the indicated time.

flections dominate over the creep deformation rates. This is usually an indication that stable crack growth has set in or instability has occurred in the specimen and creep crack growth is no longer pertinent. Negative $\dot{V}_c$ values can also be caused by history effects resulting from crack tip creep deformation when rapid acceleration in the crack growth rates occurs at the end of the test.

Returning to the data analysis, we proceed with the estimation of $C_t$. The value of $\eta$ for CT specimens is defined by equation (10.45) and is given by comparing equations (10.8) and (10.45) as:

$$\eta = \left( \frac{2}{(1-a/W)} + 0.522 \right) \frac{n}{n+1}$$

*Table 11.3 Calculations in the Solution for Example Problem 11.1*

| | Time (hr) | $\dot{V}_c$ (m/hr) | $J_p$ (J/m²) | $\dot{V}_c / \dot{V}$ | $K_{eff}$ (MPa√m) |
|---|---|---|---|---|---|
| 1 | 60 | 2.09 x 10⁻⁶ | 0 | 0.99 | 47.64 |
| 2 | 250 | 1.53 x 10⁻⁷ | 0 | 0.80 | 47.93 |
| 3 | 550 | 2.41 x 10⁻⁷ | 4.9 | 0.88 | 48.89 |
| 4 | 620 | 2.57 x 10⁻⁷ | 9.8 | 0.87 | 49.24 |
| 5 | 830 | 3.00 x 10⁻⁷ | 15.0 | 0.85 | 50.42 |
| 6 | 880 | 3.66 x 10⁻⁷ | 20.1 | 0.86 | 50.72 |
| 7 | 950 | 4.07 x 10⁻⁷ | 30.5 | 0.86 | 51.28 |
| 8 | 1000 | 4.62 x 10⁻⁷ | 41.0 | 0.86 | 51.71 |
| 9 | 1120 | 6.18 x 10⁻⁷ | 89.3 | 0.86 | 53.03 |
| 10 | 1170 | 7.00 x 10⁻⁷ | 138.3 | 0.87 | 53.73 |
| 11 | 1290 | 8.87 x 10⁻⁷ | 385.0 | 0.87 | 55.78 |
| 12 | 1340 | 1.018 x 10⁻⁶ | 615.3 | 0.87 | 56.92 |
| 13 | 1390 | 1.30 x 10⁻⁶ | 891.1 | 0.87 | 57.98 |
| 14 | 1460 | 1.644 x 10⁻⁶ | 1499 x 10³ | 0.87 | 59.84 |
| 15 | 1530 | 3.33 x 10⁻⁶ | 2.46 x 10³ | 0.87 | 62.21 |
| 16 | 1580 | 3.38 x 10⁻⁶ | 1.19 x 10⁴ | 0.65 | 78.10 |
| 17 | 1655 | 2.48 x 10⁻⁶ | 2.6 x 10⁴ | 0.40 | 95.61 |
| 18 | 1700 | -12.5 x 10⁻⁵ | 2.97 x 10⁵ | -4.81 | 253.61 |

and

$$F = \frac{K}{P} BW^{1/2} = \frac{2 + a/W}{(1-a/W)^{3/2}} f$$

where

$$f = .886 + 4.64\,(a/W) - 13.32\,(a/W)^2 + 14.72\,(a/W)^3 - 5.6\,(a/W)^4$$

$$F' = \frac{dF}{d(a/W)}$$

and

$$F'/F = \left[\frac{1}{2+a/W} + \frac{3}{2(1-a/W)}\right] + f'/f,$$

$$f = 4.64 - 26.64\ (a/W) + 44.16\ (a/W)^2 - 22.4\ (a/W)^3$$

The equation for estimating $C^*$ for CT specimens was given in equation (10.10). The values of $C^*$ and K are substituted in equation (10.25) to obtain the value of $t_T$ which can be substituted into equation (11.11) along with other parameters to obtain $C_t$. Alternatively, equation (10.46) can also be used to estimate $C_t$. All of these are identical equations for estimating $C_t$ expressed in different forms. Table 11.4 provides all of these calculated values.

*Table 11.4 Calculated Values for Example Problem 11.1*

|    | t(hrs) | $t_T$ (hrs) | $C^*$ (J/m²hr) | $C_t$ (J/m²hr) | da/dt(m/hr) |
|----|--------|-------------|----------------|----------------|-------------|
| 1  | 60     | 215         | 5.82           | 152            | $4.49 \times 10^{-7}$ |
| 2  | 250    | 200         | 6.32           | 11.8           | $1.69 \times 10^{-6}$ |
| 3  | 550    | 160         | 8.25           | 19.8           | $1.31 \times 10^{-6}$ |
| 4  | 620    | 147         | 9.08           | 21.2           | $1.56 \times 10^{-6}$ |
| 5  | 830    | 113         | 12.4           | 25.6           | $2 \times 10^{-6}$ |
| 6  | 880    | 105         | 13.5           | 31.4           | $2.2 \times 10^{-6}$ |
| 7  | 950    | 93.4        | 15.5           | 35.6           | $2.3 \times 10^{-6}$ |
| 8  | 1000   | 85.3        | 17.2           | 40.2           | $2.72 \times 10^{-6}$ |
| 9  | 1120   | 65.4        | 23.6           | 55.3           | $3.2 \times 10^{-6}$ |
| 10 | 1170   | 57.4        | 27.4           | 62.8           | $3.3 \times 10^{-6}$ |
| 11 | 1290   | 4.12        | 40.6           | 82.0           | $3.25 \times 10^{-6}$ |
| 12 | 1340   | 35.6        | 48.3           | 94.1           | $3.45 \times 10^{-6}$ |
| 13 | 1390   | 31.6        | 55.5           | 122            | $3.78 \times 10^{-6}$ |
| 14 | 1460   | 26.8        | 67.5           | 155            | $3.78 \times 10^{-6}$ |
| 15 | 1530   | 22.9        | 81.3           | 319            | $5.53 \times 10^{-6}$ |

Note that from the above table in this example problem, it takes approximately 550 hours to grow the crack 0.5mm. Until this point, the data are not considered as being under steady-state due to damage transients during the early crack growth in test. A rigorous explanation for the presence of damage transient will be provided in Section 11.2. Therefore, the first two data points must also be discarded from consideration leaving points 3 to 15 as valid data. The da/dt vs. $C_t$ behavior is plotted in Figure 11.6 and the constants b and q are obtained from the regression analysis as 4.435

x $10^{-7}$ and 0.45, respectively. To use these constants, $C_t$ must be expressed in $J/m^2hr$ and da/dt in m/hr.

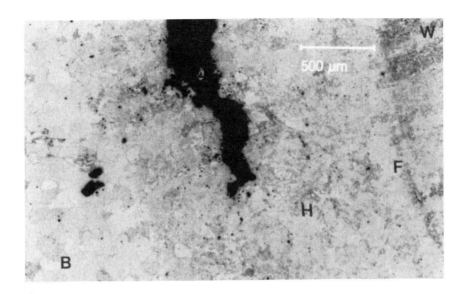

*Figure 11.6* Plot of da/dt vs. $C_t$ for the data from Example Problem 11.1. Slashed points indicate data obtained during the initial transient and are not considered valid.

### 11.1.7 Data Reduction for Creep-Brittle Materials

The data reduction for creep brittle materials is similar to the procedure for creep-ductile materials, except the da/dt is correlated with K. However, it is expected that the da/dt vs. K correlation will be valid only after passage of time equal to $t_g$ given by equation (10.71). Therefore, only data beyond $t \geq t_g$ can be considered steady-state data. If instantaneous plasticity is significant, the data can be correlated with J-integral. Figures 11.7, 10.19, and 11.8 show examples of such correlations for cold-worked 316 stainless steel [11.15], carbon steel [11.17] in cold worked and normalized condition, and a Ti-6242 alloy [11.18], respectively. Approaches for correlating crack growth rates for $t < t_g$ are being pursued currently as research topics. Some preliminary ideas are discussed in reference [11.19].

## 11.2 Microscopic Aspects of Creep Crack Growth

Figures 11.10 and 11.11 show the process of creep damage accumulation at crack tips in a Cr-Mo steel [11.20] and a model antimony-copper alloy [11.21], respectively. From these photomicrographs, it is evident that the crack tip damage accumulates in the form of cavities or voids that form preferentially on grain boundaries. In Figure 11.9, these cavities nucleate at second phase particles lying on the grain boundaries where, as in Figure 11.10, the cavities seem to nucleate unassisted at the grain boundaries because no second phase particles are associated with their presence. In the latter, nucleation is said to occur by coalescence of point defects (or vacancies) which can migrate easily on the grain boundary surface. Once a nucleated cavity reaches a critical size, it becomes energetically stable [11.22] and can then continue to grow and coalesce with neighboring cavities to form microcracks and eventually cause fracture. At crack tips, due to stress and strain concentrations, cavities nucleate and grow preferentially in the damage zone and the crack extends by the coalescence of such cavities.

**378** *Nonlinear Fracture Mechanics for Engineers*

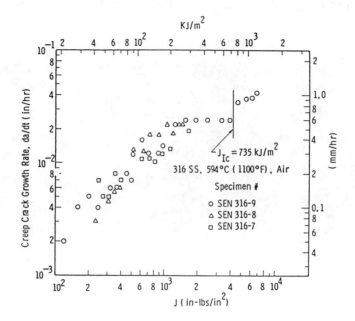

**Figure 11.7** *Creep crack growth behavior of cold-worked 316 stainless steel at 594 °C (1100 °F) (Ref. 11.15).*

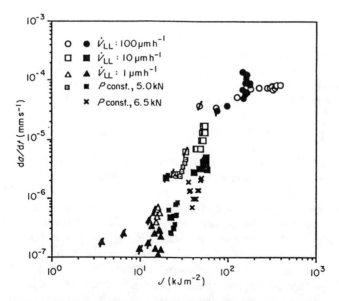

**Figure 11.8** *Creep crack growth behavior of Ti-6242 alloy (Ref. 11.17).*

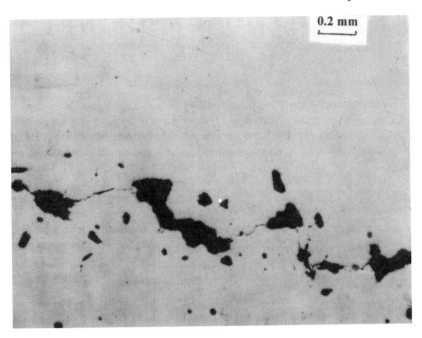

*Figure 11.9 Grain boundary voids developing on second-phase particles in front of a growing creep crack in a 1Cr-1Mo-0.25 turbine rotor steel during service at 538 °C (Ref. 11.20).*

The above physical explanations are the basis of several mathematical models for creep crack growth. These models assume that cavities form rapidly at the crack tip upon application of load at elevated temperature, either at second phase particles or by rapid nucleation. The rate of creep crack growth is governed by the kinetics of cavity growth in the crack tip environment. In this section, we will consider a simple model first proposed by Wilkinson and Vitek [11.23] with modifications by Bassami and Vitek [11.24] and by Jani and Saxena [11.25]. This model has the ability to describe several aspects of creep crack growth behavior.

Consider an array of N grain boundary cavities ahead of the crack tip as shown in Figure 11.11. These cavities have an average spacing 2b between them and the radius of the cavities are represented by $\rho_i$ where i represents the i$^{th}$ cavity from the crack tip. The rates of growth of cavities, $\dot{\rho}_i$, is given by the following equation:

$$\dot{\rho} = \phi(\rho, b)\sigma^\alpha \tag{11.13}$$

where $\alpha$ has a value which can vary with the mechanism of cavity crack growth. For example, for the classical Hull and Rimmer type of cavity growth law [11.26] in which cavities grow by diffusion and the growth is otherwise unconstrained, $\alpha = 1$. This law applies to growth of isolated cavities. For crack-like cavities which grow at a higher rate in the region of intersection of the cavity with the grain boundary, Rice and Chuang [11.27] have shown that $\alpha = 3$. If one invokes the Dyson [11.28] proposal that the rate of growth of cavities is constrained by the deformation in the surrounding grains, $\alpha = n$ for steady-state creep. In equation (11.13), $\phi$ is simply a function of average cavity spacing b and the cavity radius $\rho$. In the crack tip region, the cavities are clearly not isolated and do not appear to have crack-like shapes as seen in Figures 11.9 and 11.10. Therefore, the most likely mechanism is the constrained cavity growth of Dyson where $\alpha = n$. Rice [11.29] has analyzed this type of growth further and has proposed the following form for the function $\phi$:

$$\phi = \frac{1}{2.5}\left(\frac{2b}{\rho}\right)^2 Ad \qquad (11.14)$$

where d = grain diameter and A = pre-exponent constant for secondary creep. In the discussion in this section, we will assume secondary creep or steady-state creep to be the dominant creep deformation mode. The crack tip stresses for extensive secondary creep were derived in Chapter 10. Here, we will be

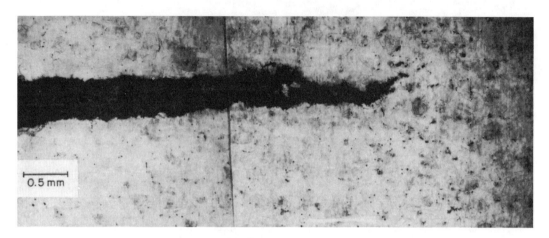

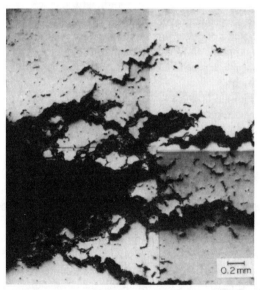

*Figure 11.10* Grain boundary cavities in front of a growing creep crack in a 1% antimony copper alloy at 400 °C. The voids in this material nucleate by coalescence of vacancies (Ref. 11.21), as compared to second phase particles as shown in the previous figure.

concerned primarily with the stress in the y-direction represented by σ (r, 0°). Assuming that the angular function in equation (10.5a) is on the order of one, we write:

$$\sigma(r,0) \approx \left( \frac{C^*}{I_n A r} \right)^{\frac{1}{n+1}} \qquad (11.15)$$

Thus, by substituting equations (11.15) and (11.14) into equation (11.13), the cavity growth rates in the crack tip region are completely defined.

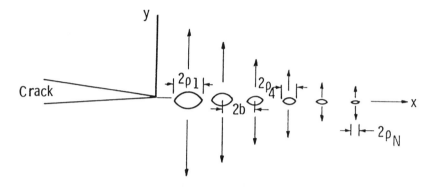

**Figure 11.11** *A one-dimensional array of grain boundary cavities in front of crack tip used to model creep crack growth (Refs. 11.23, 11.24, 11.25).*

Next, a criterion for cavity coalescence is chosen. Whenever the cavity nearest to the crack tip grows to a critical size, $\rho_c$, the crack is assumed to advance by a distance $2b$. Therefore, if $\Delta t$ is the time required for the cavity to grow from $\rho_1$ to $\rho_c$, the crack growth rate is given by:

$$\frac{da}{dt} = \frac{2b}{\Delta t} \qquad (11.16)$$

A steady-state crack growth process is established if one assumes that during the time $\Delta t$, cavity with radius $\rho_2$ grows to $\rho_1$ and the cavity with radius $\rho_m$ grows to $\rho_{m-1}$. Also, during this period, a new cavity of radius $\rho_N$ nucleates at a distance $2bN$ from the crack tip. This leads to the following set of N integral equations:

$$\Delta t = \int_{\rho_1}^{\rho_c} \frac{d\rho}{\dot{\rho}} \qquad (11.17a)$$

$$\Delta t = \int_{\rho_2}^{\rho_1} \frac{d\rho}{\dot{\rho}} \qquad (11.17b)$$

$$\Delta t = \int_{\rho_m}^{\rho_{m-1}} \frac{d\rho}{\dot{\rho}} \tag{11.17c}$$

$$\Delta t = \int_{\rho_n}^{\rho_{n-1}} \frac{d\rho}{\dot{\rho}} \tag{11.17d}$$

Substituting equation (11.15) into equation (11.13) and the result into equation (11.17), we get (assuming $\alpha = n$):

$$\Delta t = \left| \int_{\rho_{m+1}}^{\rho_m} \left( \frac{d\rho}{\phi(\rho,b)} \right) \right| \bigg/ \left( \frac{C^*}{I_n A r} \right)^{\frac{n}{n+1}}$$

or

$$\Delta t = \left[ \int_{\rho_{m+1}}^{\rho_m} r^{\frac{n}{n+1}} \frac{d\rho}{\phi(P,b)} \right] \bigg/ \left( \frac{C^*}{I_n A} \right)^{\frac{n}{n+1}} \tag{11.18}$$

Substituting $r = 2b(m+1)$, equation (11.18) becomes:

$$\frac{\Delta t}{(2b(m+1))^{n/(n+1)}} = \left[ \int_{\rho_{m+1}}^{\rho_m} \frac{d\rho}{\phi(\rho,b)} \right] \bigg/ \left( \frac{C^*}{AI_n} \right)^{\frac{n}{n+1}} \tag{11.19}$$

If we assume:

$$\int \frac{d\rho}{\phi(\rho,b)} = \Psi(\rho,b) \tag{11.20}$$

and substitute into equation (11.19) and then vary the value of m from 0 to N-1, we get a set of N integral equations as follows:

$$\left( \frac{C^*}{AI_n} \right)^{\frac{n}{n+1}} \frac{\Delta t}{(2b)^{n/(n+1)}} = \Psi_0(\rho,b) - \Psi_1(\rho,b) \tag{11.21a}$$

$$\left(\frac{C^*}{AI_n}\right)^{\frac{n}{n+1}} \frac{\Delta t}{(4b)^{n/(n+1)}} = \Psi_1(\rho,b) - \Psi_2(\rho,b) \tag{11.21b}$$

$$\left(\frac{C^*}{AI_n}\right)^{\frac{n}{n+1}} \frac{\Delta t}{(2b(m+1))^{n/(n+1)}} = \left[\Psi_m(\rho,b) - \Psi_{m+1}(\rho,b)\right] \tag{11.22a}$$

Adding equation (11.22) for m = 0 to N-1, we get:

$$\Delta t = \frac{\Psi_0 - \Psi_N}{(C^*/AI_n)^{n/(n+1)}} \sum_{m=0}^{N-1} (2b(m+1))^{n/(n+1)} \tag{11.23}$$

Substituting equation (11.23) into (11.16), we get:

$$\frac{da}{dt} = \frac{(2b)^{1-\frac{n}{n+1}}}{[\Psi_0 - \Psi_N]} \frac{(C^*/I_n A)^{n/(n+1)}}{\sum_{m=0}^{N}(m)^{n/(n+1)}} \tag{11.24}$$

Substituting equation (11.14) into (11.20) and simplifying, we get:

$$\Psi = \frac{2.5\rho^3}{3(2b)^2 A d}$$

or

$$\Psi_0 - \Psi_N = \frac{2.5}{3(2b)^2 A d}(\rho_c^3 - \rho_N^3) \tag{11.25}$$

Substituting (11.25) into (11.24) yields:

$$\frac{da}{dt} = \frac{(2b)^{\frac{2n+3}{n+1}} 3\, d\, A^{1/(n+1)}}{(2.5)(\rho_c^3 - \rho_N^3) \sum_{m=1}^{N}(m)^{n/(n+1)}} \left(\frac{C^*}{I_n}\right)^{\frac{n}{n+1}} \tag{11.26}$$

Several trends in creep crack growth behavior can be rationalized using this model. The first is that da/dt is uniquely related to C* under extensive creep conditions. This further supports the use of the fracture mechanics approach for predicting creep crack growth. It has been observed that the influence of temperature on the da/dt - C* relationship is only marginal. If we assume that the value of $n$ is relatively constant with temperature and it is A that changes rapidly, equation (11.26) predicts that the da/dt - C* relationship varies only as a function $A^{1/(1+n)}$ where $n$ is between 5 and 15. Therefore, the weak temperature dependence of the da/dt vs. C* relationship is also predicted by this model. Next, the creep crack growth exponent, $q$, is predicted to be equal to $n/(n+1)$. For $n$ values between 5 and 15, this changes from 0.83 to 0.94. The observed values of q have been in the range of 0.6 to 0.9. This can be considered a reasonable agreement considering the simplicity of the model. Also, if diffusive cavity growth or crack-like cavity growth is used, the predicted exponents can decrease significantly. For example, for purely diffusive cavity growth $n = 1$ and the predicted exponent will then be 0.5. The range from 0.5 to 0.9 certainly covers most of the observed trends.

By a casual inspection of equation (11.26), one may conclude that a smaller intercavity spacing (2b) will lower creep crack growth rates, which is opposite to what one might expect. However, we must consider that if the cavity spacing is smaller, the value of N which determines the value of the summation term in the denominator will increase because the damage zone size is controlled by the deformation properties of the material. The deformation properties, at least to a first order of approximation, can be assumed to be independent of the inclusion spacing which determines intercavity distances. If $\rho_N$ is the nucleation radius of the cavity determined by the size of the particles which participate in cavity nucleation, then we can also predict from equation (11.26) that larger grain boundary particles will lead to higher creep crack growth rates which is in agreement with what one might intuitively expect.

## 11.3 Creep Crack Growth in Weldments

Weldments are essentially composite materials consisting of a base metal (BM) and a weld metal (WM), often with different mechanical properties, and an interface region between BM and WM with considerable microstructure/property gradients. For example, the creep resistance of the BM alloy may be significantly different from that of the WM and the interface region may consist of a gradient of creep properties from the fusion region of the weld metal through the heat affected zone (HAZ) of the BM. Weldments are often subjected to post weld heat treatments (PWHT) to reduce the residual stresses in the interface region and also to homogenize the microstructure of the interface region. Such treatments can be quite expensive and therefore are used only when no other alternatives are available.

In this section, we will be discussing the creep crack growth behavior in weldments, which from a practical standpoint is extremely important, because welding is used as the preferred joining method for elevated temperature components. Also, a larger number of cracking problems in elevated temperature components involve either weldments used during fabrication or weldments used to repair damage during manufacturing or damage accumulated during service.

Current approaches for predicting creep crack growth in weldments in which the base and weld metals have dissimilar creep deformation rates are based entirely on concepts developed for homogeneous bodies. Under nominal mode I loading, for example, Liaw et al. [11.40] and Saxena et al. [11.1] have shown that the creep crack growth rates in the fusion line region correlate well with the $C_t$ parameter. As discussed in Chapter 6, linear elastic and elastic-plastic analyses of bimaterial interface cracks show that the crack tip fields and shapes and sizes of plastic zones are quite different for homogeneous bodies and bimaterials. Also, even though the nominal loading is mode I, the interface experiences a combined mode I and mode II loading. However, it was also shown that for a crack located at a mathematically sharp interface and loaded nominally in mode I, the J-integral was path-independent. By analogy, one can argue that the C*-integral will also be path-independent. Further, the crack tip stress and strain fields are also determined by J (or C* for creep). Therefore, the success in correlating da/dt with $C_t$ for fusion line region cracks is not entirely without a theoretical basis. Therefore, the approach developed for homogeneous bodies can be used, with good reason, for evaluation of properties of weldments. However, the creep crack propagation problems encountered during service and involving weldments can be considerably more complex as shown in Figure 11.12. Figure 11.12a shows the mismatch in creep deformation rates between the weld metal and base metal at 538°C (1000°F) in a 1.25Cr-0.5Mo steel weldment at a stress of 68.95 MPa (10 Ksi). Such difference in creep deformation behavior accumulates the strains in the fusion line region of the corresponding weld leading to preferential creep crack nucleation site and higher crack growth rates as shown in Figure 11.12b. In the following discussion, we will therefore adhere to laboratory testing and evaluation of weldments using specimens with cracks located at the interface and loaded nominally in mode I.

Figure 11.13 shows the creep crack growth behavior of the interface region of Cr-Mo steel weldment [11.1]. The base metal behavior of the materials used in this weldment can be approximated by the data shown in Figure 11.1. Figure 11.14 shows a comparison between the creep crack growth behavior of the base metal and the weldment. On average, crack growth rates are higher in the weldments. This is due to the strain concentrations that occur in the interface region of weldments consisting of BM and WM with dissimilar creep properties. Figure 11.15a shows a similar comparison between the weldment and BM crack growth behavior in which the WM was over-matched in terms of the creep deformation resistance for a 2.25 Cr-1Mo steel tested at 538°C [11.30]. The creep deformation behavior of the two regions is shown in Figure 11.15b. Figure 11.15c shows the crack path and its relationship to the fusion line. In this case, the crack path stays on the BM side of the interface region because the BM is the weaker of the two regions.

The post weld heat treatment (PWHT) has been shown to influence creep crack growth behavior of the heat affected zone regions of weldments. Figure 11.16 shows the creep crack growth behavior of Cr-Mo steel weldments as influenced by the impurity levels and PWHT [11.6, 11.1]. The combination of higher impurity content (sample D) with low PWHT temperature of 620°C (1148°F) resulted in the highest creep crack growth rates. Conversely, lower levels of impurity content (sample A) with high PWHT temperature of 720°C (1328°F) resulted in a creep crack growth rate comparable to those of the good welds in Figures 11.13 and 11.14. The fusion zone immediately adjacent to the weld, upon solidification, may contain significant amounts of segregation of solute elements as well as inclusions [11.31]. Segregation has been shown to have significant detrimental effects on temper embrittlement of Cr-Mo steels [11.31, 11.32]. Therefore, the cleanliness of the base steel and the weld rod as well as an effective PWHT are important in producing high quality welds for elevated temperature service. It is equally important to avoid high degrees of mismatch in the creep deformation properties of the WM and BM. In practice, the latter can be minimized but perhaps not eliminated.

Even if the base metal and weld rod contain fewer impurities, the cleanliness of the weldment cannot be guaranteed unless necessary precautions are taken during welding itself. For example, a majority of Cr-Mo steel industrial weldments are made by submerged arc welding (SMAW) process. Weldments performed with acid flux were shown to have creep-rupture strength values lower than base metal while weldments made with basic or neutral fluxes have shown creep rupture strength values comparable to the mean to lower bound base metal values [11.33]. The adverse influence of the flux was attributed to the presence of oxygen and sulfur which, upon solidification, formed high volume fractions of manganese sulfide inclusions which act as sites for creep cavitation, thereby reducing resistance to creep rupture and creep crack growth rate.

A rigorous treatment of elastic-plastic fracture and creep crack growth in welded structures is beyond the current capability of fracture mechanics. Consequently, somewhat ad hoc correlation procedures are presently available; such procedures are based on homogeneous material, nonlinear fracture mechanics concepts, all of which do not have rigorous justification. Nevertheless, experimental work has clearly demonstrated the promise of such approaches and merits further study.

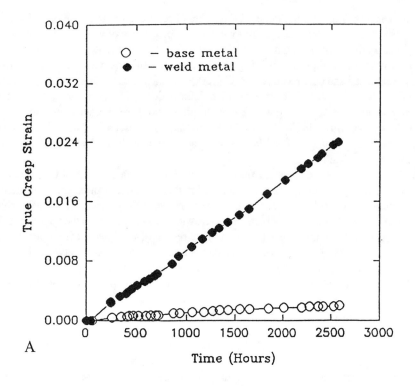

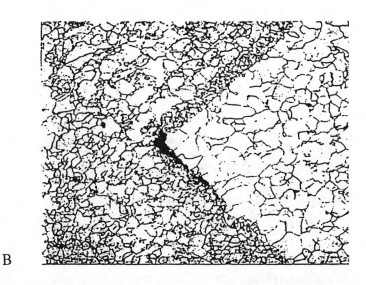

*Figure 11.12* (a) Creep deformation curves for base metal and weld metal at 538 °C for a stress level of 68.95 MPa. (b) Evolution of a creep crack in the cusp region of a "double V" weld as a result of differing creep rates from the weld used in (a) for testing.

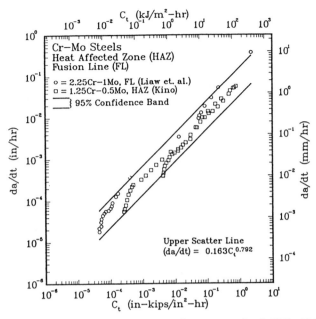

**Figure 11.13** Creep crack growth behavior of the interface region for 1.25Cr-1Mo and 2.25Cr-1Mo steel weldments (Ref. 11.4).

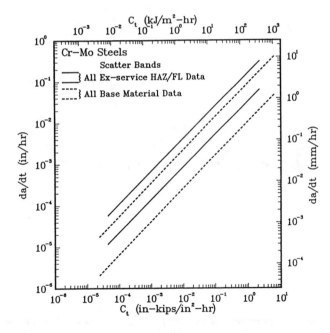

**Figure 11.14** Comparison of creep crack growth rates in Cr-Mo weldments and base metal (Ref. 11.1).

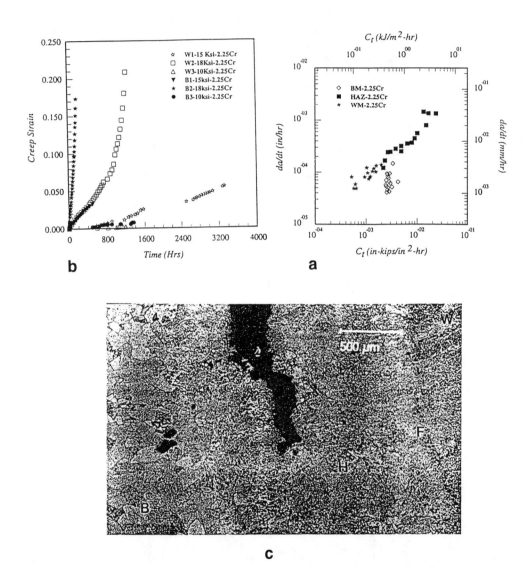

*Figure 11.15* (a) Creep crack growth behavior of a 2.25Cr-1Mo weldment from an ex-service steel; (b) the creep deformation rates in the BM and WM regions; and (c) crack path followed during creep crack growth testing relative to the fusion line (Ref. 11.30).

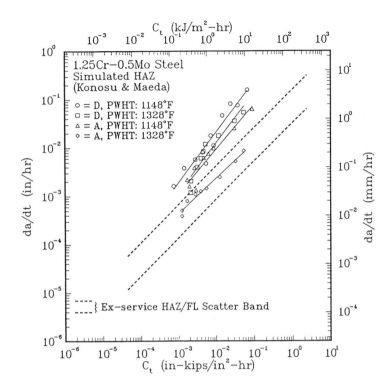

*Figure 11.16* Creep crack growth behavior in Cr-Mo steel weldments as influenced by impurity levels and post weld heat treatments (Refs. 11.6, 11.1).

## 11.4 Summary

In this chapter, the detailed test method for conducting creep crack growth tests was considered. The experimental methods used for testing creep-ductile and creep-brittle materials are very similar, but there are important differences in the analyses of the results. The data analysis method was illustrated using an example. The creep crack growth rates for creep-ductile materials correlate with $C_t$ and for creep-brittle materials with K after a transition period. The mechanisms of creep crack growth was modeled by growth and coalescence of grain boundary cavities. A mathematical model was described to predict creep crack growth based on this phenomenon. It was shown that the model was able to predict several of the commonly observed trends during creep crack growth.

The last section in the chapter dealt with the creep crack growth behavior of weldments. It was shown that if the base metal and weld metal have dissimilar creep deformation properties, the rates of creep crack growth can be faster in weldments. In addition, the microstructural gradients around the fusion zone or the interface between the base metal and weld metal can also lead to more than normal variability in the creep crack growth properties. The rates of creep crack growth were shown to be generally higher than in the base metal.

## 11.5 References

11.1 A. Saxena, J. Han, and K. Banerji, "Creep Crack Growth Behavior in Power Plant Boiler and Steam Pipe Steels", Journal of Pressure Vessel Technology, ASME, Vol. 110, 1988, pp. 137-146.

11.2 H. Riedel and V. Detampel, "Creep Crack Growth in Ductile, Creep Resistant Steels", International Journal of Fracture, Vol. 33, 1987, pp. 239-262.

11.3 H. Kino, "Electric Potential Technique for Monitoring Crack Growth and Creep Crack Growth Behavior in Various Steels", unpublished data, Mitsibushi Heavy Industries, Ltd., Dec. 1985.

11.4 P.K. Liaw, A. Saxena, and J. Schaffer, "Residual Life Prediction and Inspection Criterion for High Temperature Seam-welded Piping I - Material Properties", Engineering Fracture Mechanics, Vol. 32, 1989, pp. 709-722.

11.5 D.J. Gooch and B.L. King, "High Temperature Crack Propagation in 2.25Cr-1Mo Mannual Metal Arc Weld Metals", Proceedings of the 5th Bolton Landing Conference, Weldments: Physical Metallurgy and Failure Phenomena, eds. R.J. Christoffel et al., Aug. 1978, pp. 393-408.

11.6 S. Konosu and K. Maeda, "Creep Embrittlement Susceptibility and Creep Crack Growth Behavior in Low Alloy Steels", Nonlinear Fracture Mechanics: Vol. I - Time-Dependent Fracture, ASTM STP 995, American Society for Testing and Materials, Philadelphia, 1989, pp. 127-152.

11.7 C.H. Wells, unpublished data, Failure Analysis Associates, April 1987.

11.8 A. Saxena, "Creep Crack Growth under Non Steady-State Conditions", Fracture Mechanics: Seventeenth Volume, ASTM STP 905, American Society for Testing and Materials, Philadelphia, 1986, pp. 185-201.

11.9 A. Saxena, "Evaluation of $C^*$ for the Characterization of Creep-Crack-Growth Behavior in 304 Stainless Steel", Fracture Mechanics: Twelfth Conference, ASTM STP 700, American Society for Testing and Materials, Philadelphia, 1980, pp. 131-151.

11.10 "Standard Test Method for Measurement of Creep Crack Growth Rates in Metals: ASTM Standard 1457-92. ASTM Book of Standards, Vol. 03.01, 1992, pp. 1031-1043.

11.11 H.H. Johnson, Materials Research and Standards, Vol. 5, No. 9, 1965, pp. 442-445.

11.12 K.H. Schwalbe and D. Hellman, Journal of Testing and Evaluation, Vol. 9, 1981, pp. 218-221.

11.13 A. Saxena and S. J. Hudak, Jr., "A Review on Extension of Compliance Information on Common Crack Growth Information", International Journal of Fracture, Vol. 14, 1978, pp. 453-468.

11.14 "Standard Test Method for Measurement of Fatigue Crack Growth Rates", ASTM Standard E 647-91, ASTM Book of Standards, Vol. 03.01, 1992, pp. 674-701.

11.15 A. Saxena, H.A. Ernst, and J.D. Landes, "Creep Crack Growth Behavior in 316 Stainless Steel at 594°C", International Journal of Fracture, Vol. 23, 1983, pp. 245-257.

11.16 A. Saxena and J.D. Landes, "Characterization of Creep Crack Growth in Metals", Advances in Fracture Research, S.R. Valluri et al., editors, Pergamon Press, 1985, pp. 3977-3988.

11.17 B. Dogan, A. Saxena, and K.H. Schwalbe, "Creep Crack Growth in Creep-Brittle Ti-6242 Alloys", Materials at High Temperatures, Vol. 10, 1992, pp. 138-143.

11.18 Y. Gill, "Creep Crack Growth Characterization of SA-106C Carbon Steel", Ph.D. Thesis, Georgia Institute of Technology, Atlanta, Georgia, 1994.

11.19 D.E. Hall, D.L. McDowell, and A. Saxena, "Parameters for Characterizing Crack Growth in Creep-Brittle Materials", Submitted for publication to Fracture and Fatigue of Engineering Materials and Structures", 1996.

11.20 V.P. Swaminathan, "Advanced Rotor Forgings for High Temperature Steam Turbines Vol. I: Ingot and Forging Production", CS 4516, 1986, Electric Power Research Institute, Palo Alto.

11.21 J.T. Staley, Jr. and A. Saxena, "Mechanisms of Creep Crack Growth in 1% Sb-Cu Alloy", Acta Metallurgica, Vol. 38, 1990, pp. 897-905.

11.22 A.S. Argon, "Nucleation of Creep Cavities", Creep-Fatigue Environment Interactions, TMS-AIME, 1980, pp. 131-150.

11.23 D.S. Wilkinson and V. Vitek, "Propagation of Cracks by Cavitation--a General Theory", Acta Metallurgica, Vol. 30. 1982, pp. 1723-1732.

11.24 J.L. Bassani and V. Vitek, "Propogation of Cracks Under Creep Conditions:, Proceedings of the 9th National Congress of Applied Mechanics--Symposium on Non-linear Fracture Mechanics, 1982, pp. 127-133.

11.25 S.C. Jani and A. Saxena, "Influence of Thermal Aging on Creep Crack Growth Behavior of a Cr-Mo Steel", Effects of Load and Thermal Histories, TMS, 1987, pp. 201-220.

11.26 D. Hull and D. Rimmer, "The Growth of Grain-Boundary Voids Under Stress", Philosophical Magazine, Vol. 4, 1959, pp. 673-677.

11.27 T.J. Chuang, K.I. Kagawa, J.R. Rice, and L.B. Sills, "Non-Equilibrium Models for Diffusive Cavitation of Grain Boundaries", Acta Metallurgica, Vol. 27, 1979, pp. 265-271.

11.28 B.F. Dyson, "Constraints on Diffusional Cavity Growth Rates", Metallurgical Science, Vol. 10, 1976, pp. 349-355.

11.29 J.R. Rice, "Constraint on Diffusive Cavitation of Isolated Grain Boundary Facet in Creeping Polycrystals", Acta Metallurgica, Vol. 29, 1981, pp. 675-681.

11.30 R.H. Norris, "Creep Crack Growth Behavior in Weld Metal/Base Metal/Fusion Zone Regions in Cr-Mo Steels", Ph.D. Dissertation, Georgia Institute of Technology, 1995.

11.31 F. Masuyama, N. Nishimura, and Y. Takeda, High Temperature Technology, Vol. 8, 1990, pp. 66-70.

11.32 C.D. Lundin, K.K. Khan, S.D. Hilton, and W. Zielke, "Failure Analysis of a Service-Exposed Hot Reheat Steam Line in a Utility Power Plant", WRC Bulletin 354, Welding Research Council, 1990.

## 11.6 Exercise Problems

11.1 A standard 1T compact specimen of a 1Cr-1Mo-0.25V steel has the dimensions W = 50.8, B = 25.4mm, and is subjected to a load of 30.26 kN at 538°C. Using these same material properties, derive the creep crack growth rate behavior as a function of $C_t$ parameter from the data shown in Table 11.5.

11.2 A steam pipe is subjected to an internal pressure of 4.82 MPa at 538°C. The outside diameter of the pipe is 76.2 cm and the wall thickness is 3.81cm. The pipe contains a long radial crack on the inside surface which is 0.508 cm deep. You are given the following material data:

$$A = 2.94 \times 10^{-22} \qquad n = 8 \qquad \sigma \text{ in MPa} \qquad \dot{\varepsilon} = hr^{-1}$$

$$\frac{da}{dt} = 2.74 \times 10^{-3} (C_t)^{0.825}$$

for da/dt in m/hr and $C_t$ in Joules/m² hr

The expression for the stress intensity parameter for this configuration is given by the following equation:

$$\frac{K}{p\sqrt{t}} = F = \frac{2(\pi a/t)^{1/2}}{[1 - R_i/R_o]} \left( 1.12 + 0.405 \ (a/t) + 2.55 \ (a/t)^2 + 3.25 \ (a/t)^3 \right)$$

p = internal pressure    t = wall thickness    $R_i$ = internal pipe radius

$R_o$ = outside pipe radius

Estimate the time required to grow the crack to a depth of 1 cm during service.

11.3 In Example Problem 11.1, use the secant method in place of the seven-point incremental polynomial to reduce the data and compare the da/dt vs. $C_t$ plot.

11.4 List the advantages and disadvantages of testing CT and CCT specimens for creep crack growth testing.

11.5 Why is it necessary to discard the first 0.5mm of crack extension prior to calculating steady-state creep crack growth rates for creep-ductile materials?

*Table 11.5  Unprocessed Creep Crack Growth Data for Example Problem 11.1*

| Time (hrs) | Crack Size (mm) | Displacement (mm) |
|---|---|---|
| 0 | 26.16 | 0 |
| 20 | 26.36 | 0.254 |
| 235 | 26.52 | 0.381 |
| 475 | 26.96 | 0.508 |
| 520 | 27.12 | 0.533 |
| 570 | 27.28 | 0.588 |
| 640 | 27.58 | 0.647 |
| 690 | 27.78 | 0.658 |
| 740 | 27.94 | 0.737 |
| 810 | 28.19 | 0.864 |
| 860 | 28.44 | 0.940 |
| 910 | 28.77 | 1.067 |
| 975 | 29.31 | 1.295 |
| 1000 | 29.58 | 1.422 |
| 1025 | 29.80 | 1.562 |
| 1050 | 30.09 | 1.867 |
| 1055 | 30.43 | 2.083 |

11.6  Complete all the steps on your own and derive the Wilkinson, Vitek, and Bassani model for predicting creep crack growth. Derive an equation between the creep crack growth rate and the crack tip parameter under small-scale creep conditions.

11.7  A Cr-Mo steel has the following information provided:

inclusion spacing = 100 μm
grain diameter = 50 μm
average inclusion diameter = 5 μm

If the process zone in which the damage develops extends through five inter-inclusion spacings, $n = 8$ and $A = 1.16 \times 10^{-24}$ MPa$^{-8}$ hr$^{-1}$, estimate the da/dt vs. $C^*$ relationship for this material. Compare the predicted trend to the measured behavior in Figure 11.6 and comment on any discrepancies that you may find.

11.8 Derive an expression for estimating $C_t$ in a double-edge-notch tension (DENT) specimen from small scale to extensive creep when the load-line deflection rate is measured. Assume that the specimen is loaded with a point load, P, remote from the crack plane.

11.9 List the reasons why the concepts of nonlinear fracture mechanics developed for homogeneous materials do not directly apply to weldments for predicting creep crack growth.

11.10 Show that in a bimaterial containing a mathematically sharp interface and a crack lying on the interface, the $C^*$-integral is path-independent for a point load applied normal to the crack surface.

# CHAPTER 12

# CREEP-FATIGUE CRACK GROWTH

In the previous two chapters, we have built the analytical framework for considering time-dependent creep deformation in the crack tip region and have used the analyses to define crack tip parameters for predicting crack growth under the conditions of sustained loading. Cyclic or fatigue loading is quite prominent in several components that are operated at elevated temperature, making crack growth behavior under creep-fatigue conditions quite important. Figure 12.1 shows a schematic of a section of a steam header in which the high stress locations are marked and the load/stress vs. time history experienced at these locations during one complete operating cycle of the header is also shown [12.1]. Such duty cycles involving combination of time varying stress, periods of sustained stress, and high temperatures are typical of several elevated temperature components. Creep deformation can occur during both the sustained load period and also during the loading portion of the cycle.

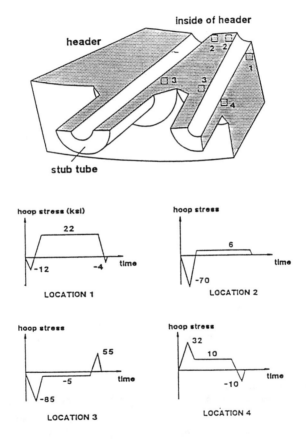

*Figure 12.1* Schematic representation of the stress vs. time histories at various locations in a steam header (Ref. 12.1).

In this chapter, we will be considering methods for predicting crack growth rates during combined creep and fatigue conditions. Some of the common loading waveforms used in the study of creep-fatigue crack growth behavior are shown in Figure 12.2. The important time parameters characterizing the load-

395

ing wave forms are, as shown in Figure 12.2, $t_r$ = rise time, $t_h$ = hold time, and $t_d$ = decay time. The amplitude of the loading waveform can be characterized by $\Delta P = P_{max} - P_{min}$, the difference between the maximum and minimum loads, and the load ratio, $R = P_{min}/P_{max}$. The loading frequency is termed $\nu$.

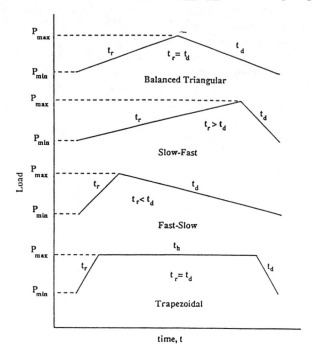

*Figure 12.2  Schematic diagrams of the loading waveforms typically employed in creep-fatigue experiments.*

## 12.1 Early Approaches for Characterizing Creep-Fatigue Crack Growth Behavior

The early approaches for characterizing the creep-fatigue crack growth behavior were based entirely on linear-elastic fracture mechanics (LEFM) and clearly met with some degree of success. In the subsequent discussion, some of these approaches will be reviewed as well as the limitations of these approaches, which will then set the stage for discussion of the more recent nonlinear fracture mechanics approaches.

### 12.1.1 LEFM Approaches

The linear elastic fracture mechanics (LEFM) approach for characterizing creep-fatigue crack growth behavior relied on $\Delta K$ for characterizing the crack growth rate per cycle, da/dN, while keeping the loading frequency and loading waveform constant. The experimental work of James [12.2, 12.3] in the early nineteen seventies which demonstrated the influence of loading frequency and waveform on da/dN vs. $\Delta K$ behavior was pioneering, and clearly the most complete at the time. Figure 12.3 shows a plot of da/dN vs. $\Delta K$ for a triangular waveform at several loading frequencies for 304 stainless steel at 538°C taken from James' early work [12.2]. The crack growth rate per cycle, da/dN increases with decreasing frequency. Figure 12.4 shows similar data demonstrating the influence of hold time in a trapezoidal loading waveform on the da/dN vs. $\Delta K$ relationship for a Cr-Mo-V steel, also at 538°C [12.4]. These data show that da/dN increases with hold time at a constant $\Delta K$. Figure 12.5 shows the influence of loading waveshape for a constant overall cycle time on the fatigue crack growth behavior of 304 stainless

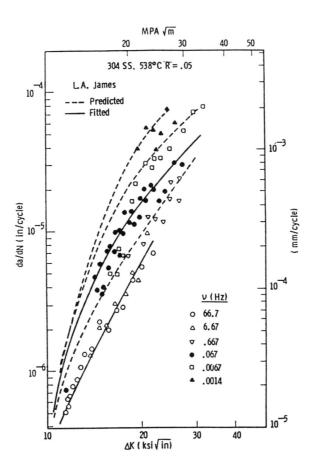

**Figure 12.3** *Fatigue crack growth behavior of 304 stainless steel as a function of loading frequency at 538 °C, James (Ref. 12.2).*

steel at 570°C [12.5]. These data show that the time-dependent damage accumulates more rapidly during the loading portion of the cycle as compared to the hold time.

The above data clearly establish the importance of time-dependent damage on the fatigue crack growth behavior at elevated temperature. Such trends can be attributed to creep damage at the crack tip, or the influence of environment, or possibly due to microstructural changes such as formation of cavities which occur due to loading at elevated temperature, or a combination of the above factors. The separation of contributions to damage due to environment (oxidation) and creep have been attempted by several researchers [12.6-12.8] by preforming elevated temperature crack growth tests at various loading frequencies and waveforms in air and in inert environments, Figure 12.6 [12.7]. These results show that the time-dependent effects, at least in some materials, are significantly enhanced by oxidation. Therefore, environmental effects must be considered in any model for predicting the fatigue crack growth behavior. The microstructural changes near the crack tip due to creep cavitation are quite often constrained by the deformation behavior in the surrounding grains. Therefore, their contribution to damage development is implicitly included when the crack growth response is correlated with the magnitude of the crack tip parameter for a given loading waveshape and frequency.

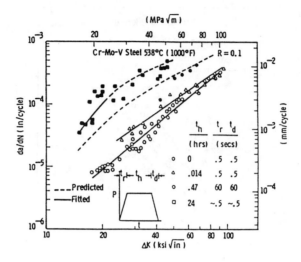

**Figure 12.4** *Fatigue crack growth behavior in Cr-Mo-V steel as a function of hold time at 538 °C (Ref. 12.4).*

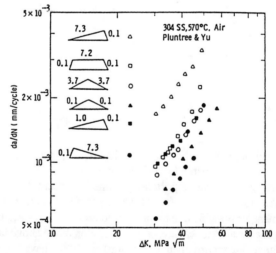

**Figure 12.5** *Influence of cyclic waveform on elevated temperature fatigue crack growth behavior of 304 stainless steel, Plumtree and Yu (Ref. 12.5).*

### 12.1.2 Limitations of the LEFM Approaches

There are several questions about the validity of the linear elastic parameter $\Delta K$ in the presence of significant creep deformation at the crack tip [12.9, 12.10]. For example, it is unlikely that $\Delta K$ remains valid for uniquely determining the fatigue crack growth behavior at low loading frequencies. There is an additional concern about the use of $\Delta K$ at elevated temperature because for a large number of high temperature ductile materials, the 0.2% yield strength decreases significantly with temperature and the influence of instantaneous plasticity becomes important. When instantaneous plasticity becomes significant, one can use the ΔJ-integral described in Chapter 9, instead of ΔK, to correlate crack growth rates. The validity of ΔJ for characterizing high temperature crack growth under plasticity conditions has been shown in several studies [12.9-12.10] and as mentioned before in detail, in Chapter 9. However, the validity of ΔJ is limited to cyclic plastic deformation and cannot be extended to include the time-dependent creep deformation which occurs if the loading frequencies are low or there are hold times involved in the loading cycles. For these conditions, time-dependent fracture mechanics parameters are needed. An extensive discussion of time-dependent fracture mechanics crack tip parameters was included in Chapters 10 and 11. In the subsequent discussion, we will build on those concepts and extend them for cyclic loading.

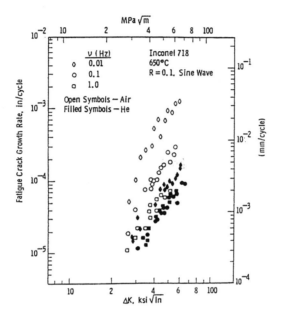

*Figure 12.6* An example of creep-fatigue-environment interactions during elevated temperature fatigue crack growth behavior, Floreen and Kane (Ref. 12.7).

## 12.2 Time-Dependent Fracture Mechanics Parameters for Creep-Fatigue Crack Growth

During a trapezoidal loading waveform, creep deformation can occur at the crack tip during hold time as well as during the loading portions of the cycle. During loading, creep deformation at the crack tip will also be accompanied by plastic and elastic deformation (instantaneous deformation) and the relative amount of creep and instantaneous deformation depends on the loading rate. For fast loading rates, creep will be negligible where as for very slow loading rates, creep can be dominant. During the hold period, creep deformation will occur causing the crack tip stresses to redistribute resulting in elastic strains. If the hold period is sufficiently long, extensive creep conditions will ultimately prevail. Therefore, to gain a complete understanding of the creep-fatigue crack tip mechanics, solutions must be obtained for a wide range of loading and material deformation conditions. However, analytical solutions for crack tip stresses are not possible for such complex conditions, thus, one must depend on the numerical approach. In this section, the results of some basic numerical studies will be first described. These results will then be used to define crack tip parameters which are suitable for describing some aspects of creep-fatigue crack growth.

### 12.2.1 Crack Tip Stresses Under Creep-Fatigue Loading

As a first step toward understanding crack tip conditions under creep-fatigue loading, Riedel [12.11] in his analysis performed in the early nineteen eighties assumed that the material deformed by elastic and power-law creep. He considered triangular loading waveforms and trapezoidal loading waveforms. For a triangular waveform without hold time, the load was increased linearly with time. The peak stress in the crack tip region was shown to depend on the loading rate. The higher the loading rate, the higher the peak stress because less time was available for creep strains to develop and redistribute the stresses. Riedel also derived that for a linearly increasing load, the transition from small-scale creep conditions to extensive creep conditions will occur at a time $t_{TC}$ given by:

$$t_{TC} = \frac{1+2n}{n+1} \frac{K^2(t_{TC})\ (1-v^2)}{C^*(t_{TC})E} \tag{12.1}$$

Recall that the form of the above equation is quite similar to equation (10.25) which estimates the transition time for constant load whereas equation (12.1) is for linearly increasing load. In the above equation, $K(t_{TC})$ and $C^*(t_{TC})$ are the values of $K$ and $C^*$ at $t = t_{TC}$, respectively. Also note that the transition time for monotonic loading is larger by a factor of $(1 + 2n)$ in comparison to the transition time for constant load.

**Example Problem 12.1**

A 50mm wide and 25.5mm thick CT specimen is to be loaded to a load of 18kN in a manner that the specimen approaches extensive creep conditions by the time the maximum load is reached. The specimen is made from a 304 stainless and the load is to be applied at a temperature of 594°C. The crack size is 25mm and the following material data are provided:

$$A = 2.13 \times 10^{-18}\ (MPa)^{-n}\ hr^{-1} \quad \text{and} \quad n = 6$$

a) What is the fastest loading time that can be used?

b) If the objective were to maintain small-scale creep conditions, what is the slowest loading time?

*Solution:*

The load levels, specimen geometry, size, and material are identical to Example Problem 10.3. Thus, we can make use of several quantities that have been previously calculated. For example, the K corresponding to the maximum load is 31.1 MPa√m and the $C^*$ is $0.139 \times 10^{-3}$ MJ/m² hr. These values also represent the $K(t_{TC})$ and $C^*(t_{TC})$ in equation (12.1). Thus, $t_{TC} = 68.38$ hours.

(a) If we allow more than $2t_{TC}$ hours for loading the specimen, extensive creep conditions can be ensured. (b) If we allow less than $0.5\ t_{TC}$ hours, small-scale conditions can be ensured.

The value of $t_{TC}$ can be used to select loading frequencies during fatigue crack growth testing at high temperatures to maintain dominantly elastic conditions. On the other hand, we can also see that if the loading rate is slow enough, there is little justification for using $\Delta K$ for characterizing fatigue crack growth rates at elevated temperature.

The next type of loading waveform used by Riedel [12.11] was a trapezoidal loading waveform in which he varied the loading rate and then held the load constant. Figure 12.7 summarizes his results in which the normalized load ($P/P_{max}$) and the normalized value of the crack tip stress ($\sigma/\sigma_{ss}$), where $\sigma_{ss}$ is

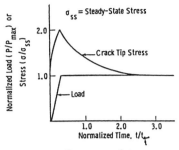

**Figure 12.7** *Normalized crack tip stress as a function of time normalized by transition time, at a fixed distance from the crack tip for fast loading followed by sustained load, from Riedel (Ref. 12.11).*

the fully relaxed, steady-state stress, are plotted against normalized time, $t/t_T$. If the loading is reasonably rapid, $t_T$ is the transition time as determined by equation (10.25). The magnitude of the peak crack tip stress depends on the loading rate. A slower loading rate will result in a lower peak crack tip stress. This would imply that at low loading rates, cycle-dependent crack growth due to fatigue will be suppressed and time-dependent crack growth due to creep may become dominant.

The above analyses were not performed for repeated cycles. During cyclic loading, considerable cyclic deformation occurs at the crack tip during repeated loading and unloading. In order to develop a clearer picture of creep-fatigue interactions, it is necessary to include the effects of cyclic deformation and also the effects of repeated cycling. A study by Yoon, Saxena, and McDowell [12.12] and another by Adefris, Saxena, and McDowell [12.13] considered such effects for trapezoidal loading waveforms for a 1.25Cr-0.5Mo steel and a 1Cr-1Mo-0.25V steel, respectively. These studies were carried out with the purpose of investigating the applicability of $C_t$ for creep-fatigue crack growth. Thus, the numerical analyses results are presented in the form of variation in $C_t$ as a function of time following the sudden application of load for several successive cycles.

Detailed finite element analyses were carried out using a code developed by Leung and McDowell [12.14, 12.15] which combines the rate-independent state variable plasticity model applicable to fatigue loading with nonlinear kinematic hardening. The total strain is partitioned into elastic, plastic, and creep. The plasticity and creep deformations are decoupled. The model has the flexibility to include both the secondary and primary creep behavior. For details about the model, the reader is referred to a paper by McDowell [12.15]. Both analyses were carried out on 50.8mm wide CT specimens assuming plane strain conditions.

We will first discuss the results of Yoon et al. [12.12] which include stable cyclic plasticity behavior and secondary creep in addition to elastic deformation. Figure 12.8 shows a plot of the numerically calculated value ot the $C_t$ parameter as a function of normalized time and Figure 12.8b shows a plot of the C-integral, equation (10.26), plotted similarly. In these analyses, the specimens were rapidly loaded to the maximum load level, the load was held for 10 seconds in one case and for 10 minutes in the other, and was then quickly removed. The C-integral directly represents the amplitude of the crack tip stress fields as was described in Section 10.2.4. The $C_t$ parameter is calculated using equations (10.40) and (10.42). In terms of the overall trends in the C-integral and $C_t$, there is not much difference. Since $C_t$ is used as a crack tip parameter for characterizing crack growth rate, we will be primarily using $C_t$ in the discussion from this point forward to characterize the crack tip conditions. We observe in Figures 12.8a and b that there is considerable influence of cyclic plasticity in the initial portion of the $C_t$ and C-integral trends. This is evident when the trends for the first cycle are compared with the trend for the elastic-secondary creep trend which can be analytically predicted using the combination of equations (10.41) and (10.42). There also is a significant difference between $C_t$ values between the first and second cycles. By comparison, the difference between the second and third cycle is marginal, indicating that the crack tip conditions quickly settle into a repeatable pattern which is followed during each cycle. The initial value

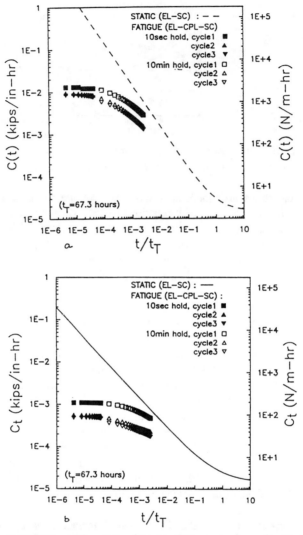

**Figure 12.8** (a) $C_t$ and (b) $C(t)$ as a function of time in fatigue analyses for 10 second and 10 minute hold times, from Yoon et al. (Ref. 12.12). The material was a 1.25Cr-0.5Mo steel at 538 °C with $A = 1.26 \times 10^{-26}$ MPa$^{-n}$ hr$^{-1}$ and $n = 10.3$.

of $C_t$ for each cycle does not appear to be significantly influenced by the hold time because it seems the same for the 10-sec and 10-minute hold-time cases.

Figures 12.9 and 12.10, respectively, show the development of the monotonic plastic zone at the end of the hold period of each cycle and the development of the cyclic plastic zone between the consecutive cycles from the 600-second hold-time case. The monotonic plastic zones are defined as the locus of points for which the magnitude of effective elastic and plastic strains are equal. The cyclic plastic zone is similarly defined by the locus of points where the ranges of the elastic and plastic effective strains are equal. The ranges are calculated based on the difference between the strain at the end of a cycle and the beginning of the hold time during the subsequent cycle. Even though the monotonic plastic zone increases cycle by cycle, the cyclic plastic zone size from the second cycle on is almost constant much like the $C_t$ and $C(t)$ response. Therefore, it is believed that the cyclic plasticity and not the monotonic plastic-

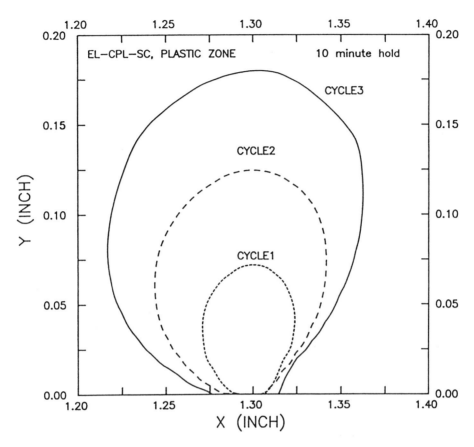

*Figure 12.9 Development of the monotonic plastic zones at the end of the hold period of each cycle for a cycle with a 10-minute hold time. The crack tip is located at 33mm (1.3in) along the x-axis, from Yoon et al. (Ref. 12.12).*

ity is governed the initial crack tip stress field. The effective stress as a function of distance from the crack is shown in Figure 12.11.

Next, we move to discuss the results of the analyses performed on 1 Cr-1Mo-0.25V steel by Adefris et al. [12.13]. In this material, it was found that in order to adequately represent the observed trends, it was necessary to account for plasticity, secondary creep, and also primary creep in the finite element analysis of the specimen. Figure 12.12 shows a plot of $C_t$ as a function of normalized time which also includes measured values of $C_t$. We note that the analysis including elastic, plastic, and only secondary creep under-predicts the $C_t$ value considerably. When primary creep is added, the results agree reasonably well with measured values of $C_t$ and also the analytical values of $C_t$ which includes primary and secondary creep but not plasticity. However, the difference between including cyclic plasticity or not including it is somewhat small. This behavior is quite different from the earlier case in which the cyclic deformation during loading and unloading was quite significant. Figure 12.13a shows a plot of $C_t$ as a function of normalized time with the time initialized to coincide with the start of each hold period. There seems to be significant differences between the first, second, and third cycles with no clear trend toward stabilization from one cycle to the next of the type observed for 12.5Cr-0.5Mo steel shown in Figure 12.8.

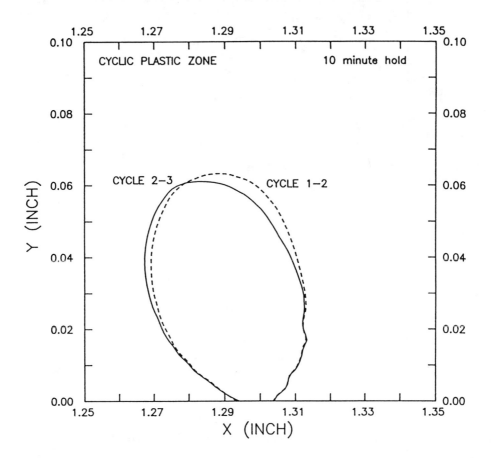

*Figure 12.10 Development of cyclic plastic zones between the end of a cycle and the beginning of the hold period of the following cycle in fatigue analysis with a 10-minute hold time for the same specimen as in Figure 12.9, from Yoon et al. (Ref. 12.12).*

In Figure 12.13b, we plot the same results as in Figure 12.13a, except the time is considered running between cycles. In other words, we set the time at the beginning of the current cycle to equal the time at the end of the previous cycle. The $C_t$ vs. time behavior falls into a single trend. This strongly implies that the loading and unloading events have no lasting influence on the crack tip stress behavior of the type seen in 1.25Cr-0.5Mo steel seen earlier. Therefore, the two steels represent very different kinds of response during creep-fatigue loading. The relative sizes of the creep, cyclic plastic zone and monotonic plastic zones are compared at the end of the third cycle in Figure 12.14. Clearly, the cyclic plastic zone is the smallest and is embedded in the other two suggesting the domination of creep over fatigue in this material. Therefore, it is not surprising that the influence of cyclic plasticity is not significant.

Figure 12.15 attempts to reconcile the two very diverse trends between a Cr-1Mo-0.25V steel and a 1.25Cr-0.5Mo steel when subjected to creep-fatigue loading. It schematically represents the comparison between the relative sizes of creep zone and the cyclic plastic zone for two types of material behavior. Figure 12.15a represents a material in which the creep zone is larger than the cyclic plastic zone,

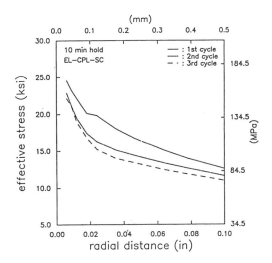

*Figure 12.11* Effective stress distribution along the 94° radial line from the crack tip at the beginning of the hold period during three successive cycles, from Yoon et al. (Ref. 12.12).

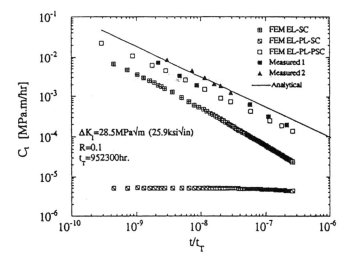

*Figure 12.12* $C_t$ as a function of normalized time for elastic-secondary creep, (EL-SC), elastic-plastic-secondary creep (EL-PL-SC), and elastic-plastic-secondary and primary creep (EL-PL-PSC) finite element analyses for 15-minute hold time tests along with experimental results from two separate tests, from Adefirs et al. (Ref. 12.13).

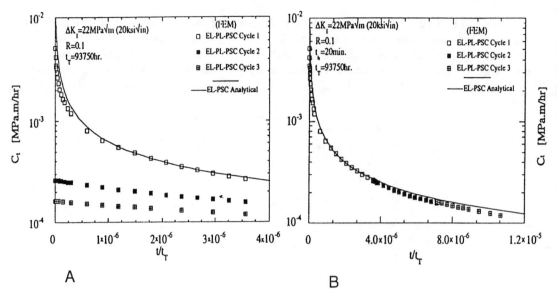

**Figure 12.13** *The creep fracture mechanics parameter $C_t$ as a function of normalized time (a) for the finite element analysis of a CT specimen of 1 Cr-1Mo-o.25V at 538 °C using elastic-plastic and primary and secondary creep for 3 successive fatigue cycles. The time is reinitialized at the beginning of each hold time (b) the same results as in (a) except the time for the second and third cycle are taken cumulatively, from Adefris et al. (Ref. 12.13).*

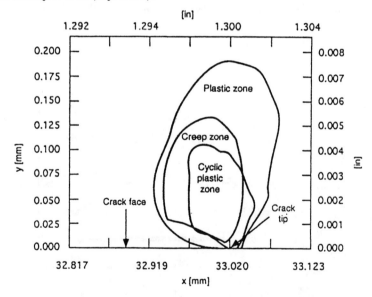

**Figure 12.14** *Relative sizes of the cyclic plastic zone, monotonic plastic zone, and creep zone at the end of the third cycle of a 20-minute hold trapezoidal loading analyses from Figures 12.12 and 12.13.*

such as in the case of 1Cr-1Mo-0.25V steel. This is expected in a material with high cyclic yield strength, therefore not only is the cyclic plastic zone smaller for the same applied $\Delta K$, the creep rates in the crack tip region are accelerated due to the high stresses. This condition can also be induced by simply increasing the length of the hold time to increase the creep zone size. Re-instatement of the stress amplitude will not occur from cycle to cycle as shown schematically in Figure 12.15a where the $C_t$ in each subsequent cycle is shown to start from the value at the end of the previous cycle. Figure 12.15b, on the other hand, represents a material whose creep zone is completely engulfed by the cyclic plastic zone as a result of either a short hold time or low cyclic yield strength of the material. Under this condition, significant reversal of creep deformation occurs and the material will behave as one in which the $C_t$ value is reinstated during each cycle.

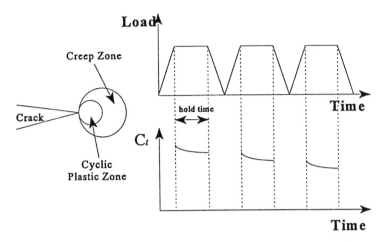

(a) Creep Zone Size >> Cyclic Plastic Zone Size

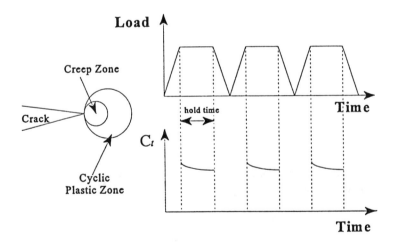

(b) Cyclic Plastic Zone Size >> Creep Zone Size

*Figure 12.15* Schematics of the change in load and $C_t$ as a function of time for trapezoidal loading at elevated temperature (a) for the condition when the creep zone size formed during the hold time is much greater than the cyclic plastic zone size; (b) for the condition when cyclic plastic zone size is much greater than creep zone size.

## 12.3 Crack Tip Parameters during Creep-Fatigue

Most successful correlations of time-dependent crack growth during trapezoidal loading waveform at elevated temperature are with time-dependent fracture mechanics parameters. These parameters are based on estimating C* values using measured load-line deflection rate. The first available evidence of this approach is in the work of Jaske and Begley [12.16] and Taira, Ohtani, and co-workers [12.17]. These proposals were for extensive creep conditions. A similar approach based on C(t)-integral was proposed by Saxena et al. [12.4, 12.10] during the same period. The latter approach also included small-scale and transition creep conditions which are quite commonly encountered in creep-fatigue. This approach was later modified in which the time-dependent crack growth rate during the hold time of a fatigue cycle is characterized by the average value of the $C_t$ parameter, $(C_t)_{avg}$, during the hold time in a given cycle [12.18]:

$$(C_t)_{avg} = \frac{1}{t_h} \int_{t(N)}^{t(N)+t_h} C_t \, dt \tag{12.2}$$

where t(N) is a function of the number of fatigue cycles, N. For materials in which the crack tip stresses are completely re-instated and insignificant creep deformation accumulates from cycle to cycle, t(N) can be chosen as 0, such as for 1.25Cr-0.5Mo steels. On the other hand, when negligible creep reversal due to cyclic plasticity occurs, such as in the case of 1Cr-1Mo-0.25V steel, t(N) is a simple linear function of N which for the $N^{th}$ cycle will be $(N - 1)t_h$. Thus, for materials which exhibit complete stress reversal each cycle:

$$(C_t)_{avg} = \frac{1}{t_h} \int_0^{t_h} C_t \, dt \tag{12.3a}$$

and for materials that do not exhibit any stress reversal:

$$(C_t)_{avg} = \frac{1}{t_h} \int_{(N-1)t_h}^{Nt_h} C_t \, dt \tag{12.3b}$$

A method to interpolate between the two limits will be presented in the next section. The $\Delta J_c$ of Taira et al. is defined in the following way:

$$\Delta J_c = \int_0^{t_h} C^* \, dt \tag{12.4}$$

From equations (12.3a) and (12.4), it is clear that $(C_t)_{avg} = \Delta J_c / t_h$ for conditions when $C_t = C^*$. Since this condition is met for extensive creep conditions, we can state that $\Delta J_c$ is a subset of the $(C_t)_{avg}$ parameter. The $(C^*)_{exp}$ parameter proposed by Webster et al. [12.19] which is based on the measured value of $C^*$-integral, is identical to $(C_t)_{avg}$ for extensive creep conditions. For small-scale creep and transition creep, it is approximately equal to $(C_t)_{avg}$, within a factor of 2, with the difference varying with the geometry of the specimen. Thus, it appears that the seemingly different approaches proposed by various researchers are not that different after all. The $(C_t)_{avg}$ approach is able to capture the essence of all other parameters. Therefore, in the subsequent discussion, it will be used more frequently to address creep-fatigue crack growth. In the next section, we will focus on methods of determining $(C_t)_{avg}$.

## 12.4 Methods of Determining $(C_t)_{avg}$

Methods of determining $(C_t)_{avg}$ include (i) those that are more suitable for test specimens in which both load and load-line deflection behavior with time are measured and (ii) those that are more suitable for components in which the deflection rates must be analytically predicted. These methods are discussed in that order.

### 12.4.1 Methods for Determining $(C_t)_{avg}$ in Test-Specimens

$(C_t)_{avg}$ for test specimens can be obtained from the following equation [12.9]:

$$(C_t)_{avg} = \frac{\Delta P \Delta V_c}{BWt_h} \frac{F'}{F} - C^* \left( (F'/F) / \eta - 1 \right) \tag{12.5}$$

The above equation is a direct adaptation from equation (10.46) with the following modifications. $\dot{V}_c$ is replaced by $\Delta V/t_h$ where $\Delta V$ is the amount of deflection accumulation during the hold time. $\Delta P$ corresponds to the amplitude of the applied load given by $P_{max} - P_{min}$. $C^*$ is calculated using the load during the hold time, $P_{max}$. Thus, the value calculated represents the average value of $C_t$ during the hold time. Since the deflection change is measured, the influence of instantaneous plasticity, primary and secondary creep are already in that measurement. If primary creep is significant, it should be included in the calculation of $C^*$ as in equation (10.58).

### 12.4.2 Analytical Methods of Determining $(C_t)_{avg}$

In deriving these equations, we will consider instantaneous cyclic plasticity and secondary creep, assuming plane strain conditions. Primary creep can be added, if necessary, following the methods outlined in Section 10.3. We will also separate the discussion for the cases of (i) complete reversal of creep deformation, (ii) no creep deformation reversal by cyclic plasticity, and (iii) a condition in between the two limiting cases.

**$(C_t)_{avg}$ for Complete Creep Reversal** We will first derive an expression for elastic, secondary creep deformation and then modify it to include instantaneous cyclic plasticity. Combining equations (12.3a), (10.41), and (10.42), we can easily show that [12.18]:

$$(C_t)_{avg} = \frac{2\alpha\beta(1-\nu^2)}{E} F_{cr}(\theta,n) \frac{\Delta K^4}{W} F'/F (EA)^{\frac{2}{n-1}} t_h^{-\frac{n-3}{n-1}} + C^* \tag{12.6}$$

The above equation can be adjusted for instantaneous cyclic plasticity by following a procedure first suggested by Kubo [12.20] and implemented by Yoon et al. [12.12]. When load is suddenly applied to the specimen and if we assume that $m = n$ (this may be an over simplification and will be discussed further), the initial stress distribution must be the same as for an elastic secondary creep material with a creep zone size, $r_c$, which is proportional to the cyclic plastic zone size, $r_p^c$. If we then assume that a time,

$t_{p1}$, is needed for the creep zone size to become equal to $r_p^c$, the crack tip stress field for elastic-plastic-creeping material at time, t, is equivalent to that of the elastic, secondary creep material at time $t + t_{pl}$. Thus, the $(C_t)_{avg}$ can be obtained by first substituting $t + t_{pl}$ for t in equation (10.42) and then substituting the result in equation (10.2a). The resulting expression for $(C_t)_{avg}$ is given by:

$$(C_t)_{avg} = \frac{2\alpha\beta(1-v^2)}{E} F_{cr}(\theta,n) \frac{\Delta K^4}{W} F'/F (EA)^{\frac{2}{n-1}} \left[ \frac{(t_n+t_{pl})^{\frac{2}{n-1}} - t_{pl}^{\frac{2}{n-1}}}{t_n} \right] + C^* \quad (12.7)$$

The value of $t_{pl}$ can be analytically estimated by relating $r_c$ and $r_p^c$ as follows:

$$r_p^c = \xi_1 \, r_c \, |_{t = t_{pl}} \quad (12.8)$$

If we designate the cyclic plastic exponent by $m'$, and relate $r_p^c$ to $m'$ and the .2% cyclic yield strength, $\sigma_0^c$, we get:

$$r_p^c = \xi_2 \left( \frac{m'-1}{m'+1} \right) \left( \frac{\Delta K}{\sigma_0^c} \right)^2 \quad (12.9)$$

Then combining equations (12.8), (12.9), and the equation for $r_c$, (10.23a), we get:

$$t_{pl} = \frac{1}{EA} \left[ \xi \left( \frac{m'-1}{m'+1} \right) \left( \frac{1}{2\sigma_0^c} \right)^2 \frac{1}{\alpha F_{cr}(90,m')} \right]^{(n-1)/2} \quad (12.10)$$

where

$$\alpha = \frac{1}{2\pi} \left( \frac{(n+1)^2}{2n\alpha_n^{n+1}} \right)^{\frac{2}{n-1}} \quad (12.11)$$

and

$$\xi = \xi_2 / \xi_1 \quad (12.12)$$

This equation shows that $t_{pl}$ is not a function of geometry or applied load and is a material constant. The value of $\xi_1$ is one for m = n but not for m≠n; its value will depend on m and n. For example, for m = 5.4 and n = 8, $\xi_1$ = 1.26 from comparing the results of the finite element analysis with equation (12.8) and $\xi_2$

= 0.69 for m' = 5.4, making $\xi = 0.54$. Using the above value of $t_{pl}$, the $(C_t)_{avg}$ was calculated for specimens in which $(C_t)_{avg}$ was also measured. Figure 12.16 shows the comparison between the predicted and experimental value of $(C_t)_{avg}$. The comparison appears to be quite satisfactory. However, for other materials, in which m' ≠ n and these values are different from 5.4 and 8, respectively, we do not have an analytical method for estimating $t_{pl}$. Perhaps $\xi = 0.547$ is a good starting point and the other parameters in equation (12.10) are also available. No verification of such an assumption either by finite element or by experiments is available. If experimental estimates of $(C_t)_{avg}$ are available, it is perhaps best to determine the value of $t_{pl}$ by comparing experimental and predicted values of $(C_t)_{avg}$.

One can also see that if small-scale-creep conditions dominate, the contribution of $C^*$ will be negligible in comparison to the first term on the right-hand side of equation (12.7). Then for a given material and hold time, one can make the argument that $(C_t)_{avg}$ and $\Delta K$ are uniquely related. Hence, if da/dN for a constant hold time were plotted as a function of $\Delta K$ as in Figure 12.4, the correlation is not surprising and does not contradict the correlation between $(da/dt)_{avg}$ and $(C_t)_{avg}$. However, a distinct trend will be observed for each hold time as also observed in Figure 12.4. Such equivalence relates this more recent approach with the early approach for characterizing creep-fatigue crack growth.

**$(C_t)_{avg}$ for No Creep Reversal** The $(C_t)_{avg}$ for this condition can be estimated by [12.13]:

$$(C_t)_{avg} = \frac{4\alpha\beta F_{cr}(\theta,n)}{(n-1)EW} (1-v^2) K^4 \frac{F'}{F} (EA)^{\frac{2}{n-1}} (Nt_h)^{-\frac{n-3}{n-1}} + C^* \qquad (12.13)$$

This equation is identical to equation (10.42) except for time which is replaced by $Nt_h$. The derivation of the equation is left as an exercise. Also, note that since $t_h$ is multiplied by N; the result for finite fatigue cycles will be much larger than $t_{pl}$. Therefore, $t_{pl}$ can be ignored as is the case when dealing with sustained loading.

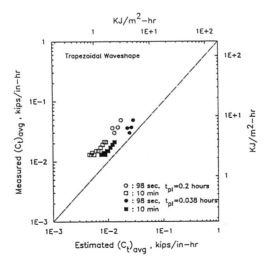

**Figure 12.16** *Comparison between the measured and estimated values of $(C_t)_{avg}$ for 1.25Cr-0.5Mo CT specimens subjected to fatigue cycles with 98-second and 10-minute hold times, from Yoon et al. (Ref. 12.12).*

**$(C_t)_{avg}$ for Partial Creep Reversal** Grover [12.21] suggested a method by which $(C_t)_{avg}$ can be estimated for situations which are in between complete reversals and no reversal at all. Figure 12.17 schematically shows the applied load vs. time in a trapezoidal loading waveform and the deflection vs. time behavior. He noted that if $\Delta V_h$ is the total hold time accumulated during the hold time and $\Delta V_R$ is the displacement range which is reversed; the latter is the difference between the displacement at the end of the hold time and the displacement at the beginning of the hold time during the subsequent cycle. Thus, the two displacements are measured immediately prior to unloading and immediately after the completion of the subsequent loading to obtain $\Delta V_R$. The creep reversal parameter, $C_R$, is given by:

$$C_R = \frac{\Delta V_R}{\Delta V_h} \qquad (12.14)$$

Then the value of $(C_t)_{avg}$ is calculated as:

$$(C_t)_{avg} = [1 - C_R]\frac{1}{t_h}\int_{(N-1)t_h}^{Nt_h} C_t\, dt + C_R \frac{1}{t_h}\int_0^{t_h} C_t\, dt \qquad (12.15)$$

As one can easily see, when $C_R = 1$, equation (12.15) reduces to equation (12.3a) and if $C_R = 0$, it reduces to equation (12.3b). Grover [12.21] used experimental data to study the variation in $C_R$ as a function of fatigue cycles. An example of this variation is shown in Figure 12.18. It seems that large variations in $C_R$ occur during the early portion of the test and during the final portion of the test, and in

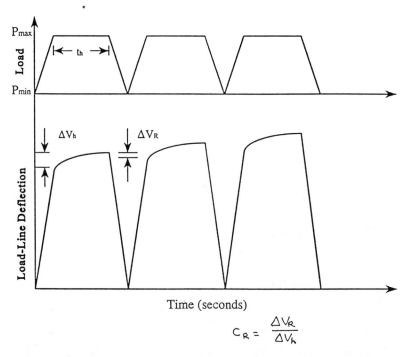

**Figure 12.17** *Definition of the creep-reversal parameter, $C_R$, and the experimental quantities required for calculating $C_R$, from Grover (Ref. 12.21).*

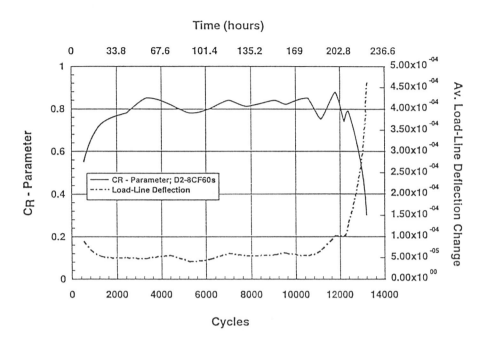

**Figure 12.18** *Variation of $C_R$ and the load-line deflection change during hold time during a typical CFCG experiment, from Grover (Ref. 12.21).*

the middle it remains approximately constant. We expect $C_R$ to be a function of the ratio of the cyclic plastic zone size to the creep zone size and the fatigue cycles, N:

$$C_R = f\left(\frac{r_p^c}{r_c}, N\right) \quad (12.16)$$

No forms for the function f have yet been proposed.

## 12.5 Experimental Methods for Characterizing Creep Crack Growth

Primarily, two methods have been used to obtain creep-fatigue crack growth rate data. The first is a traditional fracture mechanics method based on testing compact type specimens subjected to cyclic loading under load-control. In this method, CT specimens are subjected to cyclic loading of a prescribed loading wave form and frequency and a positive load ratio. The load-line displacement during the entire loading cycle is measured and the crack size is measured by the electric potential method. A schematic of such an experimental set up is shown in Figure 12.19. This method yields satisfactory results provided linear-elastic conditions can be maintained in the specimens. Incremental creep and plastic deformation can accumulate during each cycle and eventually the dominance of linear elasticity conditions cannot be assured. It has been observed [12.9, 12.21, 12.22] that if the initial a/W values in test specimens of about 0.3 are chosen, sufficient crack growth data in Cr-Mo steels and Cr-Mo-V steels can be obtained in the

414  *Nonlinear Fracture Mechanics for Engineers*

temperature range of interest before LEFM conditions are violated. It has also been found that the extensometry required in this method to accurately measure changes in the load-line displacement during a small hold period must be of a higher quality than required for creep crack growth testing. This is because the displacements during one cycle are so small that the resolution of the displacement gage has to be an order of magnitude better than used during creep crack growth testing. Also, the extensometer should be capable of reversing the direction of displacement without backlash and have the needed frequency response of typically 1 Hz.

The second method used for generating creep-fatigue crack growth data has been largely through the efforts of Japanese researchers [12.23-12.26]. In this method, hollow cylindrical specimens containing a small crack along the circumference are loaded under axial conditions. The loading is imposed in the form of prescribed displacement or stress. Crack size is monitored using electric potential method as in the previously described method. The test specimens are typically 10 to 13mm in diameter. Because of their small size, it is unlikely that significant data under dominantly elastic conditions can be obtained. As yet, to the best of our knowledge, no direct comparisons between creep-fatigue crack growth rate data from CT specimens and the hollow cylindrical specimens are available.

## 12.6 Creep-Fatigue Crack Growth Correlations

Early in this chapter, we had represented creep-fatigue crack growth rate data as da/dN vs. $\Delta K$. In the subsequent sections, we argued that such a correlation cannot always be expected and defined time-dependent fracture mechanics parameter $(C_t)_{avg}$ to correlate with the time-dependent crack growth rate during creep-fatigue loading. We even showed that in special cases $\Delta K$ and $(C_t)_{avg}$ are equivalent para-

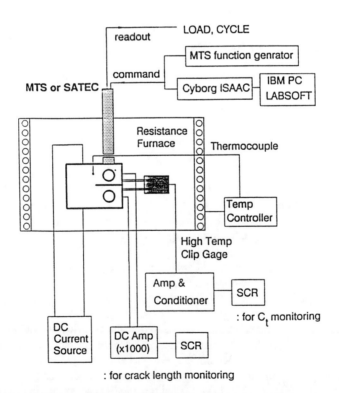

*Figure 12.19 A schematic of the test setup for creep-fatigue crack growth testing.*

meters for loading waveforms with hold times. Thus, an argument can be made that $(C_t)_{avg}$ is a unifying parameter which is inclusive of the earlier correlations between creep-fatigue crack growth rates expressed as da/dN and $\Delta K$. The use of $(da/dt)_{avg}$ implies that the crack growth rate is not uniform during the cycle. For example, during each hold period the crack growth rate, da/dt, as well as the value of $C_t$ varies with time. However, these variations cannot be measured and it is therefore necessary to average their values over the entire cycle.

The total crack growth rate during the cycle can be partitioned into a cycle dependent part and a time-dependent part. Such partitioning is given by:

$$\frac{da}{dN} = \left(\frac{da}{dN}\right)_{cycle} + \left(\frac{da}{dN}\right)_{time} \tag{12.17}$$

The cyclic portion of the crack growth rate is determined from a high frequency fatigue test with no hold period. Then, one invokes the assumption that the cycle dependent crack growth is not influenced by the stress redistribution at the crack tip due to creep deformation. To test this hypothesis, it is useful to consider data from fatigue crack growth tests with various hold times starting from 0. Figure 12.20 shows fatigue crack growth rate data for cyclic frequencies of 1 Hz and 0.083 Hz with no hold time along with data from loading waveforms that have hold times ranging from 100 seconds to 8 hours [12.27]. The loading/unloading periods during cycles with hold time were the same as for the 0.083 Hz pure fatigue cycle. It is clear that the da/dN for loading cycles with a 100 sec hold time was less than the da/dN for 0 hold time at the low end of the $\Delta K$ values for which data are available. These data clearly challenge the approach of crack growth partitioning into cycle-dependent and time-dependent components.

Another approach of considering the cycle-dependent and time-dependent crack growth rates during creep-fatigue is to assume that those mechanisms are competing mechanisms and the crack growth rate is determined by whichever of the two is greater [12.28]:

$$\frac{da}{dN} = \max\left[\left(\frac{da}{dN}\right)_{cycle}, \left(\frac{da}{dN}\right)_{time}\right] \tag{12.18}$$

The data shown in Figure 12.20 clearly favors the above approach known as the dominant damage hypothesis. The $(da/dt)_{avg}$ by the partitioning approach is given by:

$$\left(\frac{da}{dt}\right)_{avg} = \frac{1}{t_h}\left[\frac{da}{dN} - \left(\frac{da}{dN}\right)_0\right] \tag{12.19}$$

and by the dominant damage hypothesis:

$$\left(\frac{da}{dt}\right)_{avg} = \frac{1}{t_h}\left[\frac{da}{dN}\right] \tag{12.20}$$

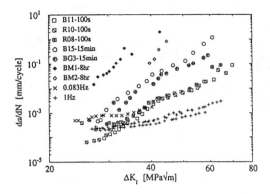

***Figure 12.20*** *Cyclic crack growth rate, da/dN, as a function of ΔK for 1Cr-1Mo-0.25V steel at 538 °C for loading waveforms with varying hold times from 0 to 8 hours. The 0 hold time tests were at loading frequencies of 1 Hz and 0.083 Hz. The loading/unloading times for the 0.083Hz match the loading/unloading times for the hold time tests (Ref. 12.27).*

The difference between the two approaches is small except in the very low ΔK range. Because of its more general applicability and also being conservative, equation (12.20) is favored for analyzing creep-fatigue data. In applications, it is more conservative to use equation (12.19), therefore, it is preferred in those cases.

### 12.6.1 Creep-Fatigue Crack Growth Rate Correlations

Figure 12.21 shows the correlation between $(da/dt)_{avg}$ and $(C_t)_{avg}$ for a 1.25 Cr-0.5Mo steel at 538°C for tests conducted with a variety of hold times ranging from 10 seconds to 24 hours and also including creep crack growth rate for the same material [12.9]. All data appear to lie in the same scatter band. These data imply that time-dependent crack growth during creep-fatigue is identical to creep crack growth. This is a very significant observation for applications of these data. The 10-second hold-time tests were completed within seventy five hours and the 24-hour hold-time tests required up to 10,000 hours for completion and typical creep crack growth tests required 2,000 hours for completion. Thus, the 10-second hold-time test can be used to predict the behavior of tests conducted over periods of 2 to 3 orders of magnitude longer. This conclusion, though very useful, must be used with caution because it may hold for only certain types of materials such as the creep-ductile materials. Figure 12.22 shows creep-fatigue crack growth rate data for 1 Cr-1Mo-0.25V steel at various hold times. For this material, the agreement between creep-fatigue data at various hold times followed a similar trend. However, the trend for creep crack growth rate was quite distinct from the creep-fatigue trend and was lower than the creep-fatigue trend [12.27]. A recent study [12.29] showed that 2.25Cr-1Mo steel at 594°C, the creep-fatigue crack growth rate data and creep crack growth rate data show very comparable trends similar to the observations on 1.25Cr-0.5Mo steels. Thus, both possibilities exist if we are considering a variety of materials.

### 12.6.2 Models for Creep-Fatigue Crack Growth

In applications, it is frequently necessary to mathematically represent crack growth rates. The following equation is used to represent the creep-fatigue crack growth rate [12.30]:

$$\frac{da}{dN} = C_0 \, (\Delta K)^{n_0} + C_1 \left[ (C_t)_{avg} \right]^q \tag{12.21}$$

where $C_0$ and $n_0$ are regression constants which represent the fast-frequency or the cycle-dependent part of the crack growth rate, and the second term has $C_1$ and $q$ as regression constants which represent the time-dependent part of the overall crack growth rate. The above equation assumes damage partitioning analysis, which as mentioned before, is preferred for applications because it is conservative.

The above model, though suitable for engineering purposes, lacks a detailed mechanistic explanation and is also focused entirely on creep-damage as the time-dependent damage mechanism. Unfortunately, no complete models currently exist which account for combined environment and creep effects on time-dependent crack growth. In the following discussion, we will consider the results from some phenomenological studies to explore the relative contributions of environment and creep to the time-dependent

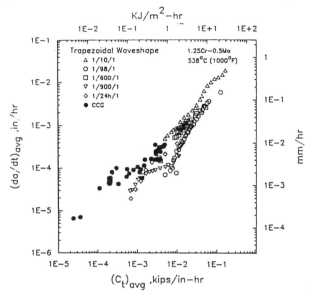

**Figure 12.21** $(da/dt)_{avg}$ *as a function of* $(C_t)_{avg}$ *for 1.25Cr-0.5Mo steel at 538 °C for hold times ranging from 10 sec to 24 hours and including creep crack growth (CCG) results (Ref. 12.9).*

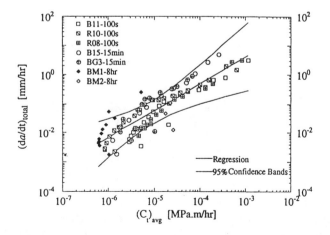

**Figure 12.22** *Same as Figure 12.21 except for 1Cr-1Mo-0.25V steel at 538 °C (Ref. 12.27).*

crack growth behavior. At the outset of this discussion, it is important to recognize that such results are extremely material dependent as well as environment (including temperature) dependent. For example, Figure 12.6 shows how cyclic crack growth rate in the frequency range of 0.01 to 1 Hz changes in air and the environment changes for Inconel 718 at 650°C [12.7]. In air, the da/dN increases very significantly for the same $\Delta K$ as the loading frequency decreases. In He, there is no significant change in the da/dN vs. $\Delta K$ behavior between loading frequencies of 1Hz and 0.1Hz and only a slight increase in da/dN was observed when the loading frequency was decreased to 0.01Hz. Thus, it is evident that the time-dependent damage in Inconel 718 at this temperature is dominated by environment.

We now turn to an example from 1.25Cr-0.5Mo steel [12.9] where it is shown that the crack growth rates in air are somewhat higher than the crack growth rates at 538°C in vacuum for a hold time of 10 sec but no significant differences were observed between the crack growth rates in air and in vacuum at higher hold times. This observation indicates that at shorter hold times (~10 sec), environment plays a significant role in determining the crack growth rate while at longer hold times (~100 sec or more), creep is dominant. Further investigations of the damage in the crack tip regions of specimens tested at various hold times were conducted. Figures 12.23a and 23b show the crack profiles of specimens tested at a hold time of 10 sec in air. The dominant crack growth appears to be transgrannular with several oxide spikes emanating from the crack surface. Figures 12.24a and 12.24b show the crack profiles and the damage ahead of the crack tip of a specimen tested at a hold time of 600 seconds clearly showing cavitation damage indicative of creep, and no oxide spikes were observed. Based on these observations, the following explanations are presented for the mechanisms of creep-fatigue crack growth in 1.25Cr-0.5Mo steels.

The explanations of the relative importance of creep and environment are presented with the help of Figure 12.25. A trapezoidal loading waveform is assumed. If parabolic type oxidation occurs at the crack tip [12.32], the time-rate of damage accumulation due to oxidation is high during the early stages of the cycle time. The oxidation damage zone may be defined as a zone including the oxide layer and surrounding oxygen-penetrated region. As time elapses, the damage rate due to oxidation decreases. The rate of damage accumulation must further decrease due to passivation from the oxide layer which forms on the fracture surface. Therefore, the effects of oxidation on da/dt are expected to be prominent only during the beginning portion of the cycle time and then saturate. On the other hand, the effects due to creep deformation and cavitation damage on da/dt does not change significantly through the cycle time. Consequently, at the early stage of the cycle time both oxidation and creep influence the time-dependent crack growth rate and after saturation of oxidation (perhaps due to passivation) the effect of creep becomes dominant. Thus, the cycle time can be divided into three regimes as shown in Figure 12.25; the oxidation dominant region (Region I), the creep dominant region (Region III) and the transition region (Region III). The times over which the various regions span is dependent on material, temperature, and perhaps also the stress level. A detailed quantitative model for creep-fatigue-environment effect must consider all these factors.

## 12.6.3 Transients During Creep-Fatigue Crack Growth

As we have explained previously, fatigue crack growth at high temperature under loading cycles which include hold time can be represented as da/dN vs. $\Delta K$ for a fixed hold time or as $(da/dt)_{avg}$ vs. $(C_t)_{avg}$ for a variety of hold times under creep dominated conditions. It was also shown that for materials and loading conditions in which the creep deformation is completely reversed during unloading, the da/dN vs. $\Delta K$ relationship is entirely consistent with the $(da/dt)_{avg}$ vs. $(C_t)_{avg}$ correlations. However, a large number of tests and material conditions do not follow this idealization. For example, in Figure 12.26, we show the da/dN vs. $\Delta K$ behavior at several hold times for a 1.25Cr-0.5Mo steel. This is the same data which was previously shown in $(da/dt)_{avg}$ vs. $(C_t)_{avg}$ plot earlier in Figure 12.21. Similar trends showing an initially decreasing da/dN with $\Delta K$ have been observed in several earlier studies [12.18,12.32]. Such derivation from a unique da/dN vs. $\Delta K$ relationship is called transient behavior.

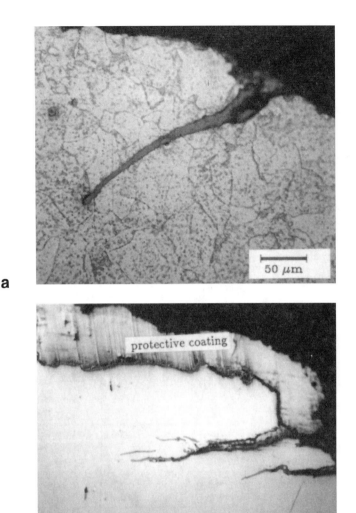

*Figure 12.23* Oxide spikes observed on the crack profiles of 1.25Cr-0.5Mo specimens tested at 538 °C under fatigue loading with hold times of 10 seconds (a) etched specimen to reveal the transgrannular crack growth and (b) unetched specimen (Ref. 12.19).

The above type of transient behavior can be normalized by simply plotting the data in terms of $(da/dt)_{avg}$ vs. $(C_t)_{avg}$. For example, when the data shown in Figure 12.26 is represented in Figure 12.21, we no longer observe the anamalous trends. The reason for this correlation is as follows. The $(C_t)_{avg}$ values from equation (12.3a) and (12.3b) were given by equations (12.7) and (12.13), respectively. Next, substituting these results into equation (12.15) gives a value of $(C_t)_{avg}$ which is both a function of $\Delta K$ and cycles, N. For a constant hold time, we assume $0 < C_R < 1$. Grover [12.21] has shown that such types of transients are nicely predicted by this model.

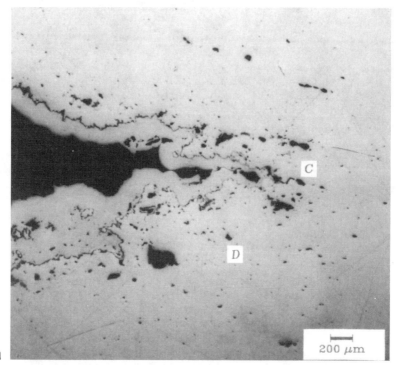

**Figure 12.24** Cavitation damage ahead of the crack tip of a specimen of 1.25Cr-0.5Mo steel tested at 538 °C with a hold time of 10 minutes (a) unetched specimen and (b) etched specimen clearly showing the intergrannular crack growth and cavitation (Ref. 12.9).

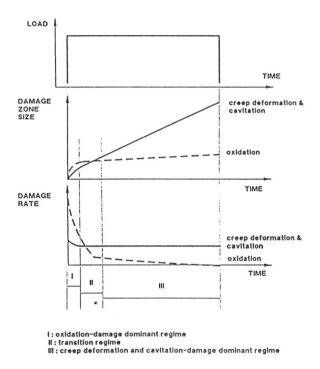

I : oxidation-damage dominant regime
II : transition regime
III : creep deformation and cavitation-damage dominant regime

*Figure 12.25 Schematic representation showing the transition of the dominant time-dependent crack growth mechanism from oxidation during the very early portion of the hold time to creep during the later portion of the hold time (Ref. 12.9).*

Another reason for transient damage condition is due to the development of steady-state damage condition in the crack tip region which evolves over a period of time. When the loads are first applied, the crack tip material is in an undamaged state. Alternatively, it could have damage which is reflective of the fatigue precracking which is typically conducted at room temperature and high frequency. Thus, the damage state changes from initial condition to the steady-state condition that is reflective of the creep-fatigue loading over some crack extension. Thus, transient crack growth response can be expected during the interim period. After sufficient crack extension has occurred, and steady-state damage conditions have been established, the normal crack growth behavior is expected. Such transients are expected to be present even if the data are correlated with $(C_t)_{avg}$.

## 12.7 Summary

When linear-elastic fracture mechanics (LEFM) conditions can be maintained such as during high loading frequencies, $\Delta K$ was shown to be an appropriate crack tip parameter for characterizing creep-fatigue crack growth rates for a constant loading frequency and waveform. As the temperature rises and the loading frequencies decrease, the validity of $\Delta K$ becomes questionable and nonlinear fracture mechanics parameters are needed. The $\Delta J$-integral and parameters such as $(C_t)_{avg}$ and $(\Delta J_c)$ are more appropriate where $\Delta K$ becomes invalid. These parameters are described in this chapter. It was shown that $\Delta J_c$ and $(C_t)_{avg}$ are equivalent under extensive creep conditions and also $\Delta K$ and $(C_t)_{avg}$ are equivalent under conditions in which the creep deformation during hold time is completely reversed by cyclic plas-

## 422  Nonlinear Fracture Mechanics for Engineers

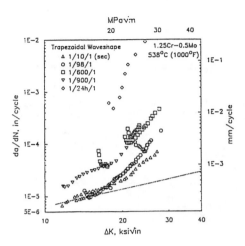

*Figure 12.26* Fatigue crack growth rates in 1.25Cr-0.5Mo steel at 538 °C as a function of ΔK for various hold times. These data are the same as ones plotted in Figure 12.21 in the form of $(da/dt)_{avg}$ vs. $(C_t)_{avg}$ (Ref. 12.9).

ticity during unloading. Thus, $(C_t)_{avg}$ emerges as the most widely applicable crack tip parameter for representing hold-time effects during creep-fatigue crack growth. Methods for estimating $(C_t)_{avg}$ in specimens and components are described in this chapter as are the test methods for characterizing creep-fatigue crack growth rates. Several examples of data correlation with ΔK and $(C_t)_{avg}$ were shown. It was found that $(C_t)_{avg}$ provides crack growth rate correlation which do not significantly vary with hold time or with slight variations in temperature, at least for some creep-ductile steels. This is a major advantage of this approach in engineering applications.

The currently available models for predicting creep-fatigue crack growth rates in components were briefly presented and their limitations were discussed. The transients in the crack growth behavior observed during the early portions of crack growth during creep-fatigue crack growth testing were also discussed.

## 12.8 References

12.1    K.B. Yoon, A. Saxena, and D.L. McDowell, "Effect of Cyclic Overload on the Crack Growth Behavior during Hold Period at Elevated Temperature", International Journal of Fracture, Vol. 59, 1993, pp. 199-211.

12.2    L.A. James, "The Effect of Frequency Upon the Fatigue-Crack Growth of Type 304 Stainless Steel at 1000F", in Stress Analysis of Growth of Cracks, ASTM STP 513, American Society for Testing and Materials, Philadelphia, 1972, pp. 218-229.

12.3    L.A. James, "Hold-Time Effects on the Elevated Temperature Fatigue-Crack Propagation of 304 Stainless Steel", Nuclear Technology, Vol. 16, 1972, pp. 521-526.

12.4    A. Saxena and J.L. Bassani, "Time-Dependent Fatigue Crack Growth Behavior of Elevated Temperature", in Fracture: Interactions of Microstructure, Mechanisms and Mechanics, TMS-AIME, Warrendale, 1984, pp. 357-383.

12.5  A. Plumtree and M. Yu, "Influence of Waveshape on High Temperature Crack Growth Behavior in Stainless Steel", in Thermal and Environmental Effects in Fatigue: Research-Design Interface, C. Jaske, S. Hudak, and M. Mayfield, editors, PVP-Vol. 71, American Society for Mechanical Engineers, 1983, pp. 13-19.

12.6  R.M. Pelloux and J.S. Huang, "Creep-Fatigue Environment Interactions in Astrolloy", in Creep-Fatigue Environment Interactions, R. Pelloux and N. Stollogg, editors, TMS-AIME, 1980, pp. 151-164.

12.7  S. Floreen and R.H. Kane, "An Investigation of Creep-Fatigue-Environment Interactions in Ni Base Super Alloy", Fatigue of Engineering Materials and Structures, Vol. 2, 1980, pp. 401-412.

12.8  A. Saxena, "A Model for Predicting Environment Enhanced Fatigue Crack Growth Behavior at High Temperatures", in Thermal and Environmental Effects in Fatigue: Research Design Interface, PVP-Vol. 71, American Society for Mechanical Engineers, 1983, pp. 171-184.

12.9  K.B. Yoon, A. Saxena, and P.K. Liaw, "Characterization of Creep-Fatigue Crack Growth Behavior under Trapezoidal Waveshape Using $C_t$-Parameter", International Journal of Fracture, Vol. 59. 1993, pp. 95-114.

12.10  A. Saxena, T.T. Shih, and R.S. Williams, "A Model for Representing the Influence of Hold Time on Fatigue Crack Growth Behavior at Elevated Temperature", in Fracture Mechanics: Thirteenth Conference, ASTM STO 743, 1981, pp. 86-99.

12.11  H. Riedel, "Crack-tip Stress Fields and Crack Growth Under Creep-Fatigue Conditions", in Elastic-Plastic Fracture: Second Symposium Vol. I - Inelastic Crack Analysis, ASTM STP 803, American Society for Testing and Materials, Philadelphia, 1983, pp. I/505-I/520.

12.12  K.B. Koon, A. Saxena, and D.L. McDowell, "Influence of Crack Tip Plasticity on Creep-Fatigue Crack Growth", in Fracture Mechanics: Twenty Second Symposium (Vol. I) ASTM STP 1131, American Society for Testing and Materials, Philadelphia, 1992, pp. 367-392.

12.13  N. Adefris, A. Saxena, and D.L. McDowell, "Creep-Fatigue Crack Growth Behavior in 1Cr-1Mo-0.25V steel. Part I: Estimation of Crack Tip Parameters", Fatigue and Fracture of Engineering Materials and Structures, Vol. 19, 1996, pp. 387-399.

12.14  C.P. Leung and D.L. McDowell, "Inclusion of Primary Creep in the Estimation of $C_t$ Parameter", International Journal of Fracture, Vol. 46, 1990, pp. 81-104.

12.15  D.L. McDowell, "A Nonlinear Kinematic Hardening Theory for Cyclic Thermo-Plasticity Theory and Thermo-Visoplasticity", International Journal of Plasticity, Vol. 8, 1992, pp. 695-728.

12.16  C.E. Jaske and J.A. Begley, "An Approach to Assessing Creep/Fatigue Crack Growth:, in Ductibility and Toughness Considerations in Elevated Temperature Service", MPC-ASME-8, American Society for Mechanical Engineers, 1978, pp. 391-409.

12.17  S.Taira, R. Ohtani, and T. Komatsu, "Application of J-integral to High Temperature Crack Propagation Part II--Fatigue Crack Propagation", Transactions of ASME, Journal of Engineering Materials Technology, Vol. 101, 1979, pp. 163-167.

12.18  A. Saxena and B. Gieseke, "Transients in Elevated Temperature Crack Growth", Proceedings of MECAMAT, International Seminar on High Temperature Fracture Mechanisms and Mechanics III, EGF-6, 1987, pp. 19-36.

12.19  K. Nishida and G.A. Webster, "Interaction Between Build Up of Local and Remote Damage on Creep Crack Growth", in Creep and Fracture of Engineering Materials and Structures, B. Wilshire and R. Evans, editors, Institute of Metals, 1990, pp. 703-711.

12.20  S. Kubo, Private communication to A. Saxena, 1988.

12.21  P.S. Grover, "Creep-Fatigue Crack Growth in Cr-Mo-V Base Material and Weldments", Ph.D. Thesis, Georgia Institute of Technology, May 1996.

12.22  A. Saxena, "Limits of Linear Elastic Fracture Mechanics in the Characterization of High Temperature Fatigue Crack Growth", ASTM STP 924, American Society for Testing and Materials, 1988. pp. 27-42.

12.23  R. Ohtani, T. Kitamura, A. Nitta, and K. Kuwabara, "High Temperature Low Cycle Fatigue Crack Propagation of Life Laws of Smooth Specimens Derived from the Crack Propagation Laws", ASTM STP 942, American Society for Testing and Materials, 1988, pp. 163-169.

12.24  K. Kuwabara, A. Nitta, T. Kitamura, and T. Ogala, "Effect of Small-Scale Creep on Crack Initiation and Propagation Under Cyclic Loading", ASTM STP 942, American Society for Testing and Materials, 1988, pp. 41-59.

12.25  R. Ohtani, T. Kitamura, and K. Yamada, "A Nonlinear Fracture Mechanics Approach to Crack Propagation in Creep-Fatigue Interaction Range", in Fracture Mechanics of Tough and Ductile Materials and its Application to Energy Related Structures, H. Liu, T. Kurio, and V. Weiss, editors, Materials Nighoff Publishers, 1981, pp. 263-270.

12.26  K. Ohji, "Fracture Mechanics Approach to Creep Fatigue Crack Growth", in Role of Fracture Mechanics in Modern Technology, Elsevier Science Publishers, Japan, 1987, pp. 131-144.

12.27  N. Adefris, A. Saxena, and D.L. McDowell, "Creep-Fatigue Crack Growth Behavior in 1Cr-1Mo-0.25V steel. Part II--Crack Growth Behavior and Models", Fatigue and Fracture of Engineering Materials and Structures, Vol. 19, 1996, pp. 401-411.

12.28  K. Ohji and S. Kubo, "Fracture Mechanics Evaluation of Crack Growth Behavior Under Creep and Creep-Fatigue Conditions", in High Temperature Fracture and Fatigue, Current Japanese Materials Research, Vol.3, Elsevier Applied Science, Tokyo, 1988, pp. 91-113.

12.29  P.S. Grover and A. Saxena, "Characterization of Creep Fatigue Crack Growth Behavior in 2.25Cr-1Mo Steel Using $(C_t)_{avg}$", International Journal of Fracture, Vol. 73, 1995, pp. 273-286.

12.30  A. Saxena, "Fracture Mechanics Approaches for Characterizing Creep-Fatigue Crack Growth", JSME International Journal, Series A, Vol. 36, No. 1, 1993, pp. 1-20.

12.31  N. Birks and G.H. Meier, "Introduction to High Temperature Oxidation of Metals", Edward Arnold, 1983.

12.32 K. Ohji and S. Kubo, "Crack Growth Behavior and Fractographs Under Creep-Fatigue Conditions Including the Near Threshold Regime", in Fractography, Publication of the Society of Materials Science, Japan, Elseveir Applied Science, 1990, pp. 105-120.

## 12.9 Exercise Problems

12.1 The creep deformation properties of 304 stainless steel at 594°C, 1.25 Cr-0.5Mo steel at 538°C and 1Cr-1Mo-0.25V steel at 538°C are given below:

|  | A (MPa$^{-n}$ x hr$^{-1}$) | n | E GPa |
|---|---|---|---|
| 304 SS (594°C) | 2.13 x 10$^{-18}$ | 6 | 172 |
| 1Cr-1Mo-0.25V (538°C) | 1.16 x 10$^{-24}$ | 8 | 162 |
| 1.25Cr-0.5M0 (538°C) | 4.49 x 10$^{-20}$ | 8 | 140.6 |

You are asked to generate creep fatigue crack growth data in the $\Delta K$ range of 15 to 25 MPa $\sqrt{m}$ for 304 SS and 1.25Cr-1Mo steel at 594°C and 538°C, respectively, and in the $\Delta K$ range of 25 MPa $\sqrt{m}$ to 40 MPa $\sqrt{m}$ for the 1Cr-1Mo-0.25V steel at 538°C using load-controlled conditions. The longest hold time is 100 seconds. Recommend the size of the compact specimen required and the starting a/W values to complete this testing. You can assume that the loading/unloading rate is very rapid.

12.2 Compare the transition times required for CT and CCT specimens of identical widths (values of W) for a 1Cr-1Mo-0.25V steel using the data provided in Exercise Problem 12.1.

12.3 If the following additional data on the plastic properties of 1Cr-1Mo-0.25V and 1.25Cr-0.5Mo steels are provided to you, estimate the value of $t_{pl}$ assuming that both materials are cyclically stable:

| Material and Test Temp (°C) | E (GPa) | .2% yield str.$\sigma_0$ (MPa) | $\varepsilon_0$ | $\alpha$ | m |
|---|---|---|---|---|---|
| 1Cr-1Mo-.25V, 538 | 162 | 473 | 2.92 x 10$^{-3}$ | 1.95 | 21.6 |
| 1.25Cr-0.5Mo, 24 | 206.8 | 211.6 | 1.02 x 10$^{-3}$ | 2.73 | 4.5 |
| 1.25Cr-0.5Mo, 538 | 140.6 | 131.0 | 0.93 x 10$^{-3}$ | 2.43 | 5.4 |

12.4 Complete all necessary steps to show that for the condition that the entire creep deformation during hold time is reversed during unloading, $(C_t)_{avg}$ is given by equation (12.6). What assumptions are implicit in equation (12.6) and how do they influence the accuracy of $(C_t)_{avg}$ for materials?

12.5 For a material deforming by elastic-secondary creep and in which the creep zone size exceeds the cyclic plastic zone size, show that the $(C_t)_{avg}$ is given by equation (12.7).

12.6 Assume dominantly small-scale-creep conditions and that in a material being tested under fatigue loading with hold time, only part of the creep deformation is reversed by cyclic plasticity during unloading. Show that the da/dN is a function of $\Delta K$ and the number of fatigue cycles (hint: you may assume that $(da/dt)_{avg}$ is correlated with $(C_t)_{avg}$ ).

12.7 Show that when creep-fatigue crack growth data are reduced by the dominant-damage hypotheses and during application the damage summation hypothesis is used, the predicted component lives will be conservative.

12.8 Construct a schematic diagram of the type shown in Figure 12.25 for Inconel 718 from the information about this material you can glean from the results provided to you in Figure 12.6. Make sure that the relative contributions to crack growth due to creep and environment are correctly represented in your figure.

12.9 Suppose you were asked to generate fatigue crack growth data with hold time for a 1Cr-1Mo-0.25V steel at 538°C using compact type specimens with a width, W. The tests are to be conducted under displacement (load-line) range control in a manner such that the displacement range is incremented as the crack grows to achieve a linear increase in $\Delta K$ with crack growth, $\Delta a$. Derive how $\Delta V$ must change with $\Delta a$ to achieve this condition. You may assume that instantaneous plasticity is negligible.

12.10 Repeat the above problem, if one cannot assume that the instantaneous plasticity can be neglected.

# CHAPTER 13

# CASE STUDIES

Fracture mechanics provides the analytical framework for developing a comprehensive methodology for predicting the potential for fracture in components. In this methodology, the magnitude and distribution of applied stresses, the resistance to fracture and crack growth of the component material, and the size, location, and geometry of defects in the component can all be simultaneously considered in predicting the crack growth and fracture behavior. This approach is also capable of determining the influence of service environment. In the previous chapters of this book we have described the physical and mathematical basis of these concepts which form the analytical framework. This chapter illustrates the generalized methodology for performing fracture mechanics analysis of engineering components through realistic examples.

## 13.1 Applications of Fracture Mechanics

Applications of fracture mechanics can be classified in several categories, but the basic methodology is common among the seemingly different applications. Depending on the need, the results of the analysis can be presented in different ways. Some broad classes of applications are briefly discussed.

### *13.1.1 Integrity Assessment of Structures and Components*

Preliminary and sometimes even final design of components is based on simple design rules which evolve from previous experience and because of their widespread use, become industry standards. They can often result in either over-conservatism or insufficient allowances for factors which contribute to fracture of components. For example, consider the widely used design rule that the maximum allowable stresses in a component not exceed 75% of the 0.2% yield strength of the material. This design criterion can be over-conservative for a material with high fracture toughness but may have an inadequate safety margin for low fracture toughness materials. In the latter case, the presence of even a small defect in a highly stressed region can cause catastrophic fracture. Fracture mechanics is used to assess the risk of fracture or the realistic safety margin in structures and components designed by such simple rules.

The risk (or safety margin) assessment consists of determining the flaw tolerance of the component using a fracture mechanics analysis. This flaw size is compared to those which can realistically exist in the component due to the fabrication process employed such as casting, welding, forging, etc., or to those that can escape detection during nondestructive inspection. Such analyses can also be used for recommending a suitable in-service maintenance and inspection program for preventing catastrophic failures. Depending on the results of such analysis, the maximum allowable stresses can either be adjusted upwards to save weight or adjusted downwards to improve the safety margin. Thus, fracture mechanics analysis provides an interactive mathematical model for evaluating a preliminary design and achieving an economically optimized final design by balancing several integrity related requirements.

### *13.1.2 Material and Process Selection*

As described in the previous chapters, fracture mechanics provides a wide variety of test methods for evaluating materials and fabrication processes. Several of these methods have been standardized by international standards writing bodies. These include fracture toughness testing [13.1-13.9], fatigue crack growth testing [13.10], dynamic fracture testing and crack arrest [13.11], creep crack growth testing [13.12], stress corrosion testing [13.13], and creep-fatigue crack growth testing [13.14-13.16]. These methods collectively provide established techniques for testing and evaluating candidate materials and fabrication processes for critical applications. However, these tests are frequently expensive and in several instances prohibitively so. Therefore, fracture mechanics test methods are not replacements for the conventional test methods such as tensile strength, hardness, charpy impact test, or other material screening tests that are often used in different industries. In fact, more than likely the preliminary material selection and process selections are made on the basis of the inexpensive tests and fracture mechanics methods are used to make final material selections.

### 13.1.3 Design or Remaining Life Prediction

Frequently when the design has already been finalized or the component is already in service, engineers are asked to predict how long the component is expected to last. From the knowledge of the magnitude and nature of the operating stresses and environment, the relevant fracture and crack growth properties and the nondestructive inspection data, it is possible to conduct a fracture mechanics analysis to predict the design life or remaining life of components already in service. Examples of such estimations will be provided among the case studies described in the later sections of this chapter.

### 13.1.4 Inspection Criterion and Interval Determination

Frequently, the nondestructive inspection techniques that are used during quality assurance or during service to detect cracks or crack-like defects in components are dictated by considerations such as the material, geometry and size of the component, the accessibility of the critical areas of the component for placing transducers, and by economics. Inspections, especially during service, can be expensive because they are often time consuming and result in long down-times, involve expensive equipment, and require highly trained professionals to perform them and to interpret results. Conservative inspection criterion and intervals can result in high rejection rates and more frequent inspections than necessary, thus, increasing the operating cost of conducting business. On the other hand, inadequate inspection criterion and insufficient inspection intervals can cause catastrophic failures with devastating consequences. Fracture mechanics analysis provides an effective analytical tool for estimating realistic inspection intervals and for determining realistic inspection criteria. Several examples of such use of fracture mechanics will be illustrated in the case studies described in this chapter.

### 13.1.5 Failure Analysis

Failure in equipment often occurs due to sudden fracture in a component. Even in electronic components such as computers, failures are often related to thermal-fatigue fractures of joints providing the electrical connection. In aerospace, automotive, and power-generation industries, the threat of failure caused by fracture of a component is always an important consideration. By conducting a thorough analysis of a failed part, important lessons can be learned about preventing similar failures in the future. The ability of fracture mechanics to relate operating stresses, material properties, and the defect sizes to fracture is extremely useful in analyzing and preventing future failures. This will be illustrated with an example in this chapter.

## 13.2 Fracture Mechanics Analysis Methodology

The information necessary for conducting fracture mechanics analysis consists of (i) identification of fracture critical regions which are usually sections of the component which are highly stressed or subjected to aggressive environment, or both; (ii) the operating conditions data such as the service environment, the parameters which determine the operating stress levels such as rotational speeds, power output, etc. and the size, shape, and location of manufacturing or service generated cracks or crack-like defects; (iii) the pertinent material data including the crack growth rates under the appropriate environment and loading conditions, fracture toughness, tensile properties, fracture appearance transition temperature (FATT), material chemistry, and microstructure.

After the above information is assembled, the next step in the fracture mechanics analysis is to calculate the magnitude of the applicable crack tip parameters, $K$, $J$, $C_t$, $C^*$ etc., using the applied stress level and expressing them as a function of the characteristic crack dimension, a. From the fracture toughness and the expressions for estimating K or J, the final crack size, $a_f$, can be estimated. The crack growth rate data can then be used to estimate the crack size at inspection, $a_i$, to meet the design life requirement. Alternatively, if $a_i$ is chosen as the largest crack that can go undetected by the chosen nondestructive inspection technique, the expected component life can be calculated. Often, $a_i$ is chosen conservatively so that the component life is also predicted conservatively. If that is the case, the predicted life should be treated merely as an inspection interval or the frequency at which inspections must be performed to detect cracks during service. If no cracks are detected during the in-service inspection, the component may be returned back to service until the next inspection becomes due, provided all other functional requirements are met.

Fracture mechanics analysis is often performed with the aid of computers because once a computer program is developed several scenarios can be examined. There are commercial computer packages available to perform fracture mechanics analysis [13.17-13.19], but, engineers very often prefer to write their own computer programs which are specific to their needs and the components they most frequently work with. Component-specific computer programs, such as PCPIPE [13.18] for integrity analysis of elevated temperature steam pipes are easy to use.

## 13.3 Case Studies

In this section, several examples are described to illustrate the use of nonlinear fracture mechanics. In each case, the assumptions are discussed to identify the limitations of the results. The examples chosen are hypothetical but do have some resemblance to actual situations except the failure analysis example which is an actual case history.

### 13.3.1 Integrity Analysis of Missile Launch Tubes

**Problem Statement** A missile launch tube has a mean diameter of 0.5m and is made from ASTM grade A537 steel which is 25mm thick. The tube is several meters long and contains several circumferential welds along its length. The welding is performed by an automatic submerged arc welding process described by the American Welding Society (AWS) specification A5.17. The process can potentially leave semi-elliptical flaws with flaw depth, a, along the thickness direction and the flaw length, 2c, along the circumferential direction. The weld metal is matched in all respects with the base metal and the weldment is stress relieved. A conservative stress calculation shows the tube can be subjected to an axial tensile stress of 125 MPa during a missile launch. You are asked to recommend an inspection procedure for the weldments. The choice is between a complete radiographic inspection, which can detect very small flaws and is capable of providing the flaw dimensions accurately, and a not very accurate method using a fiber-optic probe which scans the interior surface of the tube and provides pictures on a television screen. The latter method can miss flaws up to 20mm long along the circumference and it cannot provide any data regarding the flaw depth. The cost of the radiographic inspection is more than 50 times that of the fiber-optic probe method. Hence, it is not desirable to use radiography unless absolutely necessary. As an expert on fracture mechanics, you are asked to step in and recommend an inspection procedure. Your recommendation must be supported by rigorous analysis which you must complete within three months.

**Material Properties** The relevant material properties for avoiding fracture during the missile launch is the fracture toughness of the material in the weldment region where the flaws are most likely to be located. Also, the tensile properties are needed for estimating J. Since pieces from the welded tube are inconvenient for machining compact type (CT) specimens, we choose to acquire two 25mm thick plates of ASTM grade A537 steel and weld them together using the automatic submerged arc welding process (as shown in Figure 13.1a) following the same AWS specification used for welding the tubes. The test weldment is given the appropriate post-weld heat treatment (PWHT) to simulate the weld process of the tubes. The microstructure of the weld is characterized and the hardness values are measured in the base metal, weld metal, and the heat affected zone regions to ensure that the weld characteristics in the tube are entirely duplicated in the test material. If the results verify that the weld characteristics of our test coupon are the same as those of the missile tube material, we can proceed to machine tensile and compact type specimens from the sample. The fracture samples can be machined such that the notch plane lies along the fusion plane of the test weldment as shown in Figure 13.1b. This is chosen because the defects in the tube are also likely to be present at the interface between the weld metal and the base metal. Tensile specimens can be machined from the base metal and the weld metal regions as also shown in Figure 13.1.b. The purpose of testing tensile specimens from the two regions is to once more ensure that the weld metal properties and the base metal properties are the same.

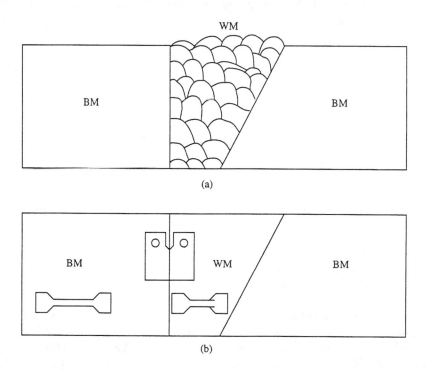

*Figure 13.1* (a) Schematic of the laboratory weld sample for conducting fracture toughness testing and (b) the layout of the test specimens to be machined from the weldment.

The following data describe the tensile and fracture toughness behavior of the weldment:

0.2% yield strength = $\sigma_o$ = 380 MPa
$m = 5.0$, $\alpha = 4.675$
$E = 205$ Gpa
$\epsilon_0 = 1.85 \times 10^{-3}$, $\nu = 0.3$
$J_{IC} = 385.5$ kJ/m$^2$

The $J_R$ curve is given by $J_R = 493.24 \times 10^3 \Delta a$, where $J_R$ is in kJ/m² and $\Delta a$ is in meters. Thus:

$$T_{mat} = \frac{dJ}{d\Delta a} \frac{E}{\sigma_o^2} = 700.2 \text{ (dimensionless)}$$

**Fracture Mechanics Analysis** The first step in analyzing the component is to assemble the expressions for estimating K and J. Since the fracture in this material is accompanied by extensive plastic deformation, we can assume that an elastic-plastic analysis will be necessary. Since the optical technique cannot suitably measure crack depth in the component, we must assume that every crack is a through-thickness crack. The expressions for estimating K and the plastic portion of the J-integral, $J_p$, for this configuration are given by [13.20-21]:

$$K = \sigma_t \sqrt{\pi R \theta} F_t \qquad (13.1)$$

where $\sigma_t$ = axial stress, R = mean radius of the pipe, $2\theta$ = crack angle as shown in Figure 13.2, and $F_t$ is given by (note that $c = \theta R$):

$$F_t = 1 + (0.125\frac{R}{t} - 0.25)^{0.25}[5.33(\frac{\theta}{\pi})^{1.5} + 18.773(\frac{\theta}{\pi})^{4.24}] \qquad (13.2)$$

t = wall thickness of the tube. The plastic part of J, $J_p$ is given by:

$$J_p = \alpha \sigma_0 \epsilon_0 R(\pi - \theta) h_1 (\frac{P}{P_o})^{m+1} \qquad (13.3)$$

where

$$P = 2\pi R t \sigma_t \qquad (13.4a)$$

and

$$P_o = 2\sigma_o Rt[\pi - \theta - 2\sin^{-1}(0.5\sin\theta)] \qquad (13.4b)$$

and $h_1$ is a function of $\theta/\pi$, m, and R/t. For R/t = 10 and m = 5, Table 13.1 lists the values of $h_1$, as a function of $\theta/\pi$ from reference [13.20].

The following expression describes the above relationship analytically:

$$h_1(\theta/\pi, 10, 5) = 7.454(\frac{\theta}{\pi}) - 20.706(\frac{\theta}{\pi})^2 + 16.21(\frac{\theta}{\pi})^3 \qquad (13.5)$$

**432** *Nonlinear Fracture Mechanics for Engineers*

Equation (13.5) can also be written in terms of the crack length, c, by noting that $\theta = c/R$ and $R = .25$ m:

$$h_1 = 9.495c - 33.601c^2 + 33.51c^3 \quad (13.5a)$$

Similarly, equation (13.2) can be written for $R/t = 10$ as:

$$F_t = 1 + 7.663c^{1.5} + 52.394c^{4.24} \quad (13.6)$$

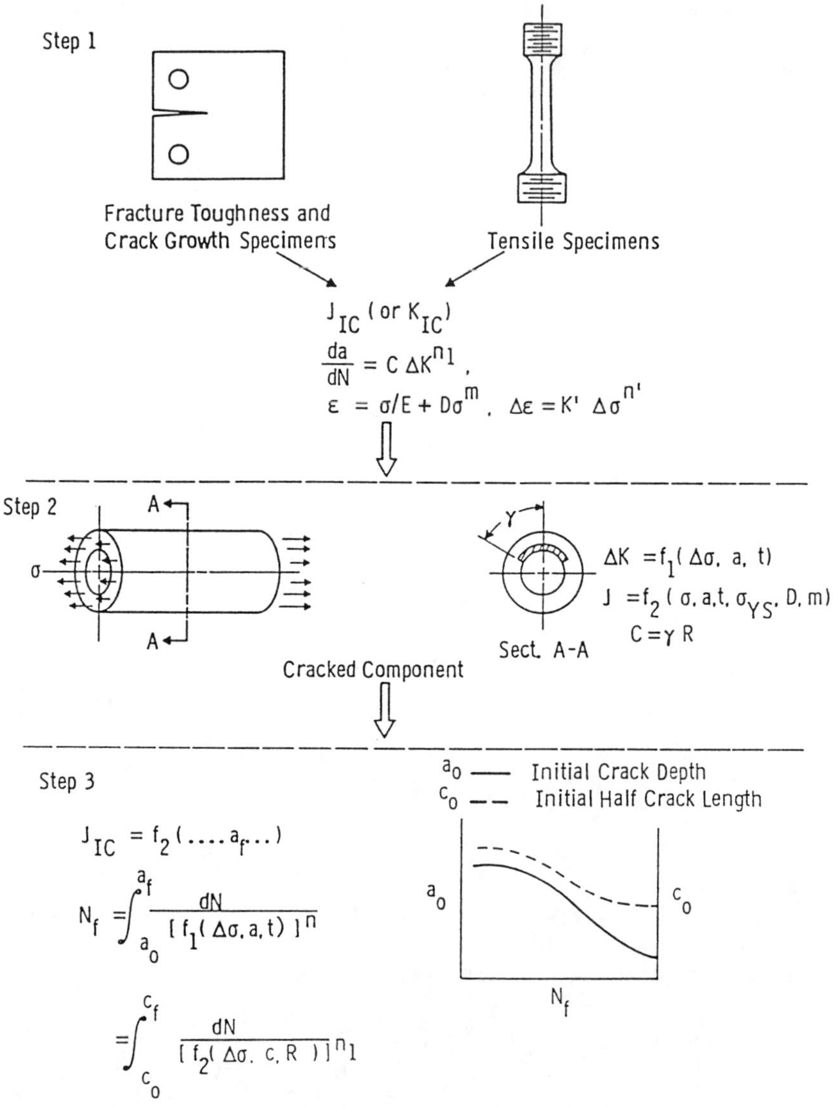

**Step 1**

Fracture Toughness and Crack Growth Specimens

Tensile Specimens

$J_{IC}$ (or $K_{IC}$)

$\frac{da}{dN} = C \Delta K^{n_1}$,

$\varepsilon = \sigma/E + D\sigma^m$, $\Delta\varepsilon = K' \Delta\sigma^{n'}$

**Step 2**

Cracked Component

Sect. A-A

$\Delta K = f_1(\Delta\sigma, a, t)$
$J = f_2(\sigma, a, t, \sigma_{YS}, D, m)$
$C = \gamma R$

**Step 3**

$J_{IC} = f_2(\ldots a_f \ldots)$

$$N_f = \int_{a_0}^{a_f} \frac{dN}{[f_1(\Delta\sigma, a, t)]^n}$$

$$= \int_{c_0}^{c_f} \frac{dN}{[f_2(\Delta\sigma, c, R)]^{n_1}}$$

$a_0$ —— Initial Crack Depth
$c_0$ --- Initial Half Crack Length

**Figure 13.2** *Fracture mechanics methodology for predicting remaining life/inspection interval and* *inspection criterion for steam pipes.*

*Table 13.1 The Values of $h_1$ as a Function of $\theta/\pi$*

| $\theta/\pi$ | $h_1$ | $\theta/\pi$ | $h_1$ |
|---|---|---|---|
|  |  | 0.3 | 0.794 |
| 0.1 | 0.518 | 0.35 | 0.759 |
| 0.15 | 0.730 | 0.40 | 0.704 |
| 0.20 | 0.800 | 0.45 | 0.640 |
| 0.25 | 0.810 | 0500 | 0.568 |

The value of J is given by:

$$J = \frac{K^2}{E}(1-v^2) + J_p \tag{13.7}$$

if we ignore the plastic zone size correction for small-scale yielding. From experience we know that when J values reach $J_{IC}$, the tube will have undergone considerable plastic deformation, therefore, the small-scale yielding value can be neglected without significantly influencing the accuracy of the calculation. However, this can be rigorously verified by comparing $J_e$ and $J_p$ at $J_{IC}$ in Table 13.2.

Substituting for K from equation (13.1) and $J_p$ from equation (13.3) into equation (13.7), we can obtain J as a function c. Table 13.2 lists these values. From this table, it is apparent that the applied J values do not exceed $J_{IC}$ until the half crack length c exceeds 0.20 m or the total crack length 2c exceeds 0.4 m (40 cm). The value of the tearing modulus remains considerably less than $T_{mat}$ even for c values of 0.3 m. Thus, it appears that the inspection with a fiber-optic device will be quite adequate for this purpose.

For c = 0.01m which is the inspection capability with the fiber-optic device, the J is much less than the $J_{IC}$ and, in fact, the value of $J_p$ is only 12.4 percent of the total J, indicating only limited plastic deformation beyond the crack tip. Therefore, even if plastic deformation in the tube is a concern to protect dimensional tolerances of the tube, a 20-mm long defect along the circumference does not pose significant problems. Hence, the recommendation in this case must be against specifying the radiographic inspection to save manufacturing cost of the missle launch tube. This recommendation does not compromise the integrity of the tube.

*Table 13.2 Values of the Various Crack Tip Parameters as a Function of Crack Size*

| c | $\theta$ | $\theta/\pi$ | $h_1$ | $F_t$ | K | $J_p$ | J | $T_{app}$ |
|---|---|---|---|---|---|---|---|---|
| (m) | (rad) |  |  |  | (MPa) | (kJ/m²) | (kJ/m²) |  |
| 0.01 | 0.04 | .0127 | .0916 | 1.007 | 22.3 | .321 | 2.528 |  |
| 0.02 | 0.08 | .0254 | .1658 | 1.0217 | 32.0 | .722 | 5.267 | .388 |
| 0.05 | 0.20 | .0637 | .3950 | 1.086 | 53.77 | 2.732 | 15.56 | .486 |
| 0.10 | 0.40 | .1274 | .6471 | 1.245 | 87.18 | 10.55 | 44.29 | .815 |
| 0.20 | 0.80 | .254 | .8231 | 1.742 | 172.52 | 111.9 | 244.02 | 2.83 |
| 0.30 | 1.20 | .382 | .7295 | 2.577 | 312.57 | 1690.8 | 2124.5 | 26.68 |

## 13.3.2 Integrity of Pipes in Nuclear Power Generating Stations

**Problem Statement** A nuclear power station is being considered for re-rating to produce ten percent more power than it is currently producing. This is being considered in conjunction with the decision to buy a new higher-capacity turbine-generator set to replace the old one. The engineers who worked in the plant when it was first commissioned 20 years ago, recall that the piping system installed to carry steam to the turbine is capable of carrying a higher load and therefore they believe that it is not necessary to replace the piping system. However, the management wants you to conduct a thorough study to ensure that the integrity of the pipes will not be compromised in any way and they also want you to estimate the inspection interval and the criterion they should adopt to get further insurance against pipe failure.

The steam pipe in question is made from austenitic stainless steel, has an outside diameter of 35.56 cm, and a wall thickness of 17.8 mm. In the new configuration, the pipe wall is subjected to an axial tensile stress of 103.4 MPa and a bending stress of 34.5 MPa. These stresses are expected to fluctuate between the maximum value and 0 at a frequency of two times a day. A piece of this pipe which was left over from the original installation is available for testing. You are asked to perform the necessary materials tests and provide an analysis for making a final decision on whether the pipes should be replaced. Figure 13.2 shows a schematic of the various steps involved in a fracture mechanics analysis. The first step involves determining the material properties, the second step involves identifying the stress intensity parameter K and the J-integral expressions for the cracked configuration of interest and the third step is the estimation of remaining life and the critical circumferential crack length for ensuring a leak-before-break.

**Material Properties** The material properties needed for the analysis include the fracture toughness and fatigue crack growth behavior. Since the fracture and fatigue crack growth behavior in this material is likely to be accompanied by significant plastic deformation, the monotonic and cyclic deformation properties may also be needed and, therefore, must also be obtained. The material properties are summarized in a paper by Liaw et al. [13.22,13.23]. The J-R curve for the material is shown in Figure 13.3 and the fatigue crack growth behavior in Figure 13.4. The relevant constants are given below:

monotonic plastic deformation constants
$\sigma_0 = 206.7$ MPa
$E = 200$ Gpa
$\epsilon_0 = 1.033 \times 10^{-3}$
$\alpha = 1.18$
$m = 5.09$
$\sigma_u = 517$ MPa (ultimate tensile strength)

cyclic deformation constants (based on equation 9.2a)
$\sigma_0^c = 310$ MPa
$m' = 7.14$
$\alpha' = 1.93 \times 10^{-3}$

fracture toughness
$J_{IC} = 300$ kJ/m$^2$
$K_{JC} = 257$ MPa$\sqrt{m}$ (value of K converted from $J_{IC}$)

fatigue crack growth
$C = 3.44 \times 10^{-13}$ for $\Delta K$ in MPa$\sqrt{m}$ and da/dN in m/cycle
$n_2 = 3.73$

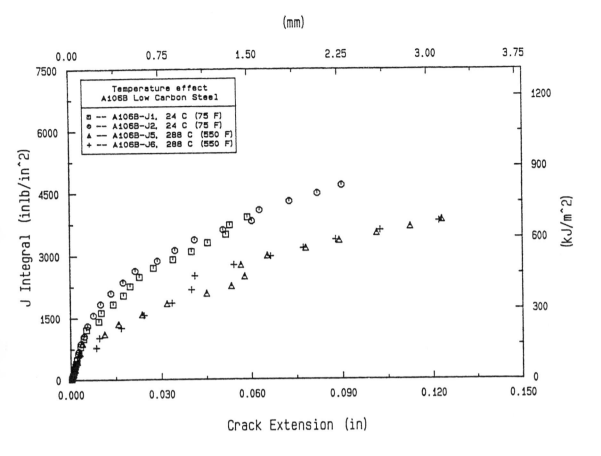

*Figure 13.3  J-R curve of the steam pipe material (Ref. 13.22).*

**Crack Models**  Figure 13.5 shows a part-through circumferential crack located on the outside surface of a pipe. During service, the crack is expected to grow in the circumferential direction as well as in the thickness direction due to axial tension and bending stresses. The growth pattern of the crack is shown by the dotted line. When the crack depth becomes equal to or greater than a critical value, the crack is expected to break through the pipe wall. At this time, rupture may occur or a leak may occur depending on the circumferential length of the flaw. The second failure mode is obviously desirable over the first and is known as the leak-before-break condition. To assure leak-before-break, it is necessary that the circumferential crack length be less than $c_{cr}$ which will cause fast fracture to occur in the circumferential direction. The fracture behavior and the fatigue crack growth behavior under elastic-plastic and fully-plastic conditions are characterized by the J-integral and the $\Delta$J-integral, respectively. It is, therefore, necessary to have expressions for estimating K and J for the pertinent flaw geometries. Figure 13.6 shows the flaw geometries for which K and J expressions are needed for predicting the crack growth and fracture behavior. Figure 13.6a shows the crack model for estimating crack growth in the circumferential direction. For crack growth in the thickness direction, a $\gamma$ value of 90° is chosen, and for crack growth in the circumferential direction, the crack is considered to be a through-the-thickness crack. The reasons for these choices are explained later. The K and J expressions for these geometries are directly taken from the work of Kumar and German [13.20] and from reference [13.21].

For a circumferentially cracked cylinder subjected to an axial load N and a bending moment M at its ends, the K for the cracked configuration shown in Figure 13.6a is given by:

$$K = \sqrt{\pi a}[\sigma_t F_1(\frac{R}{t},\frac{a}{t},\gamma) + \sigma_b F_2(\frac{R}{t},\frac{a}{t},\gamma)] \qquad (13.8)$$

where a = crack depth and:

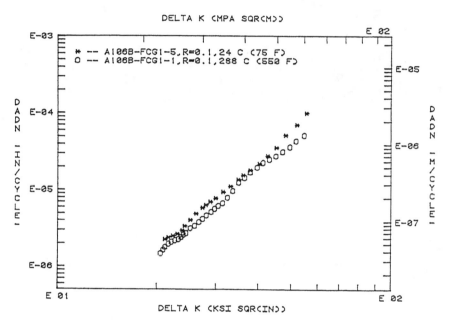

**Figure 13.4** *Fatigue crack growth behavior of the steam pipe material (Ref. 13.22).*

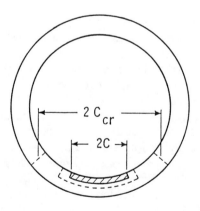

**Figure 13.5** *Schematic illustration of the leak-before-break or rupture criterion.*

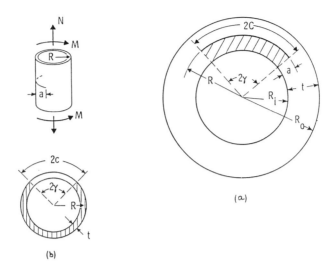

*Figure 13.6 (a) Schematic of a part-through circumferential crack in a cylinder subjected to combined tension and bending and (b) schematic of a through-thickness circumferential crack in a cylinder of constant depth.*

$$\sigma_t = \frac{N}{\pi}(R_i+R_o)t \tag{13.9}$$

$$\sigma_b = \frac{2M(R_i+R_o)}{\pi(R_o^4-R_i^4)} \tag{13.10}$$

Note that $\sigma_t$ and $\sigma_b$ are tension and bending stresses, respectively. $F_1$ and $F_2$ are calibration functions derived from finite element analyses. The $F_1$ values are available for R/t of 10 for different $\gamma$ ranging between 13.75° to 180°. The value of $\gamma = 90°$ was chosen for calculations. $F_2$ was only available for $\gamma = 45°$, therefore, those values were chosen for the calculations. These values of $F_1$ and $F_2$ are given in Tables 13.3 and 13.4, respectively. The cyclic stress intensity parameter is obtained by substituting the cyclic stress, $\Delta\sigma_t$ and $\Delta\sigma_b$ in place of $\sigma_t$ and $\sigma_b$. The expression for estimating the plastic portion of the J-integral, $J_p$, for the above configuration is given in the same reference for $0 \le \lambda \le 10$, where $\lambda$ is the tension-to-bending ratio defined later in this section.

$$J_p = \frac{1-\nu^2}{E}K^2 + Da(1-\frac{a}{t})h_1(\frac{R}{t},\frac{a}{t},\gamma,\lambda,m)[\frac{A}{A_{nc}}\frac{N}{N'_o}]^{m+1} \tag{13.11}$$

where ν is Poisson's ratio:

$$D = \frac{\alpha \varepsilon_o}{(\sigma_o)^m} \tag{13.12}$$

$$A = \pi(R_o^2 - R_i^2)$$

$$A_{nc} = \pi\left[(R_o^2 - R_i^2) + \frac{\pi - \gamma}{\pi}(R_c^2 - R_i^2)\right]$$

$$R_c = R_i + a$$

$$\lambda = \frac{\sigma_b}{\sigma_t}\frac{R_o^2}{(R_o + R_i)^2} = \frac{M}{NR}$$

$$N_o' = \frac{1}{2}\left[-\frac{\lambda N_o^2 R}{M_o} + \sqrt{\left(\frac{\lambda N_o^2 R}{M_o}\right) + 4N_o^2}\right]$$

$$M_o = 2t(R_o^2 - R_i^2)$$

$$K = \sqrt{\pi a}\left[\sigma_b F_1(a/t, \gamma)\right]$$

$$K = \sqrt{\pi a}\left[\sigma_b F_2(a/t, \gamma)\right]$$

$$N_o = \pi(R_o^2 - R_i^2)$$

For $\lambda > 10$, $J_p$ is given by:

$$J_p = \frac{(1-\nu^2)K^2}{E} + Da\left(1 - \frac{a}{t}\right)h_1\left(\frac{R}{t}, \frac{a}{t}, \lambda, m\right)\left[\frac{A}{A_{nc}}\frac{M}{M_o}\right]^{m+1} \tag{13.13}$$

where all parameters have been defined before. $h_1$ is a function of R/t, a/t, $\gamma$, $\lambda$, m, and is given in Table 13.5. For the cracked configuration shown in Figure 13.6(b), K is given by:

$$K = \sqrt{\pi c}F_3(\gamma/\pi, R/t) + \sigma_b F_4(\gamma/\pi, R/t)] \tag{13.14}$$

**Table 13.3** K-Calibration Function $F_1$ for a Part-Through Circumferential Crack in a Cylinder Subjected to Uniform Tensile Stress, $\sigma_p$ for R/t = 10 (from Refs. 13.20, 13.21)

| a/t | $F_1$ | | | | | |
| --- | --- | --- | --- | --- | --- | --- |
| | 2γ=27.5° | 45° | 90° | 180° | 270° | 360° |
| 0 | 1.12 | 1.12 | 1.12 | 1.12 | 1.12 | 1.12 |
| 0.50 | 1.446 | 1.607 | 1.749 | 1.815 | 1.818 | 1.82 |
| 0.55 | - | 1.662 | 1.852 | 1.908 | 1.911 | 1.896 |
| 0.75 | 1.472 | 1.793 | 2.245 | 2.468 | - | 2.443 |

**Table 13.4** K-Calibration Function $F_2$ for a Part-Through Circumferential Crack in a Cylinder Subjected to Pure Bending Stress, $\sigma_b$ for R/t = 10 (from Refs. 13.20, 13.21)

| a/t | $F_2$ (a/t, 45°) |
| --- | --- |
| 0 | 1.12 |
| 0.5 | 1.684 |
| 0.75 | 2.159 |

where c = $\gamma R$, $F_3$, and $F_4$ fosr R/t = 10 are:

$$F_3 = 1 + 5.3303(\gamma/\pi)^{1.5} + 18.773(\gamma/\pi)^{4.24} \quad (13.15)$$

$$F_4 = 1 + 4.5967(\gamma/\pi)^{1.5} + 2.6422(\gamma/\pi)^{4.24} \quad (13.16)$$

The $J_p$ for $0 \leq \lambda \leq 10$ is given by:

$$J_p = Dc\left(-\frac{\gamma}{\pi}\right)h_1\left(\frac{R}{t},\frac{a}{t},m,\lambda\right)\left[\frac{N}{P_o'}\right]^{m+1} \quad (13.17)$$

where:

$$P_o' = \frac{1}{2}\left[-\frac{\lambda P_o^2 R}{M_o'} + \sqrt{\left[\frac{\lambda P_o^2 R}{M_o'}\right] + 4P_o^2}\right]$$

$$P_o = 2Rt[\pi - \lambda - 2\arcsin(\frac{1}{2}\sin\lambda)] \quad (13.19)$$

$$M = \frac{4\sigma_b}{\pi R(R_o^4 - R_i^4)}$$

$$M_o' = 4R^2 t \left[ \cos\left(\frac{\lambda}{2}\right) - \frac{1}{2}\sin(\lambda) \right] \tag{13.21}$$

The $h_1$ functions are defined in Table 13.6. The value of the plastic portion of $\Delta J$, $\Delta J_p$ is obtained by replacing $\sigma_t$ and $\sigma_b$ by their cyclic values, and the constants, D and m, by their cyclic values, as will be explained. After the K/$\Delta K$, $J_p/\Delta J_p$ have been determined, the full magnitude of the J-integral can be obtained by the following relationship:

$$J = \frac{K^2}{E}(1-\nu^2) + J_p \tag{13.22}$$

**Table 13.5** $h_1$ *Functions for Part-Through Circumferential Cracks in Pipes Subjected to Tension and Bending, R/t = 10 (Ref. 13.20)*

| a/t | m | 2γ(degrees) | λ=0 | λ=0.5 | λ=∞ |
|---|---|---|---|---|---|
| 0.5 | 2 | 45 | 17.8 | - | - |
| 0.5 | 2 | 90 | 18.2 | 32.8 | 25.5 |
| 0.5 | 2 | 180 | 13.0 | - | - |
| 0.5 | 5 | 45 | 28.8 | - | - |
| 0.5 | 5 | 90 | 33.1 | 59.4 | 38.7 |
| 0.5 | 5 | 180 | 20.3 | - | - |
| 0.5 | 10 | 45 | 43.5 | - | - |
| 0.5 | 10 | 90 | 57.2 | 104.9 | 51.9 |
| 0.5 | 10 | 180 | 31.2 | - | - |
| 0.75 | 2 | 45 | 48.2 | - | - |
| 0.75 | 2 | 90 | 64.9 | 113.2 | 85.9 |
| 0.75 | 2 | 180 | 45.1 | - | - |
| 0.75 | 5 | 45 | 72.3 | - | - |
| 0.75 | 5 | 90 | 110.8 | 178.1 | 110.9 |
| 0.75 | 5 | 180 | 63.3 | - | - |
| 0.75 | 10 | 45 | 93.6 | - | - |
| 0.75 | 10 | 90 | 145.4 | 249.7 | 103.3 |
| 0.75 | 10 | 180 | 53.5 | - | - |

*Table 13.6 $h_1$ Functions for Through-Thickness Circumferential Cracks Subjected to Tension and Bending, R/t = 10 (Ref. 13.20)*

| 2γ (degrees) | m | λ=0 | 0.5 | 1.0 | 2.0 | 10 |
|---|---|---|---|---|---|---|
| 22.5 | 2 | 3.967 | 6.67 | 7.901 | 8.302 | 7.063 |
| 22.5 | 5 | 5.567 | 10.8 | 12.15 | 12.08 | 9.19 |
| 22.5 | 7 | 6.104 | 11.753 | 14.61 | 15.00 | 10.447 |
| 22.5 | 10 | 6.510 | 16.003 | 17.839 | 16.60 | 10.738 |
| 45 | 2 | 4.157 | 6.975 | 7.93 | 7.996 | 6.772 |
| 45 | 5 | 5.163 | 8.537 | 9.50 | 9.28 | 7.391 |
| 45 | 7 | 5.102 | 8.80 | 9.875 | 9.65 | 7.313 |
| 45 | 10 | 4.750 | 9.77 | 10.935 | 9.95 | 7.158 |
| 67.5 | 2 | 4.159 | 6.527 | 7.36 | 7.37 | 6.152 |
| 67.5 | 5 | 3.238 | 5.70 | 7.33 | 7.13 | 5.108 |
| 67.5 | 7 | 2.605 | 5.56 | 7.375 | 6.90 | 4.372 |
| 67.5 | 10 | 3.000 | 5.42 | 8.150 | 7.26 | 4.353 |
| 90 | 2 | 2.892 | 5.50 | 6.497 | 6.46 | 4.934 |
| 90 | 5 | 1.992 | 3.0 | 3.336 | 2.96 | 1.838 |
| 90 | 7 | 1.46 | 2.72 | 2.985 | 2.55 | 1.509 |
| 90 | 10 | 2.000 | 2.87 | 2.739 | 2.16 | 1.238 |

$\Delta J$ can be similarly obtained by substituting $\Delta K$ for K, and $\Delta J_p$ for $J_p$ in equation (13.22).

**Flow Chart of a Computer Program** Figure 13.7 shows the overall flow chart of a computer program for the fracture mechanics analysis. The first step is to receive the following input from the operator.

- Outside radius
- Pipe thickness
- Initial crack depth
- Maximum tensile stress
- Maximum bending stress
- Load ratio

The maximum bending stress is the maximum normal stress in the axial direction associated with the applied bending load. The next step in the program is to receive the material input data which includes the following:

- Ultimate tensile strength, $\sigma_u$
- Yield strength (0.2%), $\sigma_o$ (also referred to as $\sigma_{ys}$)
- Plasticity coefficient, $D = \alpha\epsilon_o/(\sigma_o)^m$
- Plasticity exponent, m
- Elastic modulus, E
- Fracture toughness:

$$K_{IC} \sqrt{\frac{J_{IC}E}{(1-v^2)}}$$

- Fatigue crack growth coef., C
- Fatigue crack growth exponent, $n_2$
- Cyclic plasticity coefficient, K'
- Cyclic plasticity exponent, n'

The constants K' and n' are cyclic fatigue constants which relate fatigue stress and strain ranges as follows:

$$\left(\frac{\Delta\sigma}{2}\right) = K'\left(\frac{\Delta\epsilon_p}{2}\right)^{n'} \quad (13.24)$$

Although the cyclic fatigue constants are often reported in the literature as values of K' and n', the following form used in equation (9.2a) is used to compute $\Delta J$:

$$\Delta\epsilon = \frac{\Delta\sigma}{E} + 2\alpha'\left(\frac{\Delta\sigma}{2\sigma_o^c}\right)^{m'} \quad (13.25)$$

By comparing the plastic parts of two forms of the cyclic stress-strain relationships, it can be easily shown that:

$$n' = 1/m' \quad (13.26)$$

and

$$K' = \sigma_o^c / (\alpha')^{1/m'} \quad (13.27)$$

and

$$D' = 2\alpha' / (2\sigma_o^c)^{m'} \quad (13.27a)$$

The values of D' and m' are used in the equations described earlier for estimating $\Delta J_p$.

Figure 13.8 shows a detailed flow chart of the subroutine for estimating the critical flaw depth at which time the flaw will break through the pipe wall, and end the life of the pipe. The critical crack depth,

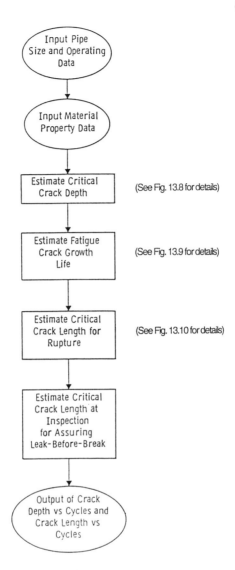

**Figure 13.7** *Main flow-chart of a computer program for analyzing the integrity of steam pipes (Ref. 13.23).*

$a_{cr}$, corresponds to the situation when the applied value of J equals to $J_{IC}$. Since the J-integral concept is based on small-scale deformation plasticity, it is important to ensure that the remaining ligament at any time is not less than 25 $(J/\sigma_f)$, where $\sigma_f = 1/2(\sigma_o + \sigma_u)$. The $a_{cr}$ value is chosen so that it meets both requirements. The determination of the critical flaw depth is an iterative calculation in which the maximum starting crack depth is limited to 0.75 t. The starting crack depth is the upper limit of crack size for which the K and J solutions described previously are valid.

Figure 13.9 shows the flow chart of the crack size vs. elapsed fatigue cycles, N(a). The number of fatigue cycles required for the crack size to reach $a_{cr}$ is called $N_f$. Thus, the remaining life, $N_R$, corresponding to any starting crack depth, a, is given by:

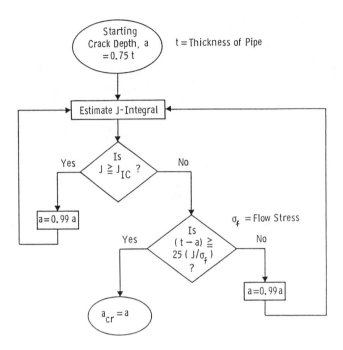

*Figure 13.8* A flow chart for estimating critical crack depth in a steam pipe (Ref. 13.23).

$$N_R = N_f - N(a) \tag{13.28}$$

The remainder of the flow chart is self-explanatory.

Figure 13.10 shows the flow chart for the calculation of the leak-before-break condition. The first step in this subroutine is to estimate the critical circumferential crack length, $c_{cr}$, for which rupture is expected to occur under normal operating conditions. This method is similar to the one for the estimation of $a_{cr}$. Subsequently, the amount of crack extension in the circumferential direction during service is estimated as follows:

$$N_f = \int_{c_o}^{c_R} \left(\frac{dc}{dN}\right) dN \tag{13.29}$$

where (dc/dN) is the circumferential fatigue crack growth rate which can be estimated from the applied value of $\Delta K$ (or $\Delta J$). In the above equation, $c_o$ is the only unknown which can be determined. Also, by replacing $N_f$ with $N(a)$, the $c_o$ corresponding to any crack depth for assuming leak-before-break can be estimated.

The last subroutine consists of the output module. The output consists a table of remaining life as a function of crack depth and the critical circumferential crack length to ensure leak-before-break.

Figure 13.11 shows a plot of crack depth vs. remaining life for bending stress of 34.5 MPa and a tensile stress of 103.4 MPa with the stress ratio of 0. The letter implies that the stress varies from 0 to its maximum value as stated for the problem. The radial crack depth and circumferential crack lengths are plotted as a function of remain life. For example, if an inspection interval of $10^4$ cycles is desired (at 2 cycles per day, it is 5000 days or 13.7 years), the critical crack depth for ensuring leak-before-break must not exceed 82.6 mm (3.25 in). Figure 13.12 shows a plot of crack depth vs. remaining life (or inspection interval) for bending stress levels of 0, 34.5 MPa, and 68.9 MPa. As expected, the remaining life decreases as the bending stress increases.

**Discussion of Results** The calculations presented above provide a relationship between initial crack size and remaining life. The initial crack sizes to meet inspection intervals of $10^4$ cycles (or 13.7 years) are quite large and are detectable by ultrasonic techniques. In fact, ultrasonic techniques are able to reliably detect cracks that are 1 mm or less deep for which the remaining life is approximately $10^5$ cycles (or 13.7 years). Therefore, if the pipes are inspected by ultrasonic techniques every 10 years, we should have a safe piping system.

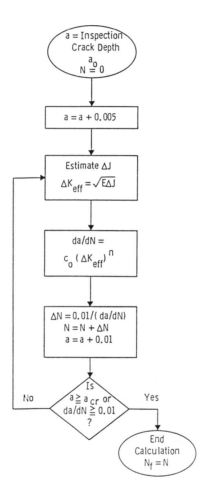

**Figure 13.9** *A flow chart for estimating crack size vs. fatigue cycles and remaining life, $N_f$. (Ref. 13.23).*

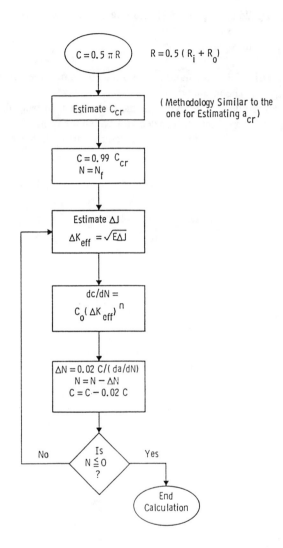

**Figure 13.10** *Flow chart for estimating critical crack length for ensuring leak-before-break conditions in the steam pipe (Ref. 13.23).*

It is worthwhile to discuss the assumptions in the analysis and their impact on the inspection criterion and interval calculation. The J-solutions available for estimating the crack growth in the radial direction are available only for $2\gamma = 90°$ for pure bending and for the bending and tension cases. The actual circumferential angles may vary over a range, hence, the J values used are only an approximation. The estimated values are expected to be an over-estimate of J (or $\Delta J$) when $2\gamma < 90°$ and an underestimate of J when $2\gamma > 90°$. However, since $2\gamma = 90°$ corresponds to a circumferential crack length of 280 mm (25% of the pipe circumference), we can limit the maximum crack length to that value. Thus, the J (or $\Delta J$) values are over-predicted leading to conservative life predictions.

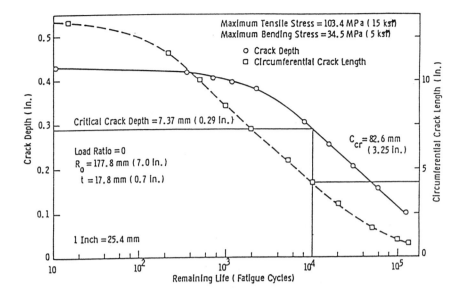

*Figure 13.11* Remaining fatigue crack growth life as a function of crack depth and the corresponding circumferential crack length for ensuring leak-before-break condition (Ref. 13.23).

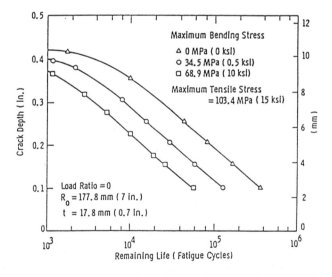

*Figure 13.12* Remaining life as a function of crack depth for a maximum tensile stress of 103.4 MPa and a bending stress ranging from 0 to 68.94 MPa (Ref. 13.23).

The J-solutions for the crack model used for estimating crack growth in the circumferential direction are for cracks which are through-wall. Since the actual cracks are not through-wall, this assumption will also lead to a conservative predictions of the leak-before-break conditions.

The bending-to-tension ratios, the a/t values, and the m values for which the actual $h_1$ functions are available are limited. For other values in the interval, $h_1$ must be interpolated. In the program which was developed, this interpolation is carried out linearly. It is not possible to determine if the errors introduced

### 448  Nonlinear Fracture Mechanics for Engineers

yield conservative or nonconservative results. However, these errors are not significant to impact the broad conclusions from this analysis on the suitability of the pipes for the intended application as well as the inspection interval and criterion.

*13.3.3 Analysis of a High Temperature Rotor Failure*

In June 1974, a steam turbine rotor at the Gallatin Plant of Tennessee Valley Authority (TVA) experienced a sudden burst during a cold start after reaching a speed of approximately 3400 RPM [13.24]. This turbine was in continuous operation (except for maintenance shutdowns) since 1957 until the time of failure and had undergone 105 cold starts and 183 warm or hot starts (288 total) [13.25]. A cold start occurs when the machine remains shut down for 72 hours or more and a warm or hot start occurs after a shutdown of less than 72 hours. Metallurgical and mechanical analyses were performed following the failure [13.24] on the rotor to understand the causes of failure. Following the initial evaluation, considerable mechanical property characterization and nondestructive inspection of the rotor pieces were also performed [13.26]. The fractogaphic observations made by Kramer and Randolph [13.24] and the stress calculations of Weiss [13.25] are first reviewed. Subsequently, the concepts of time-dependent fracture mechanics (TDFM) which were nonexistent at the time of failure and also during the period of the initial evaluation, are used to explain the causes of failure.

The fracture in the turbine rotor was traced to an originating elliptical shaped flaw located on the radial-axial plane in the bore region of the rotor under the seventh row of blades in the intermediate pressure (IP) section of the rotor, Figure 13.13. The flaw was approximately 6.45 mm (0.25 in) along the minor axis of the ellipse and 140 mm along the major axis of the ellipse. The major axis of the elliptical flaw was at a small angle with the axis of the rotor and the centroid of the flaw was at a distance of 17.8 mm from the bore surface as shown schematically in Figure 13.13. There was a secondary flaw, located on the other side of bore on the fracture plane. This flaw was 61 mm from the bore axis.

The region of the originating flaw was characterized by several clusters of long stringer type MnS inclusions. The average spacing between inclusions in the cluster was .05 mm and the average spacing between the clusters was 0.2 mm [13.24]. Surrounding the originating flaw there was a dark oxidized region followed by a shiny region where unstable fracture had clearly occurred. A close inspection of the region around the facture origin showed a gradual transition from intergrannular cracking to cleavage fracture beyond the transition region boundary. The presence of this transition region is important to the current analysis as discussed later.

The estimated steady-state temperature in the region of the originating flaw was between 413°C to 427°C. The tangential stress due to rotation in the region of the originating flaw is given in Figure 13.14 [13.25]. The stress relaxation due to creep during the 106,00 hours of operation was minimal. The combined peak thermal and centrifugal stress at the bore near the fracture origin was 537.7 MPa [13.25].

The nondestructive inspection of the rotor revealed several other flaws in the low and intermediate pressure sections of the rotor.

**Material Data**  The rotor was fabricated from a 1Cr-1Mo-0.25V, ASTM A470 class 8 forging. The material data needed to analyze the failure was yield and tensile strength, fracture toughness, creep deformation, fatigue crack growth, and creep-fatigue crack growth data at the service temperature of 427°C. The fracture toughness data were also needed in the transition region since the fracture occurred during a cold start when the metal temperature had not reached its steady-state value. These data are available from a previous study [13.26]. The creep-fatigue crack growth data in the original study were correlated in terms of $\Delta K$. These data were subsequently reanalyzed as a function of $(C_t)_{avg}$ [13.27].

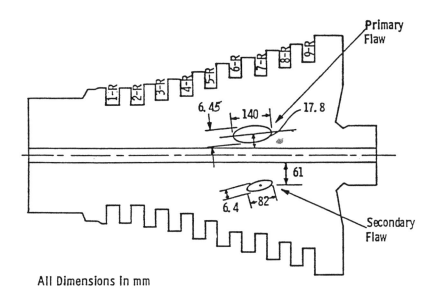

*Figure 13.13* Schematic of the intermediate pressure (IP) section of the rotor showing the size and location of the primary and secondary flaws beneath the seventh row (7-R) of blades.

Table 13.7 shows the typical tensile behavior of A470 class steel at various temperatures. Figures 13.15 and 13.16 show the variation in Charpy impact energy and the plane strain fracture toughness, $K_{IC}$, as a function of test temperature. The lower and the upper bound $K_{IC}$ values are plotted along with the mean trend.

The creep rupture data from specimens taken from the failed rotor were compared with generic data for A470 class 8 steels at 427°C [13.26]. The above study reported good agreement between the data from the failed rotor specimens and the other data. For the fracture mechanics analysis, the data needed is the creep deformation data. The secondary creep rate as function of stress at 427°C for A470 class 8 steels is plotted in Figure 13.17 [13.28]. The creep rate is given by the following equation:

$$\dot{\varepsilon} = 9.4 \times 10^{-43} \, \sigma^{13.7} \tag{13.30}$$

where $\dot{\varepsilon}$ is expressed in hr$^{-1}$ and $\sigma$ in MPa. The creep-fatigue crack growth rate data obtained on specimens from the failed rotor are plotted in Figure 13.18 along with data from other new rotor material. The data can be represented by the following equation:

450  Nonlinear Fracture Mechanics for Engineers

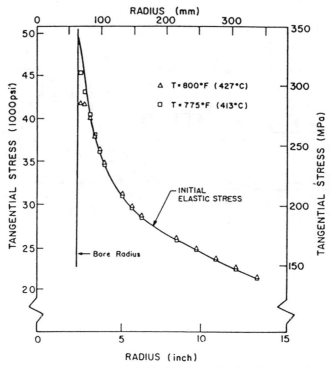

**Figure 13.14** Elastic stresses due to rotation as a function of radial distance from the center of the failed rotor for the region under the seventh row of blades (Ref. 13.25).

**Table 13.7** Tensile Properties of A470 Class 8 Steels

| Temp (°C) | 0.2% yield strength (MPa) | UTS (Mpa) | % Elong. (5 cm gage length) | Red. in area (%) |
|---|---|---|---|---|
| 24 | 623 | 775.6 | 14.2 | 39 |
| 427 | 515.7 | 624.6 | 14.2 | 53 |
| 538 | 464 | 522.6 | 17.5 | 75 |

$$\left(\frac{da}{dt}\right)_{avg} = 5.84 \times 10^{-5} (C_t)_{avg}^{0.6} \tag{13.31}$$

where $(da/dt)_{avg}$ is expected in m/hr and $(C_t)_{avg}$ in mega-Joules/m²hr (MJ/m²hr). These data were also found to be comparable to data for a generic A470 class 8 steel.

The cycle-dependent fatigue crack growth rate $(da/dN)_0$ as a function of $\Delta K$ for 427°C was also characterized using the failed rotor material and is given by [13.26]:

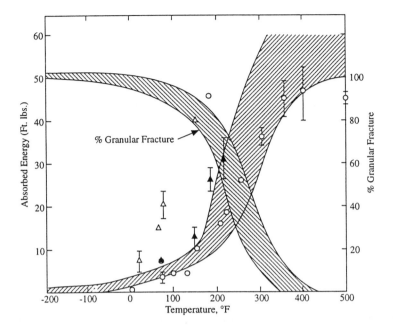

*Figure 13.15* Charpy impact energy and % granular (ductile) fracture for A470 class 8 rotor steels including some data from the Gallatin rotor material.

$$(\frac{da}{dN})_o = 1.41 \times 10^{-11} (\Delta K)^{2.7} \qquad (13.32)$$

where $(da/dN)_o$ is m/cycle and $\Delta K$ in $MPa\sqrt{m}$.

**Rotor Life Calculation** It is postulated that high strains in the ligaments between neighboring manganese sulfide (MnS) inclusions in the clusters led to the formation of the fatal elliptical flaw which was 140 mm long and 6.45 mm along the minor axis. Subsequently, this flaw grew to critical size under the creep-fatigue conditions during the remainder of the turbine life. This subcritical crack extension is responsible for the gradual transition from intergrannular cracking to cleavage fracture in the region of the flaw. We can estimate the life expended in growing the crack from the initial size of 6.45 x 140mm to the critical size and compare it to the actual life.

The first step in the crack growth analysis consists of estimating the critical crack size. Figure 13.20 shows the K-calibration function for a crack in a semi-infinite body and approaching a surface [13.29]. The K-value is given by:

$$K = \sigma\sqrt{\pi a} F_a \qquad (13.33)$$

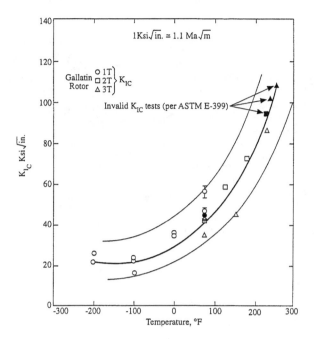

**Figure 13.16** *Fracture toughness, $K_{IC}$, as a function of temperature for the Gallatin rotor material and scatter band from other available data on generic A470 class 8 rotor steel.*

where $F_A$ is the function shown in Figure 13.19 for the crack tip labeled A which is closer to the surface. When the K at the tip A reaches $K_{IC}$, the crack is expected to have broken through the bore. Subsequently, a large surface crack will develop which can cause the catastrophic burst. Therefore, the critical crack depth, $a_c$, is given by:

$$a_c = \frac{1}{\pi}(K_{IC}/\sigma F_a)^2 \qquad (13.34)$$

Since the elliptical crack is very long, the K at the deepest point of the flaw can be represented by Equation (13.33). During a cold start, both the stress level and $K_{IC}$ vary the time. Thus, a unique combination of stress and $K_{IC}$ exist for which $a_c$ has the smallest value. The peak stress at the failure location during a cold start was estimated to be 496.3 MPa and was estimated to occur when the temperature was 110°C [13.25]. The complete relationship between stress and temperature for the failure region is shown in Figure 13.21. Next, we plot a relationship between:

$$(1/\pi)(K_{IC}/\sigma)^2 = F_A^2 a_c$$

equation (13.34), and $a_c$. We can now choose several stress-temperature combinations from Figure 13.20, determine the $K_{IC}$ range corresponding to that temperature from Figure 13.16, and then determine the pre-

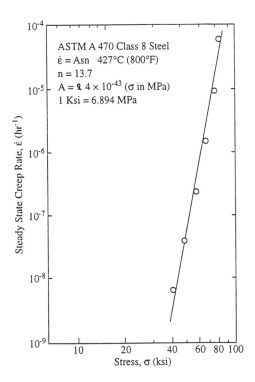

*Figure 13.17* Creep deformation behavior for A470 class 8 steels at 427 °C.

dicted $a_c$ range. These are given in Table 13.8. If the lower-bound fracture toughness behavior is assumed, the predicted $a_c$ value is 4.36 mm and it occurs at a stress of 441 MPa and a temperature of 62.2°C. If the upper-bound fracture behavior is assumed, the lowest $a_c$=10.7 mm. If we assume $a_c$ to be 4.36 mm, the crack extension by creep-fatigue will be 1.14 mm. However, if $a_c$ = 10.7 mm, the crack growth due to creep-fatigue will be 7.22 mm. On the actual rotor, there was only a very small oxidized region of intergranular crack growth between the initiating flaw and the unstable fracture which can be attributed to creep-fatigue crack growth. On that basis, we will choose the critical crack size to be 4.36 mm to estimate the crack growth life. This choice is also in agreement with the very brittle nature of the fracture that was observed. The brittleness could be the result of the high density of inclusions in the region of the fracture.

The average time between start-up and shut-down was 106,000/288 = 368 hours. It is important in this case to determine the overall level of creep deformation in the rotor. This can be done by comparing the average operating time to the transition time, $t_T$, using equation (10.25). The $C^*$ for this configuration is not known. However, an estimate of $C^*$ can be made by the expression for a center crack panel of half-wide W equal to distance from the surface to the center of the flaw. Thus, $C^*$ is given by:

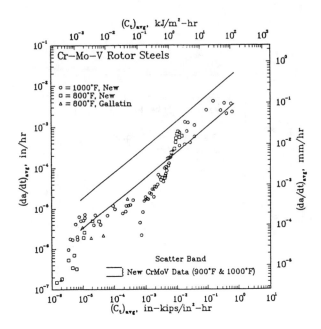

**Figure 13.18** *Creep-fatigue crack growth rate, $(da/dt)_{avg}$ as a function of the $(C_t)_{avg}$ parameter for the Gallatin rotor material and other generic A470 class 8 steel (Ref. 13.27). The scatterband represents the trend for the creep crack growth behavior.*

$$C^* = Aa(1-a/W)h_1\left(\frac{\sigma\sqrt{3}}{4(1-a/W)}\right)^{n+1} \tag{13.35}$$

The value of $h_1 \sim 3.0$ from Table 5.3a for a crack size of 3.17 mm and W = 17.8 mm and a steady-state stress value of 337.8 MPa. $C^* = 8.8 \times 10^{-12}$ MJ/m²hr. The value of K for these conditions is 33.4 MPa$\sqrt{m}$. Thus, $t_T = 4.38 \times 10^7$ hours. This value, when compared to the average operating time of 368 hours and an overall operating time of 106,000 hours makes it well within the small-scale creep regime. Thus, $(C_t)_{avg}$ for this condition can be obtained from equation (12.6) for an elastic-secondary creep material without the $C^*$ term because it is expected to be negligible for small-scale creep:

$$(C_t)_{avg} = \frac{2\alpha\beta(1-\nu^2)}{EF_{cr}(\theta,n)}\frac{\Delta K^4}{W} F'/F(EA)^{\frac{2}{n-1}} t_h^{-\frac{n-3}{n-1}}$$

$\alpha$ from equation (10.35) = 0.234 for n = 13.7, $\beta$ = 0.33 and $F_{cr}(\theta,n) \sim 0.4$, E = 179.2 × 10³ MPa, A = 9.4 × 10⁻⁴³, W = .0178 m, and $t_h$ = hold time and F and F' are the K = calibration function and its derivative, respectively. Substituting this equation into equation (13.31) and multiplying both sides by $t_h$, we get:

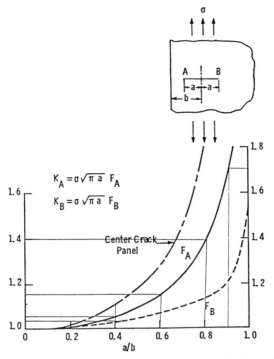

*Figure 13.19* K-calibration functions for a crack in a semi-infinite body with one tip approaching the surface. For comparison, the K-calibration of the center crack configuration in a finite plate is also shown (Tada et al., Ref. 13.29).

$$(\frac{da}{dt})_{avg}t_h = (\frac{da}{dN})_{time} = (\frac{2\alpha\beta(1-v^2)}{EF_{cr}(\theta,n)}\frac{\Delta K^4}{W}F'/F(EA)^{\frac{2}{n-1}})^{.6}(5.84 \times 10^{-5})t_h^{-\frac{n-3}{n-1}6}t_h$$

$$(\frac{da}{dN})_{time} = 6.35 \times 10^{-11}(F'/F)^{.6}\Delta K^{2.4}t_h^{.5} \tag{13.36}$$

$$F'/F = \frac{1}{2(a/W)} + \frac{F'_A}{F_A}$$

$$F = K/\sigma\sqrt{W} = (\frac{\pi a}{W})^{1/2}F_A$$

where

$$F'_A = \frac{dF_A}{d(a/W)}$$

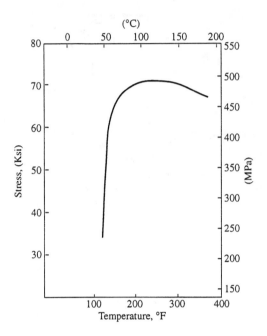

**Figure 13.20** *Tangential stress at the bore under the seventh row of blades as a function of metal temperature during a cold start of the Gallatin rotor (Ref. 13.30).*

$F_A$ in Figure 13.20 can be described by the following polynomial:

$$F_A = 1 - .63(a/W) + 5.65(a/W)^2 - 11.4(a/W)^3 + 7.619(a/W)^4$$

The following polynomial represents F'/F:

$$F'/F = \frac{0.5}{(a/W)} - .743 + 13.954(a/W) - 50.68(a/W)^2 - 65.956(a/W)^3 - 25.064(a/W)^4 \quad (13.37)$$

Thus, from equation (12.21), the total fatigue crack growth rate including the time-dependent and the cycle-dependent term is obtained by combining equations (13.32) and (13.36) as follows:

$$\frac{da}{dN} = 1.41 \times 10^{-11} (\Delta K)^{2.7} + 2.74 \times 10^{-11} (F'/F)^{0.6} \Delta K^{2.4} t_h^{.5} \quad (13.38)$$

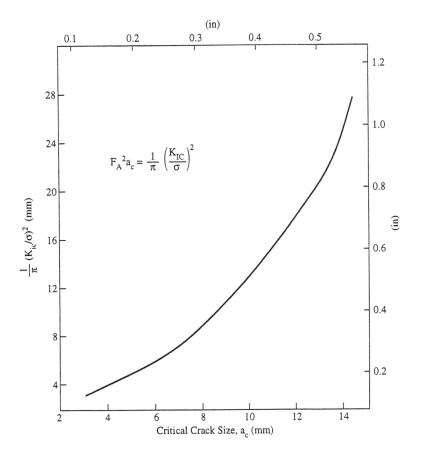

**Figure 13.21** *Critical crack size as a function of $K_{IC}$ and the applied stress.*

Substituting for $F'/F$ from equation (13.37) and $t_h = 368$ hours and for $\Delta K$ in terms of the applied stress and crack size. The resulting equation can be integrated from the initial crack size to final crack size to determine the fatigue life of the rotor. Remember that to estimate $\Delta K$ for the first term in equation (13.38) we need to use the peak stress of 496.4 MPa for cold starts and the steady-state stress value of 337.8 MPa for the warm and hot starts. To simplify the calculation, we can develop a predicted crack size vs. cycles curve for all cold starts and for all hot/warm starts, respectively, and then interpolate between them for a cold-start to hot-start ratio of 105/183. The predicted crack size, a, vs. number of start-stop cycles, N, for a starting crack size of 3.17 mm is shown in Table 13.9 for the various conditions.

From the above calculation it appears that at least during 190 of the 288 start-stop cycles, crack propagation occurred and less than 100 cycles were spent in crack formation. From the above analysis, it can be estimated that for a 40-year design life (350,000 hours or 952 start-stop cycles of operation), the initial value of crack size, a, must not exceed 0.7 mm or a total depth must not exceed 1.4 mm. This type of sensitivity is well within the capability of ultrasonic techniques where the ultrasonic inspections can be conducted through the bore of the turbine rotor. Thus, if rotors are inspected ultrasonically initially and every 5 to 10 years subsequently during service, the risk of catastrophic failures of the type experienced at this plant can be substantially reduced. This is the primary recommendation from such an analysis.

*13.3.4 Integrity Analysis of Reheat Steam Pipes*

Reheat steam pipes are used to carry steam at high pressures and high temperatures in fossil power-plants. Typical steam pressures can exceed 4.8 MPa (50 atmospheres) and the temperature can be as high as 565°C. The pipes carrying the steam are approximately 0.75 m in diameter and can be made from Cr-Mo ferritic steels with a wall thickness of 3.75 cm. Thus, creep crack growth and rupture are the dominant failure concerns in these pipes which are designed to last 30 to 40 years during service.

*Table 13.8 Results of the Critical Flaw Size Calculations in Support of the Analysis of the Turbine Rotor Failure*

| σ (MPA) | Temp °C | $K_{I_c}$ range $MPa\sqrt{m}$ | $a_c$ range(mm) |
|---|---|---|---|
| 220.6 | 49 | 44-82.5 | 9.7->14.25 |
| 303.3 | 51 | 46.2-84.7 | 6.9-13.8 |
| 372.2 | 53.3 | 47.3-86.9 | 5.05-11.7 |
| 399.8 | 55.5 | 48.4-88.0 | 4.6-10.85 |
| 420.5 | 57.7 | 49.5-90.2 | 4.4-10.70 |
| 430.9 | 60.0 | 50.6-93.5 | 4.38-10.7 |
| 441.2 | 62.2 | 51.7-96.8 | 4.36-10.9 |
| 448.11 | 64.4 | 52.8-99 | 4.4-10.95 |
| 455.0 | 66.67 | 55-101.2 | 4.6-11.05 |
| 468.8 | 75.5 | 61.6-104.5 | 5.35-11.10 |
| 482.6 | 88.9 | 66-118.8 | 5.80-12.3 |
| 496.4 | 110 | 82.5-132 | 8.40-13.6 |

*Table 13.9 Crack Size Vs. Start-Stop Cycles Predicted for the Gallatin Rotor*

| N (cycles) | a (mm) | | |
|---|---|---|---|
| | cold starts only | hot starts only | mixed |
| 0 | 3.17 | 3.17 | 3.17 |
| 50 | 3.46 | 3.44 | 3.45 |
| 90 | 3.72 | 3.68 | 3.69 |
| 190 | 4.40 | 4.34 | 4.36 |
| 290 | 5.32 | 5.13 | 5.19 |

In mid-1985 and early 1986, steam pipe bursts occurred in power plants in two unrelated incidents. Both pipes were seam welded and the failures occurred suddenly in the welds after several years of service. One of these failures resulted in loss of several lives while both caused substantial damage to the plants. In the aftermath of these failures, the management of several power plants around the world undertook extensive inspection programs to detect cracks in steam pipes to prevent other such failures from occurring. The obvious question in this case was what inspection interval criterion should be chosen to ensure integrity of these pipes. The following example demonstrates how such analysis can be performed. The example chosen is somewhat simplified to illustrate the procedure. Since the actual failures occur in welded pipes which have inhomogeneous material properties as one traverses from the weld metal to the base metal, the actual analysis can be quite complex. For additional details, the readers should consult references in the literature [13.31-13.33].

**Problem Statement** A long reheat steam pipe has an outside diameter of 0.77 m and a wall thickness of 0.0385 m. The pipe is seam welded in the axial direction. The process can leave long axial flaws on the inside surface of the pipe. The inspection technique used is capable of finding flaws that are 0.635 mm deep. The steam pressure is 4.13 MPa and the temperature is 538°C. The pipe is made from a 1.25 Cr-0.5Mo steel with the following properties at the service temperature of 538 °C:

0.2% yield strength = 131 MPa
$E = 162 \times 10^3$ MPa
$\alpha = 0.873$
$\epsilon_o = 8.08 \times 10^{-4}$
$m = 7$
$K_{IC} \approx 200$ $MPa\sqrt{m}$
$n = 10$
$A = 1.77 \times 10^{-24}$ ($\sigma$ in MPa)
Creep crack growth coefficient = $1.67 \times 10^{-3}$ ($C_t$ in MJ/m²hr and da/dt in m/hr)
creep crack growth exponent = 0.75

The system is shut down every month for various reasons. You are asked to recommend an inspection criterion and interval to prevent catastrophic failures.

**Approach** The approach consists of estimating the crack growth life starting from a defect size of 0.635 mm, the minimum inspectable size. We begin by listing the K, C*, and $C_t$ expressions for this configuration which are taken from several references [13.20, 13.21, 13.31, 13.32, 13.34].

Since radius-to-thickness ratio = 10 for this case, the expressions listed will be specifically for that case. We will model the material as an elastic-plastic-secondary creep material.

$$K = p\sqrt{b}F(a/b) \qquad (13.39)$$

where p = internal pressure, b = wall thickness, and:

$$F(a/b) = \frac{2(\pi ab)^{1/2}}{[1-(R_i/R_o)^2]} f(a/b)$$

$$f(a/b) = 1.12 + .405(a/b) + 2.556(a/b)^2 + 3.25(a/b)^3$$

$$F'(a/b) = \frac{dF}{d(a/b)}$$

$$F'/F = \frac{1}{2}(a/b) + f'/f \tag{13.40}$$

$$f'(a/b) = .405 + 5.112\,(a/b) + 9.75\,(a/b)^2$$

$$C^* = A(1-a/b)\,ah_1(a/b,n,R/b)\left(\frac{\sqrt{3}}{2}p\frac{R/b+a/b}{1-a/b}\right)^{n+1} \tag{13.41}$$

The value of $h_1$ is given in Table 5.7 for different values of m (or n). For $R/b = 10$ and $n = 10$, $h_1$ is given in Table 13.10.

**Table 13.10** Value of $h_1$ for $R/b = 10$ and $n = 10$

| a/b | $h_1$ |
|---|---|
| .125 | 9.55 |
| .250 | 6.98 |
| .500 | 2.27 |
| 0.75 | 0.787 |

For a/b -> 0, the value of $h_1$ can be derived following the procedure discussed in Chapter 5.

$$h_1(a/b \to 0) = 1.2\pi\sqrt{n}\left[\frac{R_i^2 + R_o^2}{R_i(R_i + R_o)}\right]^{n+1} \tag{13.42}$$

for $n = 10$, $h_1\,(a/b \to 0) = 22.26$. Thus, $h_1$ can be given by:

$$h_1 = 22.26 - 109.33(a/b) + 211.37\,(a/b)^2 - 137.427\,(a/b)^3 \tag{13.43}$$

$$(C_t)_{avg} = \frac{2\alpha\beta(1-\upsilon^2)}{E}F_{cr}(\theta,n)\frac{\Delta K^4}{b}F'/F(EA)^{\frac{2}{n-1}}t_h^{-\frac{n-3}{n-1}} + C^* \tag{13.44}$$

$\alpha$ for $n = 10$ from equation (10.35) = 0.258, $\beta = 0.33$, $F_{cr} = 0.4$, $\upsilon = .33$. Substituting these constants and the value of E, equation (13.44) reduces to:

$$(C_t)_{avg} = 7.73 \times 10^{-10}\Delta K^4(F'/F)t_h^{-.778} + C^*$$

Further substituting $t_h = 720$ hours:

$$(C_t)_{avg} = 4.625 \times 10^{-12} \Delta K^4 (F'/F) + C^*$$

Substituting the values of $(C_t)_{avg}$ in the equation describing the creep-fatigue crack growth rates, we can get the average value of da/dt for each start-up and shut-down cycle, and, multiplying that by 720 hours will yield crack extension per cycle as a function of crack size. The PCPIPE [13.18] computer program is designed to conduct such an analysis. The predicted remaining life is 628 cycles or $4.25 \times 10^5$ hours. At the rate of 1 cycle per month, this gives a life of 52 years. Thus, if we recommend that the pipe be inspected using ultrasonic technique with the capability to detect flaws that are 0.635 mm or deeper and if this inspection is performed every ten years, we should be able to get good assurance against catastrophic failures. A leak-before-break analysis can also be performed for this case following the approach discussed in Section 13.3.2. It can be shown that for rupture to occur, the initial crack length would have to be longer than 1.4 m. This length seems to be very long. However, in the two cases where the steam pipe rupture occurred, the lengths of the cracks were approximately 3 m. Therefore, it appears that the calculated critical crack length is in agreement with field experience, lending credibility to the analysis.

In the above formulation, primary creep effects have not been considered. Also, if the creep rates in the weld metal are different than in the base metal, the accumulation of strains can occur at a higher rate in the less deformation-resistant material. All these factors will tend to accelerate crack growth and must be considered while conducting a real analysis. Another factor which must also be considered is degradation in properties due to prolonged exposure to stress and temperature during service. Therefore, the properties used must be representative of the material in the subject piping.

## 13.4 Summary

In this chapter, a general methodology for using nonlinear fracture mechanics concepts is discussed. It is shown that fracture mechanics may be used effectively for assessing the integrity of structural components or for predicting the remaining life or design life of components. One of the most important applications of fracture mechanics is shown to be in selecting realistic inspection criterion and inspection intervals for components designed to last a long time during service.

Four realistic examples were chosen to illustrate the use of nonlinear fracture mechanics. These examples included integrity assessment, inspection criterion and interval determination, and also failure analysis. In some cases considered, the predictions from the analyses were also evaluated against service experience. Such comparisons should be performed whenever possible because they provide realistic bench-marks for the validity of the analysis. This is essential for complex components where several simplifying assumptions are frequently necessary to proceed with the analysis. Although it is possible to ensure that most of the assumptions that are made provide conservatism to the analysis, this cannot always be guaranteed. Therefore, service experience must be documented as much as possible.

## 13.5 References

13.1   E399-90, "Standard Test Method for Plane-Strain Fracture Toughness of Metallic Materials," Annual Book of ASTM Standards, Section 3, American Society for Testing and Materials, 1994, pp 407-437.

13.2   E-813-89, "Standard Test Method for $J_{Ic}$, A Measure of Fracture Toughness," Annual Book of ASTM Standards, Section 3, American Society for Testing and Materials, 1994, pp 628-642.

13.3   E1152-87, "Standard Test Method for Determining J-R Curves," Annual Book of ASTM Standards, Section 3, American Society for Testing and Materials," 994, pp 744-754.

13.4   BS 5447:1974 "Methods of Testing for Plane Strain Fracture Toughness ($K_{IC}$) of Metallic Materials," British Standards Institution, London 1974.

13.5   E561-94, "Standard Method for R-Curve Determination", Annual Book of ASTM Standards, Section 3, American Society for Testing and Materials", 1994, pp 489-501.

13.6   BS 5762:1979, "Methods for Crack Opening Displacement (COD) Testing", British Standards Institution, London 1979.

13.7   E1290-93, "Standard Test Method for Crack-Tip Opening Displacement (CTOD) Fracture Toughness Measurement," Annual Book of ASTM Standards, Section 3, American Society for Testing and Materials, 1994, pp 846-855.

13.8   E1304-89, "Standard Test Method for Plane-Strain (Chevron-Notch) Fracture Toughness of Metallic Materials, Annual Book of ASTM Standards, Section 3, American Society for Testing and Materials, 1994, pp 856-866.

13.9   "Standard Method for Measurement of Fracture Toughness" (Draft), ASTM Committee E.08 on Fatigue and Fracture Mechanics, American Society for Testing and Materials, Philadelphia, 1994.

13.10  E-647-93, "Standard Test Method for Measurement of Fatigue Crack Growth Rates," Annual Book of ASTM Standards, Section 3, American Society for Testing and Materials, 1994, pp 569-596.

13.11  E1221-88, "Standard Test Method for Determining Plane-Strain Crack Arrest Fracture Toughness, $K_{Ia}$, of Ferritic Steels," Annual Book of ASTM Standards, Section 3, American Society for Testing and Materials, 1994, pp 779-794.

13.12  E1457-92, "Standard Test Method for Measurement of Creep Crack Growth Rates in Metals", Annual Book of ASTM Standard, Section 3, American Society for Testing and Materials, 1994, pp 779-794.

13.13  E1681-95, "Standard Test Method for Determining a Threshold Stress Intensity Factor for Environment Assisted Cracking of Metallic Materials Under Constant Load," Annual Book of ASTM Standards, Section 3, American Society for Testing and Materials, 1996.

13.14  K.B. Yoon, A. Saxena, and P.K. Liaw, "Characterization of Creep-Fatigue Crack Growth Behaviors under Trapezoidal Wave Shape Using $C_t$-Parameter," International Journal of Fracture, Vol.59, 1993, pp 95-114.

13.15  P.S. Grover and A. Saxena, "Characterization of Creep-Fatigue Crack Growth Behavior in 2.25 Cr-IMo Steel Using $(C_t)_{avg}$", International Journal of Fracture, Vol. 73, 1995, pp 273-286.

13.16  A. Saxena, "Fracture Mechanics Approaches for Characterizing Creep-Fatigue Crack Growth," JSME International Journal, Series A, Vol. 36, No. 1, 1993, pp 1-20.

13.17   PC-CRACK, Structural Integrity Associates, San Jose, CA.

13.18   "PCPIPE - A Computer Program for Integrity Analysis of Steam Pipes"' Structural Integrity Associates, San Jose, CA.

13.19   NASCRAC, Engineering Mechanics Technology, Palo Alto, CA.

13.20   V. Kumar and M.D. German, "Elastic-Plastic Fracture Analysis of Through-Wall and Surface Flaws in Cylinders," EPRI Report NP-5596, Electric Power Research Institute, Palo Alto, CA, 1988.

13.21   A. Zahoor, "Ductile Fracture Handbook,"Novatech Corp., Gaithersburg, MD, EPRI Project 1757-69, October 1990.

13.22   P.K. Liaw, A. Saxena, and J. Perrin, "Life Extension Technology for Steam Pipe Systems--Part I, Development of Material Properties," Engineering Fracture Mechanics, Vol. 45, 1993, pp 759-786.

13.23   P.K. Liaw, A. Saxena, and J. Perrin, "Life Extension Technology for Steam Pipe Systems--Part II, Development of Life Prediction Methodology", Engineering Fracture Mechanics, Vol. 45, 1993, pp 787-803.

13.24   L.D. Kramer and D. Randolph, " Analysis of TVA Gallatin No. 2 Rotor Burst Part I--Metallurgical Considerations," 1976 - ASME - MPC Symposium on Creep-Fatigue Interaction, MPC-3, 1976, pp 1-24.

13.25   D.A. Weisz, "Analysis of TVA Gallatin No. 2 Rotor Burst Part II--Mechanical Analysis," 1976 ASME - MPC Symposium on Creep-Fatigue Interaction, MPC-3, 1976, pp 25-40.

13.26   G.A. Clarke, T.T Shih, and L.D. Kramer, " Evaluation of the Fracture Properties of Two 1950 Vintage CrMoV Steam Turbine Rotors," EPRI Contract RP 502, Research Report Westinghouse Research and Development Co., Pittsburgh, 1978.

13.27   K. Banerji and A. Saxena, Unpublished results, Georgia Institute of Technology, 1987.

13.28   A. Saxena, "Creep Crack Growth Under Nonsteady-State Conditions," in Fracture Mechanics: Seventeenth Volume, ASTM STP 905, American Society for Testing and Materials, 1986, pp 185-201.

13.29   H. Tada, P.C. Paris, and G.R. Irwin, "The Stress Analysis of Cracks Handbook," Del Research Co., St. Louis, MO, 1985.

13.30   W. Berry, unpublished results, 1976.

13.31   P.K. Liaw, A. Saxena, and J. Schaeffer, "Estimating Remaining Life of Elevated-Temperature Steam Pipes--Part I Material Properties," Engineering Fracture Mechanics, Vol. 32, 1989, pp 675-708.

13.32   P.K. Liaw, A. Saxena, and J. Schaeffer, " Estimating Remaining Life of Elevated-Temperature Steam Pipes-Part II. Life Estimation Methodology," Engineering Fracture Mechanics, Vol. 32, 1989, pp 709-722.

13.33   J.R. Foulds, R. Viswananthan, J.L. Landrum, and S.M. Walker, "Guidelines for the Evaluation of Seam-Welded High Energy Piping", EPRI Report, TR-104631, Electric Power Research Institute, January 1995.

13.34   V. Kumar, M.D. German, and C.F. Shih, "An Engineering Approach to Elastic-Plastic Fracture Analysis," NP-1931, Electric Power Research Institute, Palo Alto, CA, 1981.

# APPENDIX 1

# Hutchinson, Rice, Rosengren (HRR) Singular Field Quantities

*Adapted from C. F. Shih, "Tables of Hutchinson, Rice and Rosengren Singular Field Quantities,"*
*MRL E-147, Brown University, Providence, Rhode Island, June 1993*

Table A1.1  $I_m$ values as a function of m (see eq. 4.19 for polynomial expressions representing this data)

| m | $I_m$ Plane strain | $I_m$ Plane stress |
|---|---|---|
| 2 | 5.94 | 4.22 |
| 3 | 5.51 | 3.86 |
| 5 | 5.02 | 3.41 |
| 7 | 4.77 | 3.17 |
| 10 | 4.54 | 2.98 |
| 12 | 4.44 | 2.90 |
| 15 | 4.33 | 2.82 |
| 20 | 4.21 | 2.74 |
| 50 | 3.95 | 2.59 |
| 100 | 3.84 | 2.54 |
| ∞ | 3.72 | 2.49 |

Table A1.2   Angular functions in eq. 4.18 for plane strain mode I loading

| m | θ (deg) | $\hat{\sigma}_e$ | $\hat{\sigma}_{rr}$ | $\hat{\sigma}_{\theta\theta}$ | $\hat{\sigma}_{r\theta}$ | $\hat{\varepsilon}_{rr}$ | $\hat{\varepsilon}_{\theta\theta}$ | $\hat{\varepsilon}_{r\theta}$ |
|---|---|---|---|---|---|---|---|---|
| 2 | 0   | .085   | 1.5776 | 1.6758 | 0     | −.0063  | .0063   | 0 |
| 2 | 45  | .7064  | 1.535  | 1.3914 | .4014 | .0764   | −.0764  | .4255 |
| 2 | 90  | .9988  | 1.2423 | .7489  | .5212 | .3696   | −.3696  | .7808 |
| 2 | 135 | .8193  | .9074  | .1797  | .3022 | .4471   | −.4471  | .3362 |
| 2 | 180 | .2473  | .2856  | 0      | 0     | .0530   | −.0530  | 0 |
| 5 | 0   | .4621  | 1.6836 | 2.2172 | 0     | −.0183  | .0183   | 0 |
| 5 | 45  | .7652  | 1.8527 | 1.8415 | .4418 | .0030   | −.0030  | .2276 |
| 5 | 90  | .9962  | 1.2884 | 1.0895 | .5665 | .147    | −.147   | .8369 |
| 5 | 135 | .8802  | .8621  | .3317  | .4334 | .2388   | −.2388  | .3905 |
| 5 | 180 | .5819  | .6719  | 0      | 0     | .0578   | −.0578  | 0 |
| 10 | 0   | .6691  | 1.7243 | 2.4696 | 0     | .0156   | .0156   | 0 |
| 10 | 45  | .8358  | 2.0118 | 2.0534 | .4821 | −.00615 | .00615  | .1444 |
| 10 | 90  | .9959  | 1.3395 | 1.2443 | .5730 | .0688   | −.0688  | .8280 |
| 10 | 135 | .91305 | .8075  | .4158  | .4893 | .1294   | −.1294  | .324 |
| 10 | 180 | .7287  | .8414  | 0      | 0     | .0365   | −.0365  | 0 |
| 20 | 0   | .8023  | 1.7578 | 2.6841 | 0     | −.0106  | .0106   | 0 |
| 20 | 45  | .8943  | 2.1237 | 2.1913 | .5151 | −.006   | .006    | .09295 |
| 20 | 90  | .9965  | 1.3878 | 1.3428 | .5749 | .0316   | −.0316  | .8076 |
| 20 | 135 | .9414  | .7584  | .4763  | .5248 | .0670   | −.0670  | .2506 |
| 20 | 180 | .8300  | .9584  | 0      | 0     | .0209   | −.0209  | 0 |

Table A1.3  Angular functions in eq. 4.18 for plane stress mode I loading

| m | θ (deg) | $\hat{\sigma}_e$ | $\hat{\sigma}_{rr}$ | $\hat{\sigma}_{\theta\theta}$ | $\hat{\sigma}_{r\theta}$ | $\hat{\varepsilon}_{rr}$ | $\hat{\varepsilon}_{\theta\theta}$ | $\hat{\varepsilon}_{r\theta}$ |
|---|---|---|---|---|---|---|---|---|
| 2 | 0 | .9654 | .8422 | 1.0536 | 0 | .3045 | .6106 | 0 |
| 2 | 45 | .9976 | .8274 | .8393 | .3061 | .39055 | .4575 | .4226 |
| 2 | 90 | .9165 | .7709 | .3312 | .3612 | .5548 | −.0498 | .4965 |
| 2 | 135 | .6036 | .578 | .0073 | .1073 | .3467 | −.170 | .0973 |
| 2 | 180 | .4026 | −.4026 | 0 | 0 | −.1621 | .0810 | 0 |
| 5 | 0 | .9906 | .6907 | 1.1349 | 0 | .1186 | .7602 | 0 |
| 5 | 45 | .9981 | .6172 | .84115 | .3653 | .1836 | .5931 | .5095 |
| 5 | 90 | .8797 | .5677 | .2120 | .4191 | .2765 | −.043 | .3764 |
| 5 | 135 | .6625 | .5827 | −.1367 | .0151 | .1255 | .0825 | .0045 |
| 5 | 180 | .6702 | −.6702 | 0 | 0 | −.1352 | .0676 | 0 |
| 10 | 0 | .9951 | .6281 | 1.1473 | 0 | .0521 | .7971 | 0 |
| 10 | 45 | .9975 | .5166 | .8296 | .395 | .0995 | .558 | .579 |
| 10 | 90 | .8915 | .4642 | .1283 | .4554 | .1422 | −.0369 | .2429 |
| 10 | 135 | .7511 | .6054 | −.2281 | −.0478 | .0548 | −.0404 | .0054 |
| 10 | 180 | .7728 | −.7728 | 0 | 0 | −.076 | .038 | 0 |
| 20 | 0 | .9973 | .6000 | 1.1512 | 0 | .0231 | .8081 | 0 |
| 20 | 45 | .9983 | .4624 | .8223 | .4026 | .0496 | .5716 | .5839 |
| 20 | 90 | .9177 | .3968 | .0744 | .4860 | .0703 | −.0242 | .1425 |
| 20 | 135 | .834 | .6323 | .2928 | −.0893 | .0247 | −.0193 | −.0043 |
| 20 | 180 | .8496 | −.8496 | 0 | 0 | −.0384 | .0192 | 0 |

# INDEX

## A

American Society for Testing and Materials (ASTM) 62, 66, 70, 150, 154, 166, 170, 207, 256, 366
Antiplane shear 49
Arrest toughness 201-203
Asymptotic analysis 50, 88, 221, 313, 321

## B

Barenblatt model (see Dugdale - Barenblatt model)
Bend specimens (see single edge notch bend specimens)
Biaxiality ratio 29
British Standards Institute 62, 157, 171
Brittle fracture 45, 147
Budiansky-Hutchinson model 278-281

## C

C(t)-integral 323-326
$C^*$-integral 12, 311-319
Caustics 194, 209-213
Center crack tension (CCT) specimen 52, 97, 122
Cleavage fracture 147, 234-236
Cleavage toughness 147, 155-156
Closure load 274-275
Compact type (CT) specimen 52, 59, 74, 148, 151, 284
Complementary energy 24
Complementary strain energy density 25
Compliance 47, 54, 58-60, 153, 167
Constant amplitude loading 66
Constraint effects 227, 230, 239-240, 257
Continuity 236
Corner cracks 53
Crack arrest 201, 207
Crack driving force 240, 278
Crack growth resistance curves 63, 147-170
Crack mouth opening displacement (CMOD) 126-130
Crack opening angle 114-116
Crack propagation
    Corrosion-fatigue 64, 70
    Creep 363-394
    Creep-fatigue 395-426
    Fatigue 64-68, 267-303
    Stress corrosion 64, 68-70
Crack speed 189-192
Crack tip blunting 93
Crack tip opening displacement (CTOD) 12, 56, 91, 101, 109-111, 148, 157
Creep
    Activation energy 33
    Cavities 377-384
    Deformation 30-35, 309, 335
    Primary 31, 32-33, 335-345
    Rupture 31
    Secondary 31, 309-335
    Tertiary 31
Creep crack growth 363-394
Creep fatigue crack growth 395-426
Creep zone 321-323, 341
$C_t$ parameter 13, 327-335
Cyclic J (see delta J) 281-289
Cyclic stresses 268, 282
Cylinders 138-141

## D

Damage tolerance methodology 3-5, 427
Delta J 281-289
Delta K 66
Deformation theory of plasticity 29
Disk shaped compact specimen 149, 152
Displacement control 153, 272,
Divergence theorem 83
Double cantilever beam (DCB) specimen 194, 201
Double edge notch tension (DENT) specimen (also see DEN specimen) 97, 134
Driving force (see crack driving force)
Ductile fracture 45, 147, 246
Ductile-brittle transition temperature 257-260
Dugdale-Barenblatt model for plastic zone 57-58, 109
Dynamic fracture mechanics 187, 207
Dynamic fracture toughness 193, 194, 207

## E

Effective crack size 56, 126
Effective strain 26
Effective stress 26
Effective stress intensity factor 56, 373
Elastic deformation (also elasticity) 23
Elastic energy (also elastic strain energy) 23-24
Elastic energy release rate (see strain energy release rate)
Elastic stress field 50
Elastic T-stress 222-227
Elastic-plastic fracture Mechanics (EPFM) 5, 12, 81, 147
Electric potential technique 153, 366-368
Elliptical cracks 53, 449
Environmentally assisted crack growth 68-75
EPFM (see Elastic plastic fracture mechanics)
Equiaxed dimples 249
Equilibrium equations 18
Equivalent shear stress 29
Equivalent shear strain 29

## F

Failure 1-5
Failure analysis 1-5
Failure assessment diagram 184-187
Failure criterion 61, 92
Fatigue 65, 247
Fatigue crack growth 65, 267-308
Fatigue precracking 62, 149
Fractography 234-235
Fracture mechanisms
    Brittle 7, 234
    Cleavage 7, 234
    Ductile 7, 246
    Intergranular 7-8
    Transgranular 7-8
Fracture modes 48
Fracture process zone 101
Fracture toughness
    Dynamic 193, 199
    Elastic-plastic 147
    Linear elastic 60-62
    Weldments of 162-165

Frequency 65, 397
Frequency effects 70, 71, 397
Fully-plastic bodies 41, 125
Fully-plastic J-solutions 126-136

## G

Griffith's theory 45-48

## H

Heat affected zone (HAZ) 162-163, 386-387
Hinge, plastic 157
History of fracture mechanics 8-13
Hooke's law 22-23
HRR singularity 90, 313-314, 323
Hui-Riedel analysis 345
Hutchinson- Paris condition 97-101
Hydrogen embrittlement 68
Hydrostatic stress 228, 251

## I

Impact loading 194-196
Inclusions 247
Inertial effects on fracture 191
$I_m$-functions 90, Appendix
Initiation of ductile fracture 94-95, 161
Inspection interval 427, 459
Instability 175-182
Intergranular fracture 379
Irwin plastic zone correction 54-56

## J

J for cleavage fracture, $J_c$ 147, 155, 240
J for cyclic loading (also see delta J) 281-293
J-controlled fracture 100, 160
J-dominated region 97
J-estimation 107-142
J-integral 5, 81-102, 107
$J_{IC}$ 92-94, 147, 165, 252
J-Q theory 224-233
J-R curve 93, 101, 165, 248

## K

K dominated region 50
K, (see stress intensity factor )
$K_{EAC}$ 70

$K_{Ic}$ 12, 62
$K_{Iscc}$ 70
$K_p$ 63
K-R curve 63, 147

**L**

Leak before break 461
LEFM 5, 45, 70-75
Liberty ship failures 1
Limiting crack speed 189-190
Limit-load 41, 126, 128, 130, 134, 138
Load displacement diagram 47, 63-64, 113, 154, 167
Loading rate effect 191-193
LVDT 149, 366

**M**

Mixed mode 163, 164
Mott's dynamic crack analysis 189-191
Multiple specimen J testing 150

**N**

Net-section collapse 184

**O**

Opening load 281

**P**

Paris law 66, 299
Part-through crack 53
Path independent integrals
    $C^*$ 311
    $C^*_h$ 336-340
    J 84
    Delta J 281-289
Perfectly plastic materials 26, 35
Plane strain 22, 56, 57
Plane stress 22, 56, 57
Plastic collapse 57
Plastic constraint 57
Plastic zone 54
Plasticity theories 29, 98
Potential drop technique 153
Potential energy 85-86
Precracking (see fatigue precracking)

Primary creep (see creep)
Principal strain 22
Principal stresses 20
Process zone 254, 296, 298
Proportional loading 28

**Q**

Q-stress 228-230

**R**

R curve 150
Ramberg-Osgood relationship 26, 35
Rate of fatigue cracks (see Fatigue crack growth rate)
Rayleigh velocity 191
Reference stress 140-142
Reflected stress waves 198-199
Riedel-Rice analysis 320
Ritchie-Knott-Rice (RKR) model 237
River patterns 235
R-ratio 66

**S**

Secondary creep (see creep)
Short cracks 292-301
Similitude 295
Single edge notch bending (SENB) specimen 62, 97, 114, 132, 149, 150, 200
Single edge notch tension (SENT) specimen 133
Singularity 48, 50, 89
Small-scale creep 320
Small-scale yielding 54
Standards
    ASTM E 1290 157
    ASTM E1152 162
    ASTM E1221 207
    ASTM E1457 364
    ASTM E1681 70
    ASTM E399 62, 207
    ASTM E561 63
    ASTM E647 66
    ASTM E813 161
    BS 5447 62
    BS 5762 157, 161
Strain 20-22
Strain energy density 24, 82, 109

Strain hardening 26
Stress 17
Stress intensity factor (also see under K)
    Definition 48
    Expressions for K 51-54
    Relationship with $G$ 54
Stress-strain relations 23, 25, 28-29
Striations 272
Strip yield model 57-58
Subcritical crack growth 64
Surface energy 45-47
Surface flaws 65

## T

T-stress 221-224
Tearing instability theory 175
Tearing modulus 175-181
Temperature effect on
    Creep 32-34
    Creep crack growth 319, 338, 384
    Fracture toughness 452
Thermal-mechanical cracking 4
Threshold
    Environment assisted cracking 70
    Fatigue crack growth 67
Time dependent fracture mechanics 5, 12, 31
Tongues 234
Toughness (see fracture toughness)
Traction vector 19-20, 82
Transition time for creep 323
Transition time for dynamic effects 198-200

## U

Ultimate tensile strength 25
Uniform strain 25
Unloading compliance technique 150-151
Unstable crack growth 175
Upper-shelf toughness 246

## V

Velocity (see crack velocity)
Void coalescence 247
Void growth 247
Von Mises yield criterion 26

## W

Weldment testing 162-165
Westergaard stress function 48, 50
Wiebull distribution 244-246
Williams' series 10, 50, 221
Work hardening (see strain hardening)

## Y

Yield criterion 26
Yield strength 25
Young's modulus 22